Advances in Biological Science Research

Advances in Biological Science Research

A Practical Approach

Edited by

Surya Nandan Meena
Biological Oceanography Division, National Institute of Oceanography, Dona Paula, Goa, India

Milind Mohan Naik
Department of Microbiology, Goa University, Goa, India

ELSEVIER

ACADEMIC PRESS
An imprint of Elsevier

Academic Press is an imprint of Elsevier
125 London Wall, London EC2Y 5AS, United Kingdom
525 B Street, Suite 1650, San Diego, CA 92101, United States
50 Hampshire Street, 5th Floor, Cambridge, MA 02139, United States
The Boulevard, Langford Lane, Kidlington, Oxford OX5 1GB, United Kingdom

Library of Congress Cataloging-in-Publication Data
A catalog record for this book is available from the Library of Congress

British Library Cataloguing-in-Publication Data
A catalogue record for this book is available from the British Library

ISBN: 978-0-12-817497-5

For information on all Academic Press publications visit our website at
https://www.elsevier.com/books-and-journals

Publisher: Andre G. Wolff
Acquisition Editor: Linda Versteeg-buschman
Editorial Project Manager: Sandra Harron
Production Project Manager: Poulouse Joseph
Cover Designer: Vicky Pearson Esser

Typeset by TNQ Technologies

Contents

7. Biotechnological implications of hydrolytic enzymes from marine microbes

Poonam Vashist, R. Kanchana, Vazhakatt Lilly Anne Devasia, Priyanka V. Shirodkar and Usha D. Muraleedharan

8. Recent advances in bioanalytical techniques using enzymatic assay

Kanchanmala Deshpande and Geetesh K. Mishra

9. Microbial lectins: roles and applications

Hetika Kotecha and Preethi B. Poduval

20. Photocatalytic and microbial degradation of Amaranth dye

Pranay P. Morajkar, Amarja P. Naik, Sandesh T. Bugde and Bhanudas R. Naik

21. Role of nanoparticles in advanced biomedical research

R.K. Kunkalekar and Umesh B. Gawas

22. Iron-oxygen intermediates and their applications in biomimetic studies

Sunder N. Dhuri and Sarvesh S. Harmalkar

25. Advanced bioceramics

Kiran Suresh Naik

26. Production of polyhydroxyalkanoates by extremophilic microorganisms through valorization of waste materials

Bhakti B. Salgaonkar and Judith M. Bragança

Contributors

Gauri A. Achari, Department of Biological Sciences, Birla Institute of Technology and Science Pilani, KK Birla Goa Campus, Zuarinagar, India

Laurence V. Bindschedler, School of Biological Sciences, Royal Holloway University of London, Egham, Surrey, United Kingdom

Sunita Borkar, Department of Microbiology, P.E.S's R.S.N. College of Arts and Science, Goa, India

Judith M. Bragança, Department of Biological Sciences, Birla Institute of Technology and Science (BITS) Pilani, K K Birla, Goa Campus, Zuarinagar, Goa, India

Sandesh T. Bugde, Department of Chemistry, Goa University, Taleigao Plateau, Goa, India

Lakshangy S. Charya, Department of Microbiology, Goa University, Taleigao Plateau, Goa, India

Avelyno D'Costa, Department of Zoology, Goa University, Taleigao Plateau, Goa, India

Priya M. D'Costa, Department of Microbiology, Goa University, Taleigao Plateau, Goa, India

Varada S. Damare, Department of Microbiology, Goa University, Taleigao Plateau, Goa, India

Kanchanmala Deshpande, Department of Chemistry, Goa University, Taleigao Plateau, Goa, India

Shanti N. Dessai, Department of Zoology, Goa University, Taleigao Plateau, Goa, India

Vazhakatt Lilly Anne Devasia, Department of Biotechnology, Goa University, Goa, India; Present address: Department of Biotechnology, Hindustan College of Arts and Science, Padur, Kelambakkam, Chennai, India

Sunder N. Dhuri, Department of Chemistry, Goa University, Taleigao Plateau, Goa, India

James Dsouza, St. Xavier College, Mapusa, Goa, India

Samantha Fernandes, Department of Biotechnology, Goa University, Taleigao Plateau, Goa, India

Sandeep Garg, Department of Microbiology, Goa University, Taleigao Plateau, Goa, India

Umesh B. Gawas, Department of Chemistry, Dnyanprassarak Mandal's College and Research Centre, Assagao, Goa, India

Sanjeev C. Ghadi, Department of Biotechnology, Goa University, Taleigao Plateau, Goa, India

Shyamalina Haldar, Department of Microbiology, Goa University, Taleigao Plateau, Goa, India

Sarvesh S. Harmalkar, Department of Chemistry, Goa University, Taleigao Plateau, Goa, India

Md Imran, Department of Biotechnology, Goa University, Taleigao Plateau, Goa, India

Srijay Kamat, Department of Biotechnology, Goa University, Taleigao Plateau, Goa, India

R. Kanchana, Department of Biotechnology, Parvatibai Chowgule College of Arts and Science -Autonomous, Margao, Goa, India

Savita Kerkar, Department of Biotechnology, Goa University, Taleigao Plateau, Goa, India

Hetika Kotecha, Department of Biotechnology, Goa University, Taleigao Plateau, Goa, India

M.K. Praveen Kumar, Department of Zoology, Goa University, Taleigao Plateau, Goa, India

R.K. Kunkalekar, Department of Chemistry, Goa University, Taleigao Plateau, Goa, India

Mahesh S. Majik, Department of Chemistry, Goa University, Taleigao Plateau, Goa, India

Vinod K. Mandrekar, Department of Chemistry, St. Xavier's College, Mapusa, Goa, India

Kabilan Mani, Department of Biotechnology, PSG College of Technology, Coimbatore, India

Surya Nandan Meena, Biological Oceanography Division, National Institute of Oceanography, Dona Paula, Goa, India

Abhishek Mishra, Dixa Education and Research, Alto Porvorim, Goa, India

Geetesh K. Mishra, Multiscale Fluid Mechanics Lab, School of Mechanical Engineering, Sungkyunkwan University, Suwon, South Korea

Chellandi Mohandass, Biological Oceanography Division, National Institute of Oceanography, Dona Paula, Goa, India

Pranay P. Morajkar, Department of Chemistry, Goa University, Taleigao Plateau, Goa, India

Sajiya Yusuf Mujawar, Laboratory of Bacterial Genetics and Environmental Biotechnology, Department of Microbiology, Goa University, Goa, India

Usha D. Muraleedharan, Department of Biotechnology, Goa University, Goa, India

Milind Mutnale, National Centre for Polar and Ocean Research (NCPOR), Vasco-da-Gama, Goa, India

Srikanth Mutnuri, Applied and Environmental Biotechnology Laboratory, Department of Biological Sciences, Birla Institute of Technology and Science, Pilani, K K Birla Goa Campus, Zuarinagar, Goa, India

Amarja P. Naik, Department of Chemistry, Goa University, Taleigao Plateau, Goa, India

Kiran Suresh Naik, Department of Chemistry, P.E.S.'s R.S.N. College of Arts & Science, Farmagudi, Ponda, Goa, India

Milind Mohan Naik, Department of Microbiology, Goa University, Goa, India

Ravidas K. Naik, ESSO-National Centre for Polar and Ocean Research, Vasco, Goa, India

Bhanudas R. Naik, Department of Chemistry, Goa University, Taleigao Plateau, Goa, India

Prachi Parab, Department of Microbiology, Goa University, Goa, India

Chhaya Patole, Proteomics Division, National Centre for Biological Sciences, Bengaluru, India

Flory Pereira, PES's Ravi Sitaram Naik College of Arts and Science, Department of Microbiology, Ponda, Goa, India

Preethi B. Poduval, Department of Biotechnology, Goa University, Taleigao Plateau, Goa, India

Meghanath Shambhu Prabhu, Porter School of the Environment and Earth Sciences, Tel Aviv University, Tel Aviv, Israel; Applied and Environmental Biotechnology Laboratory, Department of Biological Sciences, Birla Institute of Technology and Science, Pilani, K K Birla Goa Campus, Zuarinagar, Goa, India

Neha Prabhu, Department of Microbiology, Goa University, Taleigao Plateau, Goa, India

R. Ramesh, Crop Improvement and Protection Section, ICAR-Central Coastal Agricultural Research Institute, Old Goa, India

Gouri Raut, Bioenergy division, Agharkar Research Institute, Pune, India

Ameeta RaviKumar, Institute of Bioinformatics and Biotechnology (IBB), Savitribai Phule Pune University, Pune, India

Bhakti B. Salgaonkar, Department of Biological Sciences, Birla Institute of Technology and Science (BITS) Pilani, K K Birla, Goa Campus, Zuarinagar, Goa, India

Sanika Samant, Department of Biotechnology, Goa University, Goa, India

Suvidha Samant, Dixa Education and Research, Alto Porvorim, Goa, India

Kashif Shamim, Laboratory of Bacterial Genetics and Environmental Biotechnology, Department of Microbiology, Goa University, Goa, India

Priyanka V. Shirodkar, Department of Biotechnology, Goa University, Goa, India

S.K. Shyama, Department of Zoology, Goa University, Taleigao Plateau, Goa, India

Akshaya Sridhar, Department of Biotechnology, PSG College of Technology, Coimbatore, India

Abhilash Sundarasami, Department of Biotechnology, PSG College of Technology, Coimbatore, India

Diviya Chandrakant Vainganker, Department of Microbiology, Goa University, Goa, India

Poonam Vashist, Department of Biotechnology, Goa University, Goa, India

Preface

Biological sciences are the study of life and living organisms, their life cycles, adaptations and environment. "*Advances in Biological Sciences – A Practical Approach*"describes recent progress in various rapidly growing biological sciences, such as bioinformatics, genomics, metagenomics, proteomics, enzymology, agriculture and marine microbiology, bioremediation, medicinal chemistry, and nanotechnology. This book consists of a total of thirty one chapters, each of which has been contributed by highly qualified professionals (professors, assistant professors, scientists, postdoctoral fellows, and senior research scholars) in the respective fields of research.

The chapter on bioinformatics describes the analysis and interpretation of biological experimental data using online bioinformatics tools or software. The genomics chapter describes the step-by-step strategy for identifying the gene of interest in the newly sequenced bacterial genome. Recent advances in metagenomics-based approaches have revolutionized microbial ecology science and have led to the discovery of some of the new biocatalytic molecules. In addition, metagenomics is undoubtedly the key to the discovery of secondary metabolites that can meet the urgent demand for new medicines from natural sources.

Proteomics chapters provide an overview of appropriate experimental workflows for plant proteomics, suggestions to improve the extraction and preparation of plant protein samples. Further current proteomic analysis is underway to better understand intracellular pathogen survival and disease persistence. An overview of X-ray diffraction technique the working principle, the instrumentation, and its usefulness in the structural characterization of the protein has been detailed. This section describe the various types of lectin proteins from different microbial source and also describe their role and action mechanisms. The section on enzymology explains bioanalytical techniques; principles of various enzyme inhibition assay's parameters required to optimize enzyme testing; and analytical merit characteristics for enzyme testing and its applicability to real sample analysis. Further emphasis is placed on information on routine protocols, aimed at highlighting and better understanding some of the challenges faced during the enzyme characterization/purification studies.

The section on microbiology presents an overview of microorganisms that solubilize phosphates, recent progress and future challenges with halophilic archaea as a beacon for exobiology, the use of probiotics and their mechanisms to control aquatic pathogens, and recent new research findings on the control of plant pathogenic diseases through various quorum quenching strategies. The reader will also be briefed about the advances in microbial techniques for the recovery of phosphorus in sustainable wastewater management, advances in ectomycorrhizal research methods, and techniques for the mass production of fungal arbuscular mycorrhiza.

Some chapters describe the photocatalytic and microbial degradation of amaranth dye, biodegradation of seafood waste by bacteria associated with seaweed and the use of seafood waste for the production of ethanol, the production of Polyhydroxyalkanoates by microorganisms in extreme econiches, with a particular focus on the use of cheaply available waste materials such as carbon substrates for Polyhydroxyalkanoates synthesis.

Chapters on medicinal chemistry present the various analytical techniques for the extraction, purification, and characterization of metabolites from natural sources. Special attention is focused on a discussion on case studies involving isolation of marine, microbial, and terrestrial natural products with the help of suitable examples. Another chapter discusses the advances in bioceramics, their biocompatibility, classification, application, and further research on bioceramics. Latest advancement in application of various 3D reconstructed human skin models for the screening of natural chemical compounds of cosmeceutical potential have been included.

The aquatic science section discusses the advances in sampling strategies and analysis of phytoplankton and current strategies for the isolation and preservation of ecologically important marine protists. Further, some chapters describe the role of nanoparticles in advanced biomedical research with reference to drug delivery and anticancer therapy. For better understanding of the mechanism of DNA damage as well as regions of the genome that are prone to alterations, micronucleus test and the single-cell gel electrophoresis assay techniques are described with the latest modifications.

This book describes the updates on the methodologies and protocols being used by researchers in their routine experiments of biological sciences. Greater emphasis has been given to the basic fundamentals and the latest techniques or methods in routine experiments. We have given equal importance to text and illustrations, therefore, sincere efforts have been made to support the textual clarifications and explanations with the help of flow charts, tables, and figures. It is written in a clear and concise language to enhance the self-motivation of the researchers.

The book will help graduate and postgraduate students to explore their research careers. In addition, recently updated information on various research fields and techniques in biological sciences will definitely benefit university professors, university lecturers, and scientists from different life sciences institutions worldwide.

Surya Nandan Meena
Milind Mohan Naik

Acknowledgments

A book of this nature is possible only when several diligent and hardworking minds come together with a single purpose to make a book of high standards. We editors will need a flower garden to present a flower to all those who have provided invaluable support in compiling this book from concept formulation to the present form.

We are grateful to all the authors who contributed to this book for their excellent knowledge of the multiple aspects of the subject. The contributors made the book truly exceptional and novel. We are sincerely grateful to all reviewers for sharing their valuable time and critical reviews that have really brought the book's quality to the fore. We are extremely grateful to Prof. Sanjeev Ghadi, Prof. Santosh Dubey, Prof. M. K. Janarthanam, Prof. Santosh Tilve, Prof. Prabhat Sharma, Prof. Sandeep Garg and Dr. Ram Swaroop Meena for their advice during the tenure of this work. We thank them for participating in discussions, reviewing the work, and giving us the freedom to approach them.

We would like to thank the all the staff members of the Elsevier book publication team for their direct or indirect support, in particular, Linda Versteeg-Buschman (Acquisitions Editor), Sandra Harron (Editorial Project Manager), Sandhya Narayanan (Copyright Coordinator), and Poulouse Joseph (Production Manager) for their step-by-step technical support. We appreciate the Elsevier facility in the form of an EMSS (electronic submission system for manuscripts). It is a user-friendly online tool that helps to organize the book's large content and is easy to communicate with writers and publishers in two directions.

SNM wants to dedicate efforts to his family members Shree Pana Chand Khokar (father), Smt. Kali Bai (mother), Raghu Nandan (brother), Rukamani and Chandramani (sisters), Rajkumar and Ram Swaroop (brothers in-law) for constant support and inspiration. In addition, SNM recognizes the name of his dearest daughter Bhavya Khokar (Khusi), niece Muskan and dear wife Bhavna for their unseen support in order to achieve this goal. I believe that their presence was energetic to me, and because of them, I could recover myself from the vilest time. MMN proudly acknowledges the name of his loving mother Smt. Manisha Naik and dear wife Pranaya Naik for their constant inspiration.

We would like to acknowledge Prof. Sunil Kumar Singh (Director, NIO, Goa), and Prof. Varun Sahni (Vice-Chancellor, Goa University) and

Prof YV Reddy (Registrar Goa University) for the necessary infrastructure and favorable working environment to carry out the task. It would not have been possible for us to undertake the editing of this book involving countless hours, days, and months without the financial support. So here, SNM would like to acknowledge the Department of Science and Technology, Government of India for Financial support through the postdoctoral fellowship scheme (PDF/2016/002012). MMN acknowledges the SERB-DST for financial support (Grant No. YSS/2014/000258).

Surya Nandan Meena
Milind Mohan Naik

Chapter 1

Bioinformatics methods: application toward analyses and interpretation of experimental data

Shyamalina Haldar
Department of Microbiology, Goa University, Taleigao Plateau, Goa, India

1.1 Aim of the chapter

This chapter aims to describe the tools and the techniques that are being applied globally for analysis and assessment of biological data. The chapter has been divided into three sections (nucleic acid: 1.2 and 1.3; microbial ecology: 1.4; bio statistics: 1.5).

(1) Section I deals with the bioinformatics methods applied for molecular analyses of nucleic acids. (2) Section II deals with the statistical formulae used to interpret microbial ecological data. (3) Section III describes statistical methods used to compare the biological observations to draw significant conclusions. Care has been taken to present the methods in a stepwise manner with examples for better understanding.

1.2 DNA sequencing

DNA sequencing is the process of determining the order of nucleotides within a DNA molecule. There are two methods of DNA sequencing: Maxam–Gilbert sequencing and Sanger sequencing. The former is a chemical method that chemically modifies the DNA nucleotides and subsequently cleaves the DNA backbone at the sites neighboring to the modified nucleotides [1]. However, due to technical complexity and use of hazardous chemicals, this method is not currently used for standard molecular biology.

Sanger sequencing is the method of DNA sequencing in which dideoxynucleotide phosphates (ddNTPs) are incorporated by DNA polymerase during in vitro DNA replication. Modified ddNTPs terminate DNA strand elongation since they lack a 3′-OH group required for the formation of a

Advances in Biological Science Research. https://doi.org/10.1016/B978-0-12-817497-5.00001-X

phosphodiester bond between two nucleotides, causing DNA polymerase to cease the extension of DNA. Therefore, this is called *dye-terminator sequencing*. Each of the four ddNTPs (where N = A/T/G/C) is labeled with fluorescent dyes that emit light at different wavelengths and therefore can be captured in the form of colored peaks called a *chromatogram*.

The nucleotide bases of DNA obtained from a chromatogram are converted to text-based *FASTA* format using the Applied Biosystems to FASTA converter database (www.dnabaser.com/download/Abi-to-Fasta-converter/abi-to-fasta-converter.html). A variety of free software is available for this purpose (chromas, chromaslite, etc.), which can be downloaded and installed.

1.3 Identification of organisms from nucleotide sequence

The DNA sequence obtained in FASTA format uses single-letter codes for each of the nucleotide base without mentioning the source of DNA, i.e., the name of the organism from where the DNA has been isolated. Therefore, the initial analysis of the obtained DNA sequence is to find out the source of DNA, and that is done by Basic Local Alignment Search Tool (BLAST) analysis.

1.3.1 What is BLAST?

BLAST is a program that matches the nucleotides of DNA sequences or the amino acid sequences of proteins. This helps to compare a "query sequence" (obtained from chromatogram) with a database of sequences ("subject sequences,"available on the Internet) and identify the sequences from the database that bear a resemblance to the query sequence above a definite threshold. BLAST is classified into different groups based on type of query sequence used (Table 1.1). Of these programs, nucleotide BLAST (BLASTn) and protein blast (BLASTp) are most commonly used since they directly compare the sequences without translations.

1.3.2 Methods for nucleotide BLAST

A stepwise description of nucleotide BLAST analysis is given below.

1. Open NCBI BLAST in Google.
2. Choose BLASTn.
3. Give the FASTA sequence as the query sequence (it must be minimum length of 60 nucleotides) in the blank box provided. Alternatively, you can upload the text file (.txt) or FASTA file (.fq) containing the sequences in FASTA format.
4. Adjust the parameters like database (organism: human, mouse, others [organisms other than mouse/humans]; gene: 16S gene/18S gene, chromosomal genes, etc.) from the dropdown list provided. However, if the organisms are unknown, then you can choose uncultured/environmental

TABLE 1.1 Classification of BLAST programs.

Type of BLAST	Functions
Nucleotide-nucleotide BLAST (blastn)	**Input:** DNA FASTA query **Output:** Similar DNA sequences from the DNA database that the user specifies.
Protein-protein BLAST (blastp)	**Input:** Protein query **Output:** Similar protein sequences from the protein database that the user specifies.
Position-Specific Iterative BLAST (PSI-BLAST) (blastpgp)	**Aim:** To find distant relatives of a protein. **1.** A list of all closely related proteins is created. **2.** These proteins are combined into a general "profile" sequence, which summarizes significant features present in these sequences. **3.** A query against the protein database is then run using this profile, and a larger group of proteins is found. **4.** This larger group is used to construct another profile, and the process is repeated. **Importance:** By including related proteins in the search, PSI-BLAST is much more sensitive in picking up distant evolutionary relationships than a standard protein-protein BLAST.
Nucleotide 6-frame translation-protein (blastx)	Compares the six-frame conceptual translation products of a nucleotide query sequence (both strands) against a protein sequence database.
Nucleotide 6-frame translation-nucleotide 6-frame translation (tblastx)	Translates the query nucleotide sequence in all six possible frames and compares it against the six-frame translations of a nucleotide sequence database. The purpose of tblastx is to find very distant relationships between nucleotide sequences.
Protein-nucleotide 6-frame translation (tblastn)	Compares a protein query against the all six reading frames of a nucleotide sequence database.
Megablast	Large numbers of query sequences. It concatenates many input sequences together to form a large sequence before searching the BLAST database, then postanalyzes the search results to glean individual alignments and statistical values.

sample sequences or the general nucleotide sequences. Click on the "[Save Search Strategies]" link near the top of the blast results page to save search strategies for future use. You can exclude the organisms that you don't want to be included for comparison by choosing "exclude" comment.

4. Click on BLAST.
5. The output looks like as given in Figs.1.1 and 1.2.
6. Click on each of the item (either each colored line/name of the species) to obtain the description.

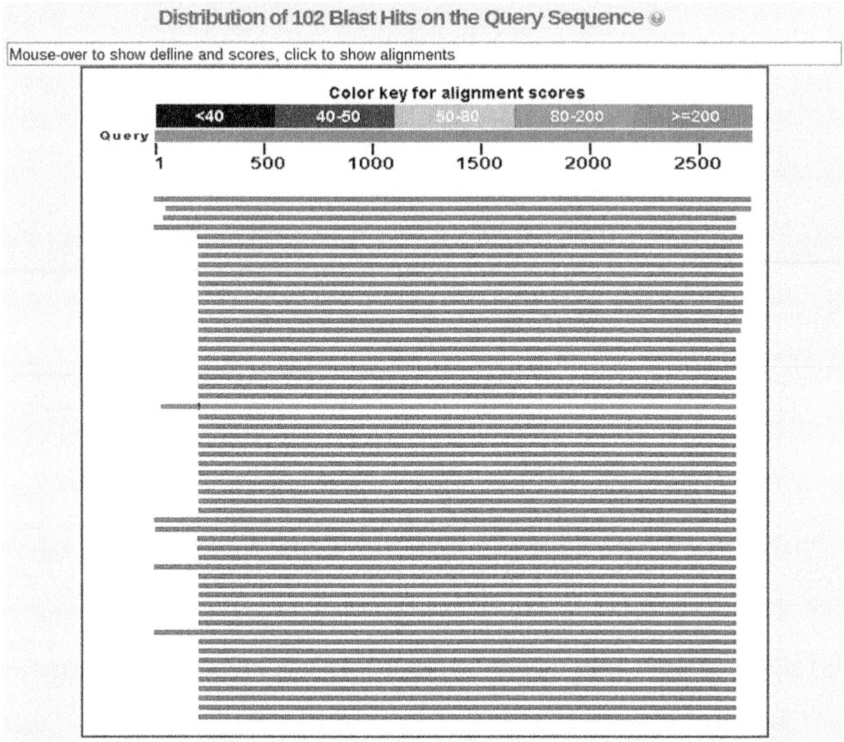

FIGURE 1.1 Distribution of BLAST hits with the subject sequences obtained for a query sequence from NCBI BLAST window. Each line denotes one subject sequence with which the query sequence has shown the similarity. Clicking on each line gives the details of the identity of the species with which the similarity is found. The *red (grey in print versions) color* indicates the hit score between the subject sequences and the query sequence to be greater than 200.

1.3.3 Interpretation of BLAST results

BLAST algorithm is based on the estimation of the similarity scores for local alignments (i.e., the most similar regions between two sequences) between the query sequence and subject sequences using specific scoring matrices, and gives the best matches ("hits") from the database as the output. The results provide the similarity score (S), query coverage (percent of the query sequence that overlaps the subject sequence), E-value, and max identity (percent similarity between the query and subject sequences over the length of the coverage area). The organisms are listed from top to bottom with decreasing similarity score (maximum identity). The score (S) is the numerical value that gives the overall quality of the alignment. Higher numbers correspond to higher

Description	Max score	Total score	Query cover	E value	Ident	Accession
Penicillium sp. Fk80 small subunit r [illegible]	390	554	46%	7e-106	97%	KC464352.1
Penicillium sp. GZU-BCECWS1-2 18S ribosomal RNA gene, partial sequence; internal transcribed s	374	521	46%	7e-101	96%	GU565133.1
Uncultured fungus clone C_D9_A 18S ribosomal RNA gene, partial sequence; internal transcribed s	364	502	46%	4e-98	95%	KC818306.1
Talaromyces verruculosus isolate AT010 18S ribosomal RNA gene, partial sequence; internal trans	364	502	46%	4e-98	95%	HQ607791.1
Penicillium aculeatum strain H23 18S ribosomal RNA gene, partial sequence; internal transcribed sp	364	508	46%	4e-98	95%	GU566285.1
Penicillium sp. C08652 18S ribosomal RNA gene, partial sequence; internal transcribed spacer 1, 5	364	508	46%	4e-98	95%	GQ996518.1
Talaromyces verruculosus ex-type CBS 388.48 18S ribosomal RNA gene, partial sequence; interna	363	515	46%	1e-97	95%	NR_103675.1

FIGURE 1.2 Species showing sequence similarity with the query sequence. The accession refers to the unique Genbank identifier for the identified species. Clicking on the accession number will provide the FASTA sequence and the details of the submission about the identified species. For description of the other parameters (score, query coverage, E-value, maximum identity), see the text.

similarity. Expect value (E) is a parameter that determines the number of expected hits to be observed by chance while searching a database for a fixed length of nucleotide. It decreases exponentially as the score (S) of the match increases. The E value denotes the random background noise, e.g., an E value of 1 for a BLAST hit means that only one match with a similar score simply by chance for the query sequence is expected to be seen in the database of the current size. The lower the E-value, or closer to zero, the more "significant" will be the match. The accession number (gb|xx99999.1|) is a unique identifier assigned to a DNA or protein sequence to track the multiple versions of that sequence record and the related sequence over time in a specific database.

1.3.4 Construction and interpretation of phylogenetic tree

The evolutionary relationships between various species and their phylogeny based upon similarities and dissimilarities in their physical or genetic characteristics is represented by a phylogenetic tree (evolutionary tree). The phyla joined together have a common ancestor phylum (Fig. 1.3). The phylogenetic tree can be constructed directly from the output window of BLAST by clicking the option "distance tree results," or it can be calculated with all the obtained FASTA sequences using the **M**olecular **E**volutionary **G**enetic **A**nalysis (MEGA) software. MEGA is free software (www.megasoftware.net) that uses different methods for phylogenomics analyses [2].

The phylogenetic tree is of two types: rooted and unrooted. The rooted tree contains "nodes" representing the common ancestor of the descendants, and the edge lengths interpret the time estimates. An unrooted tree illustrates only

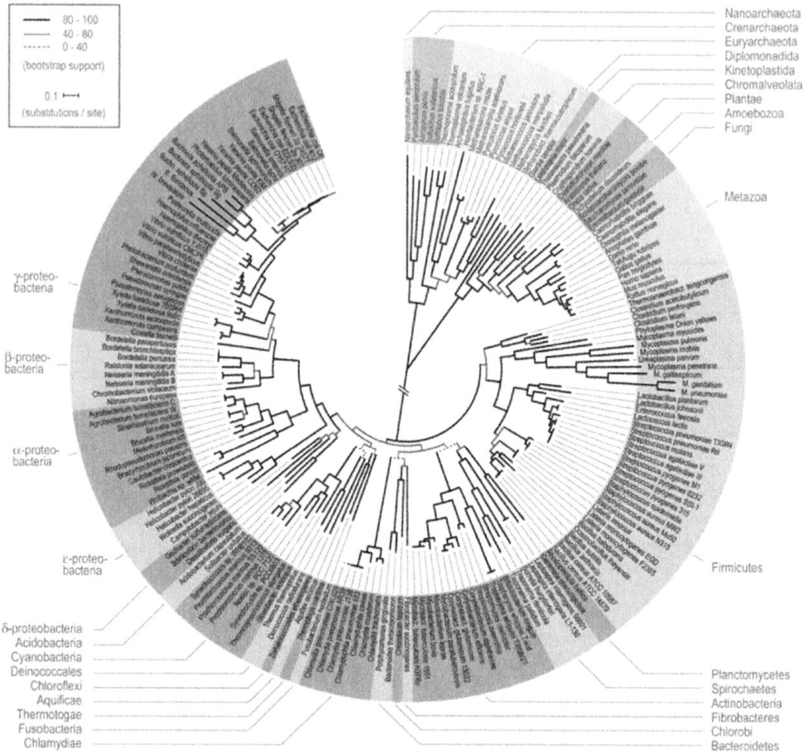

FIGURE 1.3 Phylogenetic tree (circular form) representing a common ancestor (origin) species (bacteria) with the branching arising from it showing the evolutionary and phylogenetic relationship between the different bacterial species.

the relatedness of the leaf nodes and does not require the ancestral root to be known or inferred.

The phylogenetic tree with bootstrap values calculates the redundancy of a certain character pattern among taxa. A low bootstrap value indicates claim that a certain taxon is not supported well by certain data [3].

1.3.5 Sequence deposition

Experimentally obtained DNA/protein sequences need to be deposited in the public databases for scientific references. The mandatory requirement for publication of data in a journal is the deposition of the obtained sequences in any public sequence repository. The sequences are deposited directly via online portal of the specific database or are sent via email to the respective authorities of the databases after constructing the file using the program sequin (https://www.ncbi.nlm.nih.gov/genbank/submit/opens).

1.4 Microbial ecology statistics

The principal goal of ecology is to determine the spatial and temporal diversity and abundance of organisms in a particular niche to understand the ecosystem functioning. Though the advent of technologies hold great promise to test ecological theories of quantification of microbial taxa in the environment, robust knowledge on the estimation of diversity is necessary to draw conclusions about the environmental composition. Both the cultivation-dependent (plate count methods or microscopy examinations) and the cultivation-independent (gene-based molecular analyses) require analyses of the data using various statistical parameters. A few of the statistical parameters used for the study are discussed next.

1.4.1 Species composition/species richness

The total number of different species present in a particular ecosystem is referred to as species richness (S), which is dependent on type of sampling. Increasing the area sampled increases observed species richness.

For example, the microorganisms can be grouped under different "taxa" based on their structure, biochemical properties, and sequence analyses. The species richness in a particular region (seawater; mangrove soil; sand dunes; industrial areas; etc.) will be equal to the total number of observed microbial taxa.

Statistically it is expressed by the richness estimators like Chao1 richness estimator, which is given by the formula as:

$$S_{est} = S_{obs} + \frac{(f_1)^2}{2f_2}$$

Where, S_{est} = number of species estimated, S_{obs} = number of species observed, f_1 = number of singleton taxa (taxa with only one species in that community), and f_2 = number of doubleton taxa (taxa with two species in that community). The higher number of singletons in a sample refers to higher number of undetected taxa and the Chao1 index for such cases will be high.

1.4.2 Species abundance

Abundance refers to relative representation of a species in a particular ecosystem. It takes into account the number of individuals found per taxon/group calculated by dividing the number of species from one group (ni) by the total number of species from all groups (n); usually normalized to logarithmic scale. Frequency histograms (Preston Plots) or rank-abundance diagrams ("Whittaker Plots"; Fig. 1.4) are used to represent abundance of species in a sample. The rank–abundance curve is a 2D chart with relative abundance on the *Y*-axis and the abundance rank on the *X*-axis. The highest abundant species is ranked as 1, similarly followed by 2, 3, and so on in descending order.

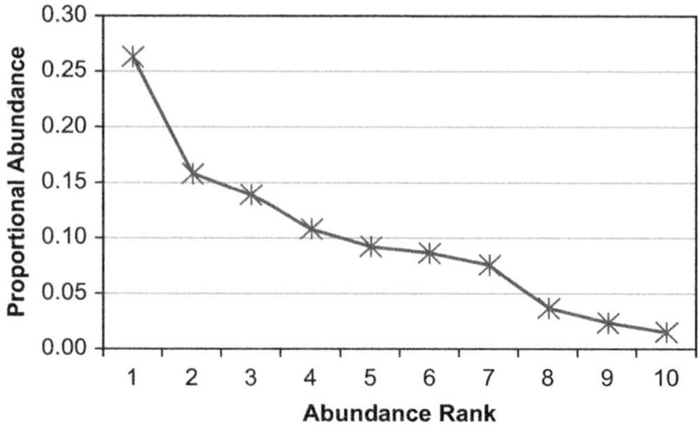

FIGURE 1.4 Rank–abundance curve showing the ranking of the species in a niche according to the abundance. The species are ranked from 1 to 10 (*X*-axis) according to the descending order of their abundance from 0.30 to 0.01 (*Y*-axis). The highest proportion of abundance (0.25) is ranked 1 while the lowest proportion of abundance (0.1) recorded is ranked 10.

Species richness and evenness together is shown in a rank–abundance curve. Species richness refers to the number of different species on the chart, i.e., how many species were ranked. The slope of the line in a logarithmic curve represents the species evenness. A steep gradient indicates low evenness or an uneven distribution of species as the high-ranking species have much higher abundances as compared to low-ranking species. The more abundant a particular species is in any system, the more dominant will be that species in that particular environment, thereby reducing the overall species diversity in the system. Hence the rank–abundance curve is also called dominance–diversity curve. Conversely, a shallow gradient rank–abundance curve indicates high evenness as the abundances of different species are similar, i.e., the proportion of species (individuals) in different groups (taxa) are similar. The following two examples illustrate the species abundance and richness from two environmental samples.

1.4.2.1 Example 1: illustration for species abundance

The data is given in Table 1.2. Here, the highest abundant taxon is *Alphaprotebacteria* containing the highest number of observed individuals and hence is given the rank 1, followed by *Betaproteobacteria* (rank 2), *Gammaproteobacteria* (rank 3), and so on until *Spirochaetes* with rank 10 containing the least number of individuals (only 6).

1.4.2.2 Example 2: comparison of species abundance with richness

The data is given in Table 1.3. Each of the four communities (A–D) in Table 1.3 has total number of individuals (N) = 30. However, the distribution

TABLE 1.2 Microscopic and biochemical studies from rhizosphere sample of a plant.

Rank	Taxa	Abundance (ni)	Relative abundance (= ni/N)
1st	*Alphaprotebacteria*	110	110/400 = 0.275
2nd	*Betaproteobacteria*	70	70/400 = 0.175
3rd	*Gammaproteobacteria*	53	53/400 = 0.132
4th	*Deltaproteobacteria*	41	41/400 = 0.102
5th	*Clostridia*	35	35/400 = 0.087
6th	*Bacillus*	33	33/400 = 0.082
7th	*Actinobacteria*	29	29/400 = 0.072
8th	*Bacteroidetes*	14	14/400 = 0.035
9th	*Cyanobacteria*	9	9/400 = 0.022
10th	*Spirochaetes*	6	6/400 = 0.015
Total	S (species richness)= 10 = total number of taxa	Total number of individuals = N = ($\sum$ni) = 400	

TABLE 1.3 Algae population in four communities.

Taxa	Community A	Community B	Community C	Community D
Chlorophyta	10	4	6	3
Rhodophyta	9	8	6	7
Glaucophyta	11	18	6	10
Chlorarachniophytes	0	0	6	4
Euglenids	0	0	6	6

of individuals under each taxon and also the total number of taxa are different in them. Both community A and B have species richness (S) = 3 as the total number of taxa is 3 while for C and D communities S = 5, thereby indicating these latter two communities to have higher species richness. However, with respect to abundance, the distribution of individuals in each taxon is highly even for community C (6 in each taxon), and all the taxa will have same ranking in rank–abundance curve. The curve will be a straight line. For communities B and D, the distribution of individuals in each taxon is less even; thereby the rank–abundance curve will mark the taxa from 1 to 3 and 1 to 5, respectively, in a descending order with respect to number of individuals beginning with *Rhodophyta* and *Glaucophyta*, respectively, as first rank. The curve will be a gradient one. However, as compared to D, the curve will be steeper for B as high-ranking *Rhodophyta* has a very high number of species compared to low-ranking taxa (*Chlorophyta* and *Glaucophyta*). As compared to B andand D, A will have more even distribution.

1.4.3 Species diversity

Species diversity is the *number of different species* represented in a given community that takes into account both the species richness and abundance. Communities that are numerically dominated by one or a few species exhibit low evenness (e.g., community B in Example 2), whereas communities where abundance is distributed equally amongst species exhibit high evenness (e.g., community C in Example 2) (Gotelli and Colwell, 2001). The diversity is expressed by one or more "indices" that quantify the species diversity. Example: Shannon index (or Shannon–Wiener) [4]; Simpson index [5] and Gini–Simpson index. During interpreting ecological terms, each of these indices corresponds to a different thing and their values are therefore not directly comparable. The Shannon index equals log(qD), where, qD = inverse of the weighted average of species proportional abundances and in practice quantifies the uncertainty in the species identity of one random individual from the dataset. The Simpson index is represented by 1/qD that refers to the probability of two random individuals in a dataset (with replacement of the first individual before taking the second) to represent the same species. The Gini–Simpson index is given by formula 1 − 1/qD, which refers to the probability of occurrence of different species by two randomly taken individuals [6–8].

The different formulae for calculation of diversity indices are given below:

1. Shannon index

$$H' = -\sum_{i=1}^{n} p_i \ln p_i$$

2. Simpson index

$$\lambda = \sum_{i=1}^{n} (p_i)^2$$

3. Gini-Simpson index

$$1 - \lambda = 1 - \sum_{i=1}^{n} (p_i)^2$$

4. Berger–Parker index refers to the highest value for pi in a particular dataset, i.e., the proportional abundance of the most abundant type. This refers to the average of the pi values when n approaches infinity, and hence equals the inverse of true diversity of order infinity.

$$\text{Here, } n = \text{total number of groups/taxa/classes}$$

$$p_i = \text{proportion of individuals in a particular group (where, } i = 1 \text{ to } n)$$
$$= \frac{\text{Number of individuals in a group}}{\text{Total number of individuals from all the groups}}.$$

e.g., p_{10} = proportion of individuals in 10th group; p_1 = proportion of individuals in first group etc.

In the Example 1 above:

p_i for *Alphaproteobacteria* = 110/400 = 0.275.
p_i for *Clostridia* = 35/400 = 0.0875.

Therefore,

$\ln(p_i)$ for Alphaproteobacteria = ln 0.275
$\ln(p_i)$ for *Clostridia* = ln (0.0875)

Accordingly,

$p_i \ln(p_i)$ for *Alphaproteobacteria* = 0.275 × ln 0.275
$p_i \ln(p_i)$ for *Clostridia* = 0.0875 × ln (0.0875)

Therefore,
$H' = -\sum \text{pi} \ln \text{pi} = -[0.275 \times \ln(0.275) + 0.0875 \times \ln(0.0875)]$ where, $i = 2$ (as we are considering two taxa as *Alphaproteobacteria* and *Clostridia*). If we would have considered all the 10 taxa from the above example, then i would be 10 and summation will be of the total products of proportion of individuals and their respective ln values from all the 10 groups.

Similarly, Simpson and Gini–Simpson index can also be calculated using the above formula for this dataset of Example 1. Berger–Parker index for this dataset will be 0.275 (i.e., 110/400, as 110 is the largest number of individuals in a group in that dataset).

The diversity of a particular or local space/region/habitat is called alpha diversity (α diversity) [9,10]. When all the species diversities from all the local regions/habitats are considered together, this is called gamma diversity (λ diversity). The ratio of gamma and alpha diversity is called true beta diversity

(β diversity), which refers to the ratio between regional and local species diversity.

$$\beta \text{ diversity} = \frac{\lambda \text{ diversity}}{\alpha \text{ diversity}}$$

In some cases, the gamma diversity is considered to be additive rather than multiplicative. Therefore, in those cases, β diversity $= \lambda$ diversity $- \alpha$ diversity. This latter type of β diversity is called absolute species turnover.

When there are two subunits, and presence-absence data are used, this can be calculated with the following equation:

$$\beta_A = (S_1 - c) + (S_2 - c)$$

Where, S_1 = the total number of species recorded in the first community, S_2 = the total number of species recorded in the second community, and c = the number of species common to both communities.

1.4.3.1 Similarity indices

The most important task is to compare the diversity and abundance of species between different samples so as to understand the similarity in species composition between different environments or under different environmental conditions. This is actually the measurement of beta diversity (between sample comparisons). There are numerous ways to visualize and analyze beta diversity, and a thorough review of multivariate techniques that are commonly used by microbial ecologists is presented by Ramette [11].

Following are a few of the statistical indices used to compute the similarities and dissimilarities between different samples with respect to species constituency [12].

1. *Jaccard index or Jaccard similarity coefficient* compares the similarity and diversity of different sample sets. It is given by the formula:

$$\frac{a}{a + b + c + \ldots\ldots z}$$

   Here, a = number of common taxa between samples; b to z = number of taxa exclusive to different samples.

   As in Example 2: The similarity coefficient between community A and B will be $= \frac{3}{3+0+0}$

   The similarity coefficient between community B and C will be $= \frac{3}{3+0+2}$

   The similarity coefficient between community C and D will be $= \frac{5}{5+0+0}$

   This is because A and B communities share all the three phyla while B and C have three common phyla. However, compared to B, the community C has two exclusive phyla that are not there in B. Similarly, C and D have all the five shared phyla and no exclusive phyla for them. *Jaccard index* can vary from 0 to 1, where 1 represents the highest similarity when all the phyla are common between all the communities.

2. *Sørensen–Dice coefficient* (Sørensen index or Dice's coefficient) is another similarity index given by the following formula.
$\frac{2a}{n_b+n_c+.......+n_z}$ where, a = common/shared taxa while n_b to n_z refers to the number of individuals in the taxa that are exclusive to the communities. *Sørensen similarity index* can vary from 0 to 1, where 1 represents the highest similarity when all the phyla are common between all the communities.

1.4.3.2 Dissimilarity indices

The distance/dissimilarity matrix is computed using either of the two methods:

1. Bray–Curtis dissimilarity [13].
2. UniFrac distances [14].

Bray–Curtis dissimilarity index between two communities is calculated by the following formula:

$$c = 1 - \frac{2w}{a+b}$$

Where, w = total number of taxa present in all the communities, a = sum of the measures of taxa in one community, and b = sum of the measures of taxa in the other community. When proportional abundance is used, a and b equal to 1 and the index collapses to 1 − w.

UniFrac distances are based on the branches of the phylogenetic tree constructed with the sequences of the species obtained from the different communities that are either shared or unique amongst samples. It depends on the quality of the input tree.

The Bray–Curtis dissimilarity matrix or UniFrac distance matrix is then used as an input for ordination and clustering analyses like principal coordinates, nonmetric multidimensional scaling, and canonical correspondence analysis (CCA) [11]. CCA is used to determine which taxa correspond with specific environmental variables.

However, presently a variety of online/downloadable software is available that can calculate all these indices, such as PAST [15] and MOTHUR [16], with the number of individuals in each group/taxa taken as the input only. The software can be freely downloaded (MAC or WINDOWS OS).

1.5 Biostatistics

The experimental observations vary between individuals as well as from time to time for an individual. However, dependable inferences cannot be drawn from mere inspection of the observed values. Hence, experimental data need to be evaluated statistically. Biostatistics helps in *systematic arrangement* of the data; *methodical comparison*; *interpretation* and *drawing of inference* from the observations. It also helps to *predict mathematically the most "probable" values* for biological properties or events. Biostatistics is applied for *designing*, *error calculation*, and *estimation of reliability and validity of the experimental methods*.

1.5.1 Sampling statistics

Mean ($\overline{X}$ for sample and μ for population) = arithmetic average of a set of scores $= \sum \frac{X_i}{n}$

X_i = individual score; n = sample size.

The sum of positive deviations of some of the scores from the mean equals to the negative deviation of the remaining scores of the sample. Therefore, $\sum(X_i - X) = 0$

$$\text{Mean deviation} = \frac{\sum|X_i - \overline{X}|}{n}$$

Standard deviation (SD) = Positive square root of the mean of squared deviations of all the scores from their mean.

$$s = \frac{\sqrt{\sum (X_i - \overline{X})^2}}{n}$$

SD is important because it helps to measure and express numerically the deviations of the scores of a sample from the mean (central value) and thereby indicates the spread or scatter of the scores around the central value. High SD indicates wide dispersion of scores.

The properties of SD are given below:

1. A change in the single score affects the value of SD
2. Addition or subtraction of a constant number from SD does not affect it. However, multiplication and division by a constant number affects SD identically.
3. If all scores have identical value, SD will be zero.
4. In a small sample size ($n < 30$), extreme scores at the two ends of the frequency distribution might be ruled out, thereby lowering the SD value. Hence, to compensate this, "degrees of freedom" [df = (n − 1)] is introduced instead of n and the new SD is called "unbiased SD" and it is given by the following formula:

$$s = \frac{\sqrt{\sum (X_i - \overline{X})^2}}{n - 1}$$

Note: The degrees of freedom (df) of a statistic is defined as the number of scores of a variable that can be altered freely in both magnitude and direction without causing any change to the values of such statistics.

Standard error (SE) = A measure of the sampling error that is the deviation of that statistic from the corresponding parameter. It is computed for numerous sample statistics like standard error of mean, proportions, SD, etc.

$$\text{Standard error } \bar{s}_X = \frac{s}{\sqrt{n}} \frac{\sqrt{N-n}}{N-1}$$

Where, n = sample size and N = population size from where the n has been drawn by simple random sampling.

1.5.2 Testing of hypothesis

To assess whether the result of any experimental data is significant, the probability (P) of that result is estimated with the help of the standard score obtained from the observed data and the probability of its random occurrence in the population using normal and t-distributions. For this, the probability of correctness of "null hypothesis" is assessed. *Null hypothesis* (H_0) proposes to nullify the hypothesis of the investigation if the observed value has evolved by chance due to random sampling or would have been false if the entire population was considered, and therefore it states that the results are not significant. On the other hand, the null hypothesis is contested by *alternative hypothesis* (H_a). Therefore, to draw any conclusion, the results of the experiments are subjected to the testing of these hypotheses.

To study the significance of difference between means of two or more groups or correlation between variables, H_0 assumes that there is no significant difference between the observed means or no significant correlation between the variables. Therefore, probability (P) is calculated for H_0, and if this estimated P-value does not exceed a particular chosen level of significance (α), the probability of correctness of H_0 is negligible. Therefore, in that case, H_0 is rejected and the observed results are considered to be significant.

For biological experiments, α is fixed at any of 0.001, 0.01, 0.02, or 0.05.

One-tailed/two-tailed t-test is performed to evaluate the significance of the difference in results between observations. The former takes into account for both the magnitude and the sign while the latter considers the magnitude only.

1.5.3 Probability distribution

Probability (P) of an incidence is the limit attained by the relative frequency of that incident in a large number of observations or trials. The relative frequency is obtained by dividing the frequency of that phenomenon by the sample size (f/n). Now, for large n the probability of the incident is expressed as a distribution of occurrence of the events between different class intervals of the

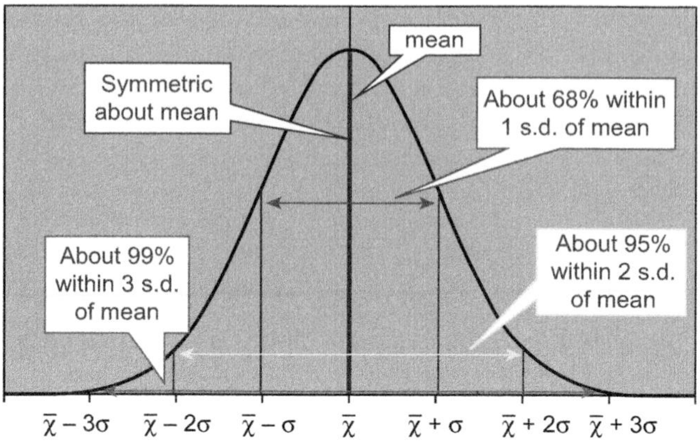

FIGURE 1.5 Normal distribution curve. The different regions within the area of the curve are marked within the figure.

given variable and is called probability distribution. This is expressed in the form of graphs by plotting scores and the probabilities of the variables along *X*-axis and *Y*-axis, respectively. This gives rise to a bell-shaped curve called a normal distribution curve if the variable is a continuous measurement variable and the n is very large ($n > 30$) (Fig. 1.5). On the other hand, the probability distribution of scores for a small sample ($n < 30$) drawn from normal distribution results in a different type of probability distribution called Student's t-distribution after the pseudonym "Student" of the discoverer, W. S. Gossett. However, t-scores vary with df; hence observed t-value must be referred to the specific t-distribution for that df.

1.5.3.1 Example

In an experiment, the mean of weights of 16 boys and 16 girls were found to be 40.3 and 37.5 kg, respectively, whereas SD values amounted to 8.15 and 6.35, respectively, for the two groups. Explain whether there the difference between the means of the weights of girls and the boys is significant or not.

Solution

According to the assumption of H_0, there is no significant difference between the means of weights of boys and girls. Whether the probability (P) for this H_0 is correct, a two-tail t-test was performed.

$$t = \frac{\text{difference of means}}{\text{SE of difference between means}}$$

$$= \frac{\overline{X}_1 - \overline{X}_2}{S\overline{X}1 - \overline{X}2} = \frac{X_1 - X_2}{\frac{\sqrt{s_1^2 + s_2^2}}{n}} = \frac{40.3 - 37.5}{\frac{\sqrt{(8.15)^2 + (6.35)^2}}{16}} = 1.084$$

$$df = 2(n-1) = 2(16-1) = 30$$

T-score (1.084) is compared with critical t-scores at df = 30 for different levels of significance from the t-table, e.g., $t_{0.05(30)} = 2.042$, $t_{0.02(30)} = 2.457$, $t_{0.01(30)} = 2.750$.

Since calculated t-score (1.084) is lower than even the critical t for 0.05 level of significance, the probability (P) for this H_0 is correct. Therefore, null hypothesis cannot be rejected and thereby there is no significant difference between the means of two groups in this study ($P > .005$).

1.6 Advanced bioinformatics tools in biological sciences

Bioinformatics is a continuous emerging field as it is utmost necessary to handle, analyze, and store the volumes of diversified data that are constantly generated worldwide. Though discussion of all the new inventions and tools is beyond the scope of this chapter, in the following section a preliminary idea is given on the basic bioinformatics tools related to nucleotide sequence and phylogenetic analyses along with the presently available databases that are used to store and retrieve information. However, the tool to be used depends on the type of analysis needed.

1.6.1 Sequence analysis

Like BLAST, ClustalW and Clustl Omega are used to match the nucleotide or protein sequences to find their evolutionary history or origin based on homology matrices [17,18]. The similarity in profile patterns for nucleotide or protein sequences is obtained by Expression Profiler and Gene Quiz [19,20]. Besides this, a wide range of bioinformatics tools are presently available that are used for primary sequence analysis like JIGSAW (to find genes and to annotate the splicing sites in the selected DNA sequences), novoSNP (to find the single nucleotide variation in the DNA sequence), WebGeSTer (to search for transcription terminator sequences to predict the termination sites of the genes during transcription), Genscan (to predict the exon-intron sites in genomic sequences), Sequerome (sequence profiling), and Softberry Tools (to annotate the genomes of eukaryots and prokaryots along with the prediction of the structure and the function of RNA and proteins; www.softberry.com). Similarly, *ab initio gene identification* is a new tool used to identify complementary DNA sequences, coding DNA sequences, and expressed sequence tags and to annotate extended genomic sequences and predict new genes. The tools are based on mathematical modeling and statistical inferences, such as dynamic programming, regression analysis, hidden Markov model, artificial neural network, clustering, and sequence mining [21].

1.6.2 Phylogenetic analysis

Molecular phylogenetics; Phylogenetic Analysis by Maximum Likelihood (both are based on maximum likelihood method), PHYLIP (phylogenetic studies), JStree (open-source library to view and edit phylogenetic trees), TreeView (to view the phylogenetic trees, with the provision of changing view); Jalview (alignment editor used for refining the alignment) are various software used to construct and interpret phylogenetic trees [22–25].

1.6.3 Sequence databases

At present, three types of bioinformatics databases are available: primary (experimental data); secondary (information from analyses of data of primary database), and composite (information from different primary databases). The primary nucleotide database includes GenBank and DNA Data Bank of Japan, storing nucleotide sequences; Rfam, storing multiple sequence alignments with RNA sequences; and European Nucleotide Archive, storing information about experimental workflows related to nucleotide sequencing. Similarly, Ensembl and PIR are the databases containing annotated genomes of eukaryotes, including human, mouse, and other vertebrates, supporting genomic and proteomic researches.

1.7 Conclusion

This chapter focused on the basic elements of bioinformatics and biostatistics with reference to nucleic acid analysis and microbial ecology studies. However, numerous bioinformatics tools are applied for analysis of amino acids, protein conformation, gene expression, lipid analysis, and very recent -omics technology, including metagenomics, metaproteomics, and metatranscriptomics. The chapter aimed to describe and explain the basic and advanced applications of statistics in interpretation of biological data with examples. With lucid language it has tried to describe the important bioinformatics tools that are used worldwide for biological data analyses. The author would consider the effort to be rewarded if the intended readers find this chapter useful to them.

References

[1] Maxam AM, Gilbert W. A new method for sequencing DNA. Proc Natl Acad Sci Unit States Am 1977;74(2):560–4.

[2] Kumar S, Stecher G, Tamura K. MEGA7: molecular evolutionary genetics analysis version 7.0 for bigger datasets. Mol Biol Evol 2016;33:1870–4.

[3] Wiesemüller B, Rothe H. Q-factor analysis as a tool for phylogenetic studies of morphometric data. Anthropol Anzeiger 2006;64(3):345–53.

[4] Shannon CE. A mathematical theory of communication. Bell Syst Tech J 1948;27:379–423.

[5] Simpson EH. Measurement of diversity. Nature 1949;163:688.

[6] Hill MO. Diversity and evenness: a unifying notation and its consequences. Ecology 1973;54:427–32.

[7] Tuomisto H. A diversity of beta diversities: straightening up a concept gone awry. Part 1. defining beta diversity as a function of alpha and gamma diversity. Ecography 2010;33:2–22.
[8] Jost L. Partitioning diversity into independent alpha and beta components. Ecology 2007;88:2427–39.
[9] Whittaker RH. Evolution and measurement of species diversity. Taxon 1972;21(2/3):213–51.
[10] Magurran AE. Measuring biological diversity. Malden, Massachusetts, USA: Blackwell Publishing; 2004. p. 264.
[11] Ramette A. Multivariate analyses in microbial ecology. FEMS Microbiol Ecol 2007;62(2):142–60.
[12] Chao A. Nonparametric estimation of the number of classes in a population. Scand J Stat 1984;11(4):265–70.
[13] Bray JR, Curtis JT. An ordination of the upland forest communities of Southern Wisconsin. Ecol Monogr 1957;27(4):325–49.
[14] Lozupone C, Knight R. UniFrac: a new phylogenetic method for comparing microbial communities. J Appl Environ Microbiol 2005;71(12):8228–35.
[15] Hammer Ø, Harper DAT, Ryan PD. PAST: paleontological statistics software package for education and data analysis. Paleontol Electron 2001;4(1):9.
[16] Schloss PD, Westcott SL, Ryabin T, Hall JR, Hartmann M, Hollister EB, Lesniewski RA, Oakley BB, Parks DH, Robinson CJ, Sahl JW, Stres B, Thallinger GG, Van Horn DJ, Weber CF. Introducing mothur: open-source, platform-independent, community-supported software for describing and comparing microbial communities. Appl Environ Microbiol 2009;75(23):7537–41.
[17] Thompson JD, Higgins DG, Gibson TJ, Clustal W. Improving the sensitivity of progressive multiple sequence alignment through sequence weighting, position-specific gap penalties and weight matrix choice. Nucleic Acids Res 1994;22:4673–80.
[18] Sievers F, Wilm A, Dineen D, Gibson TJ, Karplus K, Li W, Lopez R, McWilliam H, Remmert M, Söding J, Thompson JD, Higgins DG. Fast, scalable generation of high-quality protein multiple sequence alignments using Clustal Omega. Mol Syst Biol 2011;11(7):539.
[19] Parmigiani G, Garrett ES, Irizarry RA, Zeger SL. The analysis of gene expression data: an overview of methods and software. New York: Springer; 2003.
[20] Hoersch S, Leroy C, Brown NP, Andrade MA, Sander C. The GeneQuiz web server: protein functional analysis through the Web. Trends Biochem Sci 2000;25:33–5.
[21] Mehmood MA, Sehar U, Ahmad N. Use of bioinformatics tools in different spheres of life sciences. J Data Min Genom Proteonomics 2014;5:2.
[22] Adachi J, Hasegawa M. MOLPHY, programs for molecular phylogenetics, I: PROTML, maximum likelihood inference of protein phylogeny. Institute of Statistical Mathematics; 1992.
[23] Yang Z. PAML: a program package for phylogenetic analysis by maximum likelihood. Bioinformatics 1997;13:555–6.
[24] Retief JD. Phylogenetic analysis using PHYLIP. In: Misener S, Krawetz SA, editors. Bioinformatics methods and protocols. Methods mol biol, vol. 132. Totowa, NJ: Humana Press; 2000.
[25] Clamp M, Cuff J, Searle SM, Barton GJ. The Jalview Java alignment editor. Bioinformatics 2004;20(3):426–7.

Chapter 2

Genome sequence analysis for bioprospecting of marine bacterial polysaccharide-degrading enzymes

Md Imran, Sanjeev C. Ghadi
Department of Biotechnology, Goa University, Taleigao Plateau, Goa, India

2.1 Introduction

Complex polysaccharides (CPs) are composed of repeating units of homo/hetero monosaccharides linked by various glycosidic linkages and having diverse functional groups. CPs are widely present in plants, animals, and microorganisms. They promote structural integrity and shield organisms from predators [1]. In marine ecosystems, seaweeds and crustaceans are the main source of CPs. Additionally, the mangrove ecosystem receives a lot of plant litter rich in CPs. Agar, alginate, carrageenan, xylan, pullulan, pectin, cellulose, chitin, etc. are the predominant CPs that are widely present in marine organisms. The repeating units of monosaccharides are heavily substituted by various functional groups rendering CPs recalcitrant, and hence they are also referred to as insoluble complex polysaccharides (ICPs). Conversion of these ICPs into simpler oligosaccharides/metabolizable sugars requires the synergistic action of various polysaccharide-degrading enzymes that hydrolyzes CPs into their respective oligosaccharides/metabolizable sugars. Marine bacteria are the major sources of polysaccharide-degrading enzymes.

Polysaccharide-degrading enzymes demonstrate a wide range of applications in various industries such as paper, pulp, textiles, cosmetics, as well as in pharmacology. Therefore, various polysaccharide-degrading enzymes, such as agarase, alginate lyase, carrageenase, chitinase, and xylanase, were purified from various sources including marine bacteria. The polysaccharide-degrading enzymes producing bacteria are ubiquitous in marine ecosystems and have been reported from costal water, sediments, deep sea, seaweeds, and exoskeleton of crustacean [2—8]. Traditionally, the polysaccharide-degradation capability

Advances in Biological Science Research. https://doi.org/10.1016/B978-0-12-817497-5.00002-1

of microbes has been widely assessed by dye/precipitant-based plate assay [9]. As the various marine bacteria are reported to produce multiple polysaccharide-degrading enzymes, the dye-based plate assay becomes limited for the bioprospecting purposes [2,10−12]. However, with the advancement of technology, researchers have used genome sequencing strategy for the holistic identification of polysaccharide-degrading genes present in the bacterial genome [13,14]. Interestingly, genome analysis revealed unprecedented presence of polysaccharide-degrading genes in marine bacterial genome [13−15]. Genome study also enables understanding of mechanisms followed by polysaccharide-degrading enzymes for the hydrolysis of polysaccharides. Additionally, closer look at the marine bacterial genome revealed the organization of polysaccharide-degrading genes in polysaccharide utilization loci (PUL) for the various polysaccharides.

This chapter describes the identification of polysaccharide-degrading genes in the genome of various marine bacteria. A stepwise strategy describing the procedure for the identification of polysaccharide-degrading genes in the newly sequenced genome is presented. The importance of the carbohydrate active enzyme database (CAZy database) in the bioprospecting of polysaccharide-degrading genes/enzymes is also described. In the chapter's last section, strategies adopted for the validation of predicted polysaccharide-degrading genes are presented.

2.2 Marine polysaccharides and polysaccharide-degrading bacteria: an overview

Agar is the main constituent of the cell wall of red seaweed (Rhodophyceae). It is extracted from *Gelidium*, *Gracilaria*, and *Porphyran* spp. on industrial scale [16]. Agar is made up of agarose and agaropectin that is composed of alternating 3-O-linked β-D-galactopyranose (G) and 4-O-linked α-L-galactopyranose (L). The agar is hydrolyzed by the agarase. Based on the mode of action or cleavage pattern, agarases are classified as α-agarase and β-agarase. Although α-agarase is rare, β-agarase has been purified from several marine bacteria that have been isolated from different econiches, such as seawater, mangrove water, deep sea, seaweeds, etc. Alginate is the main structural component of cell wall of brown algae. It consists of (1−4)-linked β-D-mannuronate (M) and its C-5 epimer α-L-glucuronate (G) residues and comprises up to 40% of dry weight of seaweeds [17]. Some bacteria are also known to synthesize alginate [18,19]. Alginate lyase hydrolyzes alginate by cleaving the glycosidic bond through a β-elimination reaction mechanism [20]. Based on the substrate specificity, alginate lyases are categorized as polyM-, polyG-, and polyMG-specific lyases. The alginate lyase−producing bacteria have been reported from various marine sources including seaweeds [2,3,21−23], turban shell gut [24], sea mud [25], and deep sea sediment [4].

Carrageenan is a sulfated polysaccharide found in many red seaweeds, such as *Kappaphycus alvarezii*, *Gigartina skottsbergii*, *Chondrus crispus*, and *Eucheuma denticulatum*. The classification of carrageenans depends on the amount and the location of sulfated ester (S) as well as by the presence of 3, 6-anhydro bridges in the α-linked residues. The κ (kappa), ï (iota), and λ (lambda) carrageenan are distinguished by the presence of one-, two-, or three-ester sulfate groups per repeating disaccharide unit, respectively, and degraded by κ-carrageenase, ï-carrageenase, and λ-carrageenase, respectively. The carrageenase-producing bacteria has been isolated from seaweed [26–30], seawater [4,31], and deep sea [6]. Chitin is the most copious regenerative polymer in the oceans and is a significant source of carbon and nitrogen for marine food web. It is the main component of crustaceans as well as insect exoskeleton. Chitinases are enzymes that degrade chitin. Based on the mode of action, chitinolytic enzymes can be divided into three categories: exochitinases, demonstrating activity only for the nonreducing end of the chitin chain; endochitinases, which hydrolyze internal β-1, 4-glycoside; and β-N-acetylglucosaminidase, which cleaves N-acetyl glucosamine (GlcNAc) units sequentially from the nonreducing end of the substrate [32,33]. The chitinolytic bacteria have been isolated from shrimp/crab shell [7,8], sea sponge [34], sea sediment [35,36], and seawater [37,38].

2.3 Identification of polysaccharide-degrading genes through genome annotation

In recent years, genomes of several marine bacteria have been sequenced and annotated with special emphasis on identification of polysaccharide-degrading genes. Genome annotation studies revealed the presence of various polysaccharide-degrading genes in the genome of marine bacteria that would not be possible to identify by traditional dye/precipitant-based plate assay (Table 2.1). For example, in the genome of *Cellulophaga omnivescoria* W5C, many polysaccharide-degrading genes, including genes for agarase, neoagarobiose hydrolase, alginate lyase, carrageenase, α-glucosidase, β-glucosidase, porphyranase, α-L-arabino furanosidase, and α-L-fucosidase, were identified [39]. Similarly, genes encoding agarase, alginate lyase, xylanase, carrageenase, chitinase, amylase, cellulase, pectin/pectate lyase, and α-L-fucosidase were identified in the genome of *Flammeovirga pacifica* WPAGA1 [14]. In another marine bacterium, *Microbulbifer mangrovi* DD-13, genes for agarase, alginate lyase, carrageenase, chitinase, xylanase, amylase, pullulanase, and pectinase were identified through genome sequence annotation [13]. In the genome of *Pseudoalteromonas* sp. Strain A601, isolated from coastal sediments, multiple genes for agarase, alginate lyase, xylanase, chitinase, amylase, glucanase, and glucosidases were identified [40]. Likewise, several polysaccharide-

TABLE 2.1 Polysaccharide-degrading genes identified in the genomes of marine bacteria.

S.No.	Marine bacteria	Isolation source	Genome size (bp)	Genome completion status	Total genes	No. of polysaccharide-degrading genes	Genes related to the degradation of major polysaccharides	References
1	*Cellulophaga omnivescoria* W5C	Red seaweed	3,803,581	Draft	3334	21 (0.63%)	Agarase (04), neoagarobiose hydrolase (02), alginate lyase (01), carrageenase (05), α-glucosidase (03), β-glucosidase (02), porphyranase (02), α-L-arabino furanosidase (01), α-L-fucosidase (01)	[39]
2	*Flammeovirga pacifica* WPAGA1	Marine sediment	6,610,326	Complete	5036	98 (1.95%)	Agarase (14), alginate lyase (04), xylanase (10), carrageenase (08), chitinase (10), amylase (08), cellulase (02), pectin/pectate lyase (04), α-L-fucosidase (17)	[15]
3	*Microbulbifer mangrovi* DD-13	Mangrove water	4,528,106	Draft	3749	55 (1.47%)	Agarase (05), alginate lyase (02), xylanase (02), carrageenase (01), chitinase (04), amylase (05), pullulanase (01), pectinase/pectate lyase (03), arylsulfatase (05)	[13]

4	*Pseudoalteromonas* sp. strain A601	Coastal sediment	4,894,552	Draft	4488	33 (0.74%)	Agarase (05), alginate lyase (05), xylanase (01), chitinase (06), amylase (03), glucanase (05), glucosidases (08)	[40]
5	*Flammeovirga* sp. OC4	Seawater	8,060,000	Draft	5898	69 (1.17%)	Agarase (05), carrageenase (04), xylanase (05), mannanase (03), xylosidase (03), arylsulfatase (22), sulfatase (27)	[56]
6	*Microbulbifer elongatus* HZ11	Seawater	4,223,108	Draft	3293	17 (0.52%)	Agarase (05), alginate lyase (09), cellulase (01), amylase (02)	[44]
7	*Catenovulum* sp. Strain DS-2	Intestine of *Haliotis diversicolor*	4,572,520	Draft	4090	24 (0.59%)	β-agarase (17), α-agarase (01), κ-carrageenase (03), ï-carrageenase (03)	[57]
8	Alteromonadaceae sp. Strain G7	Coastal seawater	3,140,906	Draft	3439	89 (2.59%)	Sulfatases (50), glycoside hydrolase (17), agarase (13), β-galactosidase (08), cellulase (01)	[58]
9	*Microbulbifer* sp. CCB-MM1	Mangrove forest	3,864,326	Complete	3313	–	Cellulase, α-amylase, pullulanase, β-glucanase	[59]
10	*Cellulophaga* sp. KL-A	Decaying marine algae	3,888,471	Draft	3503	–	Carrageenase (02), β-agarase (03)	[60]

Continued

TABLE 2.1 Polysaccharide-degrading genes identified in the genomes of marine bacteria.—cont'd

S.No.	Marine bacteria	Isolation source	Genome size (bp)	Genome completion status	Total genes	No. of polysaccharide-degrading genes	Genes related to the degradation of major polysaccharides	References
11	*Saccharophagus degradans* 2-40	Decaying *Spartina alterniflora*	5,057,531	Complete	4017	–	Cellulase, mannase, xylanase, endo-β-1,3-glucanase (08), pectinase, pullulanase, agarase, alginate lyase, chitinase, amylase	[14]
12	*Alginobacter alginolytica* HZ22	Surface of *Laminaria japonica*	3,994,770	Complete	3371	–	Alginate lyase, β-agarase, β-galactosidase, β-glucosidase, endo-β-1,4-glucanase, chitinase, α-L-fucosidase, β-xylosidase, α-L-arabinofuranosidase, polygalacturonase, pectate lyase, α-amylase, α-glucosidase, β-xylosidase, enod-1,4-β-xylanase	[43]
13	*Bacillus weihaiensis* Alg07	Marine area	4,344,873	Complete	4267	–	Alginate lyase, oligo-alginate lyase, α/β-amylase, β-glucanase, pullulanase, β-glucosidase	[55]

degrading genes have been identified in the genome of *Flammeovirga* sp. OC4, *Microbulbifer elongatus* HZ11, *Catenovulum* sp. Strain DS-2, and Alteromonadaceae sp. strain G7 and are listed in Table 2.1.

2.4 Identification of polysaccharide-degrading genes in newly sequenced bacterial genome: a guide for beginners

To the best of our knowledge, genome analysis of *Saccharophagus degradans* was the first report on extensive identification and characterization of polysaccharide-degrading genes using genome annotation strategy [14]. For the sequencing of *Saccharophagus* genome, clone of genomic DNA in fosmid and plasmid were generated. During the *Saccharophagus* genome annotation, Critica and Glimmer were used for the automated prediction of gene [14]. CAZy database was used to identify the polysaccharide-degrading genes. Subsequently, next-generation sequencing (NGS) emerges and several bacterial genomes were sequenced and annotated with special emphasis to identify the polysaccharide-degrading genes (Table 2.1). The steps involved in identification of polysaccharide-degrading genes of a microorganism after whole genome sequencing may be described as follows. Briefly, genomic DNA of good quality could be isolated using the standard protocols or by using commercially available DNA isolation kit. The quantity and quality of the DNA is determined using Qubit Fluorometer and Nano-Drop, respectively. To determine the sequence, NGS library is prepared and the sequence is read on Illumina's or other appropriate sequencing platforms. The obtained sequencing reads need to be quality filtered to remove the low-quality reads with ambiguous sequences "N." The quality-filtering step can be done using FASTX-Toolkit (http://hannonlab.cshl.edu/fastx_toolkit/). The quality-filtered reads are used for assembly to obtain the scaffolds or contigs that would represent the genome of the bacteria. The assembly can be done by one of the two suitable approaches, namely, de novo assembly or reference-guided assembly. The de novo assembly is performed using SOAPdenovo assembler (http://soap.genomics.org.cn/soapdenovo.html). To do the assembly using SOAPdenovo, K-mer values needs to be optimized in order to get the best assembly. Subsequently, the genes are predicted using Glimmer or Prodigal [41,42]. Total predicted genes are subjected to BLASTx analysis against NCBI's nonredundant protein database to predict the putative functions of the proteins. Results of BLASTx analysis are imported to the excel sheet. The polysaccharide-degrading genes are later screened from the excel sheet. Additionally, the nucleotide/amino acid sequence of identified polysaccharide-degrading genes can be annotated against CAZy database for the presence of conserved domains specific to polysaccharide-degrading enzymes.

2.5 Genome sequence analysis unravels organization of polysaccharide-degrading genes as polysaccharide utilization loci

The criteria adopted for the identification of PULs was based on the proximity of polysaccharide-degrading genes and transporters homologous to SusC/SusD-like proteins encoding genes and TonB-dependent receptor/transporter genes in the genome [43]. The genes for agarase, alginate lyase, and carrageenase, along with the genes for several transporters/receptors, are observed in the vicinity of each other and constitute PULs for agar, alginate, and carrageenan in the genome of *Cellulophaga omnivescoria* W5C [39]. The agar PUL of strain W5C was comprised of 4 β-agarases, nine sulfatases, 3 β-galactosidases, and several receptor/transporter proteins. Similarly, the PULs for alginate degradation/utilization in strain W5C comprised of 3 CDS annotated as enzymes of PLs families. Additionally, for λ- and ï-carrageenan, two separate PULs were identified and each was comprised of two λ- and two ï-carrageenase-encoding genes, respectively [39]. In the genome of another bacterium, namely *Algibacter alginolytica*, 82 glycoside hydrolases and 18 polysaccharide lyases are structured in a total of 17 PULs [43]. The major PULs identified in the genome of *Algibacter alginolytica* were specific for alginate, agarose, cellulose, chondroitin, chitin, fucosides, fucoidans, rhamnogalacturonan homogalacturonan, xanthan, and xylan degradation [43].

2.6 Genome annotation: a potential tool for the elucidation of glycometabolism pathways

Genome sequencing and annotation identifies the total polysaccharide-degrading genes in a genome, which enables elucidation of glycometabolism pathways. Genome annotation studies of *Flammeovirga pacifica* WPAGA1 indicated presence of 14 β-agarase belonging to GH16, GH50, and GH86 family. Additionally, a protein belonging to GH117 was also identified. Based on the presence of these agarase genes, agar degradation pathway was elucidated in *F. pacifica* WPAGA1 [15]. Additionally, the degradation product of major polysaccharides mediated by the enzymes identified in the genome of *F. pacifica* WPAGA1 (Table 2.1) would be further metabolized following the glycometabolism pathways, such as D-galacturonate metabolic pathway, hexose metabolic pathway, pentose metabolic pathway, and D-galactose metabolic pathway were elucidated based on genome annotation data [15]. Similarly, a pathway for conversion of alginate into ethanol was predicted in *Microbulbifer elongatus* HZ11 based on computationally identified genes encoding alginate lyase and other proteins for ED pathway [44].

2.7 CAZy database: a promising tool for the classification of polysaccharide-degrading genes/enzymes identified in newly sequenced genomes

Bioprospecting of polysaccharide-degrading enzymes also includes insight into the mechanism followed by the enzymes to hydrolyze the respective polysaccharides. The mechanism followed by the polysaccharide-degrading enzymes could be easily predicted by identifying the catalytic modules in the enzyme. CAZy database is a promising tool for the detection of catalytic modules in the polysaccharide-degrading enzymes/genes. In the CAZy database, the polysaccharide-degrading enzymes belong to the glycoside hydrolase (GHs) and polysaccharide lyase (PL) families. GHs and PLs are further classified into 145 GH families and 26 PL families, respectively, in the CAZy database [45]. The newly annotated polysaccharide-degrading genes would be easily classified in GHs or PLs family by aligning them against CAZy database. The dbCAN server and carbohydrate active enzyme analysis toolkit are the potential resources for the analysis of genes against CAZy database [46,47]. Once the polysaccharide-degrading enzymes are classified in an appropriate GH/PL family, the possible mechanism followed by them would be easily predicted as the enzymes belonging to GHs family, hydrolyzes the glycosidic bond either via overall retention or via overall inversion of the anomeric configuration [48]. On the contrary, enzymes of PLs family follow β elimination mechanism to hydrolyze anionic polysaccharides [49]. Using the CAZy database, polysaccharide-degrading genes/enzymes of several marine bacteria are classified in GHs and PLs families and their possible mechanisms were predicted. For example, the agarase genes identified in the genome of *Microbulbifer mangrovi* DD-13 were classified in GH16, GH50, and GH86 families, whereas the alginate lyase of *M. mangrovi* DD-13 are reported to belong PL6 and PL17 families [13]. Additionally, carrageenase, chitinase, xylanase, and pullulanase of *M. mangrovi* DD-13 were predicted to belong to GH82, GH18, GH10, and GH13 families, respectively [13]. Similarly, the agarase genes annotated from the genome of *Microbulbifer elongatus* HZ11 belonged to GH16, GH50, and GH86 families, whereas alginate lyase was predicted to be a member of PL6, PL7, PL17, and PL18 families [44]. Likewise, following the above-described strategy, bioprospecting of polysaccharide-degrading genes from *Paraglaciecola* sp. S66 with respect to the identification of catalytic modules was achieved and genes identified for the agarase (GH16, GH50, and GH86), alginate lyase (PL6, PL7, PL14, and PL17), pectinase (GH28, GH88, GH105, PL1, PL9, PL10, and PL11), carrageenase (GH16 and GH82), and xylanase (GH10 and GH11) were classified to an appropriate family [50]. In continuation of further bioprospecting of polysaccharide-degrading genes/enzymes, identification of noncatalytic carbohydrate-binding modules (CBMs) would be a worthy step as CBMs reportedly enhance the degradative potential of cognate catalytic modules and can be considered superior over the polysaccharide-degrading enzymes whose catalytic modules are not appended with CBMs [51]. To identify the CBMs associated with cognate catalytic module of polysaccharide-degrading enzymes, again CAZy database is a promising resource. By annotating the genes against the CAZy database, CBMs have been identified in several marine bacteria [13,14,44,52].

2.8 Validation of computationally identified polysaccharide-degrading genes in the genomes of marine bacteria

Computationally identified polysaccharide-degrading genes in the genome revealed the extraordinary genomic potential of marine bacteria toward polysaccharide degradation. However, the genomic potential needs to be validated for the true potential of microbes for the polysaccharide degradation. This could be achieved by the growth studies of bacteria on various polysaccharides individually supplemented in the growth medium as a sole carbon source. Additionally, during the growth period of the bacteria, respective polysaccharide-degrading enzyme activity as well as polysaccharide-degrading enzymes mediated concomitantly produced reducing sugar can be measured. Dinitrosalicylic (DNSA) and Nelson's method have been widely used to measure the polysaccharide-degrading enzyme activity during the growth of the microbes [53,54]. Similar strategy has been adopted for the validation of the computationally identified polysaccharide-degrading genes from *M. mangrovi* DD-13 and *F. pacifica* WPAGA1 [13–15]. As the *Gracilaria* sp. and *Sargassam* sp. are majorly composed of agar and alginate, respectively, the predicated presence of genes encoding agarase and alginate lyase were validated by the growth studies of *M. mangrovi* DD-13 on *Gracilaria* and *Sargassam* powder, respectively [13]. Another strategy adopted by the researchers for the validation of computationally annotated data was studying expression of the identified genes through RT-PCR [14,55].

Acknowledgments

The authors acknowledge Goa University for providing necessary infrastructure facilities. Md. Imran sincerely acknowledges the Department of Biotechnology, Government of India, New Delhi, for financial support as DBT-JRF and DBT-SRF.

References

[1] Jonnadula R, Imran M, Poduval PB, Ghadi SC. Effect of polysaccharide admixtures on expression of multiple polysaccharide-degrading enzymes in *Microbulbifer* strain CMC-5. Biotechnol reports 2018;17:93–6.

[2] Li JW, Dong S, Song J, Li CB, Chen XL, Xie BB, Zhang YZ. Purification and characterization of a bifunctional alginate lyase from *Pseudoalteromonas* sp. SM0524. Mar Drugs 2011;9(1):109–23.

[3] Jonnadula R, Verma P, Shouche YS, Ghadi SC. Characterization of *Microbulbifer* strain CMC-5, a new biochemical variant of *Microbulbifer elongatus* type strain DSM6810T isolated from decomposing seaweeds. Curr Microbiol 2009;59(6):600–7.

[4] Kobayashi T, Uchimura K, Miyazaki M, Nogi Y, Horikoshi K. A new high-alkaline alginate lyase from a deep-sea bacterium *Agarivorans* sp. Extremophiles 2009;13(1):121–9.

[5] Khambhaty Y, Mody K, Jha B. Purification and characterization of κ-carrageenase from a novel γ-proteobacterium, *Pseudomonas elongata* (MTCC 5261) syn. *Microbulbifer elongatus* comb. Nov. Biotechnol Bioproc Eng 2007;12(6):668–75.

[6] Ohta Y, Hatada Y. A novel enzyme, λ-carrageenase, isolated from a deep-sea bacterium. J Biochem 2006;140(4):475–81.

[7] Revathi M, Saravanan R, Shanmugam A. Production and characterization of chitinase from *Vibrio* species, a head waste of shrimp *Metapenaeus dobsonii* (Miers, 1878) and chitin of *Sepiella inermis* Orbigny, 1848. Adv Biosci Biotechnol 2012;3(04):392.

[8] Annamalai N, Rajeswari MV, Vijayalakshmi S, Balasubramanian T. Purification and characterization of chitinase from *Alcaligenes faecalis* AU02 by utilizing marine wastes and its antioxidant activity. Ann Microbiol 2011;61(4):801–7.

[9] Imran M, Poduval PB, Ghadi SC. Bacterial degradation of algal polysaccharides in marine ecosystem. In: Naik MM, Dubey SK, editors. Marine pollution and microbial remediation. Singapore: Springer; 2017. p. 189–203.

[10] González JM, Mayer F, Moran MA, Hodson RE, Whitman WB. *Microbulbifer hydrolyticus* gen. nov., sp. nov., and *Marinobacterium georgiense* gen. nov., sp. nov., two marine bacteria from a lignin-rich pulp mill waste enrichment community. Int J Syst Evol Microbiol 1997;47(2):369–76.

[11] Ekborg NA, Taylor LE, Longmire AG, Henrissat B, Weiner RM, Hutcheson SW. Genomic and proteomic analyses of the agarolytic system expressed by *Saccharophagus degradans* 2–40. Appl Environ Microbiol 2006;72(5):3396–405.

[12] Vashist P, Nogi Y, Ghadi SC, Verma P, Shouche YS. *Microbulbifer mangrovi* sp. nov., a polysaccharide-degrading bacterium isolated from an Indian mangrove. Int J Syst Evol Microbiol 2013;63(7):2532–7.

[13] Imran M, Pant P, Shanbhag YP, Sawant SV, Ghadi SC. Genome sequence of *Microbulbifer mangrovi* DD-13^{T} reveals its versatility to degrade multiple polysaccharides. Mar Biotechnol 2017;19(1):116–24.

[14] Weiner RM, Taylor II LE, Henrissat B, Hauser L, Land M, Coutinho PM, Bayer EA. Complete genome sequence of the complex carbohydrate-degrading marine bacterium, *Saccharophagus degradans* strain 2-40^{T}. PLoS Genet 2008;4:e1000087.

[15] Gao B, Jin M, Li L, Qu W, Zeng R. Genome sequencing reveals the complex polysaccharide-degrading ability of novel deep-sea bacterium *Flammeovirga pacifica* WPAGA1. Front Microbiol 2017;8:600.

[16] Fu XT, Kim SM. Agarase: review of major sources, categories, purification method, enzyme characteristics and applications. Mar Drugs 2010;8(1):200–18.

[17] Matsubara Y, Kawada R, Iwasaki KI, Kimura Y, Oda T, Muramatsu T. Cloning and sequence analysis of a gene (aly PG) encoding poly (α-L-guluronate) lyase from *Corynebacterium* sp. strain ALY-1. J Biosci Bioeng 2000;89(2):199–202.

[18] Clementi F. Alginate production by *Azotobacter vinelandii*. Crit Rev Biotechnol 1997;17(4):327–61.

[19] Albrecht MT, Schiller NL. Alginate lyase (AlgL) activity is required for alginate biosynthesis in *Pseudomonas aeruginosa*. J Bacteriol 2005;187(11):3869–72.

[20] Gacesa P. Enzymic degradation of alginates. Int J Biochem 1992;24(4):545–52.

[21] Wang YH, Yu GL, Wang XM, Lv ZH, Zhao X, Wu ZH, Ji WS. Purification and characterization of alginate lyase from marine *Vibrio* sp. YWA. Acta Biochim Biophys Sin 2006;38(9):633–8.

[22] Malissard M, Chavagnat F, Duez C, Vacheron MJ, Guinand M, Michel G, Ghuysen JM. Overproduction and properties of the mannuronate alginate lyase AlxMB. FEMS Microbiol Lett 1995;126(2):105–11.

[23] Chavagnat F, Colette DUEZ, Guinand M, Potin P, Barbeyron T, Henrissat B, Ghuysen JM. Cloning, sequencing and overexpression in *Escherichia coli* of the alginate lyase-encoding aly gene of *Pseudomonas alginovora*: identification of three classes of alginate lyases. Biochem J 1996;319(2):575–83.

[24] Fu XT, Lin H, Kim SM. Purification and characterization of a Na+/K+ dependent alginate lyase from turban shell gut *Vibrio* sp. YKW-34. Enzym Microb Technol 2007;41(6):828–34.

[25] Wang Y, Guo EW, Yu WG, Han F. Purification and characterization of a new alginate lyase from a marine bacterium *Vibrio* sp. Biotechnol Lett 2013;35(5):703–8.

[26] Li S, Jia P, Wang L, Yu W, Han F. Purification and characterization of a new thermostable [kappa]-carrageenase from the marine bacterium *Pseudoalteromonas* sp. QY203. J Ocean Univ China 2013;12(1):155.

[27] Liu GL, Li Y, Chi Z, Chi ZM. Purification and characterization of κ-carrageenase from the marine bacterium *Pseudoalteromonas porphyrae* for hydrolysis of κ-carrageenan. Process Biochem 2011;46(1):265–71.

[28] Sun F, Ma Y, Wang Y, Liu Q. Purification and characterization of novel κ-carrageenase from marine *Tamlana* sp. HC4. Chin J Oceanol Limnol 2010;28(6):1139–45.

[29] Zhou MH, Ma JS, Li J, Ye HR, Huang KX, Zhao XW. A κ-carrageenase from a newly isolated Pseudoalteromonas-like bacterium, WZUC10. Biotechnol Bioproc Eng 2008;13(5):545–51.

[30] Mou H, Jiang X, Liu Z, Guan H. Structural analysis of kapp-carrageenan oligosaccharides released by carrageenase from marine *Cytophaga* MCA-2. J Food Biochem 2004;28(3):245–60.

[31] Vijayaraghavan R, Prabha VK, Rajendran S. Isolation and characterization of K-carrageenase from Bacillus subtilis. World Sci Tech 2012;2(9):13–8.

[32] Fukamizo T, Kramer KJ. Mechanism of chitin hydrolysis by the binary chitinase system in insect moulting fluid. Insect Biochem 1985;15(2):141–5.

[33] Kurita K. Controlled functionalization of the polysaccharide chitin. Prog Polym Sci 2001;26(9):1921–71.

[34] Han Y, Yang B, Zhang F, Miao X, Li Z. Characterization of antifungal chitinase from marine *Streptomyces* sp. DA11 associated with South China Sea sponge *Craniella australiensis*. Mar Biotechnol 2009;11(1):132.

[35] Annamalai N, Giji S, Arumugam M, Balasubramanian T. Purification and characterization of chitinase from *Micrococcus* sp. AG84 isolated from marine environment. Afr J Microbiol Res 2010;4(24):2822–7.

[36] Guo SH, Chen JK, Lee WC. Purification and characterization of extracellular chitinase from *Aeromonas schubertii*. Enzym Microb Technol 2004;35(6):550–6.

[37] Murao S, Kawada T, Itoh H, Oyama H, Shin T. Purification and characterization of a novel type of chitinase from *Vibrio alginolyticus* TK-22. Biosci Biotechnol Biochem 1992;56(2):368–9.

[38] Hiraga K, Shou L, Kitazawa M, Takahashi S, Shimada M, Sato R, Oda K. Isolation and characterization of chitinase from a flake-chitin degrading marine bacterium, *Aeromonas hydrophila* H-2330. Biosci Biotechnol Biochem 1997;61(1):174–6.

[39] Valdehuesa KN, Ramos KR, Moron LS, Lee I, Nisola GM, Lee WK, Chung WJ. Draft genome sequence of newly isolated agarolytic bacteria *Cellulophaga omnivescoria* sp. nov. W5C carrying several gene loci for marine polysaccharide degradation. Curr Microbiol 2018;75(7):925–33.

[40] Li J, Cheng Y, Wang D, Li J, Wang Y, Han W, Li F. Draft genome sequence of the polysaccharide-degrading marine bacterium *Pseudoalteromonas* sp. strain A601. 20 Genome Announc 2017;5(29):e00590–17.

[41] Delcher AL, Harmon D, Kasif S, White O, Salzberg SL. Improved microbial gene identification with GLIMMER. Nucleic Acids Res 1999;27:4636–41.

[42] Hyatt D, Chen GL, LoCascio PF, Land ML, Larimer FW, Hauser LJ. Prodigal: prokaryotic gene recognition and translation initiation site identification. BMC Bioinformatics 2010;11:119.

[43] Sun C, Fu GY, Zhang CY, Hu J, Xu L, Wang RJ, Su Y, Han SB, Yu XY, Cheng H, Zhang XQ. Isolation and complete genome sequence of *Algibacter alginolytica* sp. nov., a novel seaweed-degrading Bacteroidetes bacterium with diversity of putative polysaccharide utilization loci. Appl Environ Microbiol 2016. AEM-00204.

[44] Sun C, Chen YJ, Zhang XQ, Pan J, Cheng H, Wu M. Draft genome sequence of *Microbulbifer elongatus* strain HZ11, a brown seaweed degrading bacterium with potential ability to produce bioethanol from alginate. Mar Genomics 2014;18:83–5.

[45] Lombard V, Golaconda RH, Drula E, Coutinho PM, Henrissat B. The carbohydrate active enzymes database (CAZy) in 2013. Nucleic Acids Res 2014;42:D490–5.

[46] Yin Y, Mao X, Yang J, Chen X, Mao F, Xu Y. dbCAN: a web resource for automated carbohydrate-active enzyme annotation. Nucleic Acids Res 2012;40(W1):W445–51.

[47] Park BH, Karpinets TV, Syed MH, Leuze MR, Uberbacher EC. CAZymes Analysis Toolkit (CAT): web service for searching and analyzing carbohydrate-active enzymes in a newly sequenced organism using CAZy database. Glycobiology 2010;20:1574–84.

[48] Davies G, Henrissat B. Structures and mechanisms of glycosyl hydrolases. Structure 1995;3(9):853–9.

[49] Michaud P, Da Costa A, Courtois B, Courtois J. Polysaccharide lyases: recent developments as biotechnological tools. Crit Rev Biotechnol 2003;23(4):233–66.

[50] Schultz-Johansen M, Glaring MA, Bech PK, Stougaard P. Draft genome sequence of a novel marine bacterium, *Paraglaciecola* sp. strain S66, with hydrolytic activity against seaweed polysaccharides. Genome Announc 2016;4(2):e00304–16.

[51] Boraston AB, Bolam DN, Gilbert HJ, Davies GJ. Carbohydrate binding modules: fine-tuning polysaccharide recognition. Biochem J 2004;382:769–81.

[52] Boncan DA, David AM, Lluisma AO. A CAZyme-rich genome of a taxonomically novel rhodophyte-associated carrageenolytic marine bacterium. Mar Biotechnol 2018:1–21.

[53] Miller GL. Use of dinitrosalicylic acid reagent for determination of reducing sugar. Anal Chem 1959;31:426–8.

[54] Nelson N. A photometric adaptation of the somogyi method for the determination of glucose. J Biol Chem 1944;153:375–80.

[55] Zhu Y, Chen P, Bao Y, Men Y, Zeng Y, Yang J, Sun J, Sun Y. Complete genome sequence and transcriptomic analysis of a novel marine strain *Bacillus weihaiensis* reveals the mechanism of brown algae degradation. Sci Rep 2016;6:38248.

[56] Liu Y, Yi Z, Cai Y, Zeng R. Draft genome sequence of algal polysaccharides degradation bacterium, *Flammeovirga* sp. OC4. Mar genomics 2015;21:21–2.

[57] Shan D, Li X, Gu Z, Wei G, Gao Z, Shao Z. Draft genome sequence of the agar-degrading bacterium *Catenovulum* sp. strain DS-2, isolated from intestines of *Haliotis diversicolor*. Genome Announc 2014;2(2):e00144–14.

[58] Kwak MJ, Song JY, Kim BK, Chi WJ, Kwon SK, Choi S, Chang YK, Hong SK, Kim JF. Genome sequence of the agar-degrading marine bacterium *Alteromonadaceae* sp. strain G7. J Bacteriol 2012;194(24):6961–2.

[59] Moh TH, Lau NS, Furusawa1 G, Amirul AA. Complete genome sequence of *Microbulbifer* sp. CCB-MM1, a halophile isolated from Matang Mangrove Forest, Malaysia. Stand Genomic Sci 2017;12:36.

[60] Shan D, Ying J, Li X, Gao Z, Wei G, Shao Z. Draft genome sequence of the carrageenan-degrading bacterium *Cellulophaga* sp. strain KL-A, isolated from decaying marine algae. Genome Announc 2014;2(2):e00145–14.

Chapter 3

Proteomics analysis of *Mycobacterium* cells: challenges and progress

Suvidha Samant, Abhishek Mishra
Dixa Education and Research, Alto Porvorim, Goa, India

3.1 Introduction

Tuberculosis is currently one of the deadliest infectious diseases that impose a global threat to health care. According to international statistics, one-third of the world's population is infected with tuberculosis and approximately 1.3 million deaths from the disease were reported in 2015 across the globe [1]. The causal organism, *Mycobacterium tuberculosis*, has the ability to survive in the latent stage for years before reactivation and subsequently result in disease progression. The keystone of tuberculosis control efforts has been demolished by the eruption of multidrug resistance (resistance to frontline drugs) and extremely drug-resistant (resistance to all reported drugs) *M. tuberculosis*. In 2015, the World Health Organization estimated that there were about 450,000 people who developed multidrug-resistant tuberculosis, which caused 170,000 deaths across the globe [1]. Although the first-line drugs isoniazid and rifampicin potentially kill 99.0%–99.9% of *M. tuberculosis* cells in vitro within 4–7 days, resistance to both of these bactericidal drugs develops rapidly under both axenic and host microenvironment [2–5]. This implies an urgent need to discover potential antimicrobial compounds as well as new preventive methods to combat this threatening disease. More important is developing new therapies that provide sustainable and effective strategy against bacterial pathogens.

Both pathogenic and nonpathogenic *Mycobacterium* sp. are classically categorized as an obligate aerobe. However, the plasticity of genomes in this group challenges this idea by exhibiting machinery to survive in anaerobic environments, whether soil or host system. This metabolic adaptation leads to changes in the cell shape and makes cell walls thicker, which confers a formidable challenge in the isolation of total proteome [6]. These hardships

Advances in Biological Science Research. https://doi.org/10.1016/B978-0-12-817497-5.00003-3

cause variation in similar samples and create inconsistencies in proteomic analysis. Both nonpathogenic and clinically relevant mycobacteria have been studied enthusiastically using genomic and proteomic approaches. To date, 213 mycobacterial species have been described, many of which are associated with infectious diseases in humans or animals [7,8]. These species also include clinically hazardous *M. tuberculosis*, *Mycobacterium leprae*, and *Mycobacterium ulcerans*, which cause tuberculosis, leprosy, and Buruli ulcer, respectively. Unsurprisingly, the features of the metabolism, physiology, and genetic organization of *M. tuberculosis* are of the greatest interest. In 1998, Cole et al. deciphered the complete genome sequence of *M. tuberculosis*, and it is not surprising that this led to thorough study of the proteome of this organism [9]. The comprehensive proteomic approaches could achieve the existence of 97% of 4012 annotated proteins, which was predicted by whole-genome sequencing [10]. In the recent past, the genomic organization of the tuberculosis pathogen was the major focus of research groups, which resulted in genomic sequencing data for more than 10,000 *M. tuberculosis* strains. This provided the availability of diverse phenotypes and genotypes of screened strains. However, whole-genome sequencing revealed a limited applicability of comparative analysis in drug resistance and pathogenicity bacteria [11]. Thus, this approach suggests that the majority of the point mutations that categorize these strains are the mutations in the promoter regions of the genes and/or regions encoding proteins with a hypothetical function, and their role is still unknown in the physiology of mycobacteria [9]. Moreover, it is evident that as a species *M. tuberculosis* exhibits very little genomic sequence diversity compared to other *M. tuberculosis* complex (MTBC) organisms, such as *Mycobacterium africanum*, *Mycobacterium bovis*, *Mycobacterium avium*, and *M. ulcerans*. Despite limited genetic variability, members of the MTBC have been shown to exhibit vast discrepancies in a phenotypic presentation in terms of virulence, elicited immune response, and transmissibility. Therefore, a functional genomics analysis in this context is relevant to encipher the pathogen genome information by using quantitative and qualitative proteomics.

Today we have extensive protocols on rapid and cost-effective isolation of nucleic acid from different mycobacterial strains that are applied in various laboratories as mundane molecular biology techniques [12–14]. However, in the present scenario the isolation of total protein fraction is relatively difficult and so is proteomic analysis. As the cell wall in mycobacteria is resistant to acids and alkalis, it makes it hard to achieve total proteome isolation. While considering the extremely slow culture growth of the *Mycobacterium* species, the complexity of accumulation of a large bacterial mass requires a better protein isolation strategy for better protein coverage. Hence, there is the need to develop specialized methods for protein extraction and proteomic analysis of pathogen.

In 1986, Mark Wilkins introduced the term *proteome*, which combined parts of two words: *PROTEin* and *genOME* and the term *proteomics* is consequently analogous to genomics, which studies the sequence of genomic DNA and its function in the organism [15,16]. Similarly, in proteomics we study the total proteins produced by an organism/cell in a specific condition. This includes the qualitative and quantitative composition of proteins in the cell and their interaction with nucleic acids, carbohydrates, and lipids. It also includes if there is any addition or removal of prosthetic groups (e.g., carbohydrates or lipids) to the set of proteins or posttranslational modifications (e.g., carbamylation) [17–19]. It is evident that proteomics still lags behind genomics and transcriptomics due to the requirement of sophisticated instruments, less sensitivity, and protein coverage. However, the number of works that utilize proteomic methods for studying infectious agents has recently increased significantly. In the past decade, R. Aebersold's research group has made an unmatchable contribution to the development of the proteomics of tough cell wall genus, *Mycobacterium* [20,21]. A generalized model of proteomics study in *Mycobacterium* that is described in this chapter is presented in concise form in Fig. 3.1.

3.2 Proteome analysis of axenic mycobacteria

The conventional method two-dimensional gel electrophoresis coupled with MALI-TOF or ESI-MS is used for protein qualitative and quantitative analysis to study changes in the experimental sample. Contrary to its popularity, this method remains ambiguous due to its low load, poor resolution, and manual error, which results in multiple spots for a single protein or multiple proteins in a single spot [19–21] (Fig. 3.2). A majority of these problems are due to poor separation of hydrophobic, acidic, or alkaline proteins, identification and concentration of proteins, whereas functional analysis remained uncertain due to undetermined posttranslational modifications. Similarly, protein degradation, protein isoforms, and variability in tryptic digest recovery from the gel cause a further challenge in the quantitative and qualitative analysis of proteins [19,20].

However, the variations attributed to trypsin digestion can be normalized by using any other protease to achieve overlapping fragments. To compensate for these limitations of 2D-PAGE-MS, an advanced gel-free and label-free method had been developed for proteomic characterization of untreated and treated *Mycobacterium smegmatis* cells using liquid chromatography-elevated energy mass spectrometry (LC/MS^E) [7,22]. In this mode of acquisition, low- and high-collision energy are used in alternate fashion to achieve intact peptide ions and peptide product ions, respectively. This method provides a useful filter that allows the researcher to perform a protein homology search with a robust subset of peptide mass fingerprint. A total 463 peptide mass fingerprint was identified, which exhibited a $\log_{10}$ intensity concentration change that ranged from 0.19 to 6.4. An additional success can be achieved in this method by enriching peptides with careful standardization of HPLC flow rate and

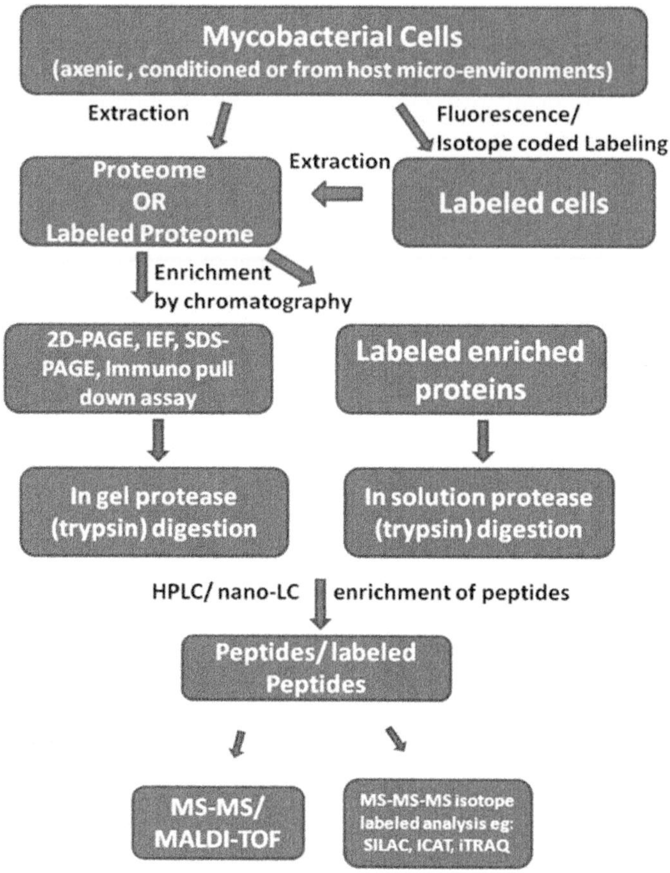

FIGURE 3.1 Generalized flowchart of proteomics operation in *Mycobacterium* cells.

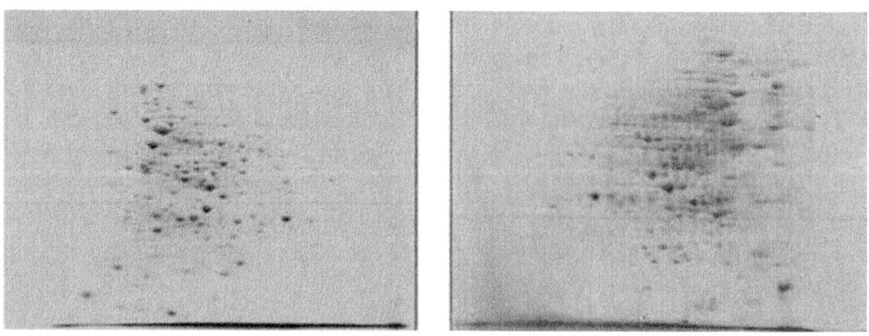

FIGURE 3.2 Variation in identical protein samples run in similar 2D-PAGE condition.

mobile phase composition during the experiment [23]. More recently the studies utilized mass spectrometry tools to investigate the proteomes of seven clinically relevant bacteria belonging to four different species, namely, *M. tuberculosis*, *M. bovis*, *M. bovis* BCG, and *M. avium*. The qualitative and quantitative analysis of proteomes reveals a varying degree of pathogenicity and virulence factors that rationalize the observed phenotypic differences. An advanced protein preparation, nano-liquid chromatography, sophisticated transcendence in mass spectrometry and data analysis by using novel bioinformatics strategies have yielded high protein coverage and was based on high confidence peptides. Through this approach, authors could identify a total of 3788 unique *M. tuberculosis* proteins out of a theoretical proteome of 4023 proteins, and they identified an average of 3290 unique proteins for each of the MTBC organisms (representing 82% of the theoretical proteomes), as well as 4250 unique *M. avium* proteins (80% of the theoretical proteome) [24].

3.3 Proteome analysis of mycobacteria-infected cells

The intracellular microenvironment protects mycobacteria from a cellular and humoral component of the host immune system. To overcome host cell defense mechanisms, pathogens manipulate the normal passage through the induction of autophagy to form a distinct replicative membranous compartment. Furthermore, *M. tuberculosis* is also able to induce vascular rupture to enrich its microenvironment by nutrients from the host cytosol and consequentially evade the host defense pathways [27]. Simultaneously, both these events favor pathogen growth in the host system. Therefore, several proteomics studies of pathogen-infected cells were initiated to collect host cell factors required to establish successful colonization and adherence at the molecular level. By using liquid chromatography-tandem MS (LC-MS/MS) approaches for proteome analysis, a higher number of peptide identification rates were achieved consistently. This allowed the quantitative and qualitative analysis of complex cellular proteomes at a different time of disease progression, as in the beginning of infection and when the full-fledged disease is caused [21]. In addition, the incorporation of stable isotope-coded labels to the proteome prior to MS analysis further improved the relative quantification of heterogeneous proteomic samples [25]. The major reason for the success of this approach is due to its ability to provide vast coverage of proteome, consistency in quantification, and high-throughput data processing and analysis.

3.4 Proteome analysis of mycobacteria-containing host vacuoles

Several recent MS studies were focused on analyzing proteome of different cell organelles of host cells infected by *M. tuberculosis* (Mtb) to elucidate the host–pathogen signaling mechanism [26]. These studies could identify wide protein coverage (1429, number of proteins identified in the study) and highlighted the inflammatory responses via pathogen pattern receptors as well as MHC-I processing and presentation. This interaction of bacterial and host factors at the intersections of pattern receptor recognition signaling pathways is analyzed in a recent review [27]. Moreover, various transcriptional factors were identified in the cytoplasm and nucleus by using SILAC [28]. A similar success is achieved by iTRAQ labeling where rare and unique proteins were identified in the endoplasmic reticulum [29], secreted exosomes [30], plasma and cell organelle membranes [31,32]. The signal strength of SILAC was robust enough to detect the cross talk between immune interaction and lipid metabolism [33]. Several reports used label-free quantitative MS approach to study phagosomes [34—38], and provided an integrated detailed insight into the cellular mechanisms of mycobacterial infection. For example, Rao et al. could describe the role of endoplasmic reticulum in phagosome biogenesis and maturation. Li et al. further demonstrated how this phagosome maturation influences antigen presentation. Similarly, membrane profiling on human dendritic cells could identify 115 proteins that were upregulated in response to heat-killed Mtb [32]. Among those host proteins, aminopeptidase N was found largely overexpressed, and the authors could demonstrate that membranous aminopeptidase N is capable of binding live bacteria and is involved in antigen presentation that impaired T cell activation to facilitate Mtb pathogenesis. Herweg et al. [34] demonstrated the effect of Mtb infection on various host organelles. Comprehensively, these studies could highlight not only the discrete components of host or pathogen signaling but also the cross talk between the orthogonally parallel signaling pathways.

3.5 Conclusion

The whole-genome analysis of the MTBC members exhibits the greatest degree of genetic conservation (99.9%), which initially led to the assumption that genetic variation amongst different strains would not be of any clinical significance. However, the advances in proteomics have opened up new approaches in studying tuberculosis by making it easier to find solutions to many complex problems, such as genetic homogeneity in clinical strains or host—pathogen interaction, or pathogen adaptations at various stages of pathogenesis. Despite the fact that proteomics lags behind genomics and transcriptomics due to limitations in instruments and insufficient sensitivity, an increasing number of studies involving proteomic approaches for the investigation of infectious agents are being published and meticulously reviewed. For example, virulence factors and their mechanisms of action, host and pathogen response to the infectious process have been described using proteomic analysis. Proteomics has made it possible to describe the unique features of various *M. tuberculosis* strains more thoroughly. A recent review that describes the scope of proteogenomics analysis of mycobacterial cells demonstrates that the genomic variability of mycobacterial strains is reflected at the proteomic level and hence the data of comparative proteomics may be helpful in understanding the phenotypic differences of different groups of bacteria, such as the degree of drug sensitivity and virulence. Whereas the proteomic studies facilitate the correct deciphering of genomic information, most mass spectrometric techniques rely on the database that is based on the forward integration of DNA sequences computationally translated as proteins. However, genome-wide analysis studies have shown that the protein annotation based on genomic data is inconsistent and not reliable [39]. This remains a challenge that needs to be addressed in the near future.

Overall we can conclude that studying tuberculosis pathogenesis at the functional level can contribute to the identification of the basic metabolic and physiological conditions essential for a successful course of infection, as well as the virulence mechanisms that allow *M. tuberculosis* to modulate the host's immune response. It is imperative that the proteins produced exclusively during the entry of mycobacteria into the host system are crucial for their survival under these conditions. Hence, these should be considered as potential targets for drug development and designing new treatment regimens, when strains of multiple types and extensive drug resistance continue to spread. Hence, studying the complete proteomic profile of mycobacteria may contribute to a better understanding of pathogen physiology and even tuberculosis treatment.

References

[1] reportWHO Report. 2015 Fact Sheet, Available online at: http://www.who.int/mediacentre/factsheets/fs104/en/.

[2] Voskuil MI, Schnappinger D, Visconti KC, Harrell MI, Dolganov GM, Sherman DR, et al. Inhibition of respiration by nitric oxide induces a *Mycobacterium tuberculosis* dormancy program. J Exp Med 2003;198:705–13.

[3] Kumar A, Toledo JC, Patel RP, Lancaster Jr JR, Steyn AJ. *Mycobacterium tuberculosis* DosS is a redox sensor and DosT is a hypoxia sensor. Proc Natl Acad Sci Unit States Am 2007;104:11568–73.

[4] Fontán P, Aris V, Ghanny S, Soteropoulos P, Smith I. Global transcriptional profile of *Mycobacterium tuberculosis* during THP-1 human macrophage infection. Infect Immun 2008;76:717–25.

[5] Shiloh MU, Manzanillo P, Cox JS. *Mycobacterium tuberculosis* senses host-derived carbon monoxide during macrophage infection. Cell Host Microbe 2008;3:323–30.

[6] Zhang Y. Persistent and dormant tubercle bacilli and latent tuberculosis. Front Biosci 2004;9:1136–56.

[7] Mishra A. Proteomic analysis of hydroxyl radical-induced resuscitation of hypoxia-induced dormant mycobacterial cells. Int J Mycobacteriology 2018;7(1):7.

[8] Tortoli E. The new mycobacteria: an update. FEMS Immunol Med Microbiol 2006;48(2):159–78.

[9] Cole ST, Brosch R, Parkhill J, Garnier T, Churcher C, Harris D, Gordon SV, Eiglmeier K, Gas S, Barry CE. Deciphering the biology of *Mycobacterium tuberculosis* from the complete genome sequence. Nature 1998;393(6685):537–44.

[10] Schubert OT, Mouritsen J, Ludwig C, Rost HL, Rosenberger G, Arthur PK, Claassen M, Campbell DS, Sun Z, Farrah T. Absolute proteome composition and dynamics during dormancy and resuscitation of *Mycobacterium tuberculosis*. Cell Host Microbe 2013;13(5):602–12.

[11] Warner DF, Mizrahi V. Complex genetics of drug resistance in *Mycobacterium tuberculosis*. Nat Genet 2013;45(10):1107–8.

[12] Akhtar S, Sarkar S, Mishra A, Sarkar D. A method to extract intact and pure RNA from mycobacteria. Anal Biochem 2011;417(2):286–8.

[13] Ghodbane R, Asmar S, Betzner M, Linet M, Pierquin J, Raoult D, Drancourt MJ. Rapid diagnosis of tuberculosis by real-time high-resolution imaging of *Mycobacterium tuberculosis* colonies. Clin Microbiol 2015;53(8):2693–6.

[14] Tyler AD, Christianson S, Knox NC, Mabon P, Wolfe J, van Domselaar G, Graham MR, Sharma MK. Comparison of sample preparation methods used for the next-generation sequencing of *Mycobacterium tuberculosis*. PLoS One 2016;11(2):e0148676.

[15] Wilkins MR, Pasquali C, Appel RD, Ou K, Golaz O, Sanchez JC, Yan JX, Gooley AA, Hughes G, Humphery- Smith I, Humphery-Smith I, et al. From proteins to proteomes: large scale protein identification by two-dimensional electrophoresis and amino acid analysis. Biotechnology 1996;14(1):61–5.

[16] James P. Protein identification in the post-genome era: the rapid rise of proteomics. Q Rev Biophys 1997;30(4):279–331.

[17] Hakkinen L, Uitto VJ, Larjava H. Cell biology of gingival wound healing. Periodontol 2000;24:127–52.

[18] Molloy MP, Witzmann FA. Proteomics: technologies and applications. Brief Funct Genomic 2002;1(1):23–39.

[19] Monteoliva L, Albar JP. Differential proteomics: an overview of gel and non-gel approaches. Brief Funct Genomic 2004;3(3):220–39.

[20] Gygi SP, Rist B, Gerber SA, Turecek F, Gelb MH, Aebersold R. Quantitative analysis of complex protein mixtures using isotope-coded affinity tags. Nat Biotechnol 1999;17(10):994–9.

[21] Schubert OT, Rost HL, Collins BC, Rosenberger G, Aebersold R. Quantitative proteomics: challenges and opportunities in basic and applied research. Nat Protoc 2017;12:1289–94.

[22] Mishra A, Sarkar D. Qualitative and quantitative proteomic analysis of Vitamin C induced changes in *Mycobacterium smegmatis*. Front Microbiol 2015;6:451.

[23] Mishra A. Commentary: superoxide generation and its involvement in the growth of *Mycobacterium smegmatis*. Front Microbiol 2017;8:1114.

[24] Peters JS, Calder B, Gonnelli G, Degroeve S, Rajaonarifara E, Mulder N, et al. Identification of quantitative proteomic differences between *Mycobacterium tuberculosis* lineages with altered virulence. Front Microbiol 2016;7.

[25] Rauniyar N, Yates III JR. Isobaric labeling-based relative quantification in shotgun proteomics. J Proteome Res 2014;13:5293–309.

[26] He Y, Li W, Liao G, Xie J. *Mycobacterium tuberculosis*-specific phagosome proteome and underlying signaling pathways. J Proteome Res 2012;11:2635–43.

[27] Mishra A, Akhtar S, Jagannath C, Khan A. Pattern recognition receptors and coordinated cellular pathways involved in tuberculosis immunopathogenesis: emerging concepts and perspectives. Mol Immunol 2017;87:240–8.

[28] Wang H, Dong D, Tang S, Chen X, Gao Q. PPE38 of *Mycobacterium marinum* triggers the cross-talk of multiple pathways involved in the host response, as revealed by subcellular quantitative proteomics. J Proteome Res 2013;12:2055–66.

[29] Saquib NM, Jamwal S, Midha MK, Verma HN, Manivel V. Quantitative proteomics and lipidomics analysis of endoplasmic reticulum of macrophage infected with *Mycobacterium tuberculosis*. Int J Proteom 2015;2015:270438. https://doi.org/10.1155/2015/270438.

[30] Diaz G, Wolfe LM, Kruh-Garcia NA, Dobos KM. Changes in the membrane-associated proteins of exosomes released from human macrophages after *Mycobacterium tuberculosis* infection. Sci Rep 2016;6:37975. https://doi.org/10.1038/srep37975.

[31] Kumar A, Jamwal S, Midha MK, Hamza B, Aggarwal S, Yadav AK, et al. Dataset generated using hyperplexing and click chemistry to monitor temporal dynamics of newly synthesized macrophage secretome post infection by Mycobacterial strains. Data Brief 2016;9:349–54. https://doi.org/10.1016/j.dib.2016.08.055.

[32] Kuo CP, Chang KS, Hsu JL, Tsai IF, Lin AB, Wei TY, et al. Analysis of the immune response of human dendritic cells to *Mycobacterium tuberculosis* by quantitative proteomics. Proteome Sci 2016;14:5. https://doi.org/10.1186/s12953-016-0095-8.

[33] Long J, Basu Roy R, Zhang YJ, Antrobus R, Du Y, Smith DL, et al. Plasma membrane profiling reveals upregulation of ABCA1 by infected macrophages leading to restriction of mycobacterial growth. Front Microbiol 2016;7:1086. https://doi.org/10.3389/fmicb.2016.01086.

[34] Herweg JA, Hansmeier N, Otto A, Geffken AC, Subbarayal P, Prusty BK, et al. Purification and proteomics of pathogen-modified vacuoles and membranes. Front Cell Infect Microbiol 2015;5:48. https://doi.org/10.3389/fcimb.2015.00048.

[35] Lee BY, Jethwaney D, Schilling B, Clemens DL, Gibson BW, Horwitz MA. The *Mycobacterium bovis* bacille Calmette-Guerin phagosome proteome. Mol Cell Proteomics 2010;9:32–53. https://doi.org/10.1074/mcp.M900396-MCP200.

[36] Li Q, Singh CR, Ma S, Price ND, Jagannath C. Label-free proteomics and systems biology analysis of mycobacterial phagosomes in dendritic cells and macrophages. J Proteome Res 2011;10:2425–39. https://doi.org/10.1021/pr101245u.

[37] Rao PK, Singh CR, Jagannath C, Li Q. A systems biology approach to study the phagosomal proteome modulated by mycobacterial infections. Int J Clin Exp Med 2009;2:233–47.

[38] Shui W, Petzold CJ, Redding A, Liu J, Pitcher A, Sheu L, et al. Organelle membrane proteomics reveals differential influence of mycobacterial lipoglycans on macrophage phagosome maturation and autophagosome accumulation. J Proteome Res 2011;10:339–48. https://doi.org/10.1021/pr100688h.

[39] Bespyatykh JA, Shitikov EA, Ilina EN. Proteomics for the investigation of mycobacteria. Acta Naturae 2017;9(1):15. 2.

Chapter 4

Plant proteomics: a guide to improve the proteome coverage

Chhaya Patole[1], Laurence V. Bindschedler[2]
[1]Proteomics Division, National Centre for Biological Sciences, Bengaluru, India; [2]School of Biological Sciences, Royal Holloway University of London, Egham, Surrey, United Kingdom

4.1 Introduction

In the last two decades, proteome analysis has gained popularity in plant science. Progress in plant proteomics is noticeable based on the increased number of publications and reviews [1–3]. The mass spectrometer is a critical aspect in proteomic experiments, and advances in the mass spectrometers, software, and sample preparation methodology have helped to push the detection limits associated with the high-dynamic range of protein conenctraions. Regardless of the analysis approach used, a meticulous experimental design and high-quality sample preparation are critical for a successful proteomic experiment.

It is vital to produce high-quality protein extracts for proteomic analysis to obtain meaningful and biological relevant liquid chromatography coupled with tandem mass spectrometry (LC-MS/MS) data. As compared to the human and other organisms, protein yield obtained from plant samples are limited due to the presence of interfering biological contaminants such as nucleic acid, polyphenols, lipids, carbohydrates, and other pigments [4,5]. Plant proteomics has come a long way from identifying a handful of proteins to the large-scale proteome profiling. This was possible due to the development of a number of novel ways to improve the sample preparation approaches to overcome difficulties inherent to the high-dynamic range of protein concentrations [5]. Using these improved approaches, it is possible to target low abundant proteins, which play a major role in signaling pathways. Large-scale proteomics could significantly contribute to our understanding of physiological mechanisms underlying plant stress tolerance [1].

Advances in Biological Science Research. https://doi.org/10.1016/B978-0-12-817497-5.00004-5

This chapter will focus on sample preparation strategies for shotgun proteomics and critical parameters to consider while designing the proteomics experiment, with a brief focus on advances in plant proteomics.

4.2 Hurdles associated with plant proteins sample preparation for mass spectrometry–based proteomics

The analytical field of proteomics has come a long way from analyzing a handful of proteins to large-scale proteome profiling. The improvement of mass spectrometers accuracy and sensitivity, together with improved nanoelectrospray (nESI) ionization sources compatible with reliable nano-chromatography separations, are probably the main factors, rendering mass spectrometry-based proteomics accessible to most biologists nowadays. Moreover, the exponential increase of available sequenced genomes allows access to genome-wide proteomics studies of an increasing number of plant species, moving from the plant model *Arabidopsis thaliana* in 2000 [6], shortly followed by rice as the first sequenced food crop [7]. Presently, 236 genomes of angiosperms have been sequenced [8].

However, to allow for large-scale analysis, the preparation of a "good" protein sample compatible with downstream mass spectrometry analysis is of prime importance. As a consequence, extensive efforts have been invested in the optimization of the preparation of plant protein samples, overcoming mainly: (1) the low protein abundance in plant tissues, (2) the requirement of mechanically breaking up cell wall matrices to access cellular content and access cell wall proteins, (3) the high-dynamic range of individual protein concentrations, and (4) the abundant presence (relative to proteins) of contaminants, such as polysaccharides, nucleic acids, or polyphenols within the tissues. All these factors are detrimental to allow for high proteome coverage in mass spectrometry–based proteomics experiments [5,9,10]. Therefore, alternative approaches have to be used to prepare plant protein samples compatible for downstream mass spectrometry analyses. The most popular scenarios are summarized in Fig. 4.1.

4.3 Primary considerations to design suitable workflows for plant proteomics

Each type of biological sample is unique and may require minor or major modifications for sample preparation from standard protocols. When designing and optimizing a proteomics experiment, it starts with the sample preparation at the protein extraction stage, while considering protein and peptide fractionations compatible for downstream mass spectrometry analysis.

The following steps need to be considered when designing the proteomics workflow, before performing the biological treatments, to allow for the right

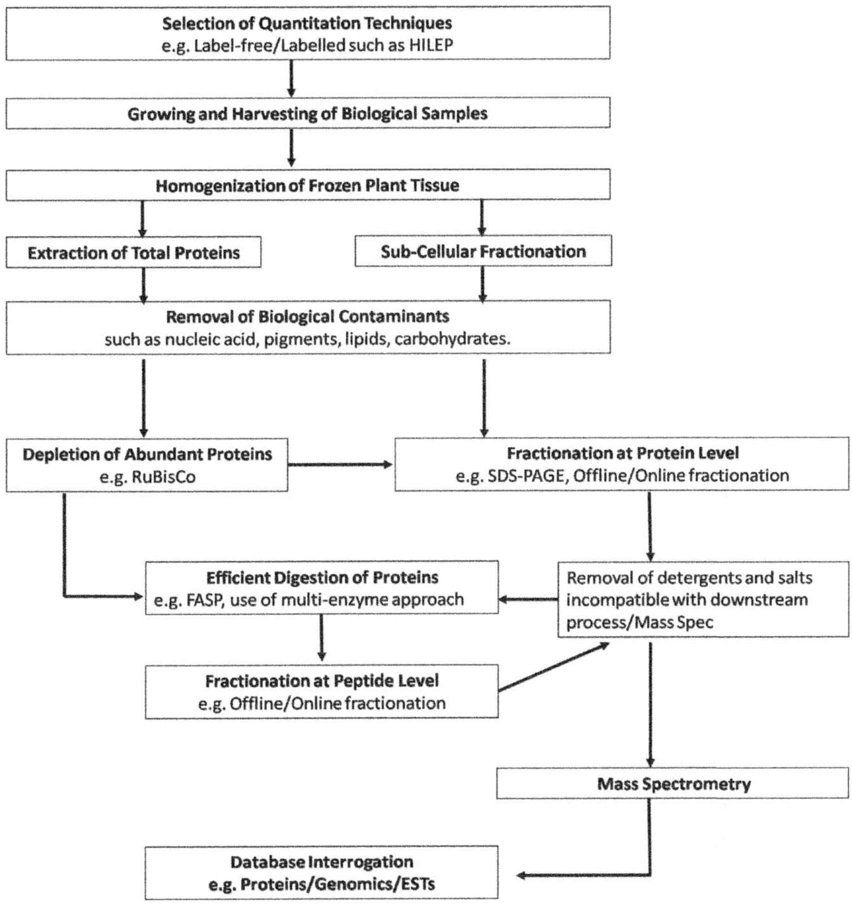

FIGURE 4.1 A systematic depiction of general workflow for proteomic analysis.

number of biological and technical repeats, to tackle the biological variation and sample-to-sample variation.

1. Consider setting up a quantitative approach rather than simply identifying the proteins.
2. For quantitative analysis, it is important to decide about the quantitative proteomic strategy to begin with, as the experimental workflow will need to be adapted depending on the relative quantification technique chosen, whether a label-free or label-based technique. Such decision will depend on the number of samples to quantify. Plant-specific metabolic labeling using "hydroponic isotope labeling of entire plants" (HILEP) [11,12] with ^{15}N salts has to be used as an alternative to "stable isotope labeling with amino acids in cell cultures" (SILAC; [13]). Plants have an internal system

to assimilate atmospheric CO_2, which compromises with the incorporation of fully labeled ^{13}C-Arginine used in SILAC [13]. HILEP is suitable for pair-wise comparisons, while isobaric chemical labeling with "isobaric tag for relative and absolute quantitation" (iTRAQ), tandem mass tag, or alike, usually at the peptide level, allows for multiplex (4 or 8) analysis [14]. Moreover, the stage at which the samples to compare are pooled together will be dependent on the type of labeling, whether metabolic or chemical (as reviewed in Ref. [14]). The stage at which samples are pooled together will dictate the number of technical replicates required for overcoming sample-to-sample variation [14]. Regarding publication guidelines, two biological replicates with three technical replicates are required for labeling based proteomics [14], while three or more biological replicates are required for label-free strategies [15,16].

3. A key for any successful experiment, whether proteomics or not, must include true independent biological repeats, where an experiment has been repeated at different times. It should also include proper reference controls (e.g., mock treatments) run in parallel with the conditions or treatments to test, to avoid interference by external uncontrolled variables, which could hinder the biological significance. It is also important to consider randomizing plant individuals for location and treatments, to avoid, for instance, variable greenhouse conditions, based on the plants location in the greenhouse. Time of the day at which samples are harvested also needs careful consideration while harvesting biological replicates in order to minimize variation due to the circadian clock.
4. At harvesting time, in order to stop biological processes or to avoid protein alteration or degradation, the biological sample has to be rapidly flash frozen in liquid nitrogen to avoid alteration of the proteome by oxidation, proteases, kinases, or phosphatases, or interference with polyphenolics following activation of polyphenol oxidases.
5. The use of internal standards such as synthetic labeled peptides, or the inclusion of a metabolic or chemical-labeled reference sample, as it has been done in super-SILAC experiments [17], are helpful for normalizing peptide amounts, avoiding technical variation occurring from sample to sample due to the quality of the nanoLC as the run spans over a period of time.
6. The selection of an appropriate protein extraction strategy very much depends on the nature of plant tissue being analyzed. The ideal extraction method should reproducibly capture the most comprehensive repertoire of proteins possible, while minimizing degradation and interference by nonproteinaceous-contaminating compounds. Grinding of the biological sample in liquid nitrogen, followed by extraction on ice with addition of proteases inhibitors, will minimize sample degradation when extracting in nondenaturing conditions. Addition of unsoluble polyvinylpolypyrrolidone (PVPP) is a requisite to remove polyphenols in order to avoid their

interference with proteins [11]. Denaturing conditions with extraction in organic solvents such as phenol [18] or TCA-acetone are often favored, as it minimizes protein degradation, while removing contaminants such as pigments, salts, phenolic compounds, and DNA [19,20], while it allows access to membrane proteins. However, denaturing conditions cannot be used if the protein structure or enzyme functionality is required for downstream analysis. The use of chaotropic agents, such as urea, are also good for membrane destabilization and protein solubilization, however, urea needs to be diluted sufficiently to allow for protein digestion when generating peptides.

a. Whatever the extraction protocol chosen, it is worth considering alternative extraction methods that may be complementary in terms of proteome coverage, when aiming to identify or quantify the largest possible number of proteins. To deepen/increase the proteome coverage, it should be considered including sample prefractionation, including subcellular fractionation, and the depletion of the most abundant proteins such as RuBisCO to increase the detectability of low-abundance proteins. Immuno depletion of RuBisCo can be performed with affinity chromatography columns coupled with anti-RuBisCO IgGs [5].

b. Most of the abundant proteins are of chloroplastic origin, or ribosomal proteins, thus organelle fractionation/depletion can, for instance, deplete for chloroplasts and improve the detection of other proteins.

c. The use of immobilized short peptide libraries (ICLL, etc.), to bind proteins to saturation, allows decreasing the relative protein amount of highly abundant proteins, thus reducing the dynamic range of protein concentrations.

However, if performing a quantitative analysis, it is important to consider the bias that such approaches might cause on protein quantitation, even if the quantification is based on relative comparison.

As stated before, it is important to have the protein extract in a solution that is suitable for protein digestion with trypsin or another protease chosen for peptide preparation. This may involve precipitating out proteins from solvent-based extraction, resuspension in a denaturing agent such as urea, but the urea concentration then needs to be reduced (below 1M), while keeping most proteins in solution to allow for protease digestion. Following protein digestion, sample cleanup to remove impurities like salts, detergents, or remaining solid particles is of prime importance for a successful downstream protein analysis. This can be done by solid phase extraction, offline, with commercially available tips packed with chromatography material, followed by online cleanup on a guard or trap column, at the start of the nLC-nESI-MS/MS run.

7. In particular, where large-scale analyses are concerned, either in terms of samples number or proteome coverage, it is important to consider time and

instrument access cost to match financial resources, while not compromising on biological and nLC–MS/MS technical repeats required. Although this will not be discussed further, the choice of the chromatographic run duration will depend on the nLC column length, the MS instrument speed, and very importantly, the sample complexity. The choice of the mass spectrometer is not discussed here, however, it is always important to consider its suitability in terms of its accuracy, sensitivity, and instrument speed while designing the nLC schedule.

8. And last, but not least, it is important to consider the statistical parameters to be used for the experimental validation, in particular, that the number of repeats (biological or technical) is often a limiting factor in proteomics analysis.

   Practical considerations and challenges associated with the various steps are discussed in more depth in the following sections.

4.3.1 Effective protein sample preparation: extraction and recovery from difficult plant samples

A lot of emphasis has been given in developing new technologies for high-throughput protein separation and identification using high-end mass spectrometers and software, while the most critical step in any proteomics study is protein extraction [21–23]. An ideal extraction protocol should reproducibly capture and solubilize all of the proteins in a given sample, while keeping the activity of proteases and other enzymes altering proteins, as well as nonproteinaceous contaminants to the minimum level [19,24]. Generally, protein extraction from plant material is more problematic as compared to tissues from other organisms as plant tissues usually have a poor protein content, while they are often rich in proteases, lipids, and phenolic compounds and need breaking up of cell walls to access cellular content [19,25]. Indeed, the prevalence of these contaminants is likely to interfere with protein stability and solubility, and hinder downstream separation [19,25].

4.3.1.1 Sample harvesting

Generally, plant samples are harvested and immediately snap frozen in liquid nitrogen to minimize protein degradation or modification, e.g., proteolytic or phosphatase activity. When the experimental aim is to describe presence or abundance of proteins within a proteome, frozen plant tissue can be processed immediately or stored at −80°C until further use; however, when subcellular fractionation, or other processes that require membrane integrity or some enzymatic function, protein complexes, fresh samples might be required [26].

4.3.1.2 Tissue homogenization and sample integrity

Tissue homogenization for cell lysis is a critical step as it not only releases proteins but also exposes these to proteases or other modifying enzymes such as phosphatases. Protein alteration must be prevented as it may change the true state of the biological sample, for instance, due to irreversible protein degradation or oxidation, reversible phosphorylation, or ubiquitination [24]. In particular, when proteins are extracted in non-denaturing conditions, to keep protein sample integrity, samples are kept on wet ice to slow down any enzymatic activity, and a cocktail of protease inhibitors (commercially available) should be added [24]. It is advisable to use a plant-specific protease inhibitor cocktail, as each inhibitor is effective against a different class of enzymes and classes of plant proteases differ from other organisms. If the phosphoproteome is being analyzed, phosphatase inhibitors should also be added [27,28]. Similarly, if studying ubiquitylated proteins, it is important to use deubiquitylating enzyme inhibitors in the extraction buffer.

4.3.1.3 Protein extraction in denaturing conditions

When membrane integrity or subcellular organelles are not required to be preserved, denaturing conditions are favored to fix the biological activity to capture the protein abundance changes at specific times and conditions. Therefore, we will focus on some of the most popular methods used to extract plant proteins in denaturing conditions, suitable for large-scale proteome analyses. There is "no single extraction" protocol or buffer system that could capture an entire proteome in a sample due to the diverse nature of proteins that vary in molecular sizes, charges, polarity or hydrophobicity, pH stability, posttranslational modifications, and cellular distribution, etc. [22]. Hence, it is essential to consider the nature of the sample tissue while deciding the extraction strategy that will favor a subproteome, e.g., native extraction will favor the identification of cytoplasmic proteins while membrane proteins will be underrepresented. With the increased access to mass spectrometry−based proteomics, progress has been made to develop and compare protein extraction protocols that could enhance crop proteomic analysis. A range of plant protein extraction protocols have been published, and the most common extraction methods are described here.

Commonly adopted plant protein extraction protocols are discussed here to give a general overview. Ten percent trichloroacetic acid (TCA) in acetone-based extraction is the most widely adopted method in the plant proteomics community [20,29]. This extraction method involves a minimum number of steps, while the simple organic solution used in the method is capable of disrupting membranes and denaturing proteins, thus preventing proteases and other enzymatic activities altering the protein constituents. At the same time, proteins precipitate out of the TCA-acetone solution, thus allowing for increasing the protein concentration of the extract, while removing acetone soluble contaminants [20]. After precipitation by centrifugation, the proteins can be resuspended in a urea-based denaturing buffer. The TCA/acetone procedure is extremely effective for young-growing vegetative tissues, whilst less effective for more mature and recalcitrant tissues. To overcome this hurdle, a modified TCA/phenol-based method is reported to be more effective for more mature tissues such as tomato, banana, and avocado fruits, which are known for their high levels of soluble cell wall polysaccharides and polyphenols [25]. Variations of the TCA-acetone method are reviewed in detail by Ref. [30].

An alternative method involves the solubilization of proteins in phenol, in presence or absence of SDS after extraction in a 0.7 M sucrose; 0.1 M KCl; 0.5 MTris-HCl, pH 7.5 and 50 mM EDTA, in presence of 2% β-mercaptoethanol added immediately before extraction to act as antioxidant. Proteins are subsequently precipitated with methanol and ammonium acetate [19]. This method is more effective for protein extractability and removal of contaminants from difficult tissues such as wood, tomato, banana, avocado fruits, and olive leaves, which contain large amounts of polyphenols [19].

Comparison of extraction methods revealed that a phenol-based extraction method outperformed methanol/chloroform and SDS-based extraction protocols in deep proteome profiling [4]. Another study author has evaluated 17 different methods to extract proteins from sunflower seeds [31]. Any traces of color in dry pellet reflects the presence of contaminants, such as pigments or oxidized phenolic compounds. In such cases, re-extraction with phenol, or any known precipitation methods, is advised, as it is likely to improve the proteome coverage [19]. Another strategy to increase the proteome coverage is the sequential extraction of tissues with different solvents. This is an effective means to generate distinct subproteomes, in order to favor the detection of low-abundance proteins.

A caveat in the use of a multistep extraction procedure is the number of fractions that require subsequent handling (e.g., re-solubilization, trypsin digestion). This might increase sample-to-sample variation, workflow complexity, and reliability, as well as increase mass spectrometer analytical time. However, if the goal is to increase the proteome coverage, adopting a fractionation strategy at the protein level, this will decrease the complexity of each subproteome, allowing for more in-depth analysis, i.e., identifying more low-abundance proteins. It is essential to adopt the extraction method with a minimum steps without compromising the protein quality.

4.3.1.4 Removal of biological contaminants and re-solubilization of proteins

Extraction of proteins from plant tissue is often complicated by the presence of other non-protein contaminants indigenous to the plant, such as organic acids, lipids, polyphenols, pigments, terpenes, etc. It is essential to remove these contaminants during the extraction procedure to obtain good-quality extracts for proteomics analysis. The quality of protein extracts is further improved by the addition of unsoluble PVPP or sodium metabisulfate during the protein extraction to remove interfering polyphenols [11,32]. Additional cleanup and concentration steps should also be carried out after the initial extraction to remove the biological contaminants. For example, an extraction protocol based on phenol is followed by chloroform methanol precipitation or ammonium acetate methanol precipitation [19,33].

4.3.2 Contaminant removal from or during protein digestion

The usage of detergents and salts is inescapable parts of the protein extraction as they accelerate the protein solubilization. Most of the detergents and non-volatile salts will interfere with enzymatic digestion and damaging the mass spectrometry analysis. Thus, detergent removal must be included in the sample preparation workflow. In the last decades, there has been major efforts on developing methods to remove the detergents with minimum protein loss [34,35]. Precipitation is one of the classical approaches to purify proteins, being first described through the application of high salt concentrations (salting out) and through the addition of organic solvents [34,36].

To optimize protein recovery through solvent precipitation, it is critical to first understand the factors controlling protein solubility. Proteins can be precipitated by causing perturbations in the solvent with respect to pH, ionic strength, and temperature [36]. In aqueous solutions, proteins adopt a structure that exposes hydrophilic regions to the surroundings, allowing the formation of a hydration layer that shields protein–protein interactions. Disruption of this hydration layer generally causes protein precipitation [37]. In proteomics, protein precipitation in acetone remains a widely adopted approach for sample purification ahead of mass spectrometry. The method is applicable for effective removal of sodium dodecyl sulfate (SDS) prior to MS analysis. The concentration of SDS should be reduced to below 0.01% prior to LC-MS analysis [38].

The degree of protein loss through acetone precipitation has been widely reported. Studies have shown that the precipitation efficiency depends on the ratio of chilled acetone and sample volume and incubation time. In 2002, Thongboonkerd [39] reported reduced recovery of urinary proteins after acetone precipitation, where they observed predominant loss of basic and hydrophobic proteins. In another study, 50% acetone was used to precipitate the proteins, which may not be optimal for recovery of all proteins [40]. For optimal recovery of complex mixtures through protein precipitation, it is generally accepted to employ four parts of cold acetone (i.e., 80% acetone) with overnight incubation of the sample in the freezer) [35,38].

Further, Crowell [41] has reported a modified version of acetone precipitation that includes acetone (80%) with 10 mM NaCl, which showed improved protein recovery over the conventional acetone method. The amount of salt required correlates with the amount of protein in the sample, as well as the intrinsic protein charge, and the dielectric strength of the solution [41]. Another precipitation technique used extensively is the methanol-chloroform precipitation. In 1984, Wessel and Flugge [42] introduced this precipitation method for soluble as well as hydrophobic proteins from dilute solutions, especially those containing membrane proteins. This method is very resilient to the presence of detergents, lipids, salt, buffers, and β-mercaptoethanol. The procedure includes addition of methanol:chloroform:water in 4:1:3 ratio to the pellet, with an additional three volumes of methanol.

Removal of detergents and salts could be carried out at protein or peptide level based on the downstream procedure. Many of the precipitation methods are employed at protein level. However, Yeung [43] demonstrated a simple method using ethyl acetate to remove interference such as octylglycoside, SDS, NP-40, and Triton X-100 from tryptic-digested samples with minimum loss of peptides. To achieve the better efficacy, the author recommended keeping the peptide volume between 50 and 100 μL and the salt concentration below 200 mM to prevent peptide loss to the organic phase during the extraction [43]. Alternative methods such as desalting spin columns, ultrafiltration, and molecular weight cut off filters were also employed to eliminate the detergents and salts from protein or tryptic peptide solutions. Desalting columns are widely available as C_{18} type resins.

4.3.3 Overcoming the high-dynamic range of protein concentrations for the discovery of low-abundant proteins

1. Enrichment and depletion strategies

   One of the constant challenge for proteomics is reduced protein identifications because of interference with high-abundance proteins. The challenge is particularly critical for plant proteomics analysis due to the presence of ribulose-1,5-bisphosphate carboxylase/oxygenase (RuBisCo) in green tissues; this abundant enzyme constitute 30%–50% of total plant protein [5,44]. Presence of RuBisCo in green plant tissues diminishes 2-DE resolution and shotgun proteomics analysis by limiting the loading capacity, or causing ion suppression during ionisation in the electrospray source. With time and the widening of the scope of proteomics, it was soon realized that there is a need to have depletion/enrichment strategies coupled with prefractionation stages before MS analysis to target the low-abundant proteins [44].

   In the past decades, a number of extraction methods with numerous depletion steps have been published, as summarized in Fig. 4.2. Kim and coworkers published the PEG-based depletion of RuBisCo from rice

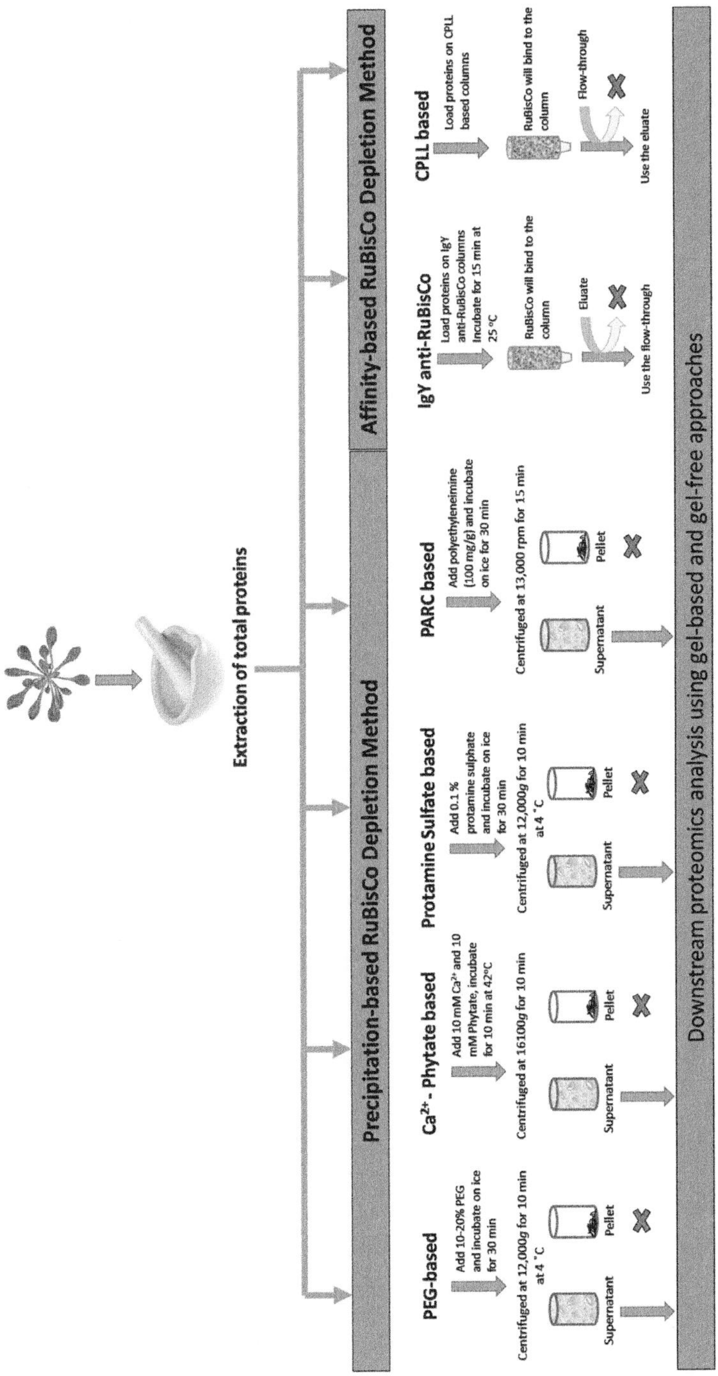

FIGURE 4.2 A summary of the methods developed for RuBisCo depletion to improve the coverage of low abundant proteins.

leaves [45]. Leaf extracts were fractionated into three fractions using different concentrations of PEG, 10%, 10%–20%, and supernatant. RuBisCo was observed in the 20% PEG precipitant, and enrichment of low-abundant proteins was observed in the final supernatant, which was further purified by acetone precipitation. This fractionation technique was able to identify at least 2600 well-separated protein spots [45]. However, systematic microsomal fractionation followed by PEG-based RuBisCo depletion to detect early signaling proteins such as protein phosphatase 2C, RNA-binding proteins, along with another 34 differentially abundant proteins, which normally escaped detection with proteomics, was based on total protein extracts in *Arabidopsis* [46]. In another study, it was shown that 16% PEG was effective to deplete the RuBisCo from *Arabidopsis* leaves [47].

PEG-based methods were laborious and time-consuming. To tackle the issue, a simple and fast technique based on Ca^{2+} and phytate protein precipitation was developed to efficiently remove the RuBisCo [48]. This method is rapid, inexpensive, and able to deplete 85% RuBisCo from soybean leaf extracts. Soluble proteins were incubated with 10 mM Ca^{2+} and 10 mM phytate for 10 min at 42°C to precipitate RuBisCo. Authors have emphasized that pH and temperature play a crucial role to efficiently remove RuBisCo. Phytate is a highly a2þ and phytate electronegative molecule [49] and strong protein binding that only occurs at pH values below the isoelectric point of a protein. Therefore a near neutral buffer at pH 6.8 containing calcium and phytate could precipitate, soybean's large subunit of RuBisCo which has an isoelectric point of 6 [48].

In 2013, a protamine sulfate precipitation method was implemented by Kim and coworkers to deplete the RuBisCO from soybean leaf [50]. For this purpose, 0.1% (w/v) protamine sulfate was found to be sufficient to deplete the large and small subunit of RuBisCo and enrich the low-abundant proteins in the supernatant fraction. Further, the efficiency of the depletion method was confirmed using Western blot, suggesting this method is able to deplete RuBisCO to below the detection limit. This method is rapid, efficient, and cost-effective, and is more specific than the previously published PEG and calcium-phytate-based methods [50]. Other methods such as polyethylenimine (PEI)-based polyethylenimine-assisted RuBisCo cleanup have also been published [44]. Basically, PEI is a positively charged polymer that can also be employed to remove acidic proteins like RuBisCo from the total soluble protein [51]. A soluble protein extraction method was developed with a nonionic detergent (50 mMTris-HCl pH 7.5, 10 mM $MgCl_2$, 150 mMNaCl, 0.1% NP-40, 1 mM PMSF) [44]. However, PEI precipitation could not selectively precipitate RuBisCo from the total protein extracted with a commercial Sigma–Aldrich kit, which includes urea and thiourea lysis buffer. The difference in selectivity might result from the change of protein charges in the sample caused by the urea and thiourea buffer.

Apart from the precipitation methods, affinity-based methods are also available, which utilize anti-RuBisCO antibodies (IgY). Commercially, Sigma–Aldrich supplies anti-RuBisCo columns under the Seppro brand name [52,53]. This affinity-based method is fast, simple, and highly specific to RuBisCO. However, the columns are expensive, limiting its wide acceptance among the plant proteomics community, especially in the developing countries. The use of affinity-based columns has significantly improved the detection of novel nitrosylated low-abundant proteins from *Brassica juncea* leaves [53].

Overall, these methods have been proved fruitful in the removal of the high-abundance proteins, which are the major hurdle in the identification of low-abundance proteins, and utilization of these techniques are, therefore, highly recommended for increasing the proteome coverage of the biological samples.

2. Online and offline fractionation techniques for peptide fractionation

Fractionation involves a series of processes for isolating a specific type of proteins/peptides from a complex mixture. Fractionation can be carried out at protein or peptide level based on the experimental design. A previously common fractionation technique employed in plant proteomics is two-dimensional electrophoresis (2-DE). Additional information regarding the 2-DE can be found in the following literature [54–58]. Although there have been improvements regarding the quality and number of spots in a 2D gel, these enhancements are still not good enough to study the entire cellular proteome [58]. Multidimensional protein identification technology (MudPIT) is generally used to reduce sample complexity and increase dynamic range and sensitivity of peptide identification by minimizing the undersampling and ion suppression problems. A majority of the MudPIT techniques are employed at peptide level, except 2-DE and strong cation exchange (SCX) which are employed at protein level [59].

The most commonly used MudPIT is the online or offline combination of SCX with reverse phase (RP) chromatography that separates peptides orthogonally based on charge and hydrophobicity, respectively [59]. In a study, iTRAQ-labeled peptides were separated into 30 fractions on a SCX column (PolySULFOETHYL, 4.6 × 100 mm, 5 μm, manufactured by PolyLC, Inc, Maryland, USA) using ATKA purifier system (GE Healthcare). SCX was coupled with RP-LC/MS, allowing the identification of 3017 phosphoproteins corresponding to 4052 phosphopeptides in maize leaves [60]. Alternatively, a parallel application of both strong anion exchange (SAX) and SCX was shown to increase the number of identified phosphopeptides since the results obtained by SCX and SAX were complementary rather than overlapping [61]. Comparison indicated that SAX has a better capability in yielding higher protein number with lower peptide load compared to SCX [61]. Unlike SCX and SAX, hydrophilic interaction liquid chromatography (HILIC) fractionates according to the peptide hydrophilicity. The more hydrophilic a peptide (e.g., phosphorylated peptides), the longer it takes to travel through the chromatography column. HILIC should be considered complementary to SCX rather than an alternative strategy, since the two techniques can lead to the identification of different sets of proteins from those in the sample [62]. Another method that is based on RP-RP resulted in 1.8- and 1.6-fold increases in peptide and protein identifications, respectively, when compared with SCX and improved protein sequence coverage [63]. RPLC−RPLC with concatenation can be automated in an offline format, which is flexible and allows multiple analyses of individual fractions to improve coverage [63].

4.3.4 Digestion of plant proteins

Shotgun proteomics requires digestion of proteins sample into peptides, which are easily ionized and detected in nESI-MS/MS analysis. Trypsin is the most commonly used protease in the field of proteomics as it efficiently and specifically cleaves at the C-terminal after lysine and arginine [64]. It generates peptides of MS-compatible size, which ionized better for MS/MS analysis due to basic residues in the C-term end, in addition to an ammonia group in the N-term, increasing the number of double charged peptides formed in the acidic conditions of the RP gradients. The restrictions in protein identification and sequence coverage imposed by the sole use of trypsin can be exemplified by several reports where multiprotease digestion has been used. Research suggested trypsin has several shortcomings as it exhibits lower cleavage efficiency toward lysine than arginine residues, thus hampering accuracy and sensitivity in peptide quantification [65]. To overcome this, different combinations of protease enzymes have been tested in combination with trypsin. The enzymes that serve as an alternative to trypsin are endoproteinases like Arg-C, Asp-N, Glu-C, and Lys-C. These provide high-cleavage efficiency and specificity. Glatter and his group [65] have reported significant improvement in the digestion strategy using LysC/trypsin, which yielded fully cleaved peptides while keeping the number of missed cleavage low and improved peptide quantification, accuracy, and sensitivity. The Lys-C enzyme shares lysine as a cleavage site with trypsin with advantage as it is tolerant to high concentration of denaturing agents such as urea (6−8 M) while trypsin becomes inactive at this condition. Compared to trypsin, Lys-C generates, on average, slightly more basic peptides [66,67].

Other serine proteases are also used in protein digestion like chymotrypsin, and the glutamyl peptidase I (Glu-C), Lys-N, and LysargiNase. The parallel use of Lys-N, Lys-C, and trypsin enabled a 72% increase in the number of detected phosphopeptides as compared to trypsin alone, whereas a trypsin replicated experiment only led to a 25% increase [68]. The data obtained from the experiments with the three different enzymes leads to a more comprehensive picture revealing 5036 redundant phosphopeptides with false discovery rate of 1% [68].

Lys-N is unique in the sense that it has enzymatic specificity for the N-terminal side of lysine, has high thermostability, and is tolerant toward denaturing agents such as urea. It is very helpful to improve digestion of insoluble proteins including membrane proteins. Lys-N can maintain its efficiency under some very harsh (denaturing) conditions (e.g., 8 M urea, 80% acetonitrile) and at temperatures as low as 4°C and up to 80°C, but is severely hampered by guanidine hydrochloride and methanol [66]. The specificities of Lys-C and Lys-N, and the sequence context of cleavages and missed cleavages, were evaluated. It was concluded that the specificity of Lys-C is substantially higher (89.7%) than that of Lys-N (71.7%), at least when incubation was done overnight. The lower specificity observed for Lys-N may be due to the presence of a contaminant peptidase in the Lys-N preparation.

In another study, six proteases, including chymotrypsin, Lys-C, Lys-N, Asp-N, Glu-C, and Arg-C, were compared. The lyophilized enzyme was dissolved in 50 mM ammonium bicarbonate at pH 8 and stored at −80°C. Bovine serum albumin (BSA) digestion was used as a first-line control for evaluating the efficiency of protease chosen for proteomics experiment. Many different peptides were identified from each proteolytic digestion, thus generating a cumulative sequence coverage of 94%. The results were found to be in line with trypsin digestion. The number of identified proteins as well as the sequence coverage were evaluated. There was an increase of 45%, which indicates high complementarily in the multiprotease approach [67].

4.3.5 Overcoming technical and biological variations

Numerous technological developments have been reported to dwell deeper into the proteome, but most poignant of these concerns is minimizing experiment-to-experiment variation to achieve reproducibility and credibility in deriving conclusions across experimental repeats. Careful experimental design must occur before any data are collected to assure success in data acquisitions and avoid unnecessary waste of time and resources [69]. To address this, researchers have to be aware of critical variables such as cellular (genetic) homogeneity, and experimental variation introduced due to the greenhouse growth conditions, sample processing, and LC-MS/MS runs [69].

An ideal proteomics experiment considers technical and biological replicates to ensure the reproducibility of the results. Biological replicates will help the user to assess whether the observed differences in the measurements exist due to the involvement of different biological conditions instead of random chance, whereas technical replicates are repeated measures from the same biological sample to assess errors introduced due the experimental techniques.

The number of biological replicates should be large enough to ensure that the proteomics experiment will have adequate power to address the question of interest while not being so large that it is inefficient in terms of time and cost [70].

Piehowski and coworkers conducted a detailed study to define the technical variations introduced at each step of a typical proteomics workflow: extraction (72%) >> instrumental variance (16%) > instrumental stability (8.4%) > digestion (3.1%) [70]. In this analysis, the largest contribution to variability came from the extraction component, which encompassed biological variation and homogenization. However, small variations could be introduced during the LC-MS/MS acquisition due to different batches of buffer, decay in performance of columns, electrospray tips, and instrument components, drifts in calibration and tuning, etc. Selection of biological samples should be randomized, so that the inference using the sample data can be generalized to the population. The use of randomization can avoid bias caused by potentially unknown systematic errors [69]. For example, a sample constituted of three to five plants can be pooled to represent the population; the order of sample processing and data acquisition should be randomized each time. In this way, potential confounding of time with biological and experimental can be avoided. By optimizing the method of sampling from different biological conditions between different experimental runs, the experimental layout can improve the efficiency of the design and analysis.

Swapping of labels is generally practiced in label-based quantitative proteomics approaches such as HILEP metabolic labeling to reduce biases introduced due to the labeled isotopes [11,71,72]. In this design, one experimental run, two labels will be randomly assigned to the biological conditions; and under another experimental repeat, the labels will be permuted between these two biological conditions to cancel confounding effects inherant to the labeling itself when evaluating the relative expression levels of proteins between different biological conditions [69,73].

A challenge for large-scale quantitative applications is maintaining reproducibility that may extend over several months or years. Use of internal standard offers a solution to this challenge. For example, generating a pool of several samples (e.g. derived from various cell cultures) labeled by SILAC yields to a whole-proteome stable isotope-labeled internal standard that can be mixed with a non-labeled tissue-derived proteome for quantification [74]. Alternatively, a pooled reference including all the samples used in that study sample was generated by labeling peptides during the tryptic digest with ^{18}O to serve as the universal reference. The labeled reference sample can then be added to each unlabeled biological sample so that each unlabeled peptide has its corresponding ^{18}O-labeled version originated from the universal reference [75].

Sigma–Aldrich, USA, in collaboration with the Association of Biomolecular Resource Facilities, Bethesda, Maryland, introduced a well-defined complex protein standard for use in the proteomics community. The product consists of 48 HPLC-purified human proteins (USP1 and USP2) in equimolar amounts. The use of standards provides insight into the effectiveness of everyday tasks prior to, and during, sample processing [75,76]. Another commercial product is known as SILuProt, a novel collection of stable isotope-labeled (SIL, or "heavy labeled") full-length proteins designed to be used as mass spectrometry internal standards for quantitative proteomics. Introduction of an internal standard at the beginning of the MS workflow minimizes variations in quantification due to protein extraction, fractionation, enrichment, proteolysis, and analysis. By increasing quantitative accuracy in the analysis of tissue proteomes, such methods should improve integration of protein expression profiling data and enhance downstream bioinformatics analyses.

4.4 Advances and applications in plant proteomics

4.4.1 Proteogenomics to help annotation of open reading frames (ORFs) in newly sequenced genomes

In shotgun proteomics, peptides are most commonly identified by matching MS/MS spectra against theoretical spectra of all candidate peptides represented in a reference protein database available from Ensembl, RefSeq, NCBI, UniProtKB/Swiss-Prot, or TrEMBL [77]. Commonly, most protein identification algorithms, such as Sequest, Mascot, and Andromeda, are using peptide identification against the available protein database [78]. Plant species such as *Arabidopsis*, rice, maize, and wheat are well studied and have an extensive and nearly complete annotated protein database. Using shotgun proteomics, 13,000 unique proteins were identified and quantified in different plant organs of *A. thaliana*, covering nearly 50% of the predicted gene models [79], whereas other plant species such as grape, barley do not have such an extensive and curated protein database. In such cases, the normal search strategy only offers limited identification of spectra for which peptides are present in the database [80]. More comprehensive approaches to identify novel proteins/peptides is termed proteogenomics. In a proteogenomics approach, novel peptides are identified by searching MS/MS spectra against customized protein sequence databases containing predicted novel protein sequences and sequence variants, transcriptomic sequence information as well as non annotated genomic sequences. Proteogenomics therefore not only provides protein-level validation of gene expression and gene model refinement but also enables the improvement of protein sequence databases [77]. Such proteogenomics strategy has been used to identify novel proteins in grape [81], barley and its pathogens [82,83], and rice [84].

4.4.2 Understanding plant development and responses to environmental clues

The way a proteome changes in response to stress is complex as it depends on a number of parameters, such as intensity of stress, plant growth stage, type of plant tissue, as well as the species and genotype [3]. Over the past few years, several proteomics studies have been conducted to understand the changes in response to different stresses in whole plants as well as in subcellular organelles such as the chloroplast, mitochondrion, nucleus, and peroxisome in response to abiotic stress in *A. thaliana* and other commercial crops. A number of stress-induced proteins have been identified in different studies [2,85]. In barley infected by the powdery mildew fungus, a number of small haustorial proteins were identified. These might play a role in controlling host metabolism and immunity in response to the pathogen [83,86]. Proteomic studies have shown modulation of proteins related to sugar metabolism and cell wall rearrangement in response to an arbuscular mycorrhizal fungus *Glomus masseae* with drought-treated wheat roots. This interaction resulted into reduced osmotic stress and maintenance of cellular integrity in wheat [87].

Recently, a large-scale proteomics that represented 24 different organ and developmental stages of *Triticum aestiivum* was undertaken. This study assisted identification of the markers of the wheat proteome, their presence in different tissues, and correlations between the abundance of functional classes of proteins [88]. Exosomes (extracellular vesicles, EV) play a central role in intercellular signaling by transporting proteins and small RNAs in response to abiotic and biotic stress. A proteomic analysi identified EVs' proteins from the apoplastic fluids of *Arabidopsis* leaves [89]. The majority of proteomics studies published sofar describe the proteome of used plant tissues containing a mixture of different cell types. Such studies failed to reveal cell-specific functions. Recently, single-cell-type proteomics studies have been performed on pollen grains, guard cells, mesophyll cells, root hairs, and trichomes [90]. High-resolution proteomic analyses will reveal novel functions taking place in single cells and thus enhance the understanding of single cell types and the heterogeneity within a single cell type or tissue.

4.5 Conclusion and future perspective

The field of proteomics continues to push the boundaries of analysis down the lower end of the biological dynamic range. Increases in instrument sensitivity and more compatible sample preparation techniques are necessary advancements for targeting the low-abundant proteins while achieving robust and reproducible analysis. It is essential to meticulously consider the different steps of proteomic sample preparation and their effect on the downstream processes such as efficient plant tissue-specific extraction methods, removal of biological and non-biological contaminants, depletion of high-abundant proteins, improving the digestion efficiency, and adopting MudPIT and other fractionation techniques to be able to identify the low-abundant proteins. Adopting the best sample preparation workflow based on the sample type will considerably improve the proteome coverage and help us to address relevant biological questions. The field of proteomics can build on the solid foundation already laid, and with some inventiveness and visionary techniques, the limitations of plant proteomics will be pushed further.

References

[1] Agrawal GK, Pedreschi R, Barkla BJ, Bindschedler LV, Cramer R, Sarkar A, et al. Translational plant proteomics: a perspective. J Proteomics 2012;75(15):4588–601.

[2] Kosová K, Vítámvás P, Urban MO, Prášil IT, Renaut J. Plant abiotic stress proteomics: the major factors determining alterations in cellular proteome. Front Plant Sci 2018;9(2):1–22.

[3] Abreu IA, Farinha AP, Negrão S, Gonçalves N, Fonseca C, Rodrigues M, et al. Coping with abiotic stress: proteome changes for crop improvement. J Proteomics 2013;93:145–68.

[4] Song G, Hsu PY, Walley JW. Assessment and refinement of sample preparation methods for deep and quantitative plant proteome profiling. Proteomics 2018;18(17):e1800220.

[5] Gupta R, Wang Y, Agrawal GK, Rakwal R, Jo IH, Bang KH, et al. Time to dig deep into the plant proteome: a hunt for low-abundance proteins. Front Plant Sci 2015;6(1):1–3.

[6] The Arabidopsis Genome Initiative. Analysis of the genome sequence of the flowering plant *Arabidopsis thaliana*. Nature 2000;408(6814):796–815.

[7] Matsumoto T, Wu J, Kanamori H, Katayose Y, Fujisawa M, Namiki N, et al. The map-based sequence of the rice genome. Nature 2005;436(7052):793–800.

[8] Chen F, Dong W, Zhang J, Guo X, Chen J, Wang Z, et al. The sequenced angiosperm genomes and genome databases. Front Plant Sci 2018;9(4):1–14.

[9] Domon B, Aebersold R. Options and considerations when selecting a quantitative proteomics strategy. Nat Biotechnol 2010;28(7):710–21.

[10] Wiśniewski JR, Zougman A, Nagaraj N, Mann M. Universal sample preparation method for proteome analysis. Nat Methods 2009;6(5):359–62.

[11] Bindschedler LV, Palmblad M, Cramer R. Hydroponic isotope labelling of entire plants (HILEP) for quantitative plant proteomics; an oxidative stress case study. Phytochemistry 2008;69(10):1962–72.

[12] Bindschedler LV, Smith CJ, Cramer R. Quantitative plant proteomics using hydroponic isotope labeling of entire plants (HILEP) [internet]. In: Ivanov AR, Lazarev AV, editors. Sample preparation in biological mass spectrometry. Dordrecht: Springer Netherlands; 2011. p. 363–80.

[13] Ong S-E, Blagoev B, Kratchmarova I, Kristensen DB, Steen H, Pandey A, et al. Stable isotope labeling by amino acids in cell culture, SILAC, as a simple and accurate approach to expression proteomics. Mol Cell Proteomics 2002;1(5):376–86.

[14] Bindschedler LV, Cramer R. Quantitative plant proteomics. Proteomics 2011;11(4):756–75.

[15] Bakalarski CE, Kirkpatrick DS. A biologist's field guide to multiplexed quantitative proteomics. Mol Cell Proteomics 2016;15(5):1489–97.

[16] Carr S, Aebersold R, Baldwin M, Burlingame A, Clauser K, Nesvizhskii A. The need for guidelines in publication of peptide and protein identification data. Mol Cell Proteomics 2004;3(6):531–3.

[17] Geiger T, Cox J, Ostasiewicz P, Wisniewski JR, Mann M. Super-SILAC mix for quantitative proteomics of human tumor tissue. Nat Methods 2010;7:383.

[18] Faurobert M, Pelpoir E, Chaib J. Phenol extraction of proteins for proteomic studies of recalcitrant plant tissues. Methods Mol Biol 2007;355:9–14.

[19] Isaacson T, Damasceno CMB, Saravanan RS, He Y, Catala C, Rose JKC. Sample extraction techniques for enhanced proteomic analysis of plant tissues. Nat Protoc 2006;1(2):769–74.

[20] Mechin V, Damerval C, Zivy M. Total protein extraction with TCA-acetone. Methods Mol Biol 2007;355:1–8.

[21] Bodzon-Kulakowska A, Bierczynska-Krzysik A, Dylag T, Drabik A, Suder P, Noga M, et al. Methods for samples preparation in proteomic research. J Chromatogr B Anal Technol Biomed Life Sci 2007;849(1–2):1–31.

[22] Feist P, Hummon AB. Proteomic Challenges: sample preparation techniques for microgram-quantity protein analysis from biological samples. Int J Mol Sci 2015;16(2):3537–63.

[23] Rose JKC, Bashir S, Giovannoni JJ, Jahn MM, Saravanan RS. Tackling the plant proteome: practical approaches, hurdles and experimental tools. Plant J 2004;39(5):715–33.

[24] Suder P, Novák P, Havlíček V, Bodzoń-Kułakowska A. General strategies for proteomic sample preparation. In: Proteomic profiling and analytical chemistry: the crossroads. 2nd ed. 2016. p. 25–49.

[25] Saravanan RS, Rose JKC. A critical evaluation of sample extraction techniques for enhanced proteomic analysis of recalcitrant plant tissues. Proteomics 2004;4(9):2522–32.

[26] Tobin AK, Bowsher CG. Subcellular fractionation of plant tissues. In: Cutler P, editor. Protein purification protocols. Totowa, NJ: Humana Press; 2004. p. 53–63.

[27] Aryal UK, Krochko JE, Ross ARS. Identification of phosphoproteins in *Arabidopsis thaliana* leaves using polyethylene glycol fractionation, immobilized metal-ion affinity chromatography, two-dimensional gel electrophoresis and mass spectrometry. J Proteome Res 2012;11(1):425–37.

[28] Benschop JJ, Mohammed S, O'Flaherty M, Heck AJ, Slijper M, Menke F. Quantitative phosphoproteomics of early elicitor signaling in Arabidopsis. Mol Cell Proteomics 2007;6(7):1198–214.

[29] Wu X, Xiong E, Wang W, Scali M, Cresti M. Universal sample preparation method integrating trichloroacetic acid/acetone precipitation with phenol extraction for crop proteomic analysis. Nat Protoc 2014;9(2):362–74.

[30] Wu X, Gong F, Wang W. Protein extraction from plant tissues for 2DE and its application in proteomic analysis. Proteomics 2014;14(6):645–58.

[31] Barbosa H de S, Quirino de Souza DL, Ferreira Koolen HH, Gozzo FC, Zezzi Arruda MA. Sample preparation focusing on plant proteomics: extraction, evaluation and identification of proteins from sunflower seeds. Anal Methods 2013;5(1):116–23.

[32] Charmont S, Jamet E, Pont-Lezica R, Canut H. Proteomic analysis of secreted proteins from *Arabidopsis thaliana* seedlings: improved recovery following removal of phenolic compounds. Phytochemistry 2005;66(4):453–61.

[33] Bindschedler LV, McGuffin LJ, Burgis TA, Spanu PD, Cramer R. Proteogenomics and in silico structural and functional annotation of the barley powdery mildew *Blumeria graminis* f. sp. hordei. Methods 2011;54(4):432–41.

[34] Burgess RR. Protein precipitation techniques. In: Methods in enzymology. 1st ed. United States: Elsevier Inc.; 2009. p. 331–42.

[35] Jiang L, He L, Fountoulakis M. Comparison of protein precipitation methods for sample preparation prior to proteomic analysis. J Chromatogr A 2004;1023(2):317–20.

[36] Englard S, Seifter S. Precipitation techniques. Methods Enzymol 1990;182:285–300.
[37] Arakawa T, Timasheff SN. Theory of protein solubility. Methods Enzymol 1985;114:49–77.
[38] Botelho D, Wall MJ, Vieira DB, Fitzsimmons S, Liu F, Doucette A. Top-down and bottom-up proteomics of SDS-containing solutions following mass-based separation. J Proteome Res 2010;9(6):2863–70.
[39] Thongboonkerd V, Gozal E, Sachleben LRJ, Arthur JM, Pierce WM, Cai J, et al. Proteomic analysis reveals alterations in the renal kallikrein pathway during hypoxia-induced hypertension. J Biol Chem 2002;277(38):34708–16.
[40] Srivastava OP, Srivastava K. Purification of gamma-crystallin from human lenses by acetone precipitation method. Curr Eye Res 1998;17(11):1074–81.
[41] Crowell AMJJ, Wall MJ, Doucette AA. Maximizing recovery of water-soluble proteins through acetone precipitation. Anal Chim Acta 2013;796:48–54.
[42] Wessel D, Flügge UI. A method for the quantitative recovery of protein in dilute solution in the presence of detergents and lipids. Anal Biochem 1984;138(1):141–3.
[43] Yeung Y-G, Stanley ER. Rapid detergent removal from peptide samples with ethyl acetate for mass spectrometry analysis. Curr Protein Pept Sci 2010;59(1). 16.12.1–16.12.5.
[44] Zhang Y, Gao P, Xing Z, Jin S, Chen Z, Liu L, et al. Application of an improved proteomics method for abundant protein Cleanup: molecular and genomic mechanisms study in plant defense. Mol Cell Proteomics 2013;12(11):3431–42.
[45] Kim ST, Cho KS, Jang YS, Kang KY. Two-dimensional electrophoretic analysis of rice proteins by polyethylene glycol fractionation for protein arrays. Electrophoresis 2001;22(10):2103–9.
[46] Widjaja I, Naumann K, Roth U, Wolf N, Mackey D, Dangl JL, et al. Combining subproteome enrichment and Rubisco depletion enables identification of low abundance proteins differentially regulated during plant defense. Proteomics 2009;9(1):138–47.
[47] Xi J, Wang X, Li S, Zhou X, Yue L, Fan J, et al. Polyethylene glycol fractionation improved detection of low-abundant proteins by two-dimensional electrophoresis analysis of plant proteome. Phytochemistry 2006;67(21):2341–8.
[48] Krishnan HB, Natarajan SS. A rapid method for depletion of Rubisco from soybean (Glycine max) leaf for proteomic analysis of lower abundance proteins. Phytochemistry 2009;70(17–18):1958–64.
[49] Lah CL, Cheryan M. Protein solubility characteristics of an ultrafiltered full-fat soybean product. J Agric Food Chem 1980;28(5):911–6.
[50] Kim YCYJ, Lee HM, Wang Y, Wu J, Kim SGST, Kang KY, et al. Depletion of abundant plant RuBisCO protein using the protamine sulfate precipitation method. Proteomics 2013;13(14):2176–9.
[51] Holler C, Zhang C. Purification of an acidic recombinant protein from transgenic tobacco. Biotechnol Bioeng 2008;99(4):902–9.
[52] Cellar NA, Kuppannan K, Langhorst ML, Ni W, Xu P, Young SA. Cross species applicability of abundant protein depletion columns for ribulose-1,5-bisphosphate carboxylase/oxygenase. J Chromatogr B Anal Technol Biomed Life Sci 2008;861(1):29–39.
[53] Sehrawat A, Abat JK, Deswal R. RuBisCO depletion improved proteome coverage of cold responsive S-nitrosylated targets in Brassica juncea. Front Plant Sci 2013;4:1–14.
[54] Pallen C, Friry-Santini C, Herouet-Guicheney C, Capt A. Technical variability of 2D gel electrophoresis – application to soybean allergens. Toxicol Reports 2014;1:734–42.
[55] Chevalier F. Highlights on the capacities of "Gel-based" proteomics. Proteome Sci 2010;8:23–33.

[56] Singh NKN, Jain N, Kumar R, Jain A, Singh NKN, Rai V. A comparative method for protein extraction and 2-D gel electrophoresis from different tissues of *Cajanus cajan*. Front Plant Sci 2015;6:606–701.

[57] Abdallah C, Dumas-Gaudot E, Renaut J, Sergeant K. Gel-based and gel-free quantitative proteomics approaches at a glance. Int J Plant Genom 2012;2012:1–17.

[58] Vadivel AKA. Gel-based proteomics in plants: time to move on from the tradition. Front Plant Sci 2015;6:369–72.

[59] Lee J, Cooper B. Alternative workflows for plant proteomic analysis. Mol Biosyst 2006;2(12):621–6.

[60] Hu X, Li N, Wu L, Li CC, Li CC, Zhang L, et al. Quantitative iTRAQ-based proteomic analysis of phosphoproteins and ABA-regulated phosphoproteins in maize leaves under osmotic stress. Sci Rep 2015;5:15626–51.

[61] Kilambi HV, Manda K, Sanivarapu H, Maurya VK. Shotgun proteomics of tomato fruits: evaluation, optimization and validation of sample preparation methods and mass spectrometric parameters. Front Plant Sci 2016;7:969–83.

[62] Longworth J, Noirel J, Pandhal J, Wright PC, Vaidyanathan S. HILIC- and SCX-based quantitative proteomics of chlamydomonas reinhardtii during nitrogen starvation induced lipid and carbohydrate accumulation. J Proteome Res 2012;11(12):5959–71.

[63] Wang Y, Yang F, Gritsenko MA, Wang Y, Clauss T, Liu T, et al. Reversed-phase chromatography with multiple fraction concatenation strategy for proteome profiling of human MCF10A cells. Proteomics 2011;11(10):2019–26.

[64] Switzar L, Giera M, Niessen WM. An overview of the available techniques and recent developments. J Proteome Res 2013;12(3):1067–77.

[65] Glatter T, Ludwig C, Ahrné E, Aebersold R, Heck AJRR, Schmidt A. Large-scale quantitative assessment of different in-solution protein digestion protocols reveals superior cleavage efficiency of tandem Lys-C/trypsin proteolysis over trypsin digestion. J Proteome Res 2012;11(11):5145–56.

[66] Taouatas N, Heck AJR, Mohammed S. Evaluation of metalloendopeptidase Lys-N protease performance under different sample handling conditions. J Proteome Res 2010;9(8):4282–8.

[67] Giansanti P, Tsiatsiani L, Low TY, Heck AJR. Six alternative proteases for mass spectrometry-based proteomics beyond trypsin. Nat Protoc 2016;11(5):993–1006.

[68] Gauci S, Helbig AO, Slijper M, Krijgsveld J, Heck AJRR, Mohammed S. Lys-N and trypsin cover complementary parts of the phosphoproteome in a refined SCX-based approach. Anal Chem 2009;81(11):4493–501.

[69] Molloy MP, Brzezinski EE, Hang J, McDowell MT, VanBogelen RA. Overcoming technical variation and biological variation in quantitative proteomics. Proteomics 2003;3(10):1912–9.

[70] Piehowski PD, Petyuk VA, Orton DJ, Xie F, Moore RJ, Ramirez-Restrepo M, et al. Sources of technical variability in quantitative LC-MS proteomics: human brain tissue sample analysis. J Proteome Res 2013;12(5):2128–37.

[71] Bantscheff M, Lemeer S, Savitski MM, Kuster B. Quantitative mass spectrometry in proteomics: critical review update from 2007 to the present. Anal Bioanal Chem 2012;404(4):939–65.

[72] Gouw JW, Krijgsveld J, Heck AJR. Quantitative proteomics by metabolic labeling of model organisms. Mol Cell Proteomics 2010;9(1):11–24.

[73] Yu F, Qiu F, Meza J. Design and statistical analysis of mass-spectrometry-based quantitative proteomics data. In: Proteomic profiling and analytical chemistry: the crossroads. 2nd ed. 2016. p. 211–37.

[74] Ong SE. Whole proteomes as internal standards in quantitative proteomics. Genome Med 2010;2(7):8–11.

[75] Tabb DL, Fernando CG, Chambers MC. MyriMatch: highly accurate tandem mass spectral peptide identification by multivariate hypergeometric analysis. J Proteome Res 2007;6(2):654–61.

[76] Falick AM, Lane WS, Lilley KS, Maccoss MJ, Phinney BS, Sherman NE, et al. ABRF-PRG07: advanced quantitative proteomics study. J Biomol Tech 2011;22(1):21–6.

[77] Nesvizhskii AI. Proteogenomics: concepts, applications, and computational strategies. Nat Methods 2014;11(11):1114–25.

[78] Paulo JA. Practical and efficient searching in proteomics: a cross engine comparison. Webmedcentral 2013;4(10):1–15.

[79] Baerenfaller K, Grossmann J, Grobei MA, Hull R, Hirsch-hoffmann M, Yalovsky S, et al. Genome-scale proteomics reveals *Arabidopsis thaliana* gene models and proteome dynamics. Science 2008;320(5878):938–41.

[80] Castellana NE, Pham V, Arnott D, Lill JR, Bafna V. Template proteogenomics: sequencing whole proteins using an imperfect database. Mol Cell Proteomics 2010;9(6):1260–70.

[81] Chapman B, Bellgard M. Plant proteogenomics: improvements to the grapevine genome annotation. Proteomics 2017;17(21):1–9.

[82] Abendroth U, Adlung N, Otto A, Grüneisen B, Becher D, Bonas U. Identification of new protein-coding genes with a potential role in the virulence of the plant pathogen Xanthomonas euvesicatoria. BMC Genomics 2017;18(1):1–14.

[83] Bindschedler LV, Burgis TA, Mills DJS, Ho JTC, Cramer R, Spanu PD. In planta proteomics and proteogenomics of the biotrophic barley fungal pathogen *Blumeria graminis* f. sp. hordei. Mol Cell Proteomics 2009;8(10):2368–81.

[84] Helmy M, Sugiyama N, Tomita M, Ishihama Y. The rice proteogenomics database OryzaPG-DB: development, expansion, and new features. Front Plant Sci 2012;3:65.

[85] Tan BC, Lim YS, Lau S-E. Proteomics in commercial crops: an overview. J Proteomics 2017;169:176–88.

[86] Bindschedler LV, Panstruga R, Spanu PD, Wildermuth MC, Panstruga R. Mildew-Omics: how global analyses aid the understanding of life and evolution of powdery mildews. Front Plant Sci 2016;7:123–34.

[87] Bernardo L, Morcia C, Carletti P, Ghizzoni R, Badeck FW, Rizza F, et al. Proteomic insight into the mitigation of wheat root drought stress by arbuscular mycorrhizae. J Proteomics 2017;169:21–32.

[88] Duncan O, Trösch J, Fenske R, Taylor NL, Millar AH. Resource: mapping the *Triticum aestivum* proteome. Plant J 2017;89(3):601–16.

[89] Rutter BD, Innes RW. Extracellular vesicles isolated from the leaf apoplast carry stress-response proteins. Plant Physiol 2017;173(1):728–41.

[90] Shaojun D, Sixue C. Single-cell-type proteomics: toward a holistic understanding of plant function. Mol Cell Proteomics 2012;11(12):1622–30.

Chapter 5

Structural analysis of proteins using X-ray diffraction technique

Umesh B. Gawas[1], Vinod K. Mandrekar[2], Mahesh S. Majik[3]
[1]*Department of Chemistry, Dnyanprassarak Mandal's College and Research Centre, Assagao, Goa, India;* [2]*Department of Chemistry, St. Xavier's College, Mapusa, Goa, India;* [3]*Department of Chemistry, Goa University, Taleigao Plateau, Goa, India*

5.1 Introduction

Proteins are large complex biomolecules that form an essential component of diet. They are indispensable for normal functioning of biological systems in all forms of life and play a central role in determining nutritional value of food. Besides their nutritional role, proteins also form a structural basis of various functional properties of foods. The functional roles of proteins result from physicochemical properties such as solubility, viscosity, water binding, gelation, cohesion/adhesion, elasticity, emulsification, and foaming. The structure of proteins determines their shape, size, amino acid sequence, hydrophobicity, hydrophilicity, charge, and ability to change in a given environment [1]. The structures of proteins are classified as: (1) primary, which refers to linear sequence of amino acid polypeptide chain, (2) secondary, in which highly regular local substructures like α-helix and β-sheet are formed due to hydrogen-bonding interactions between main chain polypeptide groups, (3) tertiary, which refers to three-dimensional (3D) form of a single protein, and (4) quaternary, are those that contain complex 3D multisubunits. Proteins form the important component of structure, function, and regulation of cells, tissues, and organs in the body. In order to understand the working of biosystems, detailed knowledge of 3D structures of proteins becomes inevitable; the folding of protein and biofunction cannot be evaluated from protein sequence alone. Also, the 3D structure is important in the study of diseases as well as development of medicines for treatment. Structure-based drug design approaches have helped in engineering new molecules with potential pharmaceutical applications [2]. Above all, the heterogeneous nature and post translational modifications are responsible for cellular localization and functional partners. The structures of

Advances in Biological Science Research. https://doi.org/10.1016/B978-0-12-817497-5.00005-7

enzyme proteins have been revealed from reactions with the substrate during enzymatic reactions. Similarly, the binding ability of hormonal proteins to their receptors has formed the basis for understanding structure and nerve transmission along the cellular membranes. X-ray crystallography has provided 3D structures of proteins, which are essential for understanding their function and activity, and hence the technique has been exploited extensively to obtain insight into the workings of numerous biological processes. In recent years, the introduction of synchrotron radiation sources, phasing by molecular replacement, and anomalous diffraction methods have facilitated the structure determination of large numbers of crystalline proteins, and the field has undergone a massive expansion worldwide. Also, the use of 3D structures of proteins in designing new drug molecules has provided thrust for the growth of protein crystallography. This chapter presents an overview of X-ray diffraction (XRD) method for crystal structure evaluation of proteins, which has contributed significantly in our understanding about different biological processes in biosystems.

5.2 Historical background

The origin of protein crystallography can be traced back from the discovery of X-rays by Conrad Röentgen (Noble Prize in Physics, 1901), and the subsequent measurements by Max von Laue (Noble Prize in Physics, 1914), who was first to observe diffraction of X-rays by atoms/ions/molecules, which revealed the wave nature of X-rays [3]. These discoveries were followed by the diffraction experiments by Henry Bragg and Laurence Bragg (Noble Prize in Physics, 1915), which proved the application of X-ray diffraction by atoms to resolve crystal structure [4]. The first application of XRD for structural analysis of biomolecules was tertiary structure of the protein myoglobin by Max Perutz and John Kendrew (Nobel Prize in Chemistry, 1962) and Francis Crick, James Watson, and Maurice Wilkins (Noble Prize in Physiology or Medicine, 1962) for molecular structure of nucleic acids [5−7]. Subsequently, several other crystallographic structures of proteins were determined using this technique. Dorothy Hodgkin (Noble Prize in Chemistry, 1964) made the remarkable contribution in the field using this method to establish the crystal structure of vitamin B_{12} and penicillin [8−13]. The crystal structure of first membrane protein, the photosynthetic reaction center, was revealed by Johann Deisenhofer, Robert Huber, and Hartmut Michel (Noble Prize in Chemistry, 1988) from XRD data [14]. Paul Boyer, John Walker, and Jens Skou (Noble Prize in Chemistry, 1997) determined the crystal structure of F1-ATPase, which has helped in elucidation of the enzymatic mechanism involved in the synthesis of ATP [15]. In the 21st century, the application of XRD for crystal structure evaluation was demonstrated by Peter Agre and Roderick MacKinnon (Noble Prize in Chemistry, 2003) for discovery and structural studies of ion channels [16]. Roger Kornberg (Noble Prize in Chemistry, 2006) determined the structure of DNA-dependent RNA polymerase [17].

The ternary structure of ribosome was established by Venkatraman Ramakrishnan, Thomas Steitz, and Ada Yonath (Noble Prize in Chemistry, 2009) using protein crystallography [18,19]. Brian Kobilka and Robert Lefkowitz (Noble Prize in Chemistry, 2012) evaluated first structure of a ligand-activated G protein-coupled receptor [20,21]. Major achievements in the development of X-ray crystallography and its application for structure evaluation directly or indirectly have been recognized by at least 14 Nobel Prizes [22].

5.3 X-ray crystallography

The word *crystallography* is defined as the science of crystals. Crystals are three-dimensional periodic arrays of atoms/molecules. The need of crystals in X-ray diffraction studies arises due to very weak scattering from individual atoms/molecules, whereas the crystals act like an amplifier increasing the scattering signal due to the multiple copies of molecules within them. X-rays are electromagnetic radiation of very short wavelength of around 1 Å ($1 \text{ Å} = 10^{-10}$ m). X-rays are generated using various laboratory sources or at synchrotrons. Synchrotrons are employed when very high-intensity and highly focused X-rays are desirable.

X-ray diffraction is a powerful nondestructive technique used for characterization of crystalline materials. The technique can be used for structure determination of single crystal as well as polycrystalline materials, wherein, it provides valuable information of structures, different phases, and preferred crystal orientations. Besides this, it also gives reliable information on structural parameters such as average grain size, crystallinity, strain, and crystal defects. X-ray diffraction occurs when electromagnetic radiation with wavelength of about 1 Å interacts with the matter inside the crystals, in particular with the electrons, resulting in bending of the wave, which is also called scattering of waves. These waves are scattered by the electrons, and each electron becomes a small X-ray source of its own. The scattered waves interact with each other, resulting in another physical phenomenon called interference. These scattered waves from all the electrons within each atom are added to each other, giving diffracted waves from each atom, which may get stronger by constructive interference or else get cancelled by destructive interference. The X-ray detector registers only stronger waves generated by constructive interference. The interaction of X-rays with matter in general is represented in Fig. 5.1. The constructive interference of scattered X-rays can be looked at using Bragg's law to determine various characteristics of the crystal as well as polycrystalline material. The interplanar distance d, for each *hkl* plane, can be calculated using Bragg's equation [23].

$$n\lambda = 2d\sin\theta$$

Where, d is the interplanar distance, n is the order of interference, λ the wavelength of X-ray target used, and θ is the angle of diffraction. Using values

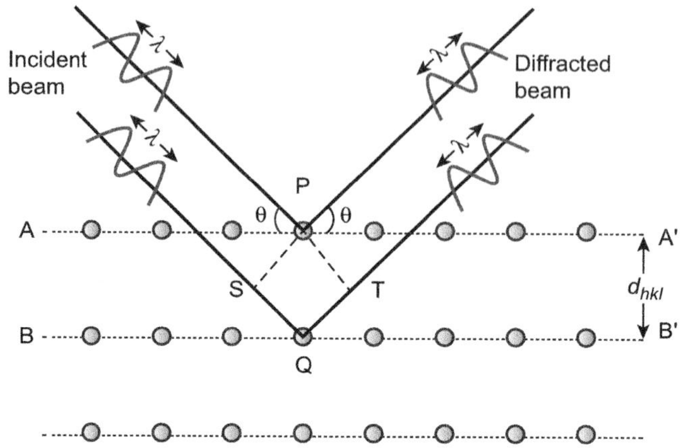

FIGURE 5.1 Schematic representation of scattering of X-rays by lattice planes.

of interplanar distance d, several other crystal parameters, such as lattice constant, X-ray density, porosity etc., can be calculated for single-crystal as well as polycrystalline material.

5.4 Protein X-ray crystallography

Protein X-ray crystallography has been exploited extensively to get insight into the workings of numerous biological processes, and the field has undergone a massive worldwide expansion during the last decade. X-ray crystallography is one of the most widely used methods for crystal structure determination of proteins. This technique accounts for about 90% of protein structures in the protein data bank (PDB). Common X-ray diffraction method requires protein macrocrystals with size 50 μm or larger. However, with the development of microfocus X-ray beams, in recent years, structural analysis of protein microcrystals with dimensions 5–10 μm has been achieved [24,25]. With the use of serial femtosecond crystallography (SFX) using X-ray free electron laser (XFEL), it is possible to study the structure of protein nanocrystals, in the range of 100 nm to 10 μm [26–30]. SFX has been found to be very useful for the structural analysis of membrane proteins since it is very difficult to grow macrocrystals for such proteins [31,32]. Membrane proteins are proteins that interact with or are part of biological membrane, including (1) integral membrane proteins, which are permanently anchored or part of the membrane, and (2) peripheral membrane proteins, which are only temporarily attached to the lipid bilayer or to other integral proteins [33,34]. It is predicted that about 25%–30% of proteins in the genome in all living species are membrane proteins, and 40%–60% of drug targets are membrane proteins [35–37]. SFX also overcomes the radiation damage problem in X-ray crystallography by the "diffraction before destruction" principle, which enables time-resolved studies

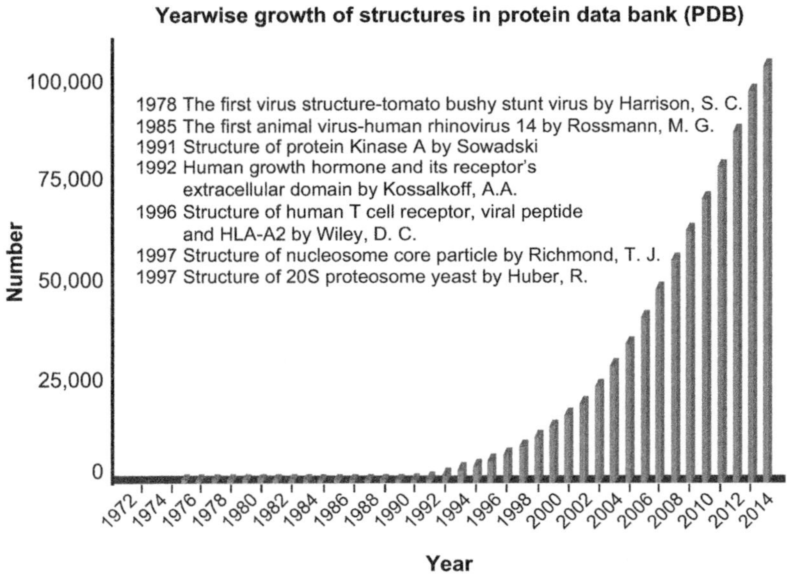

FIGURE 5.2 Graphical representation of year-wise growth of structure in protein data bank. *Source: wwPDB.*

in protein crystals [32,38–44]. With over 100,000 structures (Fig. 5.2) in PDB [45], there are less than 2000 constitute membrane proteins, of which less than 600 are unique membrane proteins [46].

The three-dimensional structures of proteins are essential for understanding protein function and activity. X-ray data of protein crystal is obtained by placing it in a monochromatic X-ray beam and the process is repeated while changing its orientation. Each exposure to X-ray radiation provides diffraction pattern and each spot on the pattern corresponds to diffracted X-ray beam emerging from the crystal, which is detected by the X-ray detector. To solve protein structure, thousands of such diffraction spots needs to be collected. The method of data collection and amount of data required depend on the type of crystal (cell dimensions and symmetry). The data collected is processed by extracting the relative intensities of diffracted X-rays. Data processing is carried out using software programs like XDS, HKL-2000, etc. In a normal setup, the X-ray beam is stationary while the crystal is rotated 1° at a time. The exposure time is normally from seconds to minutes depending on the intensity of the X-ray source. The electron density of the molecules within the crystal is calculated from intensities of X-ray diffraction spots, which in turn, gives location of atoms in the given structure. Using this information, a model of the molecule(s) in the crystal is built. The X-ray diffraction data of protein crystals is recorded using a specialized instrument for X-ray generation that employs a rotating anode. The intensity of X-rays obtained using a rotating anode is

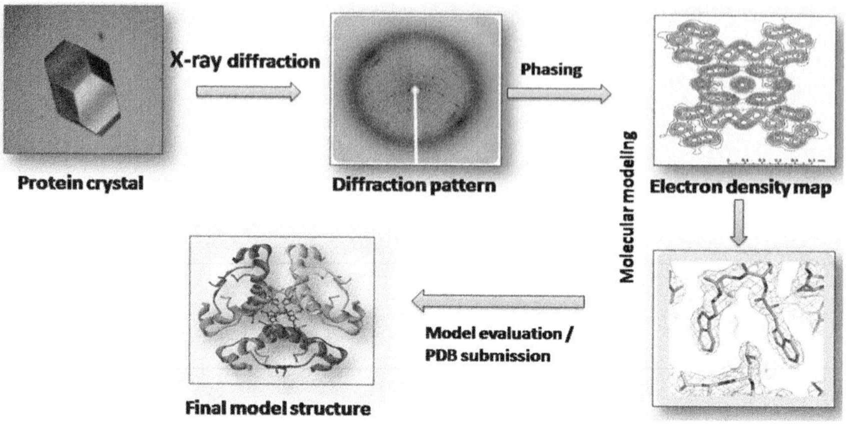

FIGURE 5.3 Schematic diagram showing the workflow for structural determination of protein by X-ray crystallography.

many times higher than used in medical X-ray machines. Higher intensity gives the better quality data for crystals, and smaller crystals can be used for data collection. Alternatively, synchrotrons can be used to obtained high-quality diffraction data. The advantage of synchrotrons is that X-rays produced at synchrotron beamlines have high intensity and have very fine focus, which allows for the diffraction measurements of very small crystals (>30–40 μm). The radiation damage caused by very high-intensity X-rays can be reduced by freezing protein crystals to cryo-temperatures before exposing them to the X-ray beam. However, before actual freezing, the protein crystals are transferred to a solution with cryoprotectant, which ensures no formation of ice crystal as ice gives its own diffraction, which will interfere in the interpretation of protein crystal data. Practically, the crystals are frozen directly while placed in X-ray beamline path and then the beam shutter is opened to allow monochromatic radiation to hit the crystal. The X-ray detector (image plate–based X-ray detector) registers the intensity of the diffracted X-rays, and intensity data is then transferred to a computer. The software program processes and corrects the data for inevitable instrumental and experimental errors and then scales them together to compare their relative intensities. The crystal characteristics such as space group symmetry and electron density map of molecule in crystal is calculated. The structural model for protein crystal is built using an electron density map. The schematic diagram of workflow for protein structure evaluation is represented in Fig. 5.3.

5.5 Advances in protein crystallography

X-ray diffraction for structural evaluation of proteins has undergone many major developments since its first application for crystal structure analysis of the protein insulin [47]. Arndt, U. et al. [48] invented rotation anode oscillator/generator, which soon became the primary X-ray diffraction data collection method for single crystals. Generation of high-intensity X-rays was achieved by continuous improvement in rotating anode generators. Rosenbaum, G. et al. [49] employed the synchrotron as a source of X-rays, and it was possible to generate X-rays with tunable wavelength and much higher X-ray flux. Phillips, W. et al. [50] developed charge-coupled detectors with synchrotron beamlines, which has made data collection much faster and more accurate. Meanwhile, shutter-free data collection was made possible by using pixel array detectors [51]. The cryogenic protection developed by Roger D [52] using flash-freezing of crystals at 100 K has made single-crystal data collection easier. Software development for the X-ray data collection and processing has seen major improvements during last 2 decades. Philips, J. et al. [53] developed multiwavelength anomalous diffraction (MAD) method, which made structure determination of single crystals easier and much faster. The use of selenomethionine incorporation into recombinant proteins and powerful synchrotron radiation made MAD the preferred method for crystal structure determination of de novo proteins. Yang, W. et al. [54] successfully employed this method to elucidate the first protein crystal structure, of Ribonuclease H. The free-R-factor method introduced by Brunger, A. [55] for structure cross-validation has prevented over refinement and became a key parameter for X-ray structure determination. Further improvement in free-R-factor method by Pannu, N. et al. [56] helped in implementation of maximum likelihood target functions in crystallographic programs. The collaborative computational project 4 (CCP4) suit, established in 1994, was a collection of a number of software programs for macromolecular structure determination methods by X-ray diffraction [57]. The structure determination packages, CNS and PHENIX, provided all the necessary programs for X-ray structure solution incorporating the refinement method of simulated annealing, X-plor [58–60]. The use of multiple isomorphous replacement (MIR) in protein structure evaluation demands generation of heavy atom derivatives of the crystals. Heavy atom agent "magic seven" prepared by Boggon, T. et al. [61] was recommended for crystals of soluble proteins, whereas the agent "Membrane's Eleven," designed by Morth, J. et al. [62] was employed for crystal structure study of membrane proteins. The phasing of larger protein complexes was possible employing polynuclear metal clusters [63]. This technique was successfully employed for structure elucidation of ribosomal subunit [64]. Besides MIR and MAD, other mainstream experimental phasing methods include single-wavelength anomalous dispersion, single isomorphous replacement, multiple isomorphous replacement with anomalous scattering, and single isomorphous replacement with anomalous scattering. Emsley, P. et al. [64] developed the automated model-building algorithm and molecular graphics COOT, which superseded the method developed by Jones, T. et al. [65] for building protein models in electron density maps. Modeling graphics has greatly enhanced the structure determination pace. It was possible to record crystallographic data of small and weakly diffracting protein crystals by intense X-rays at synchrotron facilities [66]. Neutze, R. et al. [43] developed ultrashort (femtoseconds) high-intensity X-ray pulses from free-electron lasers, which could provide structural data of crystal before its destruction by X-ray radiation. Seibert, M. et al. [67] provided proof for these studies by image construction of diffractions on individual virus particles. Chapman, H. et al. [26] have used this method for the study of the electron density map of photosystem I at 8.5 Å resolution. The first high-resolution X-ray structure of lysozyme was determined using SFX by Boutet, S. et al. [28] which was followed by de novo generation of experimental phases for lysozyme by Barends, T. et al. [68]. Further, this method has been used successfully for crystal structure determination of a precursor cathepsin B [29], photosystem II [69] and a human serotonin receptor [70]. Pawate, A. et al. [71] have used microfluidic crystallization platform for the serial time-resolved Laue diffraction analysis of macroscopic crystals of photoactive yellow protein grown on chip. The technique also avoids the manual handling of protein crystals. Maeki, M. et al. [72] have demonstrated use of cryoprotection and in situ X-ray diffraction measurement for lysozyme protein crystal using microfluidic chip. Using this technique, deterioration of crystallographic quality of protein crystals is avoided and complete dataset of protein crystal can be obtained using simple manipulation. Caballero and co-workers [73] have investigated the effects of temperature on crystal growth of glucose isomerase. Further, they have demonstrated a method to separate crystal growth and nucleation by using an ad hoc crystal growth cell and direct current. Rivera, C. et al. [74] presented novel approaches for controlling crystal growth, crystal size, and orientation by combined use of electric and magnetic field, making high-resolution X-ray crystallography easier. In addition, the very recent work by Walter, T. et al. [75] has demonstrated the use of protein thin film grown by external electric field as a template to obtain good-quality protein crystals of hen egg white lysozyme.

5.6 Case study: extended spectrum β-lactamases

X-ray crystallography has played an instrumental role in determination of three-dimensional structures of proteins. The protein data obtained from X-ray analysis, supplemented by other methods such as neutron diffraction and NMR, has helped in understanding fundamentals of structural biology, which has facilitated the development of modern medicine. The use of interactive computer graphics and molecular modeling software has assisted in rapid processing of data with high precision. β-lactam antibiotics are a class of broad spectrum antibiotics, which includes all antibiotic agents that contain a β-lactam ring in their molecular structures. This class of compounds includes derivatives of penicillin (penams), cephalosporins (cephems), monobactams, and carbapenems [76]. β-lactam antibiotics are commonly used for the prevention and treatment of bacterial infections caused by susceptible organisms. These antibiotics have been employed against infection caused by gram-positive as well as gram-negative bacteria. β-lactam antibiotics inhibit bacterial cell wall biosynthesis by targeting penicillin-binding proteins (PBPs). The binding of β-lactam antibiotics renders the PBPs chemically inert, causing bacterial death. To counter these strong β-lactam antibiotics, bacteria have evolved mechanisms to produce lactamases that are responsible for cleavage of amide bonds within the β-lactam ring [77]. β-lactamases are classified into four different classes (A,B,C, and D) according to sequence similarity. Classes A, C, and D consist of all serine-reactive hydrolases, while, class B consists of metalloenzymes, which use Zn^{2+}-bound water molecules in hydrolysis of amide bond of β-lactam ring. Class A β-lactamases were originally labeled as penicillinases, however, many β-lactamases with hydrolytic activity toward cephalosporins emerged with clinical use of oxyimino-cephalosporins. Such class A β-lactamases are called "extended spectrum of β-lactamases (ESBLs)." ESBLs were found to exhibit increased hydrolytic activity against the first-, second-, and third-generation extended spectrum cephalosporins and monobactam [78]. All class A β-lactamases use an active-site serine nucleophile to cleave the β-lactam bond of the substrate in a two-step acylation–deacylation reaction cycle leading to complete hydrolysis (Fig. 5.4). The acylation step involves: formation of precovalent substrate complex (1), general base-catalyzed nucleophilic attack on β-lactam carbonyl by the serine hydroxyl with formation of tetrahedral intermediate (2), and formation of stable acyl-enzyme adduct (3). Whereas, the deacylation step involves: general base-catalyzed attacks by a hydrolytic water molecule on acyl-enzyme adduct to form a second tetrahedral intermediate (4), formation of post covalent product complex from second tetrahedral intermediate (5), and release of the hydrolysis product.

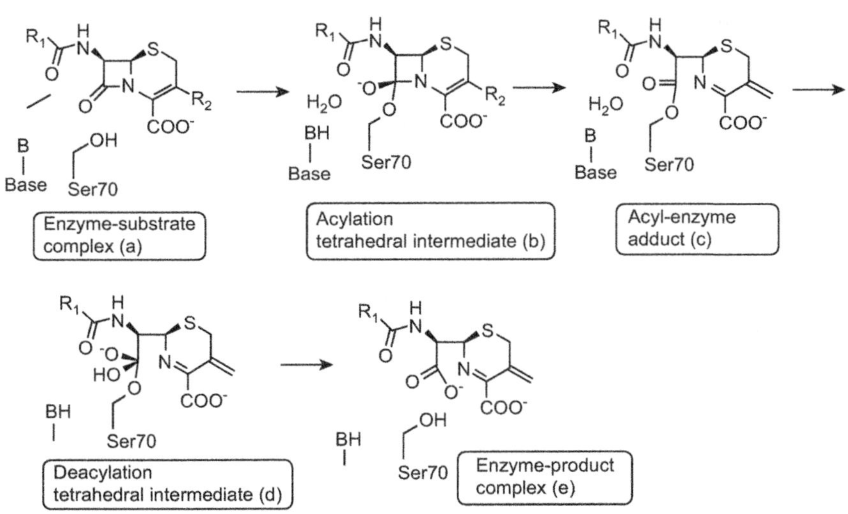

FIGURE 5.4 Schematic representation of catalytic cycle of a class A β-lactamase for a cephalosporin substrate.

Toho-1 (CTX-M-44) is CTX-M-type class A, ESBL, encoded by a plasmid in *Escherichia coli* TUH12191. It was isolated from the urine of a patient treated with β-lactam antibiotics [79]. The molecular structure of β-lactamases consists of α helical domain on the left and a mixed α/β helical domain on the right. The active site is located at the interface of these two domains. Most of the amino acid substitutions occur either in the Ω loop located at the base of the active site or in the terminal part of the B3 β-strand. The active site serine residue, Ser* (Fig. 5.5A), is located at the end of H2 α-helix [80]. The first tertiary structure of CTX-M enzyme, Toho-1 β-lactamase, was established using X-ray analysis (Fig. 5.5B). The crystal structure has revealed that there is no perceptible difference between Toho-1 and non-ESBLs in overall tertiary structures. However, the B3 β-strand and Ω loop, which are two critical components of the active site structure, may have higher flexibility in Toho-1. The amino acid residues Gly232, Asp240, and Phe160 were proposed to cause the flexibility in Toho-1 that contributes to the expansion of substrate specificity of this enzyme [81]. Furthermore, the structural analysis of the acyl-intermediate of Toho-1, complexed with antibiotics such as cefotamine, cephalothin, and benzylpenicillin, have indicated the interactions of side chain of Asn104 with all three substrates, while the side chain of Asp240 interacts only with cefotamine [82].

The crystallographic structural analyses of Toho-1 at 1.65 Å resolution by Ibuka, A. et al. [83] have revealed two alternative conformations for Ly73 side chain. The predominant conformation of Lys73 was found to be different from

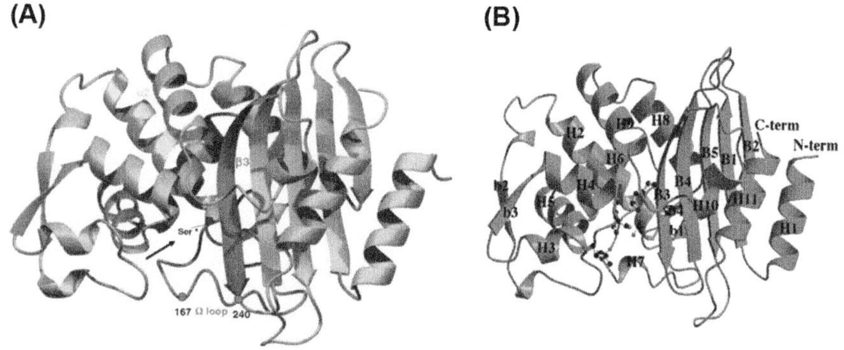

FIGURE 5.5 (A) Ribbon diagram of class A β-lactamase representing α and β domains [80], (B) Ribbon diagram of Toho-1 β-lactamase (Secondary structural elements are labeled according to the notations used for *Bacillus licheniformis* β-lactamase. The β-strands in Toho-1 are B1 (43–50), B2 (55–60), b1 (66–67), b2 (94–97), b3 (115–118), b4 (180–181), B3 (230–238), B4 (243–251), and B5 (259–266), shown in green. The helices are H1 (29–40), H2 (69–84), H3 (107–112), H4 (119–128), H5 (132–142), H6 (145–154), H7 (168–170), H8 (183–195), H9 (201–212), H10 (218–224), and H11 (276–286), shown in pink. The three regions conserved in all class A b-lactamases are the SXXK region (70–73), SXN region (130–132), and KTG region (234–236), shown as ball-and-stick models) [81].

that observed in the E166A mutant, indicating that the removal of the Glu166 side chain changes its conformation as well as interaction with Glu166. The electron density map of enzyme suggested rotation of Ser237, consequently, the hydroxyl group rotates through interaction with the carboxyl group of the substrate. Among the three sulfate ions positioned in or near the substrate binding cavity, one is tightly bound to the active site, whereas the other two are held by electrostatic interaction with two arginine residues, Arg274 and Arg276. The positively charged region between Arg274 and Arg276 is speculated to represent a pseudo-binding site of the β-lactam antibiotics involving methoxyimino group of the third-generation cephems prior to proper binding cleft for hydrolysis. The β-lactamases exhibit a strong tendency to form merohedrally twinned crystals, which affects the quality of crystals. Shimamura, T. et al. [84] used surface modification to improve the crystal quality of Toho-1 β-lactamase by removing a sulfate ion involved in packing. The twinned crystal formation was not observed for surface-modified Toho-1 (R274N/R276N) and crystal diffracted to a significantly higher resolution (~0.97 Å). Tomanicek, S. et al. [85] have employed Pd- Toho-1 β-lactamase (R274N/R276N)-benzothiophene-2-boronic acid (BZB) complex to evaluate the mechanism of hydrolysis of β-lactam bond in antibiotics. The protonated Glu166 side chain and the hydrogen-bonding network in which it participates with the inhibitor bound are consistent with Glu166 residue acting as the catalytic base in the acylation reaction mechanism, where upon binding to the BZB, the proton shuttles from Ser70 to Glu166 through the catalytic water. X-ray analysis has revealed a change in the protonation state of Glu166 due to the binding of an acylation transition state analogue and the presence of a hydrogen-bonded network is capable of facilitating breakdown of the tetrahedral intermediate. Vandavasi, V. et al. [86] have analyzed the structure of residue Lys73, which is involved in the catalytic mechanism of class A β-lactamase enzymes. The structures of both ligand-free protein and the acyl-enzyme complex of perdeuterated E166A Toho-1 β-lactamase with the antibiotic cefotaxime were studied. The structural analysis have revealed single conformation for Lys73 in ligand-free structures and two different conformations in all acyl-enzyme structures. They have predicted possible pathway for proton transfer from Lys73 to Ser130 based on existence on Lys73 closer to Ser70 in one conformer and closer to Ser130 in another conformer. Further, they have reported the structures of class A β-lactamase in an acyl enzyme complexed with aztreonam, a monobactam antibiotic, for the first time. Aztreonam is used to treat cystic fibrosis patients with chronic pulmonary infections caused by *Pseudomonas aeruginosa*. The mechanism of hydrolysis of aztreonam by CTX-M ESBLs has revealed most of the hydrogen atoms within the active site, Lys234, is fully protonated in the acyl intermediate and Lys73 is neutral, which serves as a general base during acylation step in the catalytic mechanism [87].

5.7 Conclusion

Proteins are the important constituents required for the proper functioning of almost all living organisms, and hence they are the most abundant biological macromolecules in nature. The structural analysis of proteins and other biological macromolecules always has been a very challenging task. X-ray crystallography is the most common method for determining the detailed three-dimensional structure of proteins and protein-ligand complexes. In short, the method of protein crystal structural characterization using X-ray diffraction consists of: (1) preparing high-quality protein crystals; (2) collecting diffracted X-ray signals from crystal by placing in X-ray beam; and (3) processing of diffracted X-ray signal to build a model of the atomic arrangement within the crystal. Furthermore, these models can be used for various applications, such as structure-based drug design, to understand receptor interactions and so on. It is worth noting that the protein molecules retain their function in crystalline state and interestingly their properties are not altered by crystallization. Additionally, the three 3D structures of various biological macromolecules such as proteins and nucleic acid are available in database libraries; PDB can be accessed freely with an Internet browser. http://www.rcsb.org/pdb/static.do?p=software/softwarelinks/molecular_graphics.html.

Acknowledgments

The author UBG thanks the research center of Dnyanprassarak Mandal's College (project no. DNY/CC/2014-15/03/1198) for financial support. MSM would like to acknowledge DST-SERB (New Delhi, India) (project no. YSS/2014/000776) for Young Scientist Research grant and funding. VKM would like to thank UGC (project No.F.47−1186/14/WRO) for research funding.

References

[1] Kinsella JE. Functional properties of soy proteins. J Am Oil Chem Soc 1979;56:242−58.

[2] Anderson A. The process of structure-based drug design. Chem Biol 2003;10:787−97.

[3] Friedrich W, Knipping P, von Laue M. Interferenzers cheinungen bei röntgenstrahlen. Ann Phys 1913;41:971−88.

[4] Bragg WH, Bragg WL. The structure of diamond. Proc R Soc A 1913;89:277−91.

[5] Kendrew JC, Bodo G, Dintzis HM, Parrish RG, Wyckoff H, Phillips DC. A three dimensional model of the myogloblin molecule obtained by X-ray analysis. Nature 1958;181:662−6.

[6] Kendrew JC, Dickerson RE, Strandberg BE, Hart RG, Davies DR, Phillips DC, Shore VC. Structure of myoglobin: a three dimensional Fourier synthesis at 2 Å resolution. Nature 1960;185:422−7.

[7] Watson JD, Crick FH. Molecular structure of nucleic acids; a structure for deoxyribose nucleic acid. Nature 1953;171:737−8.

[8] Carlisle CH, Hodgkin DC. The crystal structure of cholesteryl iodide. Proc R Soc A 1945;184:64−83.

[9] Abrahamsson A, Hodgkin DC, Maslen EN. The crystal structure of phenoxymethyl-penicillin. Biochem J 1963;86:514−35.

[10] Hodgkin DC, Maslen EN. The X-ray analysis of the structure of cephalosporin C. Biochem J 1961;79:393. 02.

[11] Brink C, Hodgkin DC, Lindsey J, Pickworth J, Robertson JR, White JG. X-ray crystallographic evidence on the structure of vitamin B_{12}. Nature 1954;174:1169–71.

[12] Venkatesa K, Dale D, Hodgkin DC, Nockolds CE, Moore FH, O'Connor BH. The structure of vitamin B_{12}. IX. The crystal structure of cobyric acid, Factor V 1a. Proc R Soc A 1971;323:455–80.

[13] Lenhert PG, Hodgkin DC. Structure of the 5,6-dimethyl-benzimidazolylcobamide coenzyme. Nature 1961;192:937–8.

[14] Deisenhofer J, Epp O, Miki K, Huber R, Michel H. Structure of photosynthetic reaction centre of *Rhodopseudomonas* viridis at 3 Å resolution. Nature 1985;318:618–24.

[15] Abrahams JP, Leslie AG, Lutter R, Walker JE. Structure at 2.8 Å resolution of F1-ATPase from bovine heart mitochondria. Nature 1994;370:621–8.

[16] Doyle DA, Cabral JM, Pfuetzner RA, Kuo A, Gulbis JM, Cohen SL, Chait BT, MacKinnon R. The structure of the potassium channel: molecular basis of K^+ conduction and selectivity. Science 1998;280:69–77.

[17] Cramer P, Bushnell DA, Kornberg RD. Structural basis of transcription: RNA polymerase II at 2.8 Å resolution. Science 2001;292:1863–76.

[18] Galli S. X-ray crystallography: one century of noble prizes. J Chem Educ 2009;91:2009–12.

[19] Ban N, Nissen P, Hansen J, Moore PB, Steitz TA. The complete atomic structure of large ribosomal subunit at 2.4 Å resolution. Science 2000;289:905–20.

[20] Rasmussen SG, Choi HJ, Rosenbaum HDM, Kobilka TS, Thian FS, Edwards PC, Burghammer M, Ratnala VR, Sanishvili R, Fischetti RF, Schertler GF, Weis WI, Kobilka BK. Crystal structure of the human β2 – adrenergic G protein-coupled receptor. Nature 2007;450:383–7.

[21] Cherezov V, Rosenbaum DM, Hanson MA, Rasmussen SG, Thian FS, Kobilka TS, Choi HJ, Kuhn P, Weis WI, Kobilka BK, Stevens RC. High resolution crystal structure of an engineered human β2 -adrenergic G protein – coupled receptor. Science 2007;318:1258–65.

[22] Shi Y. A glimpse of structural biology through X-ray crystallography. Cell 2014;159:995–1014.

[23] Cullity BD. Elements of X-ray diffractions. Addison Wesley Pub. Co. Inc.; 1956.

[24] Warne T, Serrano-Vega MJ, Baker JG, Moukhametzianov R, Edwards PC, Henderson R, Leslie AG, Tate CG, Schertler GF. Structure of a beta1-adrenergic G-protein-coupled receptor. Nature 2008;454:486–91.

[25] Smith JL, Fischetti RF, Yamamoto M. Micro-crystallography comes of age. Curr Opin Struct Biol 2012;22:602–12.

[26] Chapman HN, Fromme P, Barty A, et al. Femtosecond X-ray protein nanocrystallography. Nature 2011;470:73–7.

[27] Kirian RA, White TA, Holton JM, Chapman HN, et al. Structure-factor analysis of femtosecond micro-diffraction patterns from protein nanocrystals. Acta Crystallogr A: Found Crystallogr 2011;67:131–40.

[28] Boutet S, Lomb L, Williams GJ, Barends TR, et al. High-resolution protein structure determination by serial femtosecond crystallography. Science 2012;337:362–4.

[29] Redecke L, Nass K, DePonte DP, White TA, Rehders D, Barty A, et al. Natively inhibited Trypanosoma brucei cathepsin B structure determined by using an X-ray laser. Science 2013;339:227–30.

[30] Ayyer K, Yefanov OM, Oberthur D, Roy-Chowdhury S, Galli L, et al. Macromolecular diffractive imaging using imperfect crystals. Nature 2016;530:202–6.

[31] Caffrey MJ. Membrane protein crystallization. Struct Biol 2003;142:108–32.

[32] Fromme P, Spence JC. Femtosecond nanocrystallography using X-ray lasers for membrane protein structure determination. Curr Opin Struct Biol 2011;21:509–16.
[33] Johnson JE, Cornell RB. Amphitropic proteins: regulation by reversible membrane interactions (review). Mol Membr Biol 1999;16(3):217–35.
[34] Alenghat, Francis J, Golan DE. Curr Top Membr 2013;72:89–120.
[35] Yildirim MA, Goh KI, Cusick ME, Barabasi AL, Vidal M. Drug-target network. Nat Biotechnol 2007;25:1119–26.
[36] Stevens TJ, Arkin IT. Do more complex organisms have a greater proportion of membrane proteins in their genomes? Proteins Struct Funct Genet 2000;39:417–20.
[37] Krogh A, Larsson B, von Heijne G, Sonnhammer ELJ. Predicting transmembrane protein topology with a hidden Markov model: application to complete genomes. Mol Biol 2001;305:567–80.
[38] Aquila A, Hunter MS, Doak RB, Kirian RA, Fromme P, et al. Time-resolved protein nanocrystallography using an X-ray free-electron laser. Optic Express 2012;20:2706–16.
[39] Spence JC, Weierstall U, Chapman HN. X-ray lasers for structural and dynamic biology. Rep Prog Phys 2012;75:102601.
[40] Kupitz C, Basu S, Grotjohann I, Fromme R, Zatsepin NA, Rendek KN, et al. Serial time-resolved crystallography of photosystem II using a femtosecond X-ray laser. Nature 2014;513:261–5.
[41] Tenboer J, Basu S, Zatsepin N, Pande K, Milathianaki D, et al. Time-resolved serial crystallography captures high-resolution intermediates of photoactive yellow protein. Science 2014;346:1242–6.
[42] Barty A, Caleman C, Aquila A, Timneanu N, Lomb L, White TA, Andreasson J, et al. Self-terminating diffraction gates femtosecond X-ray nanocrystallography measurements. Nat Photon 2012;6:35–40.
[43] Neutze R, Wouts R, van der Spoel D, Weckert E, Hajdu J. Potential for biomolecular imaging with femtosecond X-ray pulses. Nature 2000;406:752–7.
[44] Barends TRM, Foucar L, Ardevol A, Nass K, Aquila A, Botha S, Doak RB, et al. Direct observation of ultrafast collective motions in CO myoglobin upon ligand dissociation. Science 2015;350:445–50.
[45] Burley SK, Berman HM, Kleywegt GJ, Markley JL, Nakamura H, Velankar S. Protein data bank (PDB): the single global macromolecular structure archive. Methods Mol Biol 2017;1607:627–41.
[46] White S. 2014. http://blanco.biomol.uci.edu/mpstruc/.
[47] Crowfoot D. X-ray single crystal photographs of insulin. Nature 1935;135:591–2.
[48] Arndt UW, Phillips DC. The linear diffractometer. Acta Crystallogr 1961;16:807–18.
[49] Rosenbaum G, Holmes K. Synchrotron radiation as a source for X-ray diffraction. Nature 1971;230:434–7.
[50] Philips WC, Li Y, Stanton M, Xie J, O'Mara D, Kalata K. A CCD based area detector for X-ray Crystallography using synchrotron and laboratory sources. Nucl Instrum Methods Phys Res 1993;334:621–30.
[51] Kraft P, Bergamaschi A, Broennimann CH, Dinapoli R, et al. Performance of single- photon counting PILATUS detector modules. J Synchrotron Radiat 2009;16:368–79.
[52] Rodger DW. Cryocrystallography. Structure 1994;2:1135–40.
[53] Philips JC, Hodgson KO. The use of anomalous scattering effects to phase diffraction patterns from macromolecules. Acta Crystallogr A 1980;36:856–64.
[54] Yang W, Hendrickson WA, Crouch RJ, Satow Y. Structure of ribonuclease H phased at 2 Å by MAD analysis of the selenomethionyl protein. Science 1990;249:1389–405.

[55] Brunger AT. Free R value: a novel statistical quantity for assessing the accuracy of crystal structures. Nature 1992;355:472–5.

[56] Panu NS, Read RJ. Improved structure refinement through maximum likelihood. Acta Crystallogr A 1996;52:659–68.

[57] Collaborative CP. Collaborative Computational Project, Number 4. The CCP4 suite: programs for protein crystallography. Acta Crystallogr D Biol Crystallogr 1994;50:760–3.

[58] Brunger AT, Adams PD, Clore GM, DeLano WL, Gros P, Grosse-Kunstleve RW, Jiang JS, Kuszewski J, Nilges M, Panu NS, et al. Crystallography and NMR system: a new software suite for macromolecular structure determination. Acta crystallogr D Biol Crystallogr 1998;54:905–21.

[59] Adams PD, Grosse-Kunstleve RW, Hung LW, Loerger TR, MacCoy AJ, Moriarty NW, Read RJ, Sacchettini JC, Sauter NC, Terwillger TC. PHENIX: building new software for automated crystallographic structure determination. Acta crystallogr D Biol Crystallogr 2002;58:1948–52.

[60] Brunger A, XPLOR Manual v 2.1. New Haven: Yale University.

[61] Boggon TJ, Shapiro L. Screening for phasing atoms in protein crystallography. Structure 2000;8:R143–9.

[62] Morth JP, Sorenson TL, Nissen P. Membrane's Eleven: heavy -atom derivatives of membrane-protein crystals. Acta crystallogr D Biol Crystallogr 2006;62:877–82.

[63] Dauter Z. Use of polynuclear metal clusters in protein crystallography. C R Chim 2005;8:1808–14.

[64] Emsley P, Cowtan K. Coot: model building tools for molecular graphics. Acta crystallogr D Biol Crystallogr 2004;60:2126–32.

[65] Jones TA, Zou JY, Cowan SW, Kjeldgaard M. Improved methods for building protein models in electron density maps and the location of errors in these models. Acta Crystallogr A 1991;47:110–9.

[66] Joanchimiak A. High-throughput crystallography for structural genomics. Curr Opin Struct Biol 2009;19:573–84.

[67] Seibert MM, Ekeberg T, Maia FR, Svenda M, Andreasson J, Jonsson O, Odic D, Iwan B, Rocker A, et al. Single mimivirus particle intercepted and imaged with an X-ray laser. Nature 2011;470:78–81.

[68] Barends TR, Foucar L, Botha S, Doak RB, Shoeman RL, Nass K, Koglin JE, Williams GJ, Boutet S, Messerschmidt M, Schlichting I. De novo protein crystal structure determination fro X-ray free electron laser data. Nature 2014;505:244–7.

[69] Kern J, Alonso-Mori R, Tran R, Hattne J, Gildea RJ, Echols N, Glockner C, Hellmich J, Laksmono H, et al. Simultaneous determination of femtosecond X-ray spectroscopy and diffraction of photosystem II at room temperature. Science 2013;340:491–5.

[70] Liu W, Wacker D, Gati C, Han GW, James D, Wang D, Nelson G, Weierstall U, Katritch V, et al. Serial femtosecond crystallography of G-protein coupled receptors. Science 2013;342:1521–4.

[71] Pawate AS, Šrajer V, Schieferstein J, Guha S, Henning R, Kosheleva I, Schmidt M, Ren Z, enis PJA, Perry SL. Towards time-resolved serial crystallography in a microfluidic device. Acta Crystallogr F Struct Biol Commun 2015;71:823–30.

[72] Maeki M, Pawate AS, Yamashita K, Kawamoto M, Tokeshi M, Kenis PJA, Miyazaki M. A method of cryoprotection for protein crystallography by using a microfluidic chip and its application for in situ X-ray diffraction measurements. Anal Chem 2015;87:4194–200.

[73] Caballero SM, Cruz MC, Demitri N, Polentarutti M, Romero AR, Moreno A. Glucose isomerase polymorphs obtained using an ad hoc protein crystallization temperature device and a growth cell applying an electric field. Cryst Growth Des 2016;16:1679–86.

[74] Rivera CP, Cruz MC, Escofet NE, Demitri N, Polentarutti M, Stojanoff V, Moreno A. Recent advances in the understanding of the influence of electric and magnetic fields on protein crystal growth. Cryst Growth Des 2017;17(1):135–45.

[75] Walter TK, da Gama Ferreira CF, Iulek J, Benelli EM. Use of protein thin film organized by external electric field as a template for protein crystallization. ACS Omega 2018;3(8):8683–90.

[76] Holten KB, Onusko EM. Appropriate prescribing of oral beta-lactam antibiotics. Am Fam Physician 2000;62(3):611–20.

[77] Drawz SM, Bonomo RA. Three decades of beta-lactamase inhibitors. Clin Microbiol Rev 2010;23:1–7.

[78] Knox JR. Extended-spectrum and inhibitor-resistant TEM-type beta-lactamases: mutations, specificity, and three-dimensional structure. Antimicrob Agents Chemother 1995;39:2593–601.

[79] Ishii Y, Ohno A, Taguchi H, Imajo S, Ishiguro M, Matsuzawa H. Cloning and sequence of the gene encoding a cefotaxime-hydrolyzing class A beta-lactamase isolated from *Escherichia coli.* Antimicrob Agents Chemother 1995;39:2269–75.

[80] Rossolini GM, D'Andrea MM, Mugnaioli C. The spread of CTX-M-type extended-spectrum β-lactamases. Clin Microbiol Infect 2008;14:33–41.

[81] Ibuka AS, Taguchi A, Ishiguro M, Fushinobu S, Ishii Y, Kamitori S, Okuyama K, Yamaguchi K, Konno M, Matsuzawa H. Crystal structure of the E166A mutant of extended-spectrum beta-lactamase Toho-1 at 1.8 Å resolution. J Mol Biol 1999;285:2079–87.

[82] Shimamura T, Ibuka A, Fushinobu S, Wakagi T, Ishiguro M, Ishii Y, Matsuzawa H. Acyl-intermediate structures of the extended-spectrum class A beta-lactamase, Toho-1, in complex with cefotaxime, cephalothin, and benzylpenicillin. J Biol Chem 2002;277(48):46601–8.

[83] Ibuka AS, Ishii Y, Galleni M, Ishiguro M, Yamaguchi K, Frère JM, Matsuzawa H, Sakai H. Crystal structure of extended-spectrum beta-lactamase Toho-1: insights into the molecular mechanism for catalytic reaction and substrate specificity expansion. Biochemistry 2003;42(36):10634–43.

[84] Shimamura T, Nitanai Y, Uchiyama T, Matsuzawa H. Improvement of crystal quality by surface mutations of β-lactamase Toho-1. Acta Crystallogr Sect F Struct Biol Cryst Commun 2009;65:379–82.

[85] Tomanicek SJ, Standaert RF, Weiss KL, Ostermann A, Schrader TE, Ng JD, Coates L. Neutron and X-ray crystal structures of a perdeuterated enzyme inhibitor complex reveal the catalytic proton network of the Toho-1 β-lactamase for the acylation reaction. J Biol Chem 2013;288(7):4715–22.

[86] Vandavasi VG, Weiss KL, Cooper JB, Erskine PT, Tomanicek SJ, Ostermann A, Schrader TE, Ginell SL, Coates L. Exploring the mechanism of β-lactam ring protonation in the class A β-lactamase acylation mechanism using neutron and X-ray crystallography. J Med Chem 2016;59:474–9.

[87] Vandavasi VG, Langan PS, Weiss KL, Parks JM, Cooper JB, Ginell SL, Coates L. Active-site protonation states in an acyl-enzyme intermediate of a class A β-lactamase with a monobactam substrat. Antimicrob Agents Chemother 2017;61(1). e01636–16.

Chapter 6

Technological advancements in industrial enzyme research

Vazhakatt Lilly Anne Devasia[1,#], R. Kanchana[2,*], Poonam Vashist[1], Usha D. Muraleedharan[1]
[1]*Department of Biotechnology, Goa University, Goa, India;* [2]*Department of Biotechnology, Parvatibai Chowgule College of Arts and Science - Autonomous, Margao, Goa, India;*
[#]*Present address: Department of Biotechnology, Hindustan College of Arts and Science, Padur, Kelambakkam, Chennai, India*
[*]*Corresponding author: rkr002@chowgules.ac.in.*

6.1 Introduction

Enzymes, commonly known as biocatalysts, are biomolecules that catalyze several biological as well as industrial processes. Industrial enzymes form a greener and better option in a multitude of processes that lead to the final commercial product formation. A significant number of enzymes have been applied in industries to cater to various needs in medicine, agriculture, bioremediation, biofuel, and commodity production such as textiles, detergents, pulp and paper, cosmetics, beverages, nutraceuticals, and leather. Several of these industrial enzymes are derived from recombinant microorganisms. Manufacturers often turn to molecular genetic techniques either to tailor enzymes with improved properties or to produce enzymes from microorganisms that do not survive culture conditions in laboratory or industries. Recombinant strains of microorganisms are created after judicious selection of hosts for enhanced production of enzymes with specificity and for easy downstream processing. Technological advances in molecular genetics, bioinformatics, and cell biology have thus dictated the terms and possibilities in enzyme research and production.

The field of enzyme research has seen some remarkable changes with respect to the manner of approaching the problems of screening for variants of enzymes, discovery of novel enzymes, as well as enzyme profiling and/or engineering to create "altered" enzymes that fulfill specific industrial needs. The next generation sequencing (NGS) methods have provided a platform with novel solutions in enzyme research, as in other fields. Enzymologists have

Advances in Biological Science Research. https://doi.org/10.1016/B978-0-12-817497-5.00006-9

dared to venture and dream, with the tools of gene editing, mutagenesis, and NGS in their hands.

While three-dimensional (3D) modeling and simulation methods using computer-aided discovery have undoubtedly brought the scientific community closer to perceiving the 3D structures of enzymes with exactness, they have also introduced a time-saving and hence cost-effective element [1]. Conventional methods of enzyme screening involve wet lab experiments starting with the isolation and culture of microorganisms followed by screening for the desired enzyme activity. These procedures suffer greatly with respect to time and cost involved because even after much effort being expended, isolation of an effectively novel enzyme with industrial potential might remain an unfulfilled dream. Nevertheless, the trend is fast changing of late since the metagenome is being used directly to screen for relevant enzymes, thus circumventing the intricacies of finding microbiological means for sustaining growth of organisms that are otherwise labeled "unculturable."

This chapter provides comprehensive material on technological advances in enzyme production for researchers across many related disciplines. It is conceived to take the reader through the typical processes involved in enzyme discovery and development, starting with screening methods, proceeding to engineering methods, and finally the formation of a stable product with improved enzyme efficacy. The benefits accrued by the use of newer technologies at the various stages of enzyme research are also discussed.

6.2 Enzyme discovery

With the accelerated craving for newer sources of enzymes with unique properties for efficient use, researchers are increasingly directing their attention toward the natural biodiversity as a rich source of novelty. Unassuming efforts at bioprospecting from natural sources such as mangroves, coastal habitats, corals, deep sea, salt pans, polluted soil samples, etc., have been rewarded with the unveiling of industrially useful enzymes. For instance, Kanchana [2] and Kanchana & Mesta [3] isolated a novel feather-degrading *Bacillus* sp. and *Aspergillus* sp., respectively, from poultry wastes in Goa, India, using an inexpensive feather meal broth with chicken feathers as sole carbon and nitrogen source. The enzyme keratinase, which exhibited remarkable feather-degrading ability, could be a potential industrial enzyme for the improvement of the nutritional properties of feathers (and other keratin sources) used as supplementary feedstuff. In yet another study by Kanchana [4], two novel feather-degrading fungi belonging to *Chrysosporium* and *Microsporum* genera exhibited broad pH (7.0–10.0) and temperature (30–70°C) stability as well as feather-degrading potential within a period of 96 h. A thermostable extracellular alkaline protease from *Enterobacter* sp., with prospective applications in the detergent industry, destaining of blood, and degradation of gelatinous coating of X-ray films, has been reported by Kanchana et al. [5]. A successful attempt on the production of inducible extracellular cholesterol oxidase from *Micrococcus* sp. from soil samples of Goa has also been recorded [6].

Marine and estuarine habitats, such as coastal areas and mangroves, have been subjected to scrutiny in the search for marine enzymes. For instance, two thraustochytrid protists of the genus *Thraustochytrium* isolated from coastal and mangrove habitats of Goa were studied for extracellular alkaline lipase production [7]. Another endeavor from our laboratory has led to the discovery of a novel bacterium, *Microbulbifer mangrovei*, sourced from the mangroves of Goa, which exhibited a larger number of polysaccharide-degrading activities against various insoluble complex polysaccharides than the famed *Saccharophagus degradans* [8]. Further, Devasia & Muraleedharan have reported screening a total of 32 thraustochytrid isolates for hydrolytic activities from coastal and mangrove samples, which resulted in the identification of 19 isolates with multiple polysaccharide-degrading activities including cellulase, amylase, agarase, xylanase, and pectinase, among others [9]. In addition, amylolytic activities from *Thraustochytrium* sp. have been reported [10], and Shirodkar & Muraleedharan [11] successfully produced enhanced amylolytic activity using the technique of response surface methodology.

Enzymes from extremophiles have been generating attention among researchers due to their vast power of catalysis under extreme circumstances. Certain technological breakthroughs have hastened the discovery of industrially important enzymes by providing effective means to mine the natural biodiversity. These include the integration of "omics" such as genomics, transcriptomics, proteomics, and metagenomics as biocatalyst discovery tools. The big data, generated by fast and efficient DNA sequencing technologies, has clearly emerged as a means to identify information accessible in the blueprint of the organisms.

Conventional screening methods that are either simultaneous with or are followed after the isolation of enzyme-producing strains of microorganisms have now been largely replaced by screening for enzymes from the metagenome, transcriptome, and/or proteome data. Genome data are a rich source of biodiversity in the search for novel enzymes. This has become a possibility with the replacement of the time-tested Sanger's method of DNA sequencing with rapid sequencing methods such as Illumina Solexa and SOLid, among others [12], which have therefore been perceived as a useful strategy for screening extremophiles and unculturable microbes, in particular.

Metagenome studies have led to the discovery of enzymes that hitherto went undetected. Metagenome mining, although as labor intensive as any of the conventional techniques, is considered advantageous because it is based on direct isolation of DNA from natural resources, setting aside the typical perils of isolating and culturing microorganisms. Given the fact that only about 1% of the total microbial world is culturable [13], the chances of hitting upon novel enzymes are greater when the starting material is a mixture of the DNA available from the total microbial population in an environmental source. Recently, Choi et al. [14] identified a metagenomic *Escherichia coli* clone that produced a new metagenomic enzyme, TxeA, capable of degrading the phytotoxin toxoflavin. According to Ferrer et al. [15], until the year 2015 metagenomic studies have been conducted from only 2192 sites across the globe to identify functional novelty in microorganisms. While industrial enzymes such as proteases, oxidoreductases, lipases, and acylases have had a fairly good representation in metagenomic investigations, enzymes such as nitrilases and transaminases have been poorly addressed [12,15,16].

Transcriptomes can be powerful tools in the identification of beneficial industrial biocatalysts produced by microorganisms that are grown under conditions that mimic those of industrial processes and substrates [13,15]. This is because transcriptome data are generated from mRNA transcribed from genes that are differentially expressed under the altered conditions. Additionally, enzymes that share similarities with other enzymes can be mined from an organism's transcriptome, if available in the databases. Proteome refers to the entire protein collection of an organism. Proteomic approach is used in studies related to protein expression and those concerning the function of the proteins. The proteomic and metaproteomic approaches have been applied for novel enzyme screening from cultured microorganisms as well as environmental samples. It is dependent on the identification of protein sequences present in the sample by comparison with those in the databases. Proteomics has been used for uncovering new carbohydrate-related enzyme activities [17].

The development of rapid enzyme screening methods poses a big challenge especially with the easy availability of immense amounts of sequencing and "omics" data, which is paralleled by a corresponding number of clone variants and libraries generated in a genetic engineering laboratory for novelty screening. It is reported that the choice of substrate used for functional screening of the enzyme activities plays a decisive role in the ratio of the hits versus the number of clones screened. To overcome this challenge, activity-based profiling was introduced as a new technique to promote the discovery of novel biocatalysts as well as allow the comparison of activities [18]. The technique involves the use of specific probes that are developed to bind to the active site of the proposed enzymes. The binding is followed by detection with fluorescent labels, radioactivity, or color-based changes. Activity-based proteomics has been used in the discovery and screening of serine-, cystein-, and threonine- proteases, lipases, and kinases [19]. It is therefore possible to detect activity of enzymes under physiological conditions, facilitating the study of enzymes even from unculturable microorganisms. Application of activity-based protein profiling (ABPP) has been successfully demonstrated for in vitro serine hydrolase screening from *Sulfolobus acidocaldarius*, which provided a proof-of-concept study for using ABPP in the identification and screening of enzymes from archaea [18]. The diversity of the probes for screening candidates from specific enzyme families could be integrated into a kit in the near future.

6.3 Enzyme customization

The study and the isolation of natural enzymes often prove futile as they sometimes fail to meet industry-specific requirements in terms of large-scale production, high-catalytic activity, solvent tolerance, or pH and temperature optima. However, the investment in resources and time for the discovery of novel enzymes has been substantial. The last few decades have thus created the exciting possibility of tailor-made enzymes, with the use of technologies in enzyme engineering, rational design, and directed evolution of enzymes, based upon their natural counterparts [20].

Several industrial enzymes, such as protease BYA, xylose isomerase, endo-1,4-betaxylanase II, cyclodextrin glucanotransferase, and so on, have been engineered thus far [20]. Engineered enzymes may show higher catalytic rate, enantioselectivity, thermostability, activity in the presence of solvents, or remain unaffected by salts. Enzyme engineering can involve changes either at the gene level or by protein engineering. The amino acids at the active site of enzymes can be targeted to make specific changes in the active enzyme assembly. Through computational studies, the best model has been used to synthesize enzymes that fit to the specific substrate with higher affinity [12].

Directed evolution is aimed at altering the enzyme to display desired properties of industrial potential rather than tinkering with industrial processes to suit the optimum working conditions of catalysts. Unlike rational design, determination of the structure of enzymes is not important. It involves the generation of a library of mutants followed by the identification of the mutant with the desired property from a mutated protein library using high-throughput techniques of screening of the variants. The following steps are involved: (1) choosing a parent protein, (2) generation of a mutant library, and (3) screening of proteins with the desired properties. Mutations have been introduced by several methods including random mutagenesis and DNA fragment shuffling [1]. High-throughput identification of mutants can be carried out by fluorescence-activated cell sorting (FACS) or microarray. Directed evolution has been combined with computational approach to make it more feasible in terms of applicability and practicality. In silico directed evolution is based on the theoretical knowledge of molecular interactions to predict even those properties that underlie the behavior of the enzymes but are otherwise difficult to pinpoint as being influential under practical conditions [21]. Amrein et al. [21] performed empirical valence bond calculations on amino acid substitutions of triosephosphate isomerase using computer-aided enzyme engineering to demonstrate the success with user-controlled in silico directed evolution of enzymes. In 2012, Imberger et al. [22] demonstrated an increase in the activity of cellulase CelA2 isolated from metagenomic pools to choline chloride and glycerol using directed evolution. In yet another example, temperature stability of subtilisin was improved upon by the directed evolution approach [23]. "Dr. Frances H. Arnold, one of the three recipients of the 2018

Nobel Prize in Chemistry, along with her team used directed evolution to create new enzymes that catalyze biologically unknown reactions" [24].

The rational designing of enzymes is based on their amino acid sequence information and the consequent 3D structure obtained after protein folding. Diverse enzymes with different biochemical characteristics but similar 3D structures have been found in nature [12]. The changes that have to be made in the natural amino acid sequence are logically predicted by studying and comparing such enzymes. The enzymes that catalyze different reactions and have varying substrate specificity but exhibit high similarity in their amino acid sequences therefore serve as candidates for rational design. The amino acid residues at the binding site of the enzyme are considered good targets for mutation in such designing of enzymes. The designing steps are repeated over and over on a trial-and-error basis until the best design is found. Assays are developed to functionalize the changes at every step of enzyme designing. In addition, the approach is dependent on the rapid screening of variants as well as the use of efficient techniques for determination of enzyme structures, such as X-ray diffraction and nuclear magnetic resonance spectroscopy coupled with bioinformatic tools and protein databases. Further, putative models of the structure of the enzyme so engineered have been arrived at using computational techniques [1].

6.4 Improvement of existing enzymes through mutagenic approaches

6.4.1 By site-directed mutagenesis

Progression in the area of enzyme modification has facilitated tuning of a biocatalyst to meet an industrial goal. Site-directed mutagenesis is one such device to construct novel proteins that serve as able catalysts. It involves amending of an amino acid at a precise site and evaluating the resultant mutated protein, the method thus serving as an option for those proteins whose structure and mechanisms of action are previously known. An improvement is that it takes less time in assessment since the number of variants produced is fewer. It assists in evaluating structural and functional features of particular amino acid residues in a protein. The major relevance of site-directed mutagenesis is to bring in novel properties such as improved specificity, stability, activity, and expression to the biocatalyst. A study was executed in an attempt to enhance the properties of α-galactosidase, wherein a new gene from a deep sea bacterium, *Bacillus megaterium*, was cloned and mutated. The study not only helped in improving properties of the enzyme but gave remarkable structural-functional information that also exposed means of amplified activity at the molecular level. They found a protein that holds a tunnel structure and showed how the NAD^+ (cofactor) constructed a way to the active center through this tunnel protein [25].

Site-directed mutagenesis furthermore extends its application to immobilized enzymes. In the past, immobilization was used to protect enzyme function by means of a support attached to the enzyme in an arbitrary manner. In view of the fact that orientation of the enzyme plays a major role in catalysis, researchers understood the need to immobilize the enzyme in a precise orientation with various specific residues. There have been some reports where site-directed mutagenesis had facilitated the immobilization of enzymes that could not be tethered owing to their ionic hindrances. For instance, a penicillin G acylase from *E. coli* was made to adsorb on diethylethanolamine or polyethylenimine-coated support upon addition of eight Glu residues by site-directed mutagenesis. It was repeatedly seen that enzymes dropped certain properties such as thermostability after immobilization. Site-directed mutagenesis has successfully terminated such limitations by introducing specific residues at exact sites [26].

6.4.2 By random mutagenesis

Random mutagenesis trails the unbiased move for alternative generation by mimicking the natural process. Nature needs years to evolve, by mutation or recombination, and has chosen on the basis of survival of the fittest. Random mutagenesis offers a chance to provide a mutated product in weeks by creating a library of mutants and selection of members based on the desired relevant property. It also surmounts the barrier for those proteins whose structure or catalytic mechanism is not completely known. This technique requires a competent, high-throughput screening system since the number of variants created is repeatedly high. The technique brings in unsystematic mutations in a gene, with the key objective to describe the open reading frames (ORFs), generating a diversity of variants that are subjected to screening for the relevant properties [27]. This random mutagenesis in genes is based on two methods, viz., in vitro directed evolution and gene recombination.

In vitro directed evolution/random mutagenesis may be achieved by a variety of techniques, such as chemical mutagenesis, site-saturating mutagenesis, error-prone polymerase chain reaction (PCR), and use of mutated strains, whereas the techniques that rely on gene recombination are DNA shuffling, staggered extension process, random chimeragenesis on transient templates, iterative truncation for the design of hybrid enzymes, and recombined extension on truncated templates (Fig. 6.1) [28].

At times, it has been observed that mutation causes improvement in one property while at the same time there is a compromise in other property [25]. A step forward of this restriction is to "rear" protein with the appropriate individual property and subsequently screen "progeny" for the aspiring set of properties. The engineered protein generated after mutation depends on the quality of the library. One variant (3-2G7) of subtilisin S41 (psychrophilic protease) was created by random mutagenesis, saturation mutagenesis, and

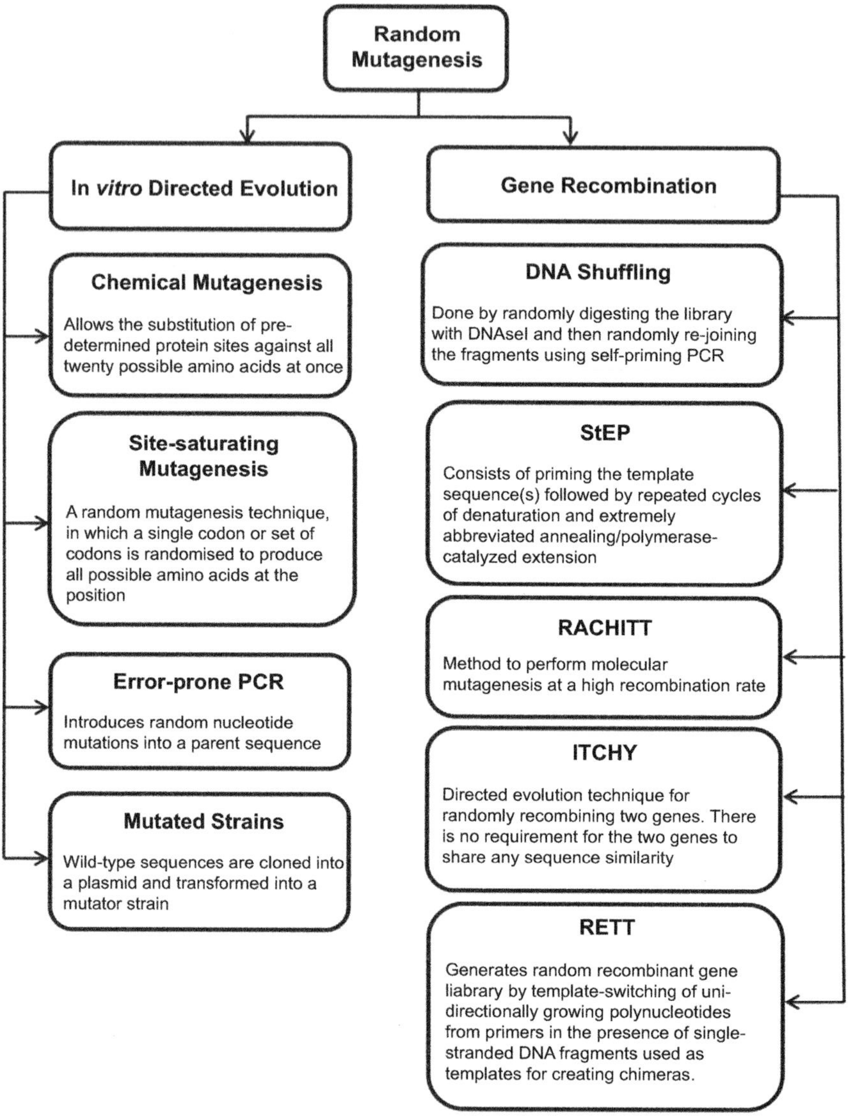

FIGURE 6.1 Techniques of Random Mutagenesis in genes.

in vitro recombination/DNA shuffling. An outstanding enhancement in temperature range was noticed without loss in its catalysis at low temperatures, and threefold higher catalytic competence was developed [29]. The DNA shuffling has also been proving to be moderately efficient in a range of applications. For example, Wang et al. [30] reported that the antifungal activity of *Lactobacillus plantarum* IMAU10014 strain was enhanced by three rounds of DNA shuffling. The mutant was efficient up to 200% compared to the wild strain, had broad antifungal spectra, and projected excellent candidature for biopreservation. Further, according to Wu et al. [31], salt tolerance in yeast was improved using two homologous Na^{+}/H^{+} antiporters from halophytes *Salicornia europaea* (SeNHX1) and *Suaeda salsa* (SsNHX1). The mutant exhibited up to 46% salt tolerance compared to the parent strain.

6.5 High-throughput screening of genetic variants for novel enzyme production

Fool-proof methods are required to screen the library of variants created by protein engineering to enact and support the rapid formation of a large number of genetic variants for enzyme discovery. The modern screening strategies applied are dependent on the strain of production, transformation efficiency, as well as assay development [32]. Microfluidics and FACS are currently used techniques in the screening for variants. Ostafe et al. [33] used FACS-based screening for cellulases by converting the products of cellulase action, viz., sugars into hydrogen peroxide, to be only used by vanadium bromoperoxide in the production of fluorescein. Besides yeasts, water-in-oil droplets of the size of bacterial cells have been developed for high-throughput screening of genetic variants of bacterial cells for novel enzyme discovery [32].

6.6 Immobilization of enzymes

The majority of enzymes, being proteins, are subject to deterioration at high temperatures, at extreme pH, and in the presence of organic solvents. An effective way to extend the stability and effectiveness of industrial enzymes is by the technique of immobilization through covalent bonding and encapsulation using polymeric substrates. Immobilization techniques serve the following primary functions: (1) preserve, if not increase, the stability of the enzyme under unfavorable conditions such as prevalent in most industrial processes; (2) conserve the relative concentration of the enzyme(s) by preventing considerable loss; and (3) permit easy recovery of the enzymes to render them reusable [20].

Covalent binding and noncovalent binding methods have been developed for immobilization of enzymes [20]. However, the search continues unabated for an ideal technology and substrates that bestow properties such as tailored pore size, cofactor regeneration, and surface functionality for enzyme immobilization.

Multiwalled carbon nanotubes (MWCNTs), magnetic or nonmagnetic nanoparticles, and quantum dots have been explored as substrates for enzyme immobilization. Carbon nanotubes are hollow nano-sized cylinders made up of single or multiple layers of graphite so as to give rise to single-walled or multiwalled carbon nanotubes, respectively. Enzymes are immobilized onto nanosized carriers by physical adsorption or covalent bonding. Enzyme immobilized nanoparticles, such as silver nanoparticles, gold nanoparticles, and silica nanoparticles, may be infused with magnetic substances that facilitate their easy separation using a magnet at the end of the process [20]. Further, magnetic nanoparticles may be engineered for suitable properties such as large pore size and surface area, stability, and catalytic activity. Yi et al. [34] used reversed micelle strategy to engineer lipase-immobilized magnetic nanoparticles containing Fe_3O_4.

Other novel composites developed for enzyme immobilization include mesoporous silica that has been popularized by its optimum pore size and mechanical stability. The mesoporous silica, when grafted with specific polymeric substances that can bind enzymes, acts as a good matrix for immobilization of enzymes. Gao et al. [35] successfully modified the mesoporous onion-like silica by coating with dopamine to act as a scaffold for *Candida* lipase, which thereby improved the stability and efficiency of the enzyme.

The entrapment method involving enclosing of enzymes in polymeric substances is one of the traditionally used methods for immobilization of industrial enzymes. A microencapsulation method has been developed recently as an alternative to bead entrapment. Microencapsulation involving cross-linking of two immiscible liquids serves to provide a thinner encapsulating membrane for easy mass transfer between the interior and exterior. A variety of technologies have been developed that involve modification of the microemulsions for improved immobilization and activation of the incorporated enzymes [36]. Addition of single-walled carbon nanotubes to the emulsions was found to further stabilize and activate immobilized lipases of *Chemobacterium viscosum* [37]. Polyethylenimine microcapsule immobilization has additionally been found to modulate enzyme activity. Su et al. [36] prepared microencapsules modified with polyethylenimine matrix and MWCN to activate and improve the activity of lipases.

Microwave irradiation has met with some degree of success in increasing the rate of the reaction. Immobilization of industrial enzymes such as papain, lipase, penicillin acylase, and horseradish peroxide has been achieved by microwave-assisted irradiation [20]. Photoimmobilization technology is yet another entrant in this field. A photoreactive polymer is used to immobilize enzymes by using UV irradiation or sunlight that generates a highly reactive species in the polymer for covalent binding with the enzymes [20].

6.7 Enzyme inhibitor studies

Conventionally, spectrophotometry (UV and fluorescence), chromatography (HPLC), and electrochemical methods have been used for screening enzyme inhibitors. However, more recently, microscale enzyme reactors that utilize capillary electrophoresis and enzyme immobilization techniques for in-house enzyme assays, determination of enzyme kinetics, and enzyme inhibitor studies have been developed [38,39].

In-line and precapillary enzyme assays based on capillary electrophoresis have been popularized in inhibitor screening studies because of fewer requirements of the samples and solvent systems, efficient and faster separation, and the ease to couple with automated detection systems. The in-line capillary electrophoretic method uses a single capillary for the entire length of time starting with sample loading and ending with detection [38]. This is different from precapillary enzyme assays wherein the capillary is used only to separate molecules. The enzyme reaction is carried out off-line, i.e., outside the capillary system and later injected into the capillary electrophoresis system. Precapillary detection has been used to determine the activity of membrane enzymes such as chlorophyllase (which is involved in chlorophyll metabolism), enzyme reaction rates, and inhibitors [39].

In-line capillary enzyme assays are divided into two types: electrophoretically mediated microanalysis (EMMA) and immobilized enzyme microreactor (IMER) [38,39]. EMMA uses the difference in electrophoretic mobility of molecules for separation of constituents in the mixture. EMMA can also be used for enzyme activity determination, stereospecificity, enzyme-mediated metabolic reactions, etc. IMERs use immobilized enzymes in the first part of the capillary while the rest of the capillary is used for separation of the metabolites [39] and are excellent tools to study enzyme assays. Scientists have prepared such IMERs to study the inhibitors of tyrosinase and trypsin from natural extracts as well as for enhancement of the loading capacity of cytochrome P-450 and glucose-6-phosphate dehydrogenase during immobilization procedures [40,41]. The same technology could be used for screening inhibitors of other enzymes of industrial significance.

6.8 Enzyme promiscuity and multifunctional enzyme studies

Some enzymes have been identified to catalyze more than one reaction. These are called multifunctional enzymes. They are promiscuous with respect to specificity, unlike their counterparts that catalyze specific reactions involving only a particular type of bond. Multifunctional enzymes are useful to the host organism that produces them. Studies on multifunctional enzymes offer critical insights into evolutionary history and phylogenetic status of several organisms. From literature reports, Cheng et al. [42] elucidated charge, polarizability, hydrophobicity, and solvent accessibility as a few of the most important character determinants of multifunctional enzymes. They constructed models in which the protein sequences were represented by vector determinants based on nine properties of the enzyme residues, including amino acid concentration, hydrophobicity, charge, surface tension, etc., to decipher information on the distribution among species and domains as well as their evolutionary significance. In a more recent study, mutations that were produced in the active-site cleft of polyethylene terephthalate (PET)-degrading enzyme, *aka* PETase, from *Ideonella sakaiensis* 201-F6 to gain insight into the functional evolution and substrate specificity, successfully led to a mutated enzyme with improved activity [43].

6.9 Sequence-dependent approach of the novel gene encoding the target enzyme/protein

Sequence-based approach discloses the details of function of the protein. This method relies on analyzing evolutionary relationships, relating phylogeny and function, identifying/detecting unknown species, and the abundance of genes. The oligonucleotides or probes are devised for the sequences encoding conserved domains of protein on the basis of consensus sequences. The target ORFs are then identified by PCR or hybridization. The 16S ribosomal RNA gene (16S rDNA) is the most popular phylogenetic marker used for identifying genome fragments obtained from precise groups of microorganisms. It, however, reveals only the phylogenetic classification of the respective bacteria and not essentially the metabolic function of the organism. For assessing the particular gene from the metagenome, the conserved gene sequences called "anchors" are used and clones are identified either by hybridization or PCR [24].

For the recognition of ORFs, similarity search algorithms (e.g., BLAST, COG, KEGG) accord ample information. Software tools are now available to question large sequence datasets such as genomes and metagenomes for the presence of gene clusters linked with biomolecules of significance. One such tool, anti SMASH (the antibiotic and Secondary Metabolite Analysis Shell), rapidly identifies and interprets secondary metabolite gene clusters from genomic sequence data. Additionally, hybridization permits the handling and screening of a large number of clones [44].

In a study by Fang et al. [45], a novel bacterial laccase gene designated lac21 was screened from a marine microbial metagenomic library of the South China Sea based on sequence screening strategy. The laccase gene had incredible potential in decolorization of azo dyes in the deficiency of redox mediators with a reasonably lower level of supplementation (15 U/L) at 20°C. Metagenomics could thus aid in the search for extremozymes that might not be discovered by traditional culture methods.

6.10 Function-based identification of the novel gene

Functional screening is a different move toward the sequence-based screening that does not require prior information of sequence. The idea of function-based screening relies on expression of a metagenomic gene of interest in a surrogate host and selection based on the phenotype. The probability to amplify the hit rate of the gene of interest can be enhanced by sample enrichment prior to building of the library, development of novel sensitive assays, and automated high-throughput screening [25].

6.11 Identification of the novel gene by sequencing techniques

The primary objective of any metagenome sequencing project would be to expose and characterize a particular microorganism in a community, linking three main aspects: (1) composition/structure of the community, their genetic and phylogenetic relations; (2) role of each member within the community; and (3) intraspecies or intrapopulation heterogeneity of the genes.

To begin with, metagenomics was confined to only diversity studies but with time it has advanced and established its application in a range of areas. The technique is being employed to unearth the practical properties of microorganisms in a community, revealing enzymes with new catalytic activity, antibiotic therapies, genes that are engaged in bioremediation, dyestuff decolorization, lignocellulosic treatment, biobleaching of paper pulp, etc. [46].

Metagenomics has bunched environmental biology, functional biology, microbial physiology, and sustainable development into one single frame. The genetic makeup of the community elucidates the type of flora, their functional role in the environment, and the effective metabolism of the species. The prosperity or deficiency of a particular species in the population in a defined area discloses the essentials of that particular environment. Understanding the function, metabolism, and succession of microorganisms helps to sustain the niche. Moreover, studies on computational enzyme docking by Karumuri et al. [47] and Singh et al. [48] also threw light on characterization of the competent substrates for the enzymes.

The analysis of sequences is executed by a series of steps with the primary objective to filter the data. The filtered data can be analyzed by:

1. Marker gene analysis, which involves comparing the metagenomic reads to a database consisting of gene families with specific genetic markers; the marker gene in the read can be identified, and reads are classified on the basis of homology to respective gene marker. There is a variety of software to taxonomically annotate the metagenomes, viz., MetaPhlAn, AMPHORA, MetaPhyler, PhyloSift, and PhylOTU.
2. Binning, a process of grouping reads to segregate them into operational taxonomic units; methods of binning are based either on compositional features or alignment or both:
   a. Compositional binning uses composition of sequences, viz., tags such as rec A, rpo B, 16S rRNA to cluster the metagenomic reads into taxons.

b. Similarity binning, which depends on the alignment of the sequences against the existing reference sequence; the aligned reads are grouped into relevant taxa. The software mainly used are MEGAN, MG-RAST, and CARMA.

c. Fragment recruitment, wherein reads are aligned to nearly identical genome sequences to construct metagenomic coverage estimates of the genome. There are numerous tools that facilitate mapping the reads, viz., MOSAIK, Genometa, SOAP, BWA, CLC, and RefCov.

3. Assembly reads that possess nearly identical sequences at their ends are linked to form a contig or complete genome. Each successive sequencing project adds to the preceding because of a collection of large data from the diverse environment. In a study conducted in acid mine drainage by Siegert et al. [49], a group of genomes belonging to unculturable strains was rebuilt using random shotgun sequencing of DNA from the biofilm. Recently, Gudenkkauf & Hewson [50] reported a broad study of the viral assemblages inhabiting marine invertebrates, revealing that different invertebrate groups anchor different assemblages.

The vast expansion of new and novel sequencing technologies and sequence searching tools has permitted extensive functional discovery of numerous microbial ecosystems. Next generation sequencing has taken over conventional sequencing methods in terms of high-throughput, low costing and allowing deeper potential into microbial community diversity. NGS has the potential for complete profiling of microbial communities from extreme samples, to uncover new species, and investigate the response of microbial populations under changing conditions. Environmental metagenomics as a field was extremely limited prior to the advent of NGS [25].

6.12 Improvement of enzymatic catalysis by microbial cell surface display

Microbial cell surface display is a powerful technique developed for any industrial or biotechnological process. Cell-surface display offers an opening to display peptides and proteins of interest on the surface of microbial cells by mingling them with the anchoring motifs. It is a synchronized expression of two or more proteins in a single-cell system that could have a combined effect on the process [51]. The protein of interest (target protein or fusion protein) is fused to the anchor protein via tethering on the cell wall and thus expressed in a host cell. The orientation of the target protein with respect to anchor protein is important for its activity. For example, in N-terminal fusion, the N-terminus of an anchor protein is fused to the C-terminus of a target protein. Conversely, in C-terminal fusion, the C-terminus of an anchor protein is fused to the N-terminus of a target protein. The technique has a broad range of biotechnological and industrial applications, including live vaccine development; a recombinant vaccine against parapoxvirus, orf virus (ORFV) that causes superficial skin lesions in infected humans and grazing animals was developed. Tan et al. [52] successfully attempted the fusion of *Echinococcus granulosus* antigen EG95 on the surface proteins of a host cell to prepare recombinants. The recombinants effectively reduced the infectivity during in vitro assay and good antibody response was observed in the inoculated sheep. In another study by Sun et al. [53], a live oral vaccine against chicken coccidiosis was developed for the first time using yeast *Saccharomyces cerevisiae* as a host strain. *Eimeria tenella* EtMic2 protein acted as a fusion protein and provided a humoral and cell-mediated immunological response. In a more recent study, Stickney et al. [54] developed nanoshuttles by engineering the exosomes and revealed their application in targeted drug delivery as well as exosome-mediated vaccine and therapy utilizing cell surface display technology. Additionally, a technique based on cell surface display was developed by Wang et al. [55] to inhibit HIV infection by using an antibody as fusion peptide together with autotransporter β-barrel domain of IgAP gene from *Neisseria gonorrhoeae*. The engineered bacteria successfully captured HIV-1 particles via surface binding and inhibited HIV-1 infection in cell culture.

6.13 Conclusion

Progression of enzymatic catalysis would require the continued mining of new enzymes from nature's vast reserves as well as the use of directed evolution or rational design techniques to close the functional gap between existing and desired properties of these catalysts. Studies highlighted herein show the power of current techniques to improve their activity, stability, and soluble expression and to change enzyme selectivity or substrate specificity as well. A few successful attempts have been made to engineer natural enzymes, but progress has still to be made to develop robust approaches. Creating novel enzymatic activity remains an elusive goal and would require the development of innovative evolutionary strategies and ultra-high-throughput screening systems to sort through immense mutant libraries. Ultimately, new directed evolution techniques, in conjunction with the continued discovery and traditional evolutionary or rational protein engineering approaches, will continue to accomplish the promise of biocatalysis in industrial, agricultural, and medical applications.

References

[1] Kotzia GA, Platis D, Axarli IA, Chronopoulou G, Karamitros C, Labrou NE. Food biochemistry and food processing. In: Simpson BK, Nollet LML, Toldrá F, Benjakul S, Paliyath G, Hui YH, editors. Biocatalysis, enzyme engineering and biotechnology. 2nd ed. Oxford: Wiley & Sons; 2012.

[2] Kanchana R. Farm waste recycling through microbial keratinases. J Appl Sci Environ Sanit 2012;7:103–8.

[3] Kanchana R, Mesta D. Native feather degradation by a keratinophilic fungus. Int J Chem Tech Res 2013;5:2947–54.

[4] Kanchana R. Utilization of biodegradable keratin containing wastes by enzymatic treatment. Int J Pharm Biol Sci 2013;4(B):117–26.

[5] Kanchana R, Jadhav S, Goletikar Y, Manerekar G. Production of alkaliphilic protease(s) by *Enterobacter* sp for application in bio-detergent formulation. Ind Biotechnol 2012;8:31–5.

[6] Kanchana R, Correia D, Sarkar S, Gawde P, Rodrigues A. Production and partial characterization of cholesterol oxidase from *Micrococcus* sp. isolated from Goa, India. Int J Appl Biol Pharmaceut Technol 2011;2:393–8.

[7] Kanchana R, Muraleedharan UD, Raghukumar S. Alkaline lipase activity from the marine protists, thraustochytrids. World J Microbiol Biotechnol 2011;27:2125–31.

[8] Vashist P, Nogi Y, Ghadi SC, Verma P, Shouche YS. *Microbulbifer mangrovi* sp. nov., a polysaccharide-degrading bacterium isolated from an Indian mangrove. Int J Syst Evol Microbiol 2013;63:2532–7.

[9] Devasia VLA, Muraleedharan UD. Polysaccharide-degrading enzymes from the marine protists, thraustochytrids. Biotechnol Bioinf Bioeng 2012;2:617–27.

[10] Shirodkar PV, Muraleedharan UD, Raghukumar S. Production of extracellular polysaccharases by the marine protists, thraustochytrids, with special reference to α-amylase activity. Int J Pharma Bio Sci 2017;8:453–62.

[11] Shirodkar PV, Muraleedharan UD. Enhanced α-amylase production by a marine protist, *Ulkenia* sp. using response surface methodology and genetic algorithm. Prep Biochem Biotechnol 2017;47:1043–9.

[12] Zhang MM, Su X, Ang EL, Zhao H. Recent advances in biocatalyst development in the pharmaceutical industry. Pharm Bioprocess 2013;1:179–96.

[13] Sturmberger L, Wallace PW, Glieder A, Birner-Gruenberger R. Synergism of proteomics and mRNA sequencing for enzyme discovery. J Biotechnol 2016;235:132–8.

[14] Choi J-E, Nguyen CM, Lee B, Park JH, Oh JY, Choi JS, Kim JC, Song JK. Isolation and characterization of a novel metagenomic enzyme capable of degrading bacterial phytotoxin toxoflavin. PLoS One 2018;13:1–14.

[15] Ferrer M, Martínez-Martínez M, Bargiela R, Streit WR, Golyshina OV, Golyshin PN. Estimating the success of enzyme bioprospecting through metagenomics: current status and future trends. Microb Biotechnol 2015;9:22–34.

[16] Gong JS, Lu ZM, Li H, Zhou ZM, Shi JS, Xu ZH. Metagenomic technology and genome mining: emerging areas for exploring novel nitrilases. Appl Microbiol Biotechnol 2013;97:6603–61.

[17] Pohl NL. Functional proteomics for the discovery of carbohydrate-related enzyme activities. Curr Opin Chem Biol 2005;9:76–81.

[18] Zweerink S, Kallnik V, Ninck S, Nickel S, Verheyen J, Wagner MBA, Feldmann I, Sickmann A, Albers SV, Bräsen C, Kaschani F, Siebers B, Kaiser M. Activity-based protein profiling as a robust method for enzyme identification and screening in extremophilic Archaea. Nat Commun 2017;8:1–12.

[19] Cravatt BF, Wright AT, Kozarich AW. Activity-based protein profiling: from enzyme chemistry to proteomic chemistry. Annu Rev Biochem 2008;77:383–414.

[20] Singh RK, Tiwari MK, Singh R, Lee JK. From protein engineering to immobilization: promising strategies for the upgrade of industrial enzymes. Int J Mol Sci 2013;14:1232–77.

[21] Amrein BA, Steffen-Munsberg F, Szeler I, Purg M, Kulkarni Y, Kamerlin SCL. CADEE: computer-aided directed evolution of enzymes. Int Union Cryst 2017;4:50–64.

[22] Ilmberger N, Meske D, Juergensen J, Schulte M, Barthen P, Rabausch U, Angelov A, Mientus M, Liebl W, Schmitz RA, Streit WR. Metagenomic cellulases highly tolerant towards the presence of ionic liquids—linking thermostability and halotolerance. Appl Microbiol Biotechnol 2012;95:135−46.

[23] Martínez R, Jakob F, Tu R, Siegert P, Maurer K-H, Schwaneberg U. Increasing activity and thermal resistance of *Bacillus gibsonii* alkaline protease (BgAP) by directed evolution. Biotechnol Bioeng 2013;110:711−20.

[24] Arnold FH. Directed evolution: bringing new chemistry to life. Angew Chem Int Ed Engl 2018;57:4143−8.

[25] Baweja M, Nain L, Kawarabayasi Y, Shukla P. Current technological improvements in enzymes toward their biotechnological applications. Front Microbiol 2016;7:1−13.

[26] Xu H, Qin Y, Huang Z, Liu Z. Characterization and site-directed mutagenesis of an α-galactosidase from the deep-sea bacterium *Bacillus megaterium.* Enzym Microb Technol 2014;56:46−52.

[27] Ramli AM, Muhammad NM, Rabu A, Munir A, Murad A, Diba FAB. Molecular cloning, expression, and biochemical characterisation of a cold-adapted novel recombinant chitinase from *Glacio zymaantarctica* PI12. Microb Cell Factories 2011;10:1−13.

[28] Sen S, Dasu V, Mandal B. Developments in directed evolution for improving enzyme functions. Appl Biochem Biotechnol 2007;143:212−23.

[29] Lillford PJ, Holt CB. In vitro uses of biological cryoprotectants. Philos Trans R Soc Lond B Biol Sci 2002;357:945−51.

[30] Wang HK, Sun Y, Chen C, Sun Z, Zhou ZC, Shen F, Zhang H, Dai Y. Genome shuffling of *Lactobacillus plantarum* for improving antifungal activity. Food Control 2013;32:341−7.

[31] Wu G, Wang G, Ji J, Li Y, Gao H, Wu J, Guan W. A chimeric vacuolar Na^+/H^+antiporter gene evolved by DNA family shuffling confers increased salt tolerance in yeast. J Biotechnol 2015;203:1−8.

[32] Wójcik M, Telzerow A, Quax WJ, Boersma YL. High-throughput screening in protein engineering: recent advances and future perspectives. Int J Mol Sci 2015;16:24918−45.

[33] Ostafe R, Prodanovic R, Commandeur U, Fischer R. Flow cytometry-based ultra-high-throughput screening assay for cellulase activity. Anal Biochem 2013;435:93−8.

[34] Yi S, Dai F, Zhao C, Si Y. A reverse micelle strategy for fabricating magnetic lipase-immobilized nanoparticles with robust enzymatic activity. Sci Rep 2017;7:9806.

[35] Gao J, Jiang Y, Lu J, Han Z, Deng J, Chen Y. Dopamine-functionalized mesoporous onion-like silica as a new matrix for immobilization of lipase *Candida* sp. 99−125. Sci Rep 2017;7:1−9.

[36] Su F, Li G, Fan Y, Yan Y. Enhanced performance of lipase *via* microcapsulation and its application in biodiesel preparation. Sci Rep 2016;6:1−12.

[37] Ghosh M, Maiti S, Dutta S, Das D, Das PK. Covalently functionalized single-walled carbon nanotubes at reverse micellar interface: a strategy to improve lipase activity. Langmuir 2012;28:1715.

[38] Scriba GKE, Belal F. Advances in capillary electrophoresis-based enzyme assays. Chromatographia 2015;78:947−70.

[39] Cheng M, Chen Z. Recent advances in screening of enzymes inhibitors based on capillary electrophoresis. J Pharm Anal 2018;8:226−33.

[40] Min W, Cui S, Wang W, Chen J, Hu Z. Capillary electrophoresis applied to screening of trypsin inhibitors using microreactor with trypsin immobilized by glutaraldehyde. Anal Biochem 2013;438:32−8.

[41] Camara MA, Tian M, Liu X, Liu X, Wang Y, Yang J, Yang L. Determination of the inhibitory effect of green tea extract on glucose-6-phosphate dehydrogenase based on multilayer capillary enzyme microreactor. Biomed Chromatogr 2016;30:1210–5.

[42] Cheng X-Y, Huang W-J, Hu S-C, Zhang H-L, Wang H, Zhang J-X, Lin H-H, Chen Y-Z, Zou Q, Ji Z-l. A global characterization and identification of multifunctional enzymes. PLoS One 2012;7:1–8.

[43] Austin HP, Allen MD, Donohoe BS, Rorrer NA, Kearns FL, Silveira RL, Pollard BC, Dominick G, Duman R, El Omari K, Mykhaylyk V, Wagner A, Michener WE, Amore A2, Skaf MS, Crowley MF, Thorne AW, Johnson CW, Woodcock HL, McGeehan JE, Beckham GT. Characterization and engineering of a plastic-degrading aromatic polyesterase. Proc Natl Acad Sci USA 2018;115:4350–7.

[44] Jackson SA, Borchert E, Gara F, Dobson ADW. Metagenomics for the discovery of novel biosurfactants of environmental interest from marine ecosystems. Curr Opin Biotechnol 2015;33:176–82.

[45] Fang ZM, Li TL, Chang F, Zhou P, Fang W, Hong YZ, Zhang XC, Peng H, Xiao YZ. A new marine bacterial laccase with chloride-enhancing, alkaline-dependent activity and dye decolorization ability. Bioresour Technol 2012;111:36–41.

[46] Kumar V, Marín-Navarro J, Shukla P. Thermostable microbial xylanases for pulp and paper industries: trends, applications and further perspectives. World J Microbiol Biotechnol 2016;32:1–10.

[47] Karumuri S, Singh PK, Shukla P. In silico analog design for terbinafine against *Trichophytonrubrum*: a preliminary study. Indian J Microbiol 2015;55:333–40.

[48] Singh PK, Joseph J, Goyal S, Grover A, Shukla P. Functional analysis of the binding model of microbial inulinases using docking and molecular dynamics simulation. J Mol Model 2016;22:1–7.

[49] Siegert MJ, Tranter M, Evans JCE, Priscu JC, Lyons WB. The hydrochemistry of Lake Vostok and the potential for life in Antarctic sub-glacial lakes. Hydrol Process 2003;17:795–814.

[50] Gudenkauf BM, Hewson I. Comparative metagenomics of viral assemblages inhabiting four phyla of marine invertebrates. Front Mar Sci 2016;3:1–12.

[51] Tanaka T, Kondo A. Cell surface engineering of industrial microorganisms for biorefining applications. Biotechnol Adv 2015;33:1403–11.

[52] Tan JL, Ueda N, Heath D, Mercer AA, Fleminga SB. Development of orf virus as a bifunctional recombinant vaccine: surface display of *Echinococcus granulosus*antigen EG95 by fusion to membrane structural proteins. Vaccine 2012;30:398–406.

[53] Sun H, Wang L, Wang T, Zhang J, Liu Q, Chen P, et al. Display of *Eimeriatenella* EtMic2 protein on the surface of *Saccharomyces cerevisiae* as a potential oral vaccine against chicken coccidiosis. Vaccine 2014;32:1869–76.

[54] Stickney Z, Losacco J, McDevitt S, Zhang Z, Lu B. Development of exosome surface display technology in living human cells. Biochem Biophys Res Commun 2016;472:53–9.

[55] Wang LX, Mellon M, Bowder D, Quinn M, Shea D, Wood C, et al. *Escherichia coli* surface display of single-chain antibody VRC01 against HIV-1 infection. Virology 2015;475:179–86.

Chapter 7

Biotechnological implications of hydrolytic enzymes from marine microbes

Poonam Vashist[1], R. Kanchana[2], Vazhakatt Lilly Anne Devasia[1,#], Priyanka V. Shirodkar[1], Usha D. Muraleedharan[1,*]

[1]*Department of Biotechnology, Goa University, Goa, India;* [2]*Department of Biotechnology, Parvatibai Chowgule College of Arts and Science -Autonomous, Margao, Goa, India;*
[#]*Present address: Department of Biotechnology, Hindustan College of Arts and Science, Padur, Kelambakkam, Chennai, India*
[*]*Corresponding author: usha@unigoa.ac.in.*

7.1 Introduction

Biocatalyst and biotechnology are intervolved terms that have camouflaged almost every socioenvironmental activity of our living style in the 21st century. To date, almost 4000 enzymes are known, out of which approximately 5% of microbial original types are used commercially [1].

The marine microbial population, while still largely unexplored because of its diversities and extreme environment, is being screened with the help of high-throughput technologies such as shotgun sequencing or pyrosequencing for deeper exploration [2], resulting in the reporting of approximately 20,000 species per liter of marine water samples [3].

Microbial enzymes have proved superior over chemical catalysts and enzymes derived from plants or animals, by virtue of their versatility, stability, enantioselectivity, reduced process time, intake of low-energy input, and cost-effectiveness, besides their nontoxic and eco-friendly characteristics [4]. Competition amongst microorganisms for space and nutrients in the marine environment is a powerful selective force that has led to the generation of multifarious enzyme systems to adapt to the complicated environments. Many of them are thus endowed with desirable features, from a general biotechnological perspective. However, these features of microbes as well as enzymes can be enhanced with the help of techno-studies such as recombinant DNA/protein engineering [5].

Advances in Biological Science Research. https://doi.org/10.1016/B978-0-12-817497-5.00007-0

Marine microorganisms, whose genetic and biochemical diversity studies are not quite out of their infancy, are of significant interest as a potential source for novel applications [6,7]. Researchers the world over have successfully isolated and characterized a variety of enzymes with novel activities from marine bacteria, actinomycetes, fungi, and other marine microorganisms, of which a few products have already found large-scale industrial applications [8–11].

Extremophilic marine microorganisms living in hydrothermal vents, sea floors, salt pans, and hot springs have been found to be rich sources of extremozymes such as proteases, cellulases, amylases, and lipases [12]. Our laboratory has had excellent success with bioprospecting of marine enzymes from several such sources. Vashist et al. reported a novel strain of *Microbulbifer* sp. isolated from a mangrove area in Divar Island, Goa, India, which exhibited more than 13 ICP degrading activities [13]. Another important avenue of interest has been the marine biodegraders classed as thraustochytrids. Kanchana et al. [14] have reported alkaline lipase activity with potential biotechnological application from these marine protists. Similarly, Devasia & Muraleedharan [15] have reported multiple polysaccharide-degrading activities from thraustochytrids for the first time. Employing response surface methodology, Shirodkar & Muraleedharan have successfully optimized a medium for commendable production of amylolytic activities from a thraustochytrid strain [16,17].

Almost 75% of all industrially applied enzymes are of the hydrolytic category, out of which proteases, carbohydrases, and lipases lead the enzyme market, with more than 70% of all enzyme sales [1]. Among these, hydrolytic enzymes resourced from bacteria hold the majority share when compared to the overall bioactive compounds discovered from marine fungi and other domains [18]. The world of enzyme demand is satisfied by about 12 major producers and 400 minor suppliers [19]. Three top companies, namely, Denmark-based Novozymes and Danisco and Switzerland-based Roche satisfy about 75% of the total enzyme demand [1]. While the world market for industrial enzymes was to the tune of $4.2 billion in 2014, an expansion at a compound annual growth rate (CAGR) of approximately 7% is predicted over the next 5-year period, escalating the figure to nearly $6.2 billion [20].

In this chapter, attention is drawn to the significant role of hydrolytic enzymes from marine microbes involved in various fields of technical applications. Also reported are aspects of modern biotechnology for improved enzyme production, enabling various industrial processes at lower energy consumption and higher efficiency, besides enhancing properties of the product while ensuring environment friendliness.

7.2 Applications of marine hydrolases

Hydrolases catalyze several reactions, including condensations, alcoholysis, and so on. The main advantages of this enzyme class are ready availability, lack of cofactor stereoselectivity, and tolerance to the addition of water-miscible solvents. In fact, the majority (about 75%) of currently used microbial enzymes in various industries fall into the hydrolytic category [21].

7.2.1 Biorefineries

Over several years, studies have been in progress in different parts of the world to find substitutes for the fractions of nonrenewable/fossil resources in use. Brazil, the United States, China, and India, covering the most-populated regions of the world at 44% of the total global population, use 45% of the total energy sources and account for 51% of the carbon dioxide release from the consumption of energy. Notwithstanding this, about 27% of primary energy is used worldwide for transportation, thus highlighting transportation fuels as notable targets for replacement with a renewable source [22].

Consistent and harmonized supply of raw materials is the most important factor in biorefinery productivity [23]. The use of aquaculture biomass (seaweeds) as a feed for biorefinery industries ensures a continuous, rapid, safer, and economic supply of feed for biorefinery industries compared to the use of lignocellulosic biomass from agricultural wastes and other biological materials. Algae have 20–80% oil content, most of which are triacylglycerols, but the robust algal cell walls present a challenge for the extraction of these lipids. Enzymatic lipid extraction has been carried out with snailase along with cellulase and lysozyme [24].

Novel and significant enzymatic properties of extremophiles and thermostable enzymes have boosted the biocatalysis in biofuel production [25]. In biorefineries for bioethanol, biogas, and biodiesel production, cellulases, hemicellulases, and ligninases are most widely used, primarily during pretreatment and saccharification of biomass. Some bacteria and fungi are capable of rapid and efficient degradation of cellulose [25,26]. Enhanced depolymerization of the cellulose fraction from lignocellulosic substrates would be favored by the use of a cocktail of enzymes such as exoglucanase, endoglucanase, β-glucosidase, cellobiase, xylanase, and other hemicellulases [27].

7.2.2 Pharmaceuticals and cosmeceuticals

Enzymes play various vital roles in the realms of pharmaceuticals, cosmetics, and diagnostics. The pharmaceutical industry in particular requires a very high degree of substrate specificity on account of the stringent routes for desired product synthesis. Coupled with the effective processing cost, this propels enzymes to advantage over chemical catalysts [4].

Enzymes are extensively being used as large-scale therapeutic drugs mainly pertaining to health concerns associated with enzymatic deficiency, digestive disorders, and in ELISA-based diagnostics [28]. Marine natural products play an important role in drug development [29]. The enzymes involved in mycothiol biosynthesis and found in the Actinobacteria, *Actinomycetes* and *Mycobacterium* are of great interest as potential targets for the production of new drugs against *Mycobacterium tuberculosis* [30].

The applications of hydrolytic enzymes in medicine are growing rapidly, as in the removal of dead skin, and enzymes such as Vibrilase (vibriolysin) obtained from the marine microorganism *Vibrio proteolyticus* have been reported for effectiveness against denatured proteins found in burnt skin [31].

Lipases have been cited as the most repeatedly used enzymes in the synthesis of alcohols, acids, lactones, and esters that are optically active [32, 33]. On account of their digestive and inflammatory properties, proteases have been testified for a wide array of applications in medicine [34]. Chitosanase catalyzes the hydrolysis of chitosan, which is used as an antimicrobial, antioxidant, and in bringing down high blood cholesterol and blood pressure levels. It has also found use in the control of arthritis, besides enhancing antitumor properties and according protection against infections [35]. Actinobacteria are considered the most important group of organisms being studied extensively for the discovery and formulations of significant drugs and other bioactive metabolites with high specificity [36]. Glucanase isolated from a marine *Bacillus* has been reported suitable for oral and other health care [37]. Oligomers obtained from carrageenan degradation have demonstrated antitumor, antiviral, anticoagulative, and other activities such as boosting immune function and digestion [38–40]. Qiu et al. reported significant advantages of the marine bacterial protease from an N1-35 strain isolated after UV mutagenesis over those from terrestrial ones [41]. Enzymes such as urease and creatinase from Actinobacteria have been used in clinical diagnostics for quantitative evaluation of diabetes and other health disorders [42,43].

Enzymes have gained increasing popularity in the cosmetic industry by virtue of their relative safety and minimal side effects. Li et al. [1] have reported the use of enzymes as free radical scavengers in preparations for hair styling and dyeing, and in toothpastes, mouthwashes, and sunscreen products. In Japan, agar-oligosaccharide is being extensively used as a moisturizing cosmetic additive and and has also been reported for good hair conditioning effects [44]. Proteases have been used in skin creams to remove dead cells and thereby effect cleaning and smoothening of the skin [45]. Papain and endoglycosidase have found wide use in toothpastes and mouthwashes, serving to whiten and remove plaque as well as other odor-causing deposits on teeth and gum tissue [46]. Lipase and papain are used in skin care products [47]. Enzymes are also routinely being incorporated in contact lens cleaners to remove proteinaceous films [48].

7.2.3 Food industry

The present rate of world average for per capita food consumption is 3000 kcal/day [19]. The estimated food consumption is expected to go up to 22.8×10^{12} kcal/day by 2020; when compared to the energy, it is roughly 16 million barrels of oil equivalent (boe) per day i.e., an approximate 70% increase by 2050 [19]. Meeting these requirements in time without compromising on food quality could be managed by enzymes application. These biocatalysts have been reported as well as used efficiently for improving food production, quality and components such as flavor, aroma, color, texture, appearance, fat modification, and in sweetener technology [1]. Application of enzymes in the food industry is principally for improving the shelf life of products, besides their stability and texture.

Agar oligosaccharides produced by the action of agarases are being used for the production of beverages, bread, and some low-calorie food [49,50]. Isolates such as *Microbulbifer* CMC-5, 10A, and LK2 have been reported for promising and extensive agarolytic activity that can be further exploited [51,52]. Agarose-derived neoagarooligosaccharides produced by β-agarase from the seaweed-decomposing bacterium *Microbulbifer* sp. strain CMC-5 showed a promising biotechnological application [52]. Similarly, esterases and lipases have been reported for betterment of bread quality by change in lipids in the dough used, in butter, cheese, and margarine flavor enhancement, in the production of crackers, pasta, etc., degumming of vegetable oils, synthesis of infant formula and nutraceuticals, increasing the concentration of polyunsaturated fatty acids in vegetable oils, as well as for improving the digestibility of natural lipids [53]. Recently Li et al. isolated a novel esterase of the hormone-sensitive lipase family, from a metagenomic library of more than 10,500 cosmid clones from the South China Sea [54]. A cold-active lipase from the marine bacterium *Janibacter* sp. strain HTCC2649 was isolated and characterized by Yuan and coworkers with sn-1/3 specificity toward mono- and di- acylglycerols [55].

Recent reviews have endorsed the application of xylanases, lipases, and amylases in the baking industry, for improving freshness and shelf life of the products, elasticity of gluten, enhancing handling, and increasing stability of the dough, as well as improved bread volumes and absorption/redistribution of water and flavor; higher content of arabinoxylooligosaccharides in bread with a positive effect on health, lighter cream crackers, improved texture, palatability, and uniformity of wafers are other benefits that have been derived [1,56,57].

Cheese, yogurt, and other milk products have been improved upon with respect to organoleptic characteristics such as aroma, flavor, color, and yield [58]. The use of enzymes (protease, catalase, lipase, esterase, lactase, etc.) in the dairy market is well recognized for reduction of allergenic properties of milk products [59].

The effectiveness of collagenase and protease has been proven in meat tenderization and further processing and protease MCP-01, an abundant extracellular serine protease produced by the deep-sea psychrophilic *Pseudoalteromonas* sp. SM9913, is reported to be one such psychrotolerant enzyme [60]. Damare et al. have described a detergent-compatible alkaline, cold-tolerant protease from a deep-sea fungus from the Central Indian Basin [61].

Enzymes have been employed in the beverage industry to control the brewing process, and to produce beer of consistently high quality. Improvement of the juice yield, aroma, and color are other positive outcomes.

Such enzymes also serve to enhance the nutritional and functional characteristics of proteins of animal and vegetable origin that are used in the process [62]. The first stage in a brewing industry for extraction of material would be to digest the plant cell wall using microbial enzymes [63]. Inulinase form *Pichia guilliermondii* OUC1 has been reported for the production of ultra-high-fructose syrups, while *Yarrowia lipolytica* OUC2 could be used for the production of inulooligosaccharides [64,65].

It has been reported very effective to use amylases, cellulases, arabinases, naringinase, pectinases, and xylanases during fruit juice/beverage processing to enhance yield, in juice clarification, for improved extraction, stability, texture, sensory characteristics, and cost/time management [1,66,67]. Tannases from *Aspergillus awamori* BTMFW032 have been used for producing coffee-flavored soft drinks and instant tea, for the clarification of fruit juice and beer, and for food detannification [68]. Many reports describe β-glucosidases as widely used during processing of fruit juices, tea, wine, and beer to effect enhancement of their organoleptic properties and essential oil content [69–72]. One specific study on β-glucosidases from *Thermotoga maritime* has been documented by Goyal et al. [73].

7.2.4 Feed industry

The continuously increasing trend in worldwide milk and meat consumption invariably escalates the demand for the feedstock for animals, and to meet the requirements of quantity and quality, the search for feed enzymes for diet formulations necessarily had to pick up the pace. Enzymes incorporated in animal feeds aid to increase the digestibility of nutrients and ensure better feed utilization [74]. The practice of supplementation of animal diets with enzymes increased exponentially from the 1980s to the subsequent decades. The world market for feed enzymes, assessed at $899.19 million in 2014, is estimated to touch the $1.3 billion mark by 2020, at a CAGR of 7.3% over the period [1]. The protein concentration in poultry feeds has been enhanced by the application of feed enzymes such as α-galactosidases, phytases, α-amylases, proteases, glucanases, xylanases, and polygalacturonases [1]. Faced with the inability of monogastric animals to digest plant-based feeds rich in hemicelluloses and cellulose, xylanase and β-glucanase have been added to the feeds to enable total degradation and digestion [75,76]. To the same end, β-glucosidases have been documented as favored additives in cellulose-based feeds for animals such as chicken and pigs [77].

7.2.5 Biopolymer industry

Involving microbial enzymes in the synthesis of biodegradable polymers appears the best solution to circumvent the increased requirement of polymers that have proved detrimental to human health and environment. The applications of biopolymers such as polyesters, polycarbonates, and polyphosphates have been reported for various biomedical purposes, e.g., orthopedic devices, tissue engineering, adhesion barriers, and controlled drug delivery [78]. The biopolymer market, which had grown substantially over the recent years, was expected to touch $3.6 billion by the end of 2018 [1].

The ever-increasing clamor for packaging materials and the concomitant environmental safety concerns could both be addressed by the flourishing biodegradable polymer industry. Lipases and laccases catalyze polymerization reactions to produce polyesters or polycarbonates and cross-links in biopolymers [79,80]. *Y. lipolytica* lipase when customized by immobilization and protein engineering has been shown to improve the performance in biopolymer synthesis [81].

7.2.6 Detergent industry

Successful integration of hydrolytic enzymes into detergent formulations is cited as the key to the production of cost-effective and environmentally gentle detergents [82]. In today's laundry detergents, enzymes such as proteases and amylases are deeply entrenched among the active ingredients. A recipe of alkaline proteases and cellulases isolated from microbial sources and that can effectively remove protein stains as well as cleave damaged cotton fibers is now a vital component of various detergents produced on a commercial scale by manufacturers such as Novozymes SA, Kao Corporation, and Genencor International [83]. Greene and his colleagues had reported cleansing properties of an alkaline protease isolated from a bacterial strain symbiotic with the gland of Deshayes of a marine shipworm [84]. *Aureobasidium pullulans* and *Bacillus mojavensis* A21 isolated from sea saltern of the China Yellow Sea and seawater, respectively, produced high yields of detergent-stable alkaline protease exhibiting excellent compatibility with a variety of commercial liquid and solid detergents [85,86].

The high specificity of enzymatic reactions indisputably mitigates the typical damages to fabrics and surfaces following the use of chemically harsh detergent components [19]. Besides, the types and ratios of enzymes in detergent mixes can be optimized to favor specific detergent applications. For instance, since dishwashing detergents would need to ensure the removal of starchy food and fat/oil deposits, they often contain varying degrees of amylase and lipase [1,19]. The industrial application of alkaline cellulases as a potential additive to laundry detergents is being actively pursued with a view to selectively contact the cellulose within the interior of fibers and facilitate soil removal from the interfibrillar spaces [87].

7.2.7 Textile industry

The use of enzymes in the textile industry is one of the most rapidly growing fields in industrial enzymology. Enzymes used in the textile industry offer various advantages as they accelerate the reaction by operating under milder conditions, providing an alternative to the polluting chemicals otherwise used. Also, they are biodegradable and easy to control, acting only on specific substrates. Current applications in the textile industry lie in the areas of biopolishing and bioscouring of fabric, antifelting of wool, softening and desizing cotton, finishing of denim and woolen items, and modification of synthetic fibers, for which the focus is primarily on hydrolases such as amylase, cellulase, protease, pectinase, and lipase/esterase [88]. There are two well-established enzyme applications in the textile industry. Firstly, in the preparatory finishing area, amylases are commonly used during the desizing process; and secondly, cellulases are used for softening in the finishing area, biostoning, and reducing of pilling propensity for cotton goods. The enzymatic desizing of cotton with α-amylases has been a state-of-the-art procedure for many decades [89]. Moreover, cellulases, pectinases, hemicellulases, lipases, and catalases are used in different cotton pretreatment and finishing processes [90]. Enzymes are emerging in a big way in the field of textile wet processing. If their cost could be managed, enzymes can be put to use in a much bigger way for textile-processing applications.

7.2.8 Leather industry

For facilitating procedures and enhancing leather quality, enzymes are required during the various stages that are integrated in the processing of leather, such as curing, soaking, liming, dehairing, bating, picking, degreasing, and tanning [91]. The usage of enzymes as substitutes to chemicals has proved efficacious in improving leather quality and in reducing environmental pollution. Lipases as well as alkaline and neutral proteases are primarily used. By removing nonfibrillar proteins during soaking and bating, alkaline proteases ensure that the finished leather is soft, supple, and pliable. Water wastage is minimized by the use of neutral and alkaline proteases during the dehairing step [92]. Removal of fats during degreasing is ensured by the addition of lipases [93]. Stainless pelt, low biological oxygen demand and chemical oxygen demand in effluents, reduced odor, and enhanced hair recovery are the benefits derived by substituting chemicals with enzymes during the liming process [1].

7.2.9 Paper and pulp industry

Aiming to diminish adverse effects on the ecosystem, the use of microbial enzymes in the paper and pulp industry has recorded steady growth. Besides saving on time and energy consumption, incorporation of enzymes cuts down on the use of chemicals during processing. Enzymes are also being used to enrich deinking and bleaching [94] and xylanases and ligninases, in particular, serve to enhance the value of the pulp by removing lignin and hemicelluloses

[95]. Raghukumar et al. had reported the potential application of crude culture filtrate of a marine fungal isolate possessing thermostable, cellulase-free alkaline xylanase activity in biobleaching of paper pulp [96]. Amylases are used by paper and pulp industries for starch coating, improving paper cleanliness, and drainage [97]. Lipases find use in deinking and enhancing pitch control, while cellulases are involved in deinking, improving softness, and drainage [97]. The role of cellulases in the bioprocess development for recycling of used printed paper is also on record [98].

7.2.10 Organic synthesis

The application of biocatalysts has invited growing attention over the past few years, and heavy demands have been placed on identifying new biocatalysts for the organic synthesis of novel compounds. The catalysis of numerous organic reactions directs the significance and high potential of this field of research. Enzymes perform an important role as biocatalysts in the synthesis of vital intermediates for the pharmaceutical and chemical industry, and new enzymatic technologies and processes have been recognized [99]. Through the last three decades, using hydrolases for the catalysis of environmentally friendly organic processes under mild reaction conditions has been well documented. The hydrolases have presented themselves as ideal tools for acceleration of synthetic transformations due to their broad substrate specificity, high stability, commercial availability, and catalytic efficiency in a wide spectrum of biocatalyzed processes. In recent years novel examples associated with nonconventional reactions catalyzed by hydrolytic enzymes have been witnessed. Amongst the biocatalysts in organic synthesis, lipases and acylases have earned great attention as promiscuous biocatalysts displaying good levels of reactivity in unique hydrolytic reactions, C—heteroatom bond formation, C—C bond formation, and oxidative processes [100]. In particular, lipases perform enantioselective hydrolytic reactions and catalyze the formation of a varied range of ester and amide bonds [101]. Furthermore, lyases are involved in the organic synthesis of acrylamide from acrylonitrile, cyanohydrins from ketones, and malic acid from fumaric acid. The use of glucose isomerase has facilitated the food and beverage industry with an annual commercial production of a multimillion ton of high-fructose corn syrup as an alternative sweetener to sucrose [19].

7.2.11 Waste treatment

Numerous enzymes are involved in the degradation of toxic pollutants and their usage in waste management has been extensive. In addition to the issues of domestic waste, industrial effluents foul the ecosystem with many lethal or toxic chemical entities. Wastewater from slaughterhouses and dairies contains high levels of proteins and fats that present low biodegradability.

Several pretreatment systems are employed to remove oil and grease to avoid many of the problems that might otherwise arise in the biological process and decrease the effectiveness of the treatment station [102].

Alone or in combinations, microbial enzymes have been employed in industrial effluent treatment by degradation or bioconversion of toxic compounds to less harmful products [103]. The enzymes involved in waste treatment are amylases, amyloglucosidases, cellulases, glucoamylases, lipases, pectinases, and proteases [19]. These enzymes are also used to convert starch to sugar, to recover additional oil from oil seeds, and to convert whey to numerous useful products [104]. A wide variety of scientific investigations are ongoing for the development of enzymatic hydrolysis processes to precede traditional biological treatment.

7.2.12 Nanoparticle synthesis

Metal nanoparticles synthesized using biomolecules are an attractive prospect owing to their variety in shapes and sizes as well as stability in colloidal solutions. The broad range of nanoparticle utility centers around their small size yet larger surface area. While a host of approaches are available for their synthesis [105], the usage of harmful radiations and chemical processes being discouraged in the current scenario, green and sustainable approaches gain favor, wherein enzymes are fast finding a place.

Silver nanoparticles (AgNPs) have been extensively used in many commercial products. They are important modules for research in biomedicine, electronics, optics, magnetics, catalysis, mechanics, energy science, and so on [106]. Gold nanoparticles (AuNPs) have huge prospects as drug carriers, in gene therapy for gene delivery, and also in optical biosensors. Pure forms of alpha-amylase have been used in the biosynthesis of AgNPs, as the enzyme reduces the silver ions, which results in the construction of stable AgNPs [107]. They are also used in the synthesis of AuNPs by reduction of tetrachloroaurate, ensuring stabilizing of the nanoparticles by capping in colloidal solution [107]. In vitro synthesis of AgNPs and AuNPs using cellulase enzyme in a single-step reaction has also been reported [108]. Enzyme-assisted green synthesis of nanoparticles is extremely dependent on the enzyme itself, the nature of the metal salt, as well as pH of the solution. The nature of capping proteins and the strength of interaction of proteins with metal nanoparticles would decide the end stability. This leads to diversity in morphologies and size control and hence monodispersity index. Enzyme-based metal nanoparticle synthesis being still in its infancy, the probability of uncovering vast applications cannot be underestimated.

7.3 Prospecting the use of hydrolytic enzymes from marine microbes

Currently, just about 5% of all chemical products are synthesized/produced using biocatalysts, and there are various doors to be yet opened to combat the numerous crises related to growing population, pollutants, or cost and time effectiveness as well as resource consumption. In recent years, the marine biological progress and enzyme bioprospecting activity have given a new source and options to humans, acquiring new potency and effectiveness especially for exploitation of bioactive compounds from marine microbial communities. As enzymes have incomparable advantages, many industries are keenly interested in adapting enzymatic methods to the requirements of their processes. Prospects are excellent for continuing to increase the usage of currently available enzymes in ongoing applications, as also in the use of novel enzymes for other purposes. Japan, Canada, Spain, Finland, Russia, and other countries have been focusing more on marine bioenzyme research.

Taken as a whole, on account of the marine biological diversity and the specificity of biological metabolism, while the study on a global scale is but just beginning, this area has huge potential for development and applications with industrial benefits.

References

[1] Li S, Yang X, Yang S, Zhu M, Wang X. Technology prospecting on enzymes: application, marketing and engineering. Comput Struct Biotechnol J 2012;2:e201209017.

[2] Caron DA, Countway PD, Jones AC, Kim DY, Schnetzer A. Marine protistan diversity. Annu Rev Mar Sci 2012;4:467–93.

[3] Sogin ML, Morrison HG, Huber JA, Welch DM, Huse SM, Neal PR, et al. Microbial diversity in the deep sea and the underexplored 'rare biosphere'. Proc Nat Acad Sci USA 2006;103:12115–20.

[4] Choi JM, Han SS, Kim HS. Industrial applications of enzyme biocatalysis: current status and future aspect. Biotechnol Adv 2015;33:1443–54.

[5] Liu L, Yang H, Shin HD. How to achieve high-level expression of microbial enzymes strategies and perspectives. Bioengineered 2013;4:212–23.

[6] Haefner B. Drugs from the deep: marine natural products as drug candidates. Drug Discov Today 2003;8:536–44.

[7] Ghosh D, Saha M, Sana B, Mukherjee J. Marine enzymes. Adv Biochem Eng Biot 2005;96:189–218.

[8] Bernan VS, Greenstein M, Maiese WM. Marine micro-organisms as a source of new natural products. Adv Appl Microbiol 1997;43:57–89.

[9] Burkholder PR, Pfister RM, Leitz FP. Production of a pyrrole antibiotic by a marine bacterium. Appl Microbiol 1966;14:649–53.

[10] David W, Michel J. Extremozymes. Curr Opin Chem Biol 1999;3:39–46.

[11] Harmsen HJM, Prieur D, Jeanthon C. Distribution of microorganisms in deep-sea hydrothermal vent chimneys investigated by whole-cell hybridization and enrichment culture of thermophilic subpopulations. Appl Environ Microbiol 1997;63:2876–83.

[12] Dalmaso GZL, Ferreira D, Vermelho AB. Marine Extremophiles: a Source of hydrolases for biotechnological applications. Mar Drugs 2015;13:1925–65. Gustafson K.

[13] Vashist P, Nogi Y, Ghadi SC, Verma P, Shouche YS. *Microbulbifer mangrovi* sp. nov., a polysaccharide-degrading bacterium isolated from an Indian mangrove. Int J Syst Evol Microbiol 2013;63:2532–7.

[14] Kanchana R, Muraleedharan UD, Raghukumar S. Alkaline lipase activity from the marine protists, thraustochytrids. World J Microbiol Biotechnol 2011;27:2125–31.
[15] Devasia VLA, Muraleedharan UD. Polysaccharide-degrading enzymes from the marine protists, thraustochytrids. Biotechnol Bioinf Bioeng 2012;2:617–27.
[16] Shirodkar PV, Muraleedharan UD. Enhanced α-amylase production by a marine protist, *Ulkenia* sp. using response surface methodology and genetic algorithm. Prep Biochem Biotechnol 2017;47:1043–9.
[17] Shirodkar PV, Muraleedharan UD, Raghukumar S. Production of extracellular polysaccharases by the marine protists, thraustochytrids, with special reference to α-amylase activity. Int J Pharma Bio Sci 2017;8:453–62.
[18] Prabavathy VR, Mathivanan N, Murugesan K. Control of blast and sheath blight diseases of rice using antifungal metabolites produced by *Streptomyces* sp. PM5. Biol Control 2006;39:313–9.
[19] Singh R, Kumar M, Mittal A, Mehta PK. Microbial enzymes: industrial progress in 21st century. 3 Biotech 2016;6:174.
[20] Industrial enzymes market by type (carbohydrases, proteases, non-starch polysaccharides & others), application (food & beverage, cleaning agents, animal feed & others), brands & by region—global trends and forecasts to 2020. www.bccresearch.com. http://www.marketsandmarkets.com/Market-Reports/industrial-enzymes-market-237327836.html. 2015.
[21] Gurung N, Ray S, Bose S, Rai V. A broader view: microbial enzymes and their relevance in industries, medicine and beyond. BioMed Res Int 2013:1–18. https://doi.org/10.1155/2013/329121.
[22] Intriago P. Marine microorganisms: perspectives for getting involved in cellulosic ethanol. Amb Express 2012;2:46–58.
[23] Cherubini F. The biorefinery concept: using biomass instead of oil for producing energy and chemicals. Energy Conserv Manag 2010;51:1412–21.
[24] Naghdi FG, González LM, Chan W, Schenk PM. Progress on lipid extraction from wet algal biomass for biodiesel production. Microb Biotechnol 2016;9:718–26.
[25] Lynd LR, Weimer PJ, Van Zyl WH, Pretorius IS. Microbial cellulose utilization: fundamentals and biotechnology. Microbiol Mol Biol Rev 2002;66:506–77.
[26] Allen F, Andreotti R, Eveleigh DE, Nystrom J. Mary Elizabeth Hickox Mandels, 90, bioenergy leader. Biotechnol Biofuels 2009;2:1–6.
[27] Chandel AK, Singh OV, Chandrasekhar G, Rao LV, Narasu ML. Key-drivers influencing the commercialization of ethanol based bio-refineries. J Commerc Biotechnol 2010;16:239–57.
[28] Mane P, Tale V. Overview of microbial therapeutic enzymes. Int J Curr Microbiol App Sci 2015;4:17–26.
[29] Debbab A, Aly AH, Lin WH, Proksch P. Bioactive compounds from marine bacteria and fungi. Microbial Biotechnol 2010;3:544–63.
[30] Vignesh S, Raja A, James RA. Marine drugs: implication and future studies. Int J Pharmacol 2011;7:22–30.
[31] Vellard M. The enzyme as drug: application of enzymes as pharmaceuticals. Curr Opin Biotechnol 2003;14:444–50.
[32] Saxena RK, Ghosh PK, Gupta R, Davidson WS, Bradoo S, Gulati R. Microbial lipases: potential biocatalysts for the future industry. Curr Sci 1999;77:101–15.
[33] Mo S, Kim JH, Cho KW. Enzymatic properties of an extracellular phospholipase C purified from a marine streptomycete. Biosci Biotechnol Biochem 2009;73:2136–7.

[34] Kumar CG, Joo HS, Koo YM, Paik SR, Chang C-S. Thermostable alkaline protease from a novel marine haloalkaliphilic *Bacillus clausii* isolate. World J Microbiol Biotechnol 2004;20:351−7.

[35] Thadathil N, Velappan SP. Recent developments in chitosanase research and its biotechnological applications: a review. Food Chem 2014;150:392−9.

[36] Das S, Ward LR, Burke C. Screening of marine *Streptomyces* spp. for potential use as probiotics in aquaculture. Aquacult 2010;305:32−41.

[37] Yahata N, Watanabe T, Nakamura Y, Yamamoto Y, Kamimiya S, Tanaka H. Structure of the gene encoding beta-1,3-glucanase A1 of *Bacillus circulans* WL-12. Gene 1990;86: 113−7.

[38] Sarwar G, Matoyoshi S, Oda H. Purification of α-carrageenan from marine Cytophaga species. Microbiol Immunol 1987;31:869−77.

[39] Roberts JN, Christopher BB, Cynthia DT, Rhonda K, Marcelino B, Peter LC, et al. Genital transmission of HPV in a mouse model is potentiated by nonoxynol-9 and inhibited by carrageenan. Nat Med 2007;13:857−61.

[40] Ngo DN, Kim MM, Kim SK. Chitin oligosaccharides inhibit oxidative stress in live cells. Carbohydr Polym 2008;74:228−34.

[41] Graham CR, David RW, Frank TR. Peptone induction and rifampin-insensitive collagenase production by *Vibrio alginolyticus*. J Bacteriol 1980;142:447−54.

[42] Dordick JS. Biocatalysts for industry. Springer Sci. Business Media; 2013, ISBN 9781475745979.

[43] Le Roes-Hill M, Prins A. Biotechnological potential of oxidative enzymes from Actinobacteria. Actinobacteria- basics and biotechnological applications. InTech. 2016. https://doi.org/10.5772/61321.

[44] Rasmussen RS, Morrissey MT. Marine biotechnology for production of food ingredients. Adv Food Nutr Res 2007;52:237−92.

[45] Cho SA, Cho JC, Han SH. Cosmetic composition containing enzyme and amino acid. Amorepacific Corporation (11/990); 2007. p. 431.

[46] Buckingham KWC. Methods for the treatment and prophylaxis of diaper rash and diaper dermatitis. The Procter & Gamble Company; 1985. Patent No US4556560 A.

[47] Diehl C. Skin treatment preparation. Crawford Healthcare Limited; 2008. Patent No. US7468195.

[48] Alfa MJ, Jackson M. A new hydrogen peroxide-based medical device detergent with germicidal properties: comparison with enzymatic cleaners. Am J Infect Contr 2001;29:168−77.

[49] Enoki T, Okuda S, Kudo Y, Takashima F, Sagawa H, Kato I. Oligosaccharides from agar inhibit pro-inflammatory mediator release by inducing hemeoxygenase 1. Biosci Biotechnol Biochem 2010;74:766−70.

[50] Ramnani P, Chitarrari R, Tuohy K, Grant J, Hotchkiss S, Philp K, et al. *In vitro* fermentation and prebiotic potential of novel low molecular weight polysaccharides derived from agar and alginate seaweeds. Anaerobe 2012;18:1−6.

[51] Ghadi S, Muraleedharan UD, Jawaid S. Screening for agarolytic bacteria and development of a novel method for *in situ* detection of agarase. J Mar Biotechnol 1997;5: 194−200.

[52] Jonnadula R, Ghadi SC. Purification and characterization of β-agarase from seaweed decomposing bacterium *Microbulbifer* sp. Strain CMC-5. Biotechnol Bioproc Eng 2011;16:513−9.

[53] Ferreira-Dias S, Sandoval G, Plou F, Valero F. The potential use of lipases in the production of fatty acid derivatives for the food and nutraceutical industries. Electron J Biotechnol 2013;16:1–38.

[54] Li PY, Ji P, Li CY, Zhang Y, Wang GL, Zhang XY. Structural basis for dimerization and catalysis of a novel esterase from the GTSAG motif subfamily of the bacterial hormone-sensitive lipase family. J Biol Chem 2014;289:19031–41.

[55] Yuan D, Lan D, Xin R, Yang B, Wang Y. Biochemical properties of a new cold-active mono- and di- acylglycerol lipase from marine member *Janibacter* sp. strain HTCC2649. Int J Mol Sci 2014;15:10554–66.

[56] Sharma M, Kumar A. Xylanases: an overview. Br Biotechnol J 2013;3:1–28.

[57] Qureshi MA, Khare AK, Pervez A. Enzymes used in dairy industries. Int J Appl Res 2015;1:523–7.

[58] Kieliszek M, Misiewicz A. Microbial transglutaminase and its application in the food industry. A review. Folia Microbiol 2014;59:241–50.

[59] Zhao GY, Zhou MY, Zhao HL, Chen XL, Xie BB, Zhang XY, et al. Tenderization effect of cold-adapted collagenolytic protease MCP01 on beef meat at low temperature and its mechanism. Food Chem 2012;134:1738–44.

[60] Law BA. The nature of enzymes and their action in foods. In: Whitehurst RJ, Law BA, editors. Enzyme in food technology. Sheffileld: Sheffield Academic Press; 2002. p. 1–18.

[61] Damare S, Raghukumar C, Muraleedharan UD, Raghukumar S. Deep-sea fungi as a source of alkaline and cold-tolerant proteases. Enzym Microb Technol 2006;39:172–81.

[62] Karlund A, Moor U, Sandell M, Karjalainen RO. The impact of harvesting, storage and processing factors on health-promoting phytochemicals in berries and fruits. Processes 2014;2:596–624.

[63] Kumar S. Role of enzymes in fruit juice processing and its quality enhancement. Adv Appl Sci Res 2015;6:114–24.

[64] Guo M. Functional foods: principles and technology. Cambridge: Woodhead Publishing; 2009. https://doi.org/10.1533/9781845696078.

[65] Lima DM, Fernandes P, Nascimento DS, Ribeiro RCLF, de Assis SAA. Fructose syrup: a biotechnology asset. Food Technol Biotechnol 2011;49:424–34.

[66] Garg G, Singh A, Kaur A, Mahajan R. Microbial pectinases: an ecofriendly tool of nature for industries. 3 Biotech 2016;6:47–59.

[67] Molkabadi EZ, Hamidi-Esfahani Z, Sahari MA, HoseinAzizi M. A new native source of tannase producer, *Penicillium* sp. EZ-ZH190: characterization of the enzyme. Iran J Biotechnol 2013;11:244–50.

[68] Choct M. Enzymes for the feed industry: past, present and future. World Poult Sci J 2006;62:5–15.

[69] Kang W, Xu Y, Qin L, Wang Y. Effects of different β-D-Glycosidases on bound aroma compounds in muscat grape determined by HS-SPME and GC-MS. J Inst Brew 2010;116:70–7.

[70] Mojsov K. Use of enzymes in wine making: a review. Int J Market Technol 2013;3:112–27.

[71] Wang Y, Zhang C, Li J, Xu Y. Different influences of β-glucosidases on volatile compounds and anthocyanins of Cabernet Gernischt and possible reason. Food Chem 2013;140:245–54.

[72] Su E, Xia T, Gao L, Dai O, Zhang Z. Immobilization of β glucosidase and its aroma increasing effect on tea beverage. Food Bioprod Process 2010;88:83–9.

[73] Goyal K, Selvakumar P, Hayashi K. Characterization of a thermostable β-glucosidase (BglB) from *Thermotogamaritima* showing transglycosylation activity. J Mol Catal B Enzym 2001;15:45–53.

[74] Bhat MK. Cellulases and related enzymes in biotechnology. Biotechnol Adv 2000;18:355–83.

[75] Kirk O, Borchert TV, Fuglsang CC. Industrial enzyme applications. Curr Opin Biotechnol 2002;13:345–51.

[76] Ulery BD, Nair LS, Laurencin CT. Biomedical applications of biodegradable polymers. J Polym Sci B Polym Phys 2011;49:832–64.

[77] Zhang Z, Marquardt RR, Wang G, Guenter W, Crow GH, Han Z, et al. A simple model for predicting the response of chicks to dietary enzyme supplementation. J Anim Sci 1996;74:394–402.

[78] Kobayashi S. Lipase-catalyzed polyester synthesis - a green polymer chemistry. Proc Jpn Acad Ser B Phys Biol Sci 2010;86:338–65.

[79] Gurung N, Ray S, Bose S, Rai V. A broader view: microbial enzymes and their relevance in industries, medicine, and beyond. BioMed Res Int 2013. https://doi.org/10.1155/2013/329121. Article ID 329121.

[80] Yan J, Han B, Gui X, Wang G, Xu L, Yan Y, et al. Engineering *Yarrowiali polytica* to simultaneously produce lipase and single cell protein from agro-industrial wastes for feed. Sci Rep 2018;8:758.

[81] Dewan SS. Global markets for enzymes in industrial applications. MA, USA: BCC Research: Wellesly; 2017.

[82] Kumar D, Savitri, Thakur N, Verma R, Bhalla TC. Microbial proteases and application as laundry detergent additive. Res J Microbiol 2008;3:661–72.

[83] Singh A, Kuhad RC, Ward OP. Industrial application of microbial cellulases. In: Kuhad RC, Singh A, editors. Lignocellulose Biotechnology: future prospects. New Delhi, India: I.K. International Publishing House; 2007. p. 345–58.

[84] Greene RV, Grifin HL, Cottal MA. Utility of alkaline protease from marine shipworm bacterium in industrial cleansing applications. Biotechnol Lett 1996;18:759–64.

[85] Chi ZM, Ma C, Wang P, Li HF. Optimization of medium and cultivation conditions for alkaline protease production by the marine yeast *Aureobasidium pullulans*. Bioresour Technol 2007;98:534–8.

[86] Haddar A, Agrebi R, Bougatef A, Hmidet N, Sellami-Kamoun A, Nasri M. Two detergent stable alkaline serine-proteases from *Bacillus mojavensis* A21: purification, characterization and potential application as a laundry detergent additive. Bioresour Technol 2009;100:3366–73.

[87] Araujo R, Casal M, Cavaco-Paulo A. Application of enzymes for textiles fibers processing. Biocatal Biotechnol 2008;26:332–49.

[88] Marcher D, Hagen HA, Castelli S. ITB Veredlung 1993;39:20–32.

[89] Meyer-Stork LS. Maschen Ind 2002;52:32–40.

[90] Mojsov K. Applications of enzymes in the textile industry: a review. In: 2nd international congress: engineering, ecology and materials in the processing industry. Jahorina, Bosnia and Herzegovina: Tehnol. Oski Fakultet Zvornik; 2011. p. 230–9.

[91] Rao MB, Tanksale AM, Ghatge MS, Deshpande VV. Molecular and biotechnological aspects of microbial proteases. Microbiol Mol Biol Rev 1998;62:597–635.

[92] Choudhary RB, Jana AK, Jha MK. Enzyme technology applications in leather processing. Ind J Chem Technol 2004;11:659–71.

[93] Srivastava N, Singh P. Degradation of toxic pollutants from pulp & paper mill effluent. Discovery 2015;40:221–7.
[94] Maijala P, Kleen M, Westin C, Poppius-Levlin K, Herranen K, Lehto JH, et al. Biomechanical pulping of softwood with enzymes and white-rot fungus *Physisporinus rivulosus*. Enzym Microb Technol 2008;43:169–77.
[95] Kuhad RC, Gupta R, Singh A. Microbial cellulases and their industrial applications. Enzym Res 2011:1–10. https://doi.org/10.4061/2011/280696.
[96] Raghukumar C, Muraleedharan U, Gaud VR, Mishra RJ. Xylanases of marine fungi of potential use for biobleaching of paper pulp. J Ind Microbiol Biotechnol 2004;31:433–41.
[97] Kirk TK, Jeffries TW. Role of microbial enzymes for pulp and paper processing. In: ACS symposium series: Am. Chem. Soc. Washington, DC; 1996. p. 2–14.
[98] Patrick K. Enzyme technology improves efficiency, cost, safety of stickies removal program. Pap Age 2004;120:22–5.
[99] Eugene AL, Uchechukwu OC, Mariagoretti UO. A review on biological catalysts in organic synthesis. Int J Adv Eng Res Appl 2016;2:296–321.
[100] Busto E, Gotor-Fernández V, Gotor V. Hydrolases: catalytically promiscuous enzymes for non-conventional reactions in organic synthesis. Chem Soc Rev 2010;39:4504–23.
[101] Davis BG, Boyer V. Biocatalysis and enzymes in organic synthesis. Nat Prod Rep 2001;18:618–40.
[102] Cammarota MC, Freire DMG. A review on hydrolytic enzymes in the treatment of wastewater with high oil and grease content. Bioresour Technol 2006;97:2195–210.
[103] Pandey D, Singh R, Chand D. An improved bioprocess for synthesis of acetohydroxamic acids using DTT (dithiothreitol) treated resting cells of *Bacillus* sp. APB-6. Bioresour Technol 2011;102:6579–86.
[104] Kalia VC, Rashmi, Lal S, Gupta MN. Using enzymes for oil recovery from edible seeds. J Sci Ind Res 2001;60:298–310.
[105] Iravani S, Korbekandi H, Mirmohammadi SV, Zolfaghari B. Synthesis of silver nanoparticles: chemical, physical and biological methods. Res Pharm Sci. 2014;9:385–406.
[106] Borase HP, Salunke BK, Salunkhe RB, Patil CD, Hallsworth JE, Kim BS, et al. Plant extract: a promising biomatrix for ecofriendly, controlled synthesis of silver nanoparticles. Appl Biochem Biotechnol 2014;173:1–29.
[107] Khan I, Duttaa JR, Ganesanb R. Enzymes' action on materials: recent trends. J Cell Biotechnol 2016;1:131–44.
[108] Mishra A, Sardar M. Cellulase assisted synthesis of nano-silver and gold: application as immobilization matrix for biocatalysis. Int J Biol Macromol 2015;77:105–13.

Further reading

[1] Kamm B, Kamm M. Principles of biorefineries. Appl Microbiol Biotechnol 2004;64: 137–45.

Chapter 8

Recent advances in bioanalytical techniques using enzymatic assay

Kanchanmala Deshpande[1], Geetesh K. Mishra[2]
[1]*Department of Chemistry, Goa University, Taleigao Plateau, Goa, India;* [2]*Multiscale Fluid Mechanics Lab, School of Mechanical Engineering, Sungkyunkwan University, Suwon, South Korea*

8.1 Introduction

Rapid analysis of a large number of samples with good accuracy, precision, and sensitivity has always remained a great challenge. Apart from the above-mentioned analytical parameters, interference due to matrix and subsequent low recoveries increases the difficulty level of analysis. Routinely used techniques that have gained acceptance have shown promise in their use. The standard analytical techniques that are used for the detection of various analytes form environmental or food samples are based on chromatographic techniques such as gas chromatography (GC) and liquid chromatography (LC), coupled with mass spectrometry (MS), such as GC-MS/MS and LC-MS/MS [1]. These methods are highly precise and have the potential of being automated. However, they still suffer from drawbacks such as tedious sample preparation, requirement of large amount of organic solvent, high level of sophistication required, and difficulty for screening a large number of samples within a short time. Several bioanalytical techniques have been developed over the years to minimize the problems stated herein. These techniques can quantify analytes of interest even at trace concentrations with good reliability [2]. Reported bioanalytical techniques can analyze a majority of analytes found in a wide range of matrices [2] but are tremendously challenging sometimes due to the variety of substances in biological samples, the complex molecular structures, and time-dependent concentrations. Most of the bioanalytical protocols require isolation of components from natural environments and separations of complex mixtures. Biosensors, as a discrete bioanalytical tool, are able to measure analytes selectively, often in a natural matrix, without prior separation of multicomponent samples and often producing quantitative data within minutes [3,4].

Advances in Biological Science Research. https://doi.org/10.1016/B978-0-12-817497-5.00008-2

In the past few years, the effects of human activities on environmental quality have been both extensive and diverse in the extent to which they disturb the ecosystem. Work in recent years has observed an emergent need to examine environmental pollutants/food contaminants and to keep a constant watch on them. These contaminants may have mild to harsh effects, which may be visible immediately or only after a long time. Some of the contaminants have lethal effects and could lead to a public health crisis [5]. Therefore, screening of toxic pollutants on a routine basis is gaining more consideration these days. Other than environmental pollutants and food contaminants, emerging areas of concern include reported high amounts of food allergens and pharmaceutical drugs such as antibiotics [6,7]. The presence of various food allergens and antibiotics in food elements is a challenge for food safety management. Thus their monitoring in food is very important for human consumption; even the minute presence of these residues can trigger an allergic reaction in hypersensitive individuals [8]. To address these important issues concerning environmental and food safety, several biosensors have been developed to help in monitoring environment and human health. Herein, we address the significance, basics, and application of biosensor systems for food and environmental monitoring. A practical solution for optimum performance of biosensors is also discussed here.

8.1.1 Why biosensors?

Defining human exposure to various toxic chemicals is an enormous task. Hence, development of novel analytical tools that not only quantify the concentration of toxic chemicals at ultra-trace level but also estimate the biological damage resulting due to their exposure is the need of the hour [9]. These toxic chemicals are reported to be carcinogenic in nature and hinder important enzymatic functions [10]. Considering these points, development of novel tools for predicting concentration and its impact at ultra-trace level is of utmost importance. Over conventional analytical techniques, biosensors excel in various aspects. The significant aspects are high-throughput ability, versatility, capability for speciation studies, minimal use of organic solvents, and suitability for point-of-care testing or field monitoring [11]. These significant advantages have led to the development and commercial success of biosensors [12]. In view of this, over the past few decades, numerous biosensors have been developed for the detection of ions, small molecules of organic or inorganic nature, and various analytes of biological origin such as proteins, deoxyribonucleic acids, and so on [13].

8.1.2 Emergence of biosensors

Biosensors represent new analytical devices that appear to be an analyst's dream. Technically, a biosensor is a probe that integrates a biological component with an electronic transducer, thereby converting a biochemical signal into a quantifiable electrical response [14]. Biosensors are characterized by a high level of specificity generated by the biocomponent, which specifically reacts with a given analyte or substrate. The combination of this specificity, with a sensitive transducer, gives to biosensors their unique characteristics for the detection of a variety of analytes, even when they occur in complex matrices [2]. It is noteworthy that bioassays having high specificity and capability for analysis of single or multianalyte are often referred as biosensors although most of these systems do not have integrated transducer systems. A schematic representing various components of a biosensor is shown in Fig. 8.1.

A plethora of literature is available that addresses the development and application of biosensors for medical diagnostics [15], food quality assurance [16], environmental monitoring [17,18], industrial process control, and to biological warfare agent detection [19]. In recent years advances in biosensor development has been observed in the area of its miniaturization, its in situ measurement capability, which ultimately leads to its commercialization [15]. By 2022, the global biosensor market will be valued at approximately USD 27.06 billion with home-use health-monitoring devices being dominant.

8.2 Classification of biosensors

Biosensors classification can be performed in different ways. The type of biological signaling mechanism utilized for biosensing can be one of the approaches, whereas the type of signal transduction system employed can also be the basis of its classification. Biological signaling achieved by biosensors

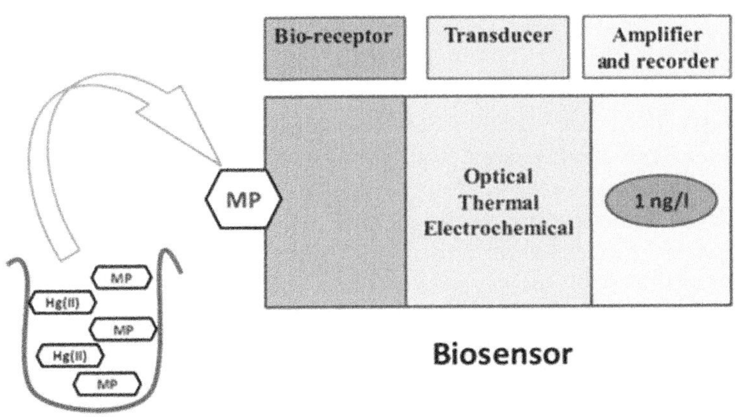

FIGURE 8.1 Schematic representing biosensor components. [Figure represent interaction of MP, i.e., methyl parathion (pollutant) with bioreceptor and subsequent steps in sensing].

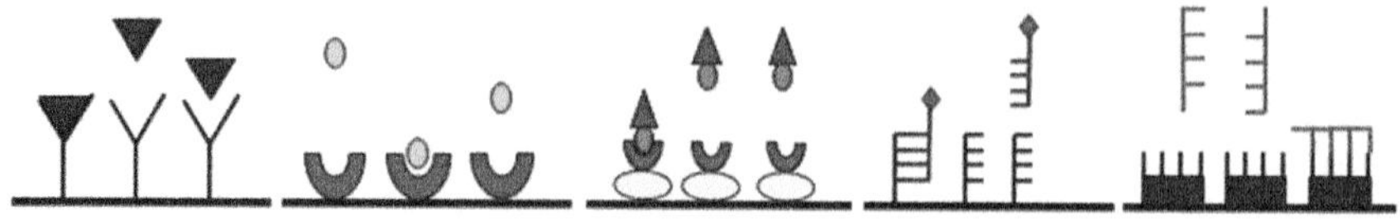

FIGURE 8.2 Type of biocomponent commonly used in biosensors (here the biocomponent is shown in immobilized form for simplicity).

can be due to various biocomponents. Some of the significant biocomponents exploited for biosensor application include antibody/antigen, enzymes, whole cells, nucleic acids, and biomimetic materials (aptamers). Various biocomponents explored in biosensors are compiled in Fig. 8.2.

When biosensors are classified according to their method of signal transduction, i.e., by measuring the change that occurs in the bioreceptor reaction, they are named as optical, thermal, electrochemical, and piezoelectric. Optical transducers respond to an analyte by undergoing a change in their optical properties such as absorption, fluorescence, luminescence, reflectance, emission, or a change in an interferometric pattern. Optical transducers represent the largest and fastest growing area in biosensor technology due to their high amplification capacity [20–24]. Some bioreceptor reactions cause changes in heat, i.e., molar enthalpy. Thermal biosensors measure this change in heat [25–28]. A third class includes electrochemical reactions, which are the most commonly used detectors. These devices measure the current produced from oxidation and reduction reactions. Concentration of the electroactive species present can be predicted form this current [29]. A fourth class includes measurement of changes in mass caused by chemical binding to small piezoelectric crystals. These are popularly known as mass sensitive biosensors.

8.2.1 Enzyme biosensor

Biosensors that utilize enzymes as the recognition elements represent the most extensively studied area. The high specificity of enzyme-substrate interactions, and the usually high turnover rates of biocatalysts, open the way to sensitive and specific enzyme-based biosensor devices development [30,31]. Each enzyme has an active site or sites containing functional groups. The interaction between the active site of enzyme and analyte is highly specific. The products of this interaction can be detected at much lower limits than with other normal interactions. Enzyme-based assay is an extremely broad field that impacts on many major industrial sectors such as the pharmaceutical, health care, food, and agricultural industries, as well as environmental monitoring [30].

Two different approaches can be identified for determining an analyte by use of an enzymatic biosensor: (1) if the enzyme metabolizes the analyte, the analyte can be determined by measuring the enzymatic product, and (2) if the analyte inhibits the enzyme, the decrease of enzymatic product can be measured and correlated to the analyte concentration. In the latter case, the device is designated as a "biosensor based on enzyme inhibition" or "inhibition biosensor." Among the reported enzyme-based biosensors, the inhibition based one comprises the major portion.

8.2.1.1 Enzyme inhibition biosensor

A large percentage of environmental pollutants and food toxicants are known to act as enzyme inhibitors, resulting in the development detection methods based on this property. The binding of an inhibitor can stop a substrate from entering the enzyme's active site and/or hinder enzyme catalysis. Inhibitor binding could be either reversible or irreversible. Irreversible inhibitors usually react with the enzyme and change it chemically. These inhibitors modify key amino acid residues required for enzymatic activity; in contrast, reversible inhibitors bind noncovalently and different types of inhibition result depending on whether these inhibitors bind the enzyme, the enzyme-substrate complex, or both [32].

It is well known that the response of a biosensor to the addition of a substrate is determined by the concentration of the product (P) of the enzymatic reaction on the surface of the sensor. The reaction is controlled by the rate of two simultaneous processes, i.e., the enzymatic conversion of the substrate (S) and the diffusion of the product from the enzyme layer. For this conversion, factors such as pH, ionic strength, temperature, enzyme concentration, and substrate concentration play significant roles. Inhibitor (I) selectively inhibits the activity of certain enzymes, thus affecting enzyme activity and the product concentration. Schematic representation of enzyme inhibition is depicted in Fig. 8.3. In the biosensors based on enzyme inhibition, the quantification of the inhibitor concentration is carried out by measuring the enzyme activity before and after exposure of the biocomponent at the target analyte [30,33]. The percentage of inhibition is calculated as follows,

$$I\% = (\mathrm{A0} - \mathrm{AI})/\mathrm{A0} \times 100$$

Where I% = percentage of inhibition; A0 = the enzyme activity before exposure to the analyte; AI = the enzyme activity after exposure to the analyte. Here, the percentage of inhibited enzyme (I%) that results after exposure to the inhibitor is quantitatively related to the inhibitor (i.e., analyte) concentration provided incubation time is constant.

The choice of enzyme/analyte system is based on the specific reaction between functional moieties present on the active site of enzyme and active

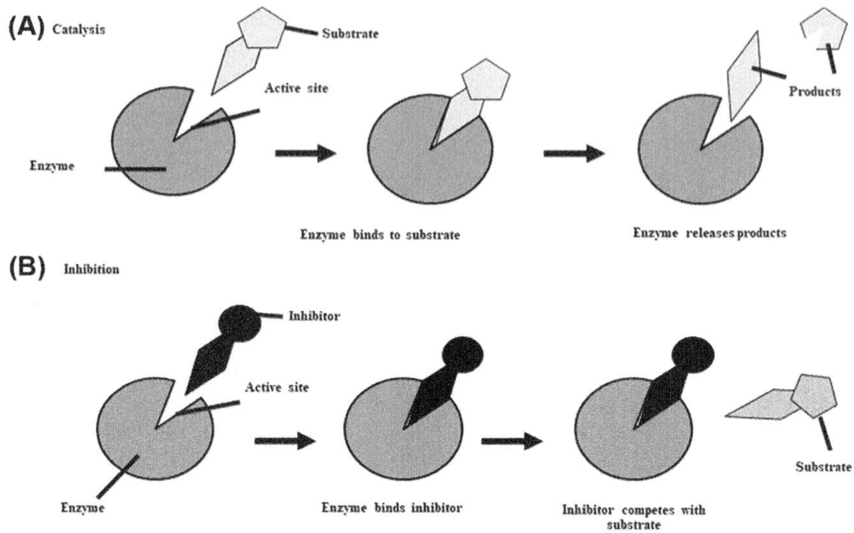

FIGURE 8.3 Schematic representing principles behind enzyme inhibition biosensor.

functional group of analyte. Most of the toxic compounds block the active site, thereby inhibiting normal enzyme function. The enzyme inhibitor reaction is often complex and has been reviewed in the current literature [34–36]. For quantification, a calibration plot is prepared using a range of known concentrations of standard analyte solution in contact with enzyme solution. The difference in the enzyme activity before and after contact is calculated. The steady-state response corrected for a background signal versus the difference in the enzyme activity is plotted against analyte concentration or its logarithm. The analytical figures of merit for assay such as dynamic range, linear range, limit of detection (LOD) and limit of quantification (LOQ) were calculated from the plot. Such calibration plots are used for quantification of unknown concentration [30]. Enzyme inhibition–based biosensors are fascinating to researchers where various enzymes such as glucose oxidase, urease, tyrosinase, cholinesterase, alcohol oxidase, and peroxidase have been utilized as biocomponents [36–38].

8.2.2 Overcoming limitations in enzyme-based biosensors

Although enzyme-catalyzed or enzyme-inhibition-based biosensors show considerable promise, there are a great number of challenges and limitations in exploiting these systems for potential applications. The majority of enzymes are fairly unstable at room temperature in their free form, due to which their application in commercial sensors or kits is often hampered. The impact is visible where lack of sensor stability under storage and operational conditions, difficulty in reuse of enzyme and recovery of product are observed [34]. Also, a protein's sequence and interactions between residues in the protein core are naturally not fully optimized and only achieve the minimum requirements for proper functioning. This situation leaves plenty of room for improvement.

In order to make enzyme utilization in the most effective way, different methods have been put into practice; immobilization is one of them [30,39]. The term "immobilized enzymes" refers to enzymes physically confined or localized in a certain defined region of space with retention of their catalytic activities, and which can be used repeatedly and continuously. Enzyme immobilization offers several advantages over using free enzymes in solution phase. With immobilization, enzyme stability improves. While working with organic solvents, under varied pH range or with different temperature conditions enzymes retain their stability. The added advantage is that with this increased stability, enzyme availability to substrate or analyte increases, which ultimately shows greater turnover over a considerable period of time. The majority of the reported work on enzyme immobilization can be carried out by physical and chemical approaches. Weak interactions between support and enzyme are the characteristics of physical enzyme immobilization, whereas covalent bond formation between enzyme and support are the characteristics of chemical immobilization [27,39,40]. The other significant modes of enzyme immobilization include entrapment and cross-linking.

For effective enzyme immobilization, efficiency of several natural and synthetic supports has been assessed. Inert polymers and inorganic materials are usually used as carrier matrices. Apart from being affordable, an ideal matrix must encompass characteristics like affinity for enzyme. This is assured by the presence of the specific active groups on the carrier, which enable the generation of the enzyme—carrier interactions. However, if absent, the interactions can be tuned by applying intermediate agents (carrier modifiers). Application of enzyme immobilization has been reviewed by several researchers. Combined approaches of proteins/enzymes with nanoparticles have been discussed in detailed by Behera et al. and Pal et al. [41,42]. Significance of enzyme—carrier interactions, especially in terms of adsorption as mode of immobilization has been reviewed by Jesionowski et al. [42]. In this paper he reviewed various carriers utilized for the immobilization with and without the intermediate agents, comparative study of methods of adsorption on different types of the carriers, and a few examples where immobilized enzymes are employed as catalysts in practical applications. Deshpande et al. [3] have demonstrated that zirconia, a white crystalline solid, is also an attractive for enzyme immobilization than the other common materials such as carbon nanotubes, graphite, silica, and magnetic nanoparticles.

Sometimes absence of a functional group required for effective interaction between enzyme and carrier is observed. Surface modification is the best choice in these cases. Here, by chemical modifiers the covalent linkage is possible between enzyme and carrier. The typical common modifying agents reported for enzyme immobilization include glutaraldehyde [27,28]. Glutaraldehyde is a choice because it possesses high affinity to bacteria, fungi, and proteins. Structurally it contains two reactive aldehyde groups with a five-atom carbon chain which serves as a spacer for enzymes, making their active sites more easily accessible for the substrates. Among the other significant modifiers used for enzyme immobilization are 3-aminopropyltrimethoxysilane and 3-aminopropyltriethoxysilane [43,44]. The latter two compounds interact more strongly with the carrier surface, which is due to the presence of three methoxy or ethoxy groups in their molecules.

Among the different enzyme inhibition assays acetylcholinesterase and butyl cholinesterase share a major role. Reported approaches for stabilization of cholinesterases include AChE with mixture of glucose, trehalose & gelatin, albumin, trehalose & gelatin, BSA, lysozyme & gelatin and combination of sucrose, polygalacturonic acid & dextran sulfate, dextrose in protein standard solution [22,24]. Other significant inhibition-based reported work includes alkaline phosphatase (ALP from calf intestine) immobilized on silica [45].

8.2.3 Application of enzyme biosensor

In recent years the research on the application of biosensors mainly has been focused on four major areas. Fig. 8.4 represents these significant areas in schematics. Among these areas, the potential of enzymatic assays for food and environmental analysis in a high-throughput context has not yet been much assessed. High-throughput analysis can be defined as the implementation of assays in the wells of microplates in combination with liquid handling

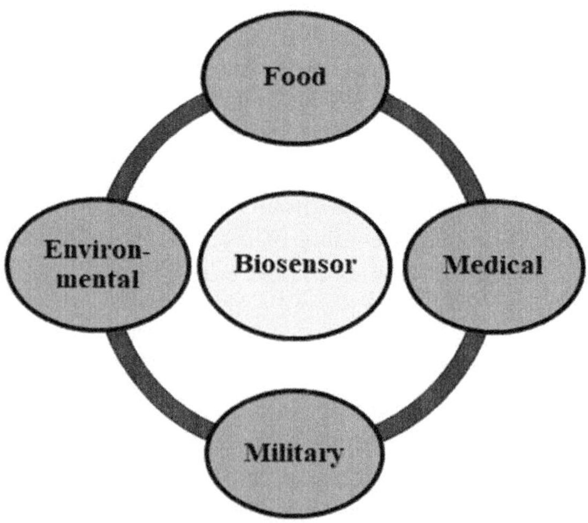

FIGURE 8.4 Current biosensor application areas.

robotics. Such a multianalyte bioanalytical system may result in significant cost reduction with negligible crossover possibilities.

8.3 Enzyme biosensors for environmental monitoring

Enzyme-inhibition-based biosensors show the potential to complement a number of emerging screening and monitoring methods for environmental applications. Potential application areas include: laboratory screening, field screening, or continuous and in situ field monitoring. Analytical tasks associated with site characterization primarily involve the identification of listed contaminants and mapping of the spatial distribution of the compounds of concern. Frequent and repetitive analysis at specific locations for particular compounds of interest is also an important step in remediation. Because enzyme-inhibition-based biosensors show the potential to operate continuously at remote or in situ locations, these devices could be particularly well suited for this task [46]. Portability and rapid analysis makes enzyme-inhibition-based biosensors a promising tool for analyzing various compartments in environment such as soil and water [46].

In the area of water quality monitoring, there is reported enzyme-based biosensor focus on screening of toxic components on a routine basis and their monitoring during remediation phase. Among the reported analytes, toxic components such as pesticides and heavy metals contribute a significant share. Since all these toxic components act as enzyme inhibitors, biosensors based on them are reported to show excellent sensitivity up to ppt level. Considerable number of published works are devoted to biosensors used for the determination of bioavailable metals in soil samples [47]. Here, simple aqueous extract of soils and sediment samples provide information on the bioavailability of the pollutants; this type of information is often more relevant for assessing the potential damage caused by a substance. Air samples have also been analyzed directly with biosensors although the number of devices developed and applied with this objective is limited.

To understand the physiological impact of pesticides in the environment, food safety, and quality control enzyme-based biosensors play a crucial role. Acetylcholinesterase (AChE)/Butyrylcholinesterase (BuCHE) inhibition-based biosensors are reported to determine trace levels of pesticide [3,35,46,51]. Over the last decade or two, AChE/BuChE-inhibition-based biosensors have developed tremendously. Among the different AchE-based sensor Xia and Xiangyou reported multiple organophosphorus compounds such as phoxim, dichlorvos, omethoate, and trichlorfon with detection range 0.01−10.00 ng/L [48]. Multielement analysis and high-throughput analysis have always remained areas of interest for commercial success of biosensors [49]. On the lab scale, the need for higher throughput has driven the shift from microwell plates with 96 wells to those with 384 or 1536 wells. The traditional 96-well format has proved inadequate and is being replaced by microplate with larger numbers of smaller wells with volumes that range from milliliters to nanoliters. This 4–16-fold increase in the sample throughput with decrease in size means that fewer reagents are needed per assay, and the resulting cost savings has evidently been beneficial. With miniaturized high-throughput assays, highly toxic compounds such as Hg(II) can be analyzed with little μL total assay volume. Microwell plate with automation can analyze thousands of samples within hours or can analyze multiple samples within a very short time. In addition, inhibition studies are performed in closer proximity, which ultimately increases reliability of results. Advancement of high-throughput analysis on various platforms has been successfully demonstrated by various scientists. The wide range of platforms reported for this work includes metal or metal oxide chips, nanoparticles, paper strips, textile sample, glass fiber filters, and even human skin [3,10,19,21–24,46,50]. The interesting fact about disposable strips includes the absence or decrease of the color indicating the AChE inhibition. The biosensor is able to detect organophosphate and carbamate pesticides with good detection limits (methomyl $= 6.16 \times 10^{-4}$ mM and profenofos $= 0.27$ mM) and rapid response times ($\sim$5 min). In this method quantification is done by naked eyes (0.0001 and 0.1 ppm of organophosphorus pesticides) making the method cost effective. On chip platform, effective chemical enzyme mobilization and its utility for sensitive mercury analysis in deep well has also been reported [50].

8.4 Enzyme biosensors for food quality monitoring

The food industry needs appropriate analytical methods for monitoring food quality and food processing. Detection of chemical and biological contaminants in food commodity is of supreme importance because, unlike impurities of a physical nature, they cannot be seen visually. Therefore, it is essential to advance the biosensor development for the analysis of food quality, since they have proven to be an extremely viable alternative to traditional analytical techniques [51]. However, very few biosensors play a prominent role in monitoring food quality or food processing. Considerable efforts have been made to develop enzyme-based biosensors that are inexpensive, reliable, and robust enough to operate under realistic conditions [52]. Advantages of enzyme-based biosensors in food quality monitoring are abundant. Trace level quantification, high precision, low sample volume and analysis time, and reusability of biosensors are some of the advantages [53]. In the area of food quality monitoring, most of the enzyme-based biosensors for food security are largely concentrated on the analysis of food contaminants, allergens, toxins, pathogens, food additives, and adulterants. In recent years, numerous transducers have been explored utilizing several enzymatic assays for the development of biosensing techniques applied for food quality monitoring. Among various biosensors, application of thermal biosensors in food analysis was studied extensively. A large number of enzymatic reactions have been reported using thermal biosensors, i.e., enzyme thermistor, amid them detection of fructose, choline, and urea were reported in different food stuffs [25]. A highly selective, interference-free biosensor for the measurement of fructose in real syrup samples was developed. The assay was based on the phosphorylation of d(-)fructose to fructose-6-phosphate by hexokinase and subsequent conversion of fructose-6-phosphate to fructose-1,6-biphosphate by fructose-6-phosphate-kinase. Moreover, determination of free and total choline in commercially available milk, milk powder, and dietary supplements was reported using thermal biosensors. The choline biosensor was based on the choline oxidase and catalase enzymes coimmobilized on controlled pore glass (CPG). The heat liberated in the bienzyme reaction was proportional to choline concentration under the optimized condition [27]. Furthermore, a simple, economical, and highly stable thermal biosensor for analysis of urea in adulterated milk was reported by Mishra et al. The biosensor is comprised of immobilized enzyme urease on CPG, which selectively hydrolyzes urea present in the milk sample. The biosensor exhibited lower detection limit and an excellent dynamic range of urea detection in milk [4]. In other reported transducers utilizing enzyme assays, an amperometric biosensor for vitamin C was reported based on the immobilization of ascorbate oxidase into a biocompatible sandwich-type composite film. Content of ascorbic acid in commercial juices was determined by the developed biosensor and characterized by a very good bioelectrocatalytic performance toward the oxidation of vitamin C in solution [54].

As modern-day society requires sensitive, accurate, and express methods of food safety monitoring, the growing field of enzymatic biosensors represents an answer to this demand. Although most of the earlier reported biosensor systems have been tested on buffered solutions, more biosensors that can be applied to real sample analysis have been reported in recent years. In this perspective, biosensors for potential environmental and food applications continue to show improvements in areas such as genetic modification of enzymes and improvements in immobilization of recognition element. Compiled works on recent trends in enzyme-based biosensors are presented in Table 8.1.

8.5 Future prospects and conclusions

The biosensors-based bioanalytical techniques utilizing enzymatic reactions are useful as an alarm or general toxicity indicator for the fast identification of food or environmental contaminants. The market trends showed about 10.4% growth in the development of biosensors for various applications for food and environmental safety, which also leads to improved human health. In contrast with clinical applications, biosensing techniques for food and environmental safety are still in their early stages and facing many difficulties due to inherent features of environmental and food matrices where sensitivity and selectivity toward a specific analyte are the key issues.

TABLE 8.1 Compiled literature on advances in enzyme- based biosensors.

Enzyme	Analyte	Matrix	LOD/Range	Ref.
Acetylcholinesterase	Paraoxon	Milk	1 μg/L	[2]
Urease	Urea	Urine	10 −1000 mM	[4]
Alcohol oxidase	Hg(II)	Water	5−500 ng/L	[10]
Organophosphorus hydrolase	Paraoxon	Gas	90−300 mg/L	[19]
Horseradish peroxidase	Parathion	Fruit	0.001 −500 μg/L	[21]
Butyrylcholinesterase	Paraoxon	Milk	0.005 −50 μg/L	[22]
Hexokinase and fructose-6-phosphate-kinase	Fructose	Syrup	0.5−6.0 mM	[25]
Alkaline phosphatase	Paraoxon	Water	0.01 −500 μg/L	[45]

Sensitivity improvement has always remained the topmost priority of enzyme-based biosensors. In practice, for quantification of toxic components in environmental or food samples highly sensitive enzyme-inhibition-based biosensors are required. To achieve such sensitivity, nanomaterials or nano-composites tailored with enzymes is a preferred approach. The tailor-made nanocomposites with versatile nanostructures provide more sensitive and flexible analysis in complex matrices with increased analytical performance [41,42]. New strategies for popularizing enzyme-based biosensors for field application or commercial market include disposable paper-based sensors, reusable sensor strips, and microfluidic devices in the form of lab on chip where enzyme immobilization plays a significant role [55]. These different platforms that are very promising are low in cost, miniaturized, user friendly, and meet the needs of on-site detection of environmental and food samples. To increase sensitivity, future work should focus on clarifying the mechanisms of interaction between nanomaterials and enzymes using novel properties to fabricate a new generation of biosensors. Emerging technologies such as lab-on-a-chip microdevices and nanosensors or wearable point-of-use screening tools for environmental and food safety applications offer opportunities for the construction of a new generation of biosensors with much better performance [19,56,57]. The second important issue is high selectivity. Most biosensors reported in the literature work very well in laboratories, however, they may meet serious problems in tests in real samples or in the field. As a result, it is essential to develop novel surface modification approaches in order to avoid nonspecific adsorption at surfaces [47]. Third, another important issue is multianalysis on a single platform. Multiplexing is critical for saving assay time, which is especially important for assays performed in laboratories. Multiarrays can perform a large number of assays for multiple sample analytes [57,58]. An ideal miniaturized enzyme-based biosensor will be one that is integrated and highly automated. Current lab-on-a-chip technologies (microfluidics) offer a solution toward this goal. We can expect that successful biosensors in the future may incorporate all these features, and can conveniently detect minute targets within a short period. Sensing formats such as lateral flow chambers, disposable electrodes, and colorimetric paper strips are also projected as potential alternatives for facilitating the on-field quantification of potential analytes in a cost-effective manner. Further, development of wireless-networking-equipped biosensors that can be implemented onsite for real-time monitoring of pollutants will also help expand the dimension of research in this field [38]. Overall, for the successful development of biosensors, effective combination of biosensing and biofabrication should be used. For this, different transducers can be utilized [59].

In practice, biosensor development is a contineous process and there is always room for improvement in the desigh and utilizing various components in emerging biosensor. With the addition of newer food toxicants and environmental contaminents that may be threat for humans and the overall ecosystem, the need for fast and accurate biosensor will be of utmost importance. Biosensors today, requires the trust of potential consumer, keeping in mind that the accepance of new efforts is the best indicator of the sucess and achievement for an emerging technology.

References

[1] Grimalt S, Dehouck P. Review of analytical methods for the determination of pesticide residues in grapes. J Chromatogr A 2016;1433:1–23.

[2] Zhang Y, Muench S, Schulze H, Perz R, Yang B, Schmid RD, et al. Disposable biosensor test for organophosphate and carbamate insecticides in milk. J Agric Food Chem 2005;53:5110–5.

[3] Deshpande K, Mishra RK, Bhand S. Determination of methyl parathion in water and its removal on zirconia using optical enzyme assay. Appl Biochem Biotechnol 2011;164(6):906–17.

[4] Mishra GK, Sharma A, Deshpande K, Bhand S. Flow injection analysis biosensor for urea analysis in urine using enzyme thermistor. Appl Biochem Biotechnol 2014;174:998–1009.

[5] Hayat A, Marty JL. Aptamer based electrochemical sensors for emerging environmental pollutants. Front Chem 2014;2:41.

[6] Mishra GK, Sharma V, Mishra RK. Electrochemical aptasensors for food and environmental safeguarding: a review. Biosensors 2018;8:28.

[7] Wang W, Han J, Wu Y, Yuan F, Chen Y, Ge Y. Simultaneous detection of eight food allergens using optical thin-film biosensor chips. J Agric Food Chem 2011;59(13):6889–94.

[8] Mishra GK, Sharma A, Bhand S. Ultrasensitive detection of streptomycin using flow injection analysis-electrochemical quartz crystal nanobalance (FIA-EQCN) biosensor. Biosens Bioelectron 2015;67:532–9.

[9] Ramanathan K, Gary R, Apostol A, Rogers K. A Fluorescence based assay for DNA damage induced by radiation chemical mutagens and enzymes. Curr Appl Phys 2003;3:99–106.

[10] Deshpande K, Mishra RK, Bhand S. A High sensitivity microformat chemiluminescence enzyme inhibition assay for determination of Hg(II). Sensors 2010;10:6377–94.

[11] Prieto-Simón B, Campa's M, Andreescu S, Marty JL. Trends in flow-based biosensing systems for pesticide assessment. Sensors 2006;6:1161–86.

[12] Kim B-M, Abd El-Aty AM, Hwang T-E, Jin L-T, Kim Y-S, Shim J-H. Development of an Acetyl cholinesterase-based detection kit for the determination of organophosphorus and carbamate pesticide residues in agricultural samples. Bull Korean Chem Soc 2007;28:929–35.

[13] Dorst B, Mehta J, Bekaert K, Rouah-Martin E, Coen W, Dubruel P, et al. Recent advances in recognition elements of food and environmental biosensors: a review. Biosens Bioelectron 2010;26:1178–94.

[14] Scheller F, Schubert F. Biosensors: techniques and instrumentation in analytical chemistry, vol. 11. Netherland: Elsevier Science Publishers B.V.; 1992. p. 1–7.

[15] Newman JD, Turner AP. Home blood glucose biosensors: a commercial perspective. Biosens Bioelectron 2005;20:2435–53.

[16] Neethirajan S, Ragavan V, Weng X, Chand R. Review biosensors for sustainable food engineering: challenges and perspectives. Biosensors 2018;8:23.

[17] Mozaz S, Marco M, Alda M, Barcelo D. Biosensors for environmental applications: future development trends. Pure Appl Chem 2004;76:723–52.

[18] Selid P, Xu H, Collins E, Collins M, Zhao J. Sensing mercury for biomedical and environmental monitoring. Sensors 2009;9:5446–59.

[19] Mishra RK, Martín A, Nakagawa T, Barfidokht A, Lu X, Sempionatto JR, et al. Detection of vapor-phase organophosphate threats using wearable conformable integrated epidermal and textile wireless biosensor systems. Biosens Bioelectron 2018;101:227–34.

[20] Borisov M, Wolfbeis O. Optical biosensors. Chem Rev 2008;8:423–61.

[21] Chouhan RS, Vinayaka AC, Thakur MS. Chemiluminescence based technique for the detection of methyl parathion in water and fruit beverages. Anal methods 2010;2:924–8.

[22] Mishra RK, Deshpande K, Bhand S. A high-throughput enzyme assay for organophosphate residues in milk. Sensors 2010;10:11274–86.

[23] Verma N, Singh M. Biosensors for heavy metals. Biometals 2005;18:121–9.

[24] Weetall HH, Mishra NN, Mahfouz A, Rogers KR. An approach for screening cholinesterase inhibitors in drinking water using an immobilized enzyme assay. Anal Lett 2004;37:1297–305.

[25] Bhand S, Soundararajan S, Wärnmark IS, Milea JS, Dey ES, Yakovleva M, et al. Fructose-selective calorimetric biosensor in flow injection analysis. Anal Chim Acta 2010;668:13–8.

[26] Danielsson B, Mattiasson B, Mosbach K. Enzyme thermistor devices and their analytical applications. Appl Biochem Bioeng 1981;3:97–143.

[27] Deshpande K, Danielsson B, Bhand S. Flow injection analysis of choline in milk and dietary supplements using an enzyme thermistor. Chemical Sensors 2011a;1(1):16.

[28] Mishra GK, Mishra RK, Bhand S. Flow injection analysis biosensor for urea analysis in adulterated milk using enzyme thermistor. Biosens Bioelectron 2010;26(4):1560–4.

[29] Kanungo L, Bacher G, Bhand S. Flow-based impedimetric immunosensor for aflatoxin analysis in milk products. Appl Biochem Biotechnol 2014;174:1157–65.

[30] Amine A, Mohammadi H, Bourais I, Palleschi G. Enzyme inhibition-based biosensors for food safety and environmental monitoring. Biosens Bioelectron 2006;21:1405–23.

[31] Harrad L, Bourais I, Mohammadi H, Amine A. Recent advances in electrochemical biosensors based on enzyme inhibition for clinical and pharmaceutical applications. Sensors 2018;18:164.

[32] Fukuto RT. Mechanism of action of organophosphorus and carbamate insecticides. Environ Health Perspect 1990;87:245–54.

[33] Arduini F, Amine A, Moscone D, Palleschi G. Reversible enzyme inhibition-based biosensors: applications and analytical improvement through diagnostic inhibition. Anal Lett 2009;42:1258–93.

[34] Luque de Castro MD, Herrera MC. Enzyme inhibition-based biosensors and biosensing systems: questionable analytical devices. Biosens Bioelectron 2003;18:279–94.

[35] Pundir CS, Chauhan N. Acetylcholinesterase inhibition-based biosensors for pesticide determination: a review. Anal Biochem 2012;429:19–31.

[36] Upadhyay L, Verma N. Enzyme inhibition based biosensors: a review. Anal Lett 2013;46(2):225–41.

[37] Bucur B, Munteanu FD, Marty JL, Vasilescu A. Advances in enzyme-based biosensors for pesticide detection. Biosensors 2018;8:27.

[38] Mehta J, Bhardwaj SK, Bhardwaj N, Paul AK, Kumar P, Kim K-H, et al. Progress in the biosensing techniques for trace-level heavy metals. Biotechnol Adv 2016;34:47–60.

[39] Datta S, Christena LR, Rani Y, Rajaram S. Enzyme immobilization: an overview on techniques and support materials. Biotech 2013;3:1–9.

[40] Ahmad R, Sardar M. Enzyme immobilization: an overview on nanoparticles as immobilization matrix. Biochem Anal Biochem 2015;4:2.

[41] Pal S, Sharma MK, Danielsson B, Willander M, Chatterjee R, Bhand S. A miniaturized nanobiosensor for choline analysis. Biosens Bioelectron 2014;54:558–64.

[42] Jesionowski T, Zdarta J, Krajewska B. Enzyme immobilization by adsorption: a review. Adsorption 2014;20:801–21.

[43] Zhou G, Fung KK, Wong LW, Chen Y, Renneberg R, Yang S. Immobilization of glucose oxidase on rod-like and vesicle like mesoporous silica for enhancing current responses of glucose biosensors. Talanta 2011;84:659–65.

[44] Falahti M, Saboury AA, Ma'mani L, Shafiee A, Rafieepour A. The effect of functionalization of mesoporous silica nanoparticles on the interaction and stability of confined enzyme. Int J Biol Macromol 2012;50:1048–54.

[45] Mishra RK, Mishra GK, Dharma Teja V, Danielsson B, Bhand S. Visual colorimetric dual readout bioassay for determination of pesticide residues in drinking water. Chemical Sensors 2013;3:12.

[46] Kumar J, Jha SK, D'Souza SF. Optical microbial biosensor for detection of methyl parathion pesticide using *flavobacterium Sp.* whole cells adsorbed on glass fiber filters as disposable biocomponent. Biosens Bioelectron 2006;21:2100–5.

[47] Rodriguez-Mozaz S, Lopez A, Barceló DMD. Biosensors as useful tools for environmental analysis and monitoring. Anal Bioanal Chem 2006;386:1025–41.

[48] Xia S, Xiangyou W. Acetylcholinesterase biosensor based on prussian-blue-modified electrode for detecting organophosphorus pesticide. Biosens Bioelectron 2010;25:2611–4.

[49] Mishra A, Kumar J, Melo JS. An optical microplate biosensor for the detection of methyl parathion pesticide using a biohybrid of *Sphingomonas* sp. cells-silica nanoparticles. Biosens Bioelectron 2017;87:332–8.

[50] Deshpande K, Mishra RK, pal S, Danielsson B, Williander M, Bhand S. A novel on chip analysis of dissolved Hg(II) in drinking water nanotech, vol. 3; 2010. p. 133–6.

[51] Da Costa Silva LM, Santos VPS, Salgado AM, Pereira KS. Biosensors for contaminants monitoring in food and environment for human and environmental health. In: State of the art in biosensors – environ med Appl; 2013. p. 152–68.

[52] Luong JHT, Groom CA, Male KB. The potential role of biosensors in the food and drink industries. Biosens Bioelectron 1991;6:547–54.

[53] Velasco-garcía MN, Mottram T. Biosensor technology addressing agricultural problems. Biosyst Eng 2003;84:1–12.

[54] Wen Y, Xu J, Liu M, Li D, He H. Amperometric vitamin C biosensor based on the immobilization of ascorbate oxidase into the biocompatible sandwich-type composite film. Appl Biochem Biotechnol 2012;167:2023–38.

[55] D'Souza SF. Immobilization and stabilization of biomaterials for biosensor applications. Appl Biochem Biotechnol 2001;96:225–38.

[56] Lee TMH. Over-the-counter biosensors: past, present, and future. Sensors 2008;8:5535–59.

[57] Mishra RK, Hubble LJ, Martín A, Kumar R, Barfidokht A, Kim J, et al. Wearable flexible and stretchable glove biosensor for on-site detection of organophosphorus chemical threats. ACS Sens 2017;2(4):553–61.

[58] Bhand S, Surugiu I, Dzgoev A, Ramanathan K, Sundaram PV, Danielsson B. Immunoarrays for multianalyte analysis of chlorotriazines. Talanta 2005;65(2):331–6.

[59] Vigneshvar S, Sudhakumari VS, Senthilkumaran B, Prakash H. Recent advances in biosensor technology for potential applications – an overview. Front. Bioeng. Biotechnol 2016;4:11.

Further reading

[1] Andreescu S, Marty JL. Twenty years research in cholinesterase biosensors: from basic research to practical applications. Biomol Eng 2006;23:1–15.

[2] Nigam VK, Shukla P. Enzyme based biosensors for detection of environmental pollutants–a review. J Microbiol Biotechnol 2015;25(11):1773–81.

[3] Ovalle M, Stoytcheva M, Zlatev R, Valdez B. Electrochemical study of rat brain acetylcholinesterase inhibition by chlorofos: kinetic aspects and analytical applications. Electrochim Acta 2009;55:516–20.

[4] Badawy ME, El-Aswad AF. Bioactive paper sensor on the acetylcholinesterase for the rapid detection of organophosphate and carbamate pesticides. Int J Anal Chem 2014;10:1–8.

Chapter 9

Microbial lectins: roles and applications

Hetika Kotecha, Preethi B. Poduval*
Department of Biotechnology, Goa University, Taleigao Plateau, Goa, India
**Corresponding author.*

9.1 Introduction

Lectins are proteins known to bind to sugar moieties on a cell surface and are widely known for their cell adhesion properties. These adhesins are known to bind to glycan receptor surfaces via carbohydrate-recognition domains, which are also known as C-type lectins, while some lectins tend to bind to internal sequences of oligosaccharide chains [1]. Their binding to sugars gives an agglutination reaction, and this phenomenon is seen to have many applications as discussed in this chapter.

Most of the reports on lectins are derived from plants such as *Glycine max*, *Ricinus communis*, and *Moringa oleifera* [2–4]. Animal and microbial lectins have been studied only recently for a wide range of applications in bioremediation, bioflocculation, and probiotics [5]. Microbial lectins are comprised of lectins obtained from fungi, bacteria, protozoa, and viruses. It was observed that only a few lectins were isolated since the 1970s, however, the importance of microbial lectins was realized later due to continuous research in this field, which led to extensive studies on lectins from microorganisms [1,5,6]. Alfred Gottschalk in the early 1950s was the first to identify a microbial lectin. This lectin was seen to be of viral origin and was isolated from the influenza virus [1]. Sharon et al. were the first to study bacterial lectins in the 1970s [6]. The main function of microbial lectins includes attachment to host cells, which is a prerequisite condition to cause infection. Lectins are beneficial to microbes in assisting their adherence to the cell surface. The prevention of this function could lead to the curtailing of several microbial diseases in humans [7,8].

Lectins were primarily known only for their adhesion properties. However, studies have shown that this property was the basis of many future applications and approaches in biomedical sciences [1,8]. They are used as carriers of chemotherapeutic agents as a part of cancer therapy, purification of proteins by chromatography, in bioremediation of heavy metals, and bioflocculation in the

Advances in Biological Science Research. https://doi.org/10.1016/B978-0-12-817497-5.00009-4

treatment of wastes [9]. Lectin-carbohydrate-mediated interactions act as stimuli for neutrophils to reach to the site of infection and act on the injury causing inflammation in humans [10].

Microbial lectins are nonimmunogenic in nature and are classified on the basis of their structure, location, and specificity to carbohydrates (Table 9.1). The ability of lectins to selectively bind to carbohydrates and glycoconjugates, causing agglutination of cells, renders lectins as unique proteins and distinguishes them from other macromolecules [14]. Antibodies are similar in structure, whereas lectins differ in amino acid sequence, three-dimensional structure, and molecular weight. This makes microbial lectin research fairly complex as very little work has been done concerning their classification, function, and organization [19].

Certain biomolecules in which hemagglutination is based entirely due to sugar activity are defined as lectins [20,21] (Fig. 9.1). Goldstein et al. stated that lectins are sugar-binding proteins or glycoproteins of nonimmune origin that agglutinate and/or precipitate glycoconjugates. In their studies, Goldstein and coworkers stated that the specificity of lectins is defined on the basis of monosaccharides or simple oligosaccharides that hamper or restrain lectin-mediated agglutination reactions [22]. Lectins are those proteins exhibiting a nonimmunoglobulin nature, capable of specific detection and reversible binding to carbohydrate moieties of complex carbohydrates, without varying the covalent structure of any of the recognized glycosyl ligands [23].

Lectins contain a carbohydrate-binding site, metal-binding site, or additional hydrophobic site along with a carbohydrate-containing site(s) capable of modulating the activity of carbohydrate-binding sites [24]. Landsteiner discovered human blood group polymorphism and named it as ABO system. Human blood group O contains H-antigen, which is a disaccharide L-fuc-α-1, 2-D-Gal-β1. Blood group A consists of N-acetyl-D-galactosamine linked in 1, 3 manner to H-antigen, whereas blood group B has D-galactose in a similar way [25]. Most lectins agglutinate red blood cells of all human blood groups, at roughly the same dilution with different blood types and are called pan-agglutinins that are broadly specific lectins [24]. As a result, this property of agglutination is exploited for the screening of lectins from other proteins. Thus, lectins have a huge potential in the field of life sciences, and this has led to its increasing importance.

9.2 Roles and mechanism of lectin action

Lectins belong to a class of natural proteins/glycoproteins that specifically and selectively bind to carbohydrate moiety on the cell surface in a noncovalent interaction [26]. This interaction has been known for more than 100 years, however, exact studies on the mechanism of lectin production as well as advantages that lectins provide to microorganisms are still in infancy stage. This nonimmunogenic glycoprotein recognizes and selectively binds to complex

TABLE 9.1 Common lectin families and their functions.

Lectin family	Location (cell)	Characteristic ligands	Functions	References
Calnexin	Endoplasmic reticulum (ER)	Glc_1Man_9 and $GlcNAc_2$	Compromises ER chaperone system, assists in protein folding, and possesses a binding site for ATP and Ca^{2+}	[11]
Chitinase-like lectins (Chi-lectins)	Extracellular	Chitooligosaccharides	Expresses YKL-40 in arthritic cartilage (exact role unclear)	[12]
C-type lectins	Cell membrane and extracellular	DC-SIGN (dendritic cell-specific intercellular adhesion molecule-3-grabbing non-integrin)/CD209 and DC-SIGNR	Immunity against infection and cell adhesions	[13]
F-box lectins	Cytoplasm	$GlcNAc_2$	Involved in degradation of proteins	[14]
Ficolins	Plasma and on mucosal surfaces	GlcNAc	Innate immunity (defense system) and complement activation	[15]
Galectins	Cytoplasm and extracellular	Mac-2, L-14 etc	Cell to extracellular matrix interactions; pro- and antiinflammatory functions	[16]
L-type lectins	ER, Golgi body and ER-Golgi intermediate compartment (ERGIC)	Multiple	Glycoprotein sorting and trafficking	[17]
P-type lectins	In the secretory pathway	Man6-phosphate	Recognizes phosphorylated mannose residues and generation of functional lysosomes within the cells of higher eukaryotes	[18]
M-type lectins	ER	Man8	ER-related glycoprotein degradation	[17]

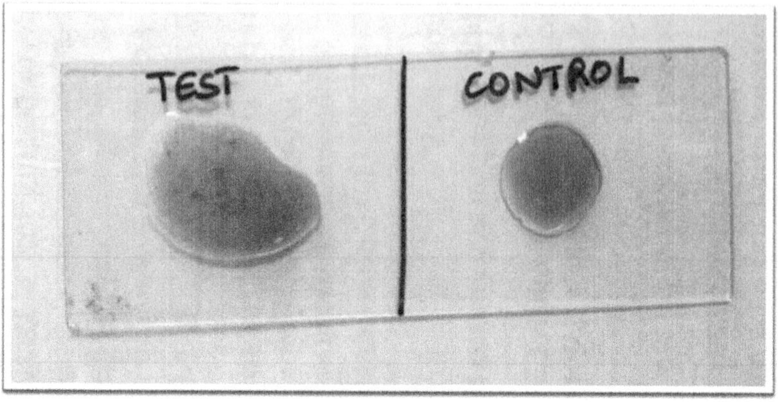

FIGURE 9.1 Lectin detection using hemagglutination. Test – PbS1 bacterial culture grown in minimal growth medium is mixed with human blood. Control – Sterile minimal growth medium mixed with human blood.

cell-bound carbohydrates. Pathogenic bacteria exploit this property to recognize and subsequently adhere to human tissue [27,28].

1. The *Escherichia coli* contain mannose-specific lectins of pili or fimbriae, which are subunits of the filament that are present on the surface of bacteria as appendages [1,29]. Type I mannose-specific lectins and fimbriae of Type P (of 17 kDa) of *E. coli* are specific for Galα4Gal, whereas β galactosides exhibit specificity toward Type II fimbriae of oral actinomycetes [30].

The major role of microbial lectins is that of adhesion, where bacteria adhere to carbohydrates-glycocalyx on the cellular surface when it is in contact for colonization [31]. This bacterial adhesion to the surface of the cells escalates with time to form a systematic structure popularly known as a *biofilm*. Biofilms help bacteria to attach irreversibly and firmly to other bacteria, leading to the production of a carbohydrate mucous layer that maintains the integrity of the biofilm [32]. The lectin-mediated biofilm formation enables bacteria to communicate via adhesion. The coordination of bacteria via biofilm adhesion is termed as quorum sensing [33]. Staining of bacteria having Type I fimbriae with fluorescent intercalating dye and a reference mannose solution for competitive binding to microchip gave a fluorescence signal that was quantified to estimate amounts of surface-bound bacteria. Self-assembled monolayer (SAM) is a more-flexible carbohydrate surface formation on a golden substratum. Bacteria with Type I fimbriae were stained with a fluorescent dye, which adhered to mannosylated SAM. The adhesion inhibition assay was inhibited by adding mannoside solution [34].

2. Lectins isolated from *Acetobacter baumannii* sp. AB119 of 30 kDa were found to be antimicrobial and exhibited antitumor activity, as this lectin

inhibited the biofilm formation for both gram-positive and gram-negative bacteria. The anticancer property of sp. AB119 was tested by MTT [3-(4, 5-dimethylthiazol-2-yl)-2, 5-diphenyltetrazolium bromide] assay that inhibited the proliferation of HeLa tumor cells at IC_{50} (inhibition concentration) of 10 mm for 24 h. The antibiofilm action of this lectin is helpful in curtailing multidrug-resistant pathogenic infection [35].

3. Bacteria secrete lectins to inhibit other microbes from colonization, benefiting them in the race for nutrients and space. A lectin like bacteriocin L1pA consisting of two β-lectin domains, secreted by gram-negative *Proteobacteria*, eliminated other bacteria by contact-dependent killing without cellular uptake [36].
4. A freshwater bloom-forming cyanobacteria, *Microcystic aeruginosa*, producing a lectin known as "*microvirin*," is a mannan-binding lectin that is responsible for cell-cell recognition and attachment. This lectin affects the colony formation, providing *M. aeruginosa* a competitive advantage over other species of phytoplankton [37].
5. Jancarikova et al. characterized a novel lectin from *Photorhabdus asymbiotica*, a species belonging to the category of bioluminescent bacteria. Analysis of *P. asymbiotica* genome revealed novel fucose-binding lectin designated as PHL that could be important for early stage infection of insects and humans. PHL lectin was genetically engineered and overexpressed in *E. coli.* Recombinant protein showed antimicrobial, phenol oxidase activity as well as the production of reactive oxidation species, highlighting the role of lectins in host-pathogen interaction. The PHL was purified by SDS-PAGE and the molecular weight was determined to be 40.18 kDa monomer [38].

The specificity of bacterial lectins to carbohydrates present on different cells of the human body is a causative factor leading to infection. This specificity could be studied and exploited to formulate antiadhesive drugs. However, bacteria and viruses exhibit different types of lectins and bind selectively to various carbohydrates for adhesion. Therefore, it is a highly challenging task for researchers to develop antiadhesion therapy [31].

6. Mitogenic activity of lectins has been reported [39]. The mitogenic activity of microbial lectins is stimulated by binding of T cell receptor complex to the ligand. The cells then undergo rapid mitosis by signaling pathway. The growth and division of T lymphocytes are stimulated by cytokine production (IL-2), which in turn is induced by the interaction of cell receptor with fungal lectins of *Rhizoctonia bataticola* [40]. There are very few reports of microfungal lectins from *Aspergillus nidulans*, *Cephalosporium*, and *R. bataticola.* These lectins have mitogenic potential, which is exploited in glycobiology, oncology, and histochemistry. Along with fungi, even bacterial species belonging to *Pseudomonas* genera contain lectins that bind to a mannose receptor on the cell surface, in turn inducing T cell

proliferation. Mannose-specific lectins aid in the initiation of infection and lectin-mediated phagocytosis [41,42].

A study conducted by Singh et al. showed the mitogenic activity exhibited by a mucin-binding microfungal lectin from *A. nidulans.* The lectins were purified using ion exchange and gel filtration chromatography and concentrated using ultracentrifugation. Concentrated lectins were fractionated on a Sephadex column. The fractions exhibiting hemagglutination were further tested for mitogenic potential on male Swiss albino mice splenocytes using MTT assay. A dose-dependent rise in cell concentration was observed with increase in lectin concentration, proving that fungal lectins have a mitogenic potential that proves to be valuable in studying biochemical changes in immune cells [43].

7. Out of the reported fungal lectins, only 15% belong to microfungi or molds, whereas 3% are reported from *Saccharomyces cerevisiae* or yeasts [44]. Some of them are involved in human pathogenesis. Paracoccin is a GlcNac-binding lectin present in *Paracoccidioides brasiliensis.* It binds to the laminin of extracellular matrix during an infection. This binding stimulates the release of tumor necrotic factor (TNFα) as well as nitric oxide that are known to be important mediators in causing a fungal infection [45].
8. Out of viruses, lectins of influenza virus are studied most elaborately. The lectins of influenza virus bind to carbohydrate moieties containing sialic acid present on the host cell surface. The binding of viral lectins with glycans present on the surface of the cell leads to interaction of the virus by endocytosis, causing pH-dependent fusion of viral envelope with the cell membrane of the host. This fusion ultimately leads to the release of viral genetic material (in this case) RNA into the cytosol. The abundance of receptors enhances adherence of viral particles on glycans of the epithelial membrane and human beings [1]. On the other hand, human immunodeficiency virus (HIV)-blocking lectins from a green alga have also been identified. Lectins that inhibit HIV viral infectivity have been studied for mechanisms involved in the HIV inactivation [46]. However, anti-HIV mechanism of lectins still has to be investigated, and since green algae are not microscopic, this role of lectins is out of the scope of this chapter.
9. Viral lectins present in *Cotesia plutellae*, a wasp, were found to be responsible for immune suppression of its host—the diamond black moth. The viral lectin was designated as CpBV, whose gene was cloned and the amino acid sequence exhibited 80% homology with bracovirus lectin. CpBV contained only one carbohydrate recognition domain and was classified into C-type lectin [47].

10. Lectins secreted by a social amoeba that aggregates to form fruiting *Dictyostelium discoideum* have been reported to modulate microbiota. Wild strains of *D. discoideum* secrete lectins that mediate bacterial carriages are called carrier strains. These bind to bacteria such as *Klebsiella* via secreted surface lectins. The bacteria enter without being digested and provide a source of nutrition. This amoeba-lectin-mediated internalization leads to horizontal gene transfer and provides a useful microbiome homeostasis model [48]. Also, a parasitic amoebozoan, *Entamoeba histolytica*, expresses a lectin-adhesin of 260 kDa that binds to terminal Gal/GalNac residues on glycoproteins and glycolipids. The heterodimeric lectin is useful in parasitic attachment, invasion, and cellular lysis of intestinal epithelium [1].

9.3 Applications of microbial lectins

Lectins from microorganisms have wide applications in many fields such as biomedical, microbial identification, bioremediation, bioflocculation etc. The applications of microbial lectins are described in brief as follows:

9.3.1 Lectins in diagnostics

In diagnostic microbiology, lectins can be used to identify the infectious organism that causes tissue damage. This is because bacterial cells possess certain lipopolysaccharides that are specifically bound to particular carbohydrate glycan, forming the use of lectins in microbial identification [49]. Use of lectins in the field of diagnostics has many advantages over other diagnostic techniques, such as their purification is homogenous and they are stable in common biological buffers. Additionally, lectins can be detected in tissue slices, and the process of freeze-drying does not involve loss of lectin activity. Also, no specific and/or sophisticated instruments are required for use of lectins as a diagnostic tool [50].

9.3.2 Lectins in bioremediation

Lectins have the capacity to cluster cells and then bind them together. This is due to the fact that microbial glycol conjugates react with their specific lectins [51]. This property can be used as an important tool in the process of bioremediation. *S. cerevisiae* was used for the removal of Cu^{+2} from highly contaminated areas. The yeast cells were found to aggregate together due to the presence of lectins present in the cells, and it was observed that it was likely for heavy metals to occupy lectin-binding sites causing increase in biosorption of metals [52].

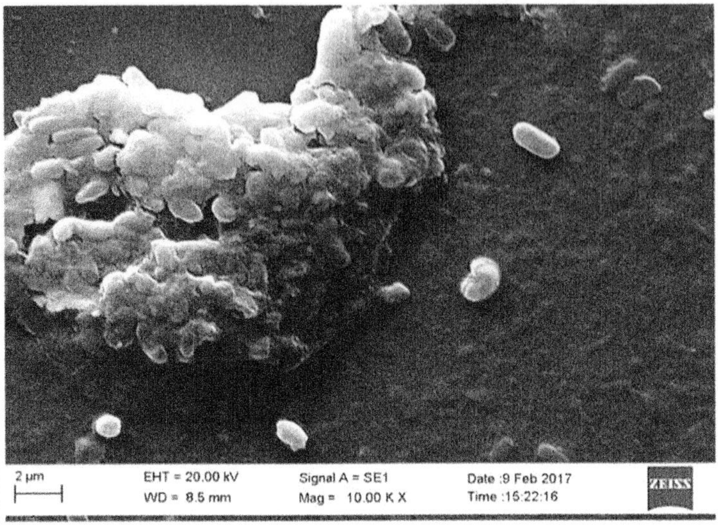

FIGURE 9.2 SEM image of CuB1 lectin-mediated aggregation.

Previously, studies revealed that the polysaccharide-binding activity of lectins is dependent on the presence of metals in their protein structure [53,54]. The decrease in concentration of a particular heavy metal validated the bioremediating ability of few screened bacterial isolates screened from a heavy metal contaminated sediment sample of Dabal, South Goa, India (15°18′40.47″N and 74°9′29.84″E), namely PbS1 and CuB1 (unpublished data). The decrease in the heavy metal concentration can be attributed to the binding of metal ions to membrane-bound lectins on the cell surface [55–57].

In comparison to most other methods employed for biosorption of heavy metals, bacterial lectins work much faster [58]. This property can be exploited for large-scale applications in bioremediation. The presence of production of a large amount of lectins resulted in the attachment of the bacterial cells to each other as lectins bind to surface polysaccharides present on the capsule or cell membrane. CuB1 cells were found closely associated with each other (Fig. 9.2).

9.3.3 Lectins in bioflocculation

Cell adhesion produces a biofilm during the process of bioremediation and bioflocculation [59,60]. Park et al. set up a laboratory-scale system consisting of an activated sludge. This was used to check the efficiency and ease of the process using lectins. It was seen that lectins help in the process of bioflocculation and also tend to increase the rate of the process to some extent [9]. Murthy observed that some proteins take part in the formation of an extracellular matrix that is lectin-like and may be one of the causes leading to bioflocculation [61].

9.3.4 Lectins in fluorescent staining

We know that the technique of fluorescent staining has gained more importance in recent years due to easy handling, higher specificity and also to avoid the long radioactive permission procedure [62]. Lectin-based fluorescent staining recently has been used to detect a particular species from a mixture of different microbes [63].

Sizemore et al. used a similar technique with modifications to selectively stain moderately thermophilic and acidophilic mining bacteria in mixed cultures that also contain *Thiobacillus ferrooxidans*. The mechanism for selective staining was binding of the wheat germ agglutinin to the exposed n-acetyl glucosamine residues of the peptidoglycan layer of the gram-positive bacteria only [64].

9.3.5 Lectin and probiotics

Probiotics have been beneficial to humans to a great extent. Bacteria such as *Lactobacilli* produce lectins, known as probiotic bacterial lectins (PBLs). The PBLs are responsible for inter- and intrapopulation relationship in the gut of humans between the bacteria and its host [65]. PBLs were tested against clinical microbial pathogens such as *Candida* and *Staphylococcus*, and results demonstrated that PBLs showed growth inhibition of various strains along with its proteolysis [66,67].

9.4 Conclusion

Lectins are ubiquitous and are produced by animals, plants, as well as microorganisms. Recent advancements in research carried out on lectins from microbes such as bacteria, viruses, fungi, and protozoa suggest the importance of lectins, and with advancements in molecular biology techniques such as genome sequencing, more microbial lectins are expected to be analyzed. As microbial lectins are involved in host-pathogen interactions that may lead to severe infections in humans, the use of specific lectins as a drug target may lead to promising research in diagnostic microbiology. The structure, specificity, and affinity of any microbial lectin toward a particular glycan on the host cell are characteristics of lectin-cellular interaction. Reports on lectins isolated from microbes and the strategies to study lectin-carbohydrate interactions suggest that lectins play important roles that are covered in this chapter. Microbial lectins play major roles such as communication in bacteria, an antimicrobial activity that provides microbes with a competitive advantage for nutrition and space, cellular recognition, as well as infection of host cells, making lectins the focus of interest for biologists, particularly glycobiologists. Applications of microbial lectins include their use in diagnostics, bioremediation, bioflocculation, fluorescent staining, and probiotic studies. Although applications have been realized for microbial lectins, available information on the underlying mechanisms is limited, providing scope for tremendous research in this area.

References

[1] Nizet V, Varki A, Aebi M. Microbial lectins: hemagglutinins, adhesins, and toxins. Essentials of glycobiology. 3rd ed., vol 37. NY, USA: Cold Spring Harbor Laboratory Press; 2017.

[2] Khatun S, Khan MM, Ashraduzzaman M, Pervin F, Bari L, Absar N. Antibacterial activity and cytotoxicity of three lectins purified from drumstick (*Moringa oleifera* Lam.) leaves. J Bio Sci 2009;17:89–94.

[3] Lis H, Sharon NA. The biochemistry of plant lectins (phytohemagglutinins). Annu Rev Biochem 1973;42(1):541–74.

[4] Nicolson GL, Blaustein J, Etzler ME. Characterization of two plant lectins from *Ricinus communis* and their quantitative interaction with a murine lymphoma. Biochemistry 1974;13(1):196–204.

[5] Slifkin M, Doyle RJ. Lectins and their application to clinical microbiology. Clin Microbiol Rev 1990;3(3):197–218.

[6] Esko JD, Sharon N. Microbial lectins: hemagglutinins, adhesins, and toxins. Cold Spring Harbor Laboratory Press; 2009.

[7] Ofek I, Doyle RJ. Principles of bacterial adhesion. Bacterial adhesion to cells and tissues. Boston: Springer; 1994. p. 1–15.

[8] Sharon N, Ofek I. Safe as mother's milk: carbohydrates as future anti-adhesion drugs for bacterial diseases. Glycoconj J 2000;17(7–9):659–64.

[9] Park C, Novak JT. Characterization of lectins and bacterial adhesins in activated sludge flocs. Water Environ Res 2009;81(8):755–64.

[10] Bevilacqua MP, Pober JS, Majeau GR, Cotran RS, Gimbrone MA. Interleukin 1 (IL-1) induces biosynthesis and cell surface expression of procoagulant activity in human vascular endothelial cells. J Exp Med 1984;160(2):618–23.

[11] Williams DB. Beyond lectins: the calnexin/calreticulin chaperone system of the endoplasmic reticulum. J Cell Sci 2006;119(4):615–23.

[12] Bigg HF, Wait R, Rowan AD, Cawston TE. The mammalian chitinase-like lectin, YKL-40, binds specifically to type I collagen and modulates the rate of type I collagen fibril formation. J Biol Chem 2006;281:21082–95.

[13] Mitchell DA, Fadden AJ, Drickamer K. A novel mechanism of carbohydrate recognition by the C-type lectins DC-SIGN and DC-SIGNR subunit organization and binding to multivalent ligands. J Biol Chem 2001;276(31):28939–45.

[14] Lakhtin V, Lakhtin M, Alyoshkin V. Lectins of living organisms. The overview. Anaerobe 2011;17(6):452–5.

[15] Holmskov U, Thiel S, Jensenius JC. Collections and ficolins: humoral lectins of the innate immune defense. Annu Rev Immunol 2003;21(1):547–78.

[16] Ochieng J, Furtak V, Lukyanov P. Extracellular functions of galectin-3. Glycoconj J 2002;19(7–9):527–35.

[17] Drickamer K, Taylor ME. Biology of animal lectins. Annu Rev Cell Biol 1993;9(1):237–64.

[18] Dahms NM, Hancock MK. P-type lectins. Biochim Biophys Acta 2002;1572(2–3):317–40.

[19] Santos AF, da Silva MD, Napoleão TH, Paiva PM, Correia MT, Coelho LC. Lectins: function, structure, biological properties and potential applications. Curr Top Pept Protein Res 2014;15:41–62.
[20] Peumans WJ, Van Damme EJ. Lectins as plant defense proteins. Plant physiol 1995;109(2):347.
[21] Rouf R, Tiralongo E, Krahl A, Maes K, Spaan L, Wolf S, May TW, Tiralongo J. Comparative study of hemagglutination and lectin activity in Australian medicinal mushrooms (higher basidiomycetes). Int J Med Mushrooms 2011;13(6).
[22] Goldstein IJ. What should be called a lectin? Nature 1980;285:66.
[23] Kocourek J, Horejsi V. A note on the recent discussion on definition of the term 'lectin'. Lectins: biology, biochemistry. Clin Biochem 1983;3:3–6.
[24] Singh RS, Bhari R. Current status of microbial lectins in biomedical research. Adv Ind Biotechnol 2013:315–61.
[25] Landsteiner K. Note the antifermantative, lytic and agglutinating activity of blood serum and lymph. Landsteiner. Centralblatt of Bacteriol. 1900;27:357–62.
[26] Michael T, Smith CM. Lectins probe molecular films in biofouling: characterization of early films on non-living and living surfaces. Mar Ecol Prog Ser 1995;119(1):229–36.
[27] Imberty A, Varrot A. Microbial recognition of human cell surface glycoconjugates. Curr Opin Struct Biol 2008;18(5):567–76.
[28] Pieters RJ. Carbohydrate mediated bacterial adhesion. Bacterial Adhesion 2011:227–40.
[29] Duguid JP, Gillies RR. Non-flagellar filamentous appendages (fimbriae) and haemagglutinating activity in dysentery bacilli. J Gen Microbiol 1956;15:6.
[30] Goldstein IJ, Hayes CE. The lectins: carbohydrate-binding proteins of plants and animals. Adv Carbohydr Chem Biochem 1978;35:127–340.
[31] Hartmann M, Lindhorst TK. The bacterial lectin FimH, a target for drug discovery–carbohydrate inhibitors of type 1 fimbriae mediated bacterial adhesion. Eur J Org Chem 2011;20(21):3583–609.
[32] Blaser MJ. An endangered species in the stomach. Sci Am 2005;292(2):38–45.
[33] Mack D, Davies AP, Harris LG, Knobloch JK, Rohde H. Staphylococcus epidermidis biofilms: functional molecules, relation to virulence, and vaccine potential. Glycoscience and Microbial Adhesion 2008:157–82.
[34] Qian X, Metallo SJ, Choi IS, Wu H, Liang MN, Whitesides GM. Arrays of self-assembled monolayers for studying inhibition of bacterial adhesion. Anal Chem 2002;74(8):1805–10.
[35] Alyousef AA, Alqasim A, Aloahd MS. Isolation and characterization of lectin with antibacterial, antibiofilm and antiproliferative activities from Acinetobacter baumannii of environmental origin. J Appl Microbiol 2018;124(5):1139–46.
[36] Ghequire MG, Swings T, Michiels J, Buchanan SK, De Mot R. Hitting with a BAM: selective killing by lectin-like bacteriocins. mBio 2018 2;9(2). e02138-17.
[37] Kehr JC, Zilliges Y, Springer A, Disney MD, Ratner DD, Bouchier C, Seeberger PH, De Marsac NT, Dittmann E. A mannan binding lectin is involved in cell–cell attachment in a toxic strain of *Microcystis aeruginosa*. Mol Microbiol 2006;59(3):893–906.
[38] Jančaříková G, Houser J, Dobeš P, Demo G, Hyršl P, Wimmerová M. Characterization of novel bangle lectin from Photorhabdus asymbiotica with dual sugar-binding specificity and its effect on host immunity. PLoS Pathog 2017;13(8). e1006564.
[39] Kilpatrick DC. Mechanisms and assessment of lectin-mediated mitogenesis. Mol Biotechnol 1999;11(1):55.

[40] Pujari R, Nagre NN, Chachadi VB, Inamdar SR, Swamy BM, Shastry P. Rhizoctonia bataticola lectin (RBL) induces mitogenesis and cytokine production in human PBMC via p38 MAPK and STAT-5 signaling pathways. Biochim Biophys Acta 2010;1800(12):1268–75.

[41] Abraham SN, Sun D, Dale JB, Beachey EH. Conservation of the D-mannose-adhesion protein among type 1 fimbriated members of the family Enterobacteriaceae. Nature 1988;336(6200):682.

[42] Sharon N. Carbohydrates as recognition determinants in phagocytosis and in lectin mediated killing of target cells. Biol Cell 1984;51(2):239–45.

[43] Singh RS, Bhari R, Singh J, Tiwary AK. Purification and characterization of a mucin-binding mycelial lectin from Aspergillus nidulans with potent mitogenic activity. World J Microbiol Biotechnol 2011;27(3):547–54.

[44] Singh RS, Bhari R, Kaur HP. Characteristics of yeast lectins and their role in cell–cell interactions. Biotechnol Adv 2011;29(6):726–31.

[45] Coltri KC, Casabona-Fortunato AS, Gennari-Cardoso ML, Pinzan CF, Ruas LP, Mariano VS, Martinez R, Rosa JC, Panunto-Castelo A, Roque-Barreira MC. Paracoccin, a GlcNAc-binding lectin from *Paracoccidioides brasiliensis*, binds to laminin and induces TNF-α production by macrophages. Microb Infect 2006;8(3):704–13.

[46] Sato Y, Hirayama M, Morimoto K, Yamamoto N, Okuyama S, Hori K. High mannose-binding lectin with preference for the cluster of α1-2 mannose from the green alga Boodlea coacta is a potent entry inhibitor of HIV-1 and influenza viruses. J Biol Chem 2011:M110.

[47] Lee S, Nalini M, Kim Y. A viral lectin encoded in Cotesia plutellae bracovirus and its immunosuppressive effect on host hemocytes. Comp Biochem Physiol Mol Integr Physiol 2008;149(4):351–61.

[48] Dinh C, Farinholt T, Hirose S, Zhuchenko O, Kuspa A. Lectins modulate the microbiota of social amoebae. Science 2018 27;361(6400):402–6.

[49] McSweegan EF, Pistole TG. Interaction of the lectin limulin with capsular polysaccharides from *Neisseria meningitidis* and *Escherichia coli*. Biochem Biophys Res Commun 1982;106(4):1390–7.

[50] Doyle R, Keller K. Lectins in diagnostic microbiology. Eur J Clin Microbiol 1984;3(1):4–9.

[51] Tanti B, Buragohain AK. Differentiation of petroleum hydrocarbon-degrading *Pseudomonas* spp. based on lectin binding of cell extracts in an agglutination assay. Indian J Biotechnol 2010;9(1):74–9.

[52] Soares EV, De Coninck G, Duarte F, Soares HM. Use of *Saccharomyces cerevisiae* for $Cu2^{+}$ removal from solution: the advantages of using a flocculent strain. Biotechnol Lett 2002;24(8):663–6.

[53] Abhilash J, Dileep KV, Palanimuthu M, Geethanandan K, Sadasivan C, Haridas M. Metal ions in sugar binding, sugar specificity and structural stability of *Spatholobus parviflorus* seed lectin. J Mol Model 2013;19(8):3271–8.

[54] Garcia-Pino A, Buts L, Wyns L, Loris R. Interplay between metal binding and cis/trans isomerization in legume lectins: structural and thermodynamic study of P. angolensis lectin. J Mol Biol 2006;361(1):153–67.

[55] Malik A. Metal bioremediation through growing cells. Environ Int 2004;30(2):261–78.

[56] Tabak HH, Lens P, van Hullebusch ED, Dejonghe W. Developments in bioremediation of soils and sediments polluted with metals and radionuclides–1. Microbial processes and mechanisms affecting bioremediation of metal contamination and influencing metal toxicity and transport. Rev Environ Sci Biotechnol 2005;4(3):115–56.

[57] Walton BT, Guthrie EA, Hoylman AM. Toxicant degradation in the rhizosphere 1994;563(2):11–26.
[58] Guo H, Luo S, Chen L, Xiao X, Xi Q, Wei W, Zeng G, Liu C, Wan Y, Chen J, He Y. Bioremediation of heavy metals by growing hyperaccumulaor endophytic bacterium *Bacillus* sp. L14. Bioresour Technol 2010;101(22):8599–605.
[59] Singh R, Paul D, Jain RK. Biofilms: implications in bioremediation. Trends Microbiol 2006;14(9):389–97.
[60] Verstrepen KJ, Klis FM. Flocculation, adhesion and biofilm formation in yeasts. Mol Microbiol 2006;60(1):5–15.
[61] Murthy SN. Bioflocculation: implications for activated sludge properties and wastewater treatment (Doctoral dissertation). Virginia Tech; 1998.
[62] Ballal R, Cheema A, Ahmad W, Rosen EM, Tapas S. Fluorescent oligonucleotides can serve as suitable alternatives to radiolabeled oligonucleotides. J Biomol Tech 2009;20(4):190.
[63] Fife DJ, Bruhn DF, Miller KS, Stoner DL. Evaluation of a fluorescent lectin-based staining technique for some acidophilic mining bacteria. Appl Environ Microbiol 2000;66(5):2208–10.
[64] Sizemore RK, Caldwell JJ, Kendrick AS. Alternate gram staining technique using a fluorescent lectin. Appl Environ Microbiol 1990;56(7):2245–7.
[65] Lakhtin M, Lakhtin Bajrakova A, Aleshkin A, Afanasiev S, Aleshkin V. Lectin systems imitating probiotics: potential and prospects for biotechnology and medical microbiology. Probiotics 2012;32:417.
[66] Lakhtin M, Alyoshkin V, Lakhtin V, Afanasyev S, Pozhalostina L, Pospelova V. Probiotic lactobacillus and bifidobacterial lectins against Candida albicans and *Staphylococcus aureus* clinical strains: new class of the pathogen biofilm destructors. Probiotics Antimicrob Proteins 2010;2(3):186–96.
[67] Lakhtin MV, Aleshkin VA, Lakhtin VM, Nesvizhskiĭ I, Afanas' ev SS, Pospelova VV. The role of lectins from probiotic microorganisms in sustaining the macroorganism. Vestn Ross Akad Med Nauk 2010;(2):3–8.

Further reading

[1] Disney MD, Seeberger PH. Carbohydrate arrays as tools for the glycomics revolution. Drug Discov Today Targets 2004 1;3(4):151–8.

Chapter 10

Biodegradation of seafood waste by seaweed-associated bacteria and application of seafood waste for ethanol production

Sanika Samant[1], Milind Mohan Naik[2,*], Diviya Chandrakant Vainganka[2], Sajiya Yusuf Mujawar[4], Prachi Parab[2], Surya Nandan Meena[3]

[1]*Department of Biotechnology, Goa University, Goa, India;* [2]*Department of Microbiology, Goa University, Goa, India;* [3]*Biological Oceanography Division, National Institute of Oceanography, Dona Paula, Goa, India;* [4]*Laboratory of Bacterial Genetics and Environmental Biotechnology, Department of Microbiology, Goa University, Goa, India*

[*]*Corresponding author: milind@unigoa.ac.in; milindnaik4@gmail.com.*

10.1 Introduction

In recent years, seafood waste has increased tremendously, since during the processing of prawns, shrimps, and other shellfish mostly the meat is utilized while the shells, bones, and head portions are thrown as wastes into marine waters [1]. Fish production around the globe has increased tremendously and reached 174 million metric tons in 2017. India is the third largest producer of fishery around the globe and hence produces an enormous amount of fish waste [2]. India generates >2 metric million tons of waste during fish processing, of which 300,000 tons contribute to visceral waste alone [3]. Commercial processing of fish generates a significant amount of waste, which includes viscera, fins, scales, and bones [4]. This huge amount of discards (fishery waste) including solid wastes along with wastewater resulting from fishery processing are unutilized and usually disposed of in landfills, or dumped near shore and into the ocean without any pretreatment, causing environmental pollution, thus severely impacting aquatic biota health ailments [5,6]. Although these wastes are biodegradable, the process is very slow. This results in accumulation of fishery waste over time and pollutes coastal and marine environments due to bad odors and secretion of biogenic amines, thereby affecting marine life [7]. Due to

Advances in Biological Science Research. https://doi.org/10.1016/B978-0-12-817497-5.00010-0

foul odors, seafood waste in the marine environment attracts flies, insects, rodents, and other vermin, creating an unhygienic atmosphere. Fish waste is classified as certified waste because it is comprised of high organic content and is thus even more costly to dispose of [4]. In fish processing industries, acid, alkali, and heat treatments are used to degrade shell waste that is hazardous to the environment [8]. Hence, biodegradation of seafood waste using microorganisms is important as they can be used for polluted environment reclamation without harming natural biota. Bioremediation is ecofriendly and cost-effective as compared to physicochemical methods. Therefore, treatment of prawn shell, shellfish, and fish waste using marine bacteria is crucial for preservation and sustainable development of the marine environment.

The fish processing procedure involves removal of fish bones, scales, heads, and internal organs. Therefore, during the processing of seafood, a large amount of shell and scale waste is discarded from fish markets, seafood restaurants, fish-processing industries, and kitchens. Fish scales consist of protein, calcium phosphate, calcium carbonate, magnesium carbonate, chitin, and pigments [9]. Generally, crustacean shells consist mainly of 30%–50% calcium carbonate, 30%–40% protein, and 20%–30% chitin and calcium phosphate [4,10,11]. Shells also contain carotenoid pigments and a trace amount of lipids. The content of shell components varies with different species and seasons [12]. Therefore, to degrade seafood waste, bacteria possessing protease, chitinase activities, and phosphate and calcium carbonate solubilization properties will be of great importance. There are very few reports on the total degradation of seafood waste by bacteria. Microorganisms possessing proteolytic activity have been applied for the deproteinization of chemically demineralized shells [13]. Purification of chitin from shrimp wastes using microbial deproteination and decalcification activities has been demonstrated [7]. *Lactobacillus plantarum* and *Pseudomonas aeruginosa* were used for deproteinization and demineralization of crab shell and shrimp waste [1]. Two bacterial cultures, *Exiguobacterium acetylicum* and *Bacillus cereus*, were studied for their ability to decompose shrimp shell waste [14]. A *P. aeruginosa* strain, K-187, isolated from soil (in Taiwan) showing protease and chitinase activities when cultured in medium containing shrimp and crab shell wastes as sole carbon sources has been reported [15]. *Serratia marcescens* FS-3 strain exhibiting strong protease activity was isolated from soil toward the seaside of a southwestern region of Korea and was used for degradation of crab (*Chionoecetes opilio*) shell wastes [16].

Most of the above-mentioned studies demonstrate the use of chemicals in the seafood waste treatment process along with the use of microbial enzymes or microorganisms. Also, detail study on complete degradation of crab shell, prawn shell, and fish scale using bacteria has not yet been undertaken. Therefore, the current study focuses on isolation of seaweed-associated bacteria possessing the ability to degrade seafood waste such as scales, crab shell, and prawn shell waste by producing organic acids and hydrolytic enzymes and their application in bioremediation of seafood waste. Here we also discuss studies on sustainable use of seafood waste for ethanol production.

10.2 Materials and methods

10.2.1 Collection of marine seaweed samples

Live and healthy seaweed samples (*Ulva* sp.) were collected from rocky intertidal regions of Anjuna Beach, Goa, India using sterile forceps in sterile petri plates. The samples were transported to the laboratory immediately under cool conditions for further analysis. *Ulva* sp. samples were processed for isolation of associated bacteria within 24 h.

10.2.2 Enrichment of *Ulva*-associated bacteria

For enrichment of *Ulva* sp.—associated bacteria, seaweed sample was rinsed gently two or three times with sterilized seawater to wash off sand particles and loosely bound bacteria. In order to isolate firmly associated bacteria, *Ulva* sp. was aseptically cut into 5-cm-long pieces and two pieces were inoculated into 50 mL sterile Zobell marine broth (ZMB) in 150-mL Erlenmeyer flask and incubated on shaker for 48 h at room temperature (RT, 28°C ± 2) with constant shaking at 150 rpm to enrich seaweed-associated bacteria.

10.2.3 Isolation of calcium carbonate solubilizing marine *Ulva*-associated bacteria

From enriched ZMB, seaweed pieces were removed aseptically and the broth was serially diluted up to 10^{-8} using sterile saline (2%) and spread plated (0.1 mL) on seawater-based agar containing 1% $CaCO_3$ and 0.4% glucose (pH 7). The plates were then incubated at room RT (28°C ± 2) for 48—72 h. Morphologically different calcium carbonate—solubilizing bacterial colonies showing a highest zone of clearance on agar were selected and purified for further study. These calcium carbonate—solubilizing bacterial cultures were maintained by regular subculturing on Zobell marine agar (HiMedia Laboratories) and stored at 4°C.

10.2.4 Investigating seafood waste (fish, crab, prawn waste) utilizing potential of selected calcium carbonate—solubilizing bacteria

10.2.4.1 Preparation of crab/prawn shell and fish scale powder

Crab shells, prawn shells, and fish scales were obtained from local markets. They were washed with distilled water, sun-dried for 1 week, and ground into a fine powder using an electronic mixer. This powder was then used as fish/crab/prawn waste for degradation studies.

10.2.4.2 Microbial utilization of seafood waste as a sole source of carbon

Seawater-based media was prepared in three different 50 mL conical flasks and 1% sterile crab shell, prawn shell, fish scale powder was added separately into three flasks as a sole carbon and nitrogen source (1.5% agar was added). Media was sterilized and poured into plates. Bacterial isolates were streaked on plates and incubated at RT (28°C ± 2) for 10 days. Plates were observed for bacterial growth.

Calcium-solubilizing bacterial isolates, which also showed the ability to utilize seafood waste as a sole carbon source, were further tested for their potential to produce protease, cellulase, chitinase, agarase, and phosphate-solubilizing activities.

10.2.5 Agarase production by marine *Ulva* sp.—associated bacteria

Bacterial isolates were plated on seawater-based agar medium (2% agar without any other added carbon source) and incubated at RT (28°C ± 2) for 24—48 h. Colonies were observed for depression in agar plates. Agarase production was also tested by flooding plates with Lugol's iodine and observing zone of clearance around colony [17]. Agarase activity was tested to rule out agar utilization by bacteria. The absence of agarase activity confirms the ability of organisms to utilize seafood waste as a sole carbon source.

10.2.6 Production of protease by *Ulva* sp.—associated bacteria

Ulva sp.—associated bacteria, which were found to be utilizing seafood waste as a sole source of carbon, were streaked on skim milk agar (HiMedia Laboratories) plates and incubated for 4 days at RT (28°C ± 2). Positive protease activity was indicated by a clear zone surrounding the bacterial streak/growth.

10.2.7 Phosphate solubilization by acid-producing *Ulva* sp.—associated bacteria

Bacterial isolates were streaked on seawater-based Pikovskaya's agar plates containing 0.05% bromothymol blue (Merck) and incubated at RT (28°C ± 2) for 72 h. Zone of clearance and yellow coloration around bacterial streak indicates phosphate solubilization due to acid production [18].

10.2.8 Cellulase production by *Ulva* sp.—associated bacteria

Bacterial isolates were streaked on seawater-based carboxymethyl cellulose agar (HiMedia Laboratories) and incubated at RT (28°C ± 2). After 4 days of incubation, plates were flooded with 1% Congo red solution (w/v) and allowed to stand for 10 min. Excess stain was poured out gently and plates were flooded with 1M NaCl solution (destaining). After incubating for 10 min, excess NaCl solution was poured out. This step was repeated two times to wash off excess Congo red stain. Cellulase activity was indicated by a zone of clearance along the bacterial streak [19], whereas the rest of the plate stained dark red.

10.2.9 Production of chitinase by *Ulva* sp.—associated bacteria

Chitinase production by *Ulva* sp.—associated bacteria was checked by streaking on seawater-based agar plates containing 1% colloidal chitin as a sole carbon source and incubated at RT (28°C ± 2) for 4 days. Plates were flooded with 1% Congo red solution (w/v) and were allowed to stand for 10 min. Excess Congo red was discarded gently and plates were flooded with 1M NaCl solution (destaining). Excess NaCl solution was poured out gently after incubating for 15 min. This step was repeated twice to wash off excess Congo red stain. Chitinase activity was indicated by a zone of clearance along streak [20], whereas the rest of the plate stained dark red.

Bacterial isolates exhibiting calcium carbonate solubilizing, protease, chitinase, and cellulase activities were selected for the treatment of fish and shellfish waste by preparing consortia.

10.2.10 Degradation of fish/crab/prawn waste using microbial consortia developed using *Ulva* sp.—associated bacteria

Development of microbial consortia for seafood waste degradation is of utmost importance since improved degradation is achieved using microbial consortia as compared to individual isolates. Before using selected bacterial isolates as consortia for seafood degradation study, these isolates were tested by the cross-inhibition test. Test isolates were inoculated as a line in the center on the surface of Zobell marine agar and incubated at RT (28°C ± 2) for 20 h. Other isolates were inoculated as a perpendicular line to the test isolates, and plates were incubated at RT (28°C ± 2) for 48 h. Positive results were indicated by a zone of inhibition of growth of other isolates. This test was repeated for all the isolates. If bacterial isolates inhibit each other, then they cannot be used in consortia for degradation of seafood waste.

Three selected bacterial isolates were inoculated separately into 50 mL seawater-based broth containing 1% crab shell powder as a sole source of carbon for 4 days at RT (28°C ± 2) with shaking at 150 rpm. Then, 0.1 mL culture broth from each flask was added as inoculum into 100 mL seawater-based media in Erlenmeyer flask (250 mL) containing 2% crab shell powder and incubated at RT (28°C ± 2) with constant shaking at 150 rpm for 4 days. The test was performed in triplicate. Control was maintained containing seawater-based broth supplemented with 2% crab shell powder without inoculum. After 4 days of incubation, 2 mL culture broth was collected under the aseptic condition and centrifuged at 8000 rpm for 10 min. The supernatant was taken in another tube and used for determining reducing sugars by DNSA method released during degradation of seafood waste [21]. Similar degradation study by consortia was repeated with prawn shell and fish scales in triplicate.

10.2.11 Identification of seaweed-associated bacteria

Three selected seaweed-associated bacteria having the potential of degrading seafood waste were identified by performing biochemical tests and referring to *Bergey's Manual of Systematic Bacteriology* [22] and also by using 16S rDNA gene sequence. Gene coding for 16S rRNA was amplified with universal eubacterial primers: 27F (5′ AGAGTTTGATCMTGGCTCAG 3′) and 1492R (5′ TACGGYTACCTTGTTACGACTT 3′). 16S rRNA sequence data were compared with GenBank database using BLAST.

10.3 Results and discussion

Nine morphologically different calcium carbonate–solubilizing seaweed-associated bacterial isolates showing a highest zone of clearance on agar were selected and purified for further studies. These nine bacterial isolates were designated as PM1, PM2, PM3, PM4, PM5, PM6, PM7, PM8, and PM9 (Fig. 10.1). Only three out of nine bacterial isolates—PM1, PM6, and PM9—were able to utilize crab shells, prawn shells, and fish scale powder as a sole carbon source in seawater-based agar after 10 days of incubation. Also, all three bacterial isolates didn't show growth on seawater-based agar media (agar

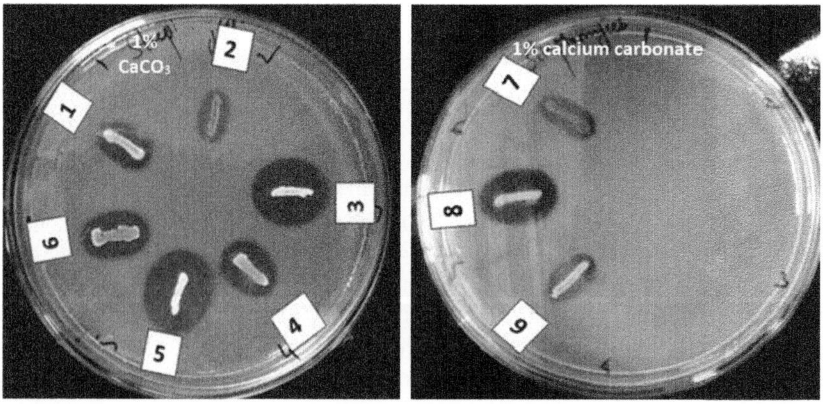

FIGURE 10.1 Calcium carbonate–solubilizing seaweed-associated bacterial isolates showing zone of clearance around colonies when streaked on seawater-based agar comprising 1% $CaCO_3$ and 0.4% glucose.

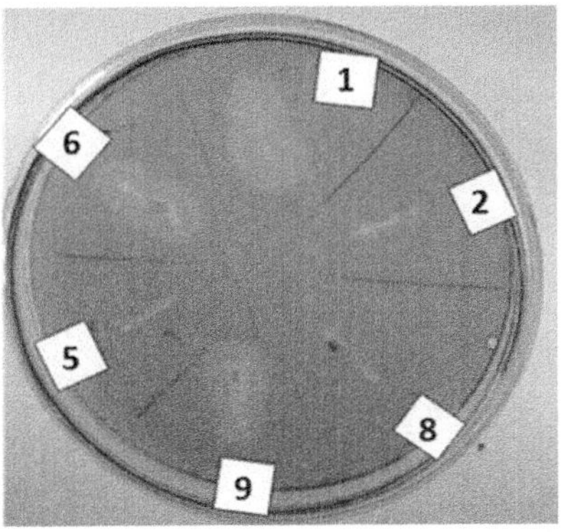

FIGURE 10.2 Seaweed-associated bacteria showing cellulase activity on seawater-based agar plates comprising 1% carboxymethyl cellulose (CMC) as a sole source of carbon.

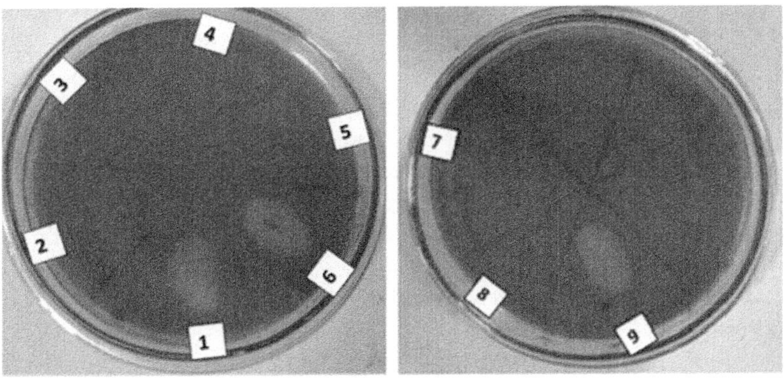

FIGURE 10.3 Seaweed-associated bacteria showing chitinase activity on seawater-based agar plates containing 1% colloidal chitin as a sole carbon source.

as the sole source of carbon), therefore, bacterial isolates PM1, PM6, and PM9 were selected for further studies since the absence of agarase activity confirms that organisms only utilize seafood waste as a sole source of carbon.

All three isolates PM1, PM6, and PM9 were found to be positive for protease, cellulase (Fig. 10.2), and chitinase (Fig. 10.3) activity and also could solubilize phosphate (Fig. 10.4) in seawater-based agar. Crustacean shells consist mainly of 30%–40% protein, 30%–50% calcium carbonate, and 20%–30% chitin and calcium phosphate [9–11]. They also contain carotenoid pigments and a trace amount of lipid resides. Seafood wastes are rich in organic contents such as protein, bioactive peptides, collagen, gelatin, calcium carbonate, and lipid, making the disposal process more complicated and expensive [4]. The content of shell components varies with different species

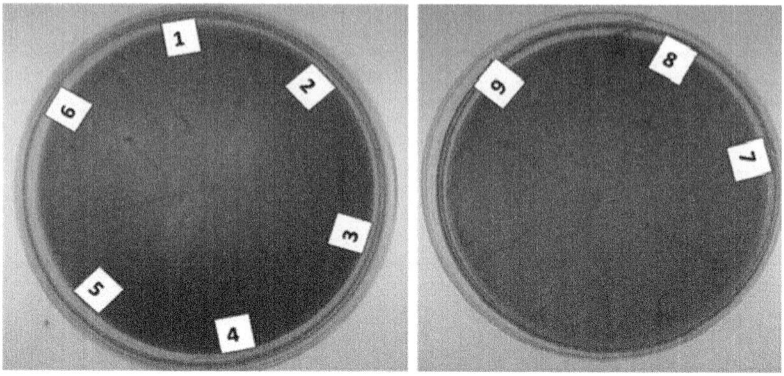

FIGURE 10.4 Seaweed-associated bacteria showing phosphate-solubilizing activity on seawater-based Pikovskaya's agar plates containing 0.05% bromothymol blue.

and seasons [12]. Therefore, to degrade seafood waste bacteria (PM1, PM6, and PM9), which have protease, chitinase activities, phosphate solubilizing and calcium carbonate solubilization properties are of great importance.

Seaweed-associated selected bacterial isolates didn't show any cross-inhibition activity with each other therefore were selected to develop microbial consortia to degrade seafood waste. Microbial consortia (PM1, PM6, and PM9) was developed to enhance degradation of seafood waste, which was evident from the amount of reducing sugars released from crab shell/prawn and fish scales (Fig. 10.5) and was found to be 310 ± 8 μg/mL, 245 ± 14 μg/mL, and 180 ± 15 μg/mL, respectively, after 4 days of incubation. These results confirmed that bacterial consortia have very high seafood waste degradation activity and can be used for bioremediation of seafood waste before discharging into marine waters and also marine sites already polluted with seafood waste. Based on morphology and biochemical tests, bacterial

FIGURE 10.5 Seaweed-associated bacteria, when used as microbial consortia (PM1, PM6, and PM9), showed degradation of seafood waste (crab shell, prawn shell, and fish scales).

isolates PM1, PM6, and PM9 were identified as *Bacillus* sp. *Brevibacterium* sp. and *Vibrio* sp., respectively, and through 16S rRNA sequencing, bacterial isolate PM6 was further confirmed as *Brevibacterium iodinum* (accession number MG971400).

Fish production generates a huge amount of solid waste in the form of whole fish waste, fish heads, tails, skin, viscera, bones, blood liver, guts, and some muscle tissue along with wastewater composed of liquid waste produced during fish processing [6]. Improper disposal of seafood wastes generated by fishery-processing industries signifies an increasing environmental and health problem [8]. The intertidal region is exposed to fish, prawn, and crab waste as they get washed toward the shoreline when dumped in the sea during the wave currents. Biodegradation of waste is perhaps the most lucrative and environmentally friendly procedure for waste utilization since chemical treatment method can, in turn, add harmful chemicals (HCl, HNO_3, H_2SO_4, CH_3COOH, and HCOOH) to the environment [9,23]. Fishery waste being collected is in mixed type form near/along the shoreline inhabited by seaweeds, and bacteria associated with seaweeds are adapted to possess enzymes protease, chitinase, phosphate solubilization, and calcium solubilization activity. These properties make them efficient in biodegradation and treatment of fisheries waste containing a mixture of proteins, cellulose, chitin, and minerals, etc. In this study, we have reported for the first time isolation of seaweed-associated bacteria (*Bacillus* sp., *Brevibacterium* sp., *Vibrio* sp.) from the intertidal region of Goa, India, to degrade fish waste efficiently without using any chemical degradation step. Also, first-time detailed studies regarding enzymes (protease, chitinase, and cellulase) and organic acids (demineralization) produced by seafood waste degrading marine bacteria are now underway. Use of marine bacterial consortia for degradation of seafood waste is an ecofriendly and cost-effective method as compared to chemical method. In the near future, genes encoding protease, cellulase, and chitinase from these seaweed-associated bacteria will be used to genetically modify *Escherichia coli* for enhanced degradation of crab shell/prawn shell and fish scale waste as an advance in seafood waste management.

10.4 Application of seafood waste for bioethanol production

The main aim of waste management is to develop advanced biotechnology to biodegrade waste and for sustainable production of biofuel without harming the environment [24]. Application of seafood waste for bioethanol (biofuel) production is a very innovative and ecofriendly concept. The present biotechnology-based concept uses marine bacteria to break down crab shell/fish scales/prawn shell by utilizing them as the sole source of carbon and nutrients and break them down into monomer sugars. The sugars thus produced during degradation of seafood waste can be used to produce bioethanol in the future, using *Saccharomyces cerevisiae* in a profitable, sustainable, and environmentally friendly manner. Biofuels (ethanol) do not contribute much toward environmental pollution and thus are beneficial over current fuels.

The rapid decline in the world's oil reserves is the main reason behind increasing interest in biofuels as a substitute for fossil fuels. Also, at present ethanol production is mainly done by yeast fermentation (*S. cerevisiae*) by using plant raw material containing very high levels of sugar. Use of plant raw material for biofuel production is economically costly, environmentally damaging, and requires a large cultivable area, therefore, we need an alternative source of raw material [24]. Since a large amount of seafood waste is generated every day, we can use this waste for biofuel production. Here the seafood waste can be first degraded using bacteria possessing calcium carbonate solubilization, cellulase, protease, and chitinase activity to release sugars. The sugars thus released can be used for bioethanol production by fermentation using *S. cerevisiae*. Sugar *N*-acetyl-D-glucosamine (GlcNAc) is the monomer of chitin and released during degradation of seafood waste. Inokuma et al. (2016) used *Scheffersomyces* (*Pichia*) *stipitis* strains for ethanol production by using GlcNAc as the sole carbon source [25]. *S. stipitis* NBRC1687, 10007 and 10063 strains gave 81%, 75% and 82% ethanol yield, respectively, after consuming 50 g/L GlcNAc at 30°C for 96 h. Not much research work has been done in this area to date, and therefore there is great potential for researchers to develop advanced biotechnological methods/processes to efficiently use seafood waste for ethanol (biofuel) production in a sustainable way without damaging the environment.

Acknowledgments

Dr. Milind Naik thanks SERB-DST project (File Number: YSS/2014/000258) and Dr. Shyamalina Haldar, postdoctoral fellow, Department of Microbiology, Goa University.

References

[1] Pal J, Verma HM, Munka VK, Maurya SK, Roy D, Kumar J. Biological method of chitin extraction from shrimp waste an eco-friendly low-cost technology and its advanced application. Int J Fish and Aquat 2014;1(6):104–7.

[2] Varun TK, Senani S, Jayapal N, Chikkerur J, Roy S, Tekulapally VB, et al. Extraction of chitosan and its oligomers from shrimp shell waste, their characterization and antimicrobial effect. Vet World 2017;10(2):170.

[3] Mahendrakar NS. Aqua feeds and meat quality of cultured fish. New Delhi: Biotech Consort India Ltd.; 2000. p. 26–30.

[4] Kumaran E, Mahalakshmipriya A, Rajan S. Effect of fish waste based Bacillus protease in silver recovery from waste X-ray films. Int J Curr Microb Appl Sci 2013;2(3):49–56.

[5] Bozzano A, Sardà F. Fishery discard consumption rate and scavenging activity in the Northwestern Mediterranean Sea. ICES J Mar Sci 2002;59(1):15–28.

[6] Rebah FB, Miled N. Fish processing wastes for microbial enzyme production: a review. 3 Biotech 2013;3(4):255–65.

[7] Xu Y, Gallert C, Winter J. Chitin purification from shrimp wastes by microbial deproteination and decalcification. Appl Microbiol Biotechnol 2008;79(4):687–97.

[8] Arvanitoyannis IS, Kassaveti A. Fish industry waste: treatments, environmental impacts, current and potential uses. Int J Food Sci Technol 2008;43(4):726–45.

[9] Younes I, Rinaudo M. Chitin and chitosan preparation from marine sources. Structure, properties and applications. Mar Drugs 2015;13(3):1133–74.

[10] Knorr D. Use of chitinous polymers in food: a challenge for food research and development. Food Technol 1984;38:85–97.

[11] Arbia W, Arbia L, Adour L, Amrane A. Chitin extraction from crustacean shells using biological methods-a review. Food Technol Biotechnol 2013;51(1):12–25.

[12] Cho YI, No HK, Meyers SP. Physicochemical characteristics and functional properties of various commercial chitin and chitosan products. J Agric Food Chem 1998;46(9):3839–43.

[13] Jung WJ, Jo GH, Kuk JH, Kim KY, Park RD. Extraction of chitin from red crab shell waste by co fermentation with *Lactobacillus paracasei* subsp. *tolerans* KCTC-3074 and *Serratia marcescens* FS-3. Appl Microbiol Biotechnol 2006;71(2):234.

[14] Sorokulova I, Krumnow A, Globa L, Vodyanoy V. Efficient decomposition of shrimp shell waste using *Bacillus cereus* and *Exiguobacterium acetylicum*. J Ind Microbiol Biotechnol 2009;36(8):1123–6.

[15] Wang SL, Chio SH. Deproteinization of shrimp and crab shell with the protease of *Pseudomonas aeruginosa* K-187. Enzym Microb Technol 1998;22(7):629–33.

[16] Jo GH, Jung WJ, Kuk JH, Oh KT, Kim YJ, Park RD. Screening of protease-producing *Serratia marcescens* FS-3 and its application to deproteinization of crab shell waste for chitin extraction. Carbohydr Polym 2008;74(3):504–8.

[17] Imran Md, Poduval PB, Ghadi SC. Bacterial degradation of algal polysaccharides in marine ecosystem. In: Naik MM, Dubey SK, editors. Marine poll microb remed; 2017. p. 196–202.

[18] Mondal D, Islam MS, Hoque MF, Hossain MK, Islam MK, Hossin MS, Ahsan SM. Isolation and screening of potential phosphate solubilizing bacteria (Psb) from tidal saline soils of Bangladesh. Octa J Environ Res 2016;4(3).

[19] Kasana RC, Salwan R, Dhar H, Dutt S, Gulati A. A rapid and easy method for the detection of microbial cellulases on agar plates using Gram's iodine. Curr Microbiol 2008;57(5):503–7.

[20] Krithika S, Chellaram C. Isolation, screening, and characterization of chitinase producing bacteria from marine wastes. Int J Pharm Pharmaceut Sci 2016;8(5):34–6.

[21] Monreal J, Reese ET. The chitinase of *Serratia marcescens*. Can J Microbiol 1969;15(7):689–96.

[22] Garrity GM, Brenner DJ, Kreig NR, Staley JT. Bergey's manual of systematic bacteriology, vol. 2. New York, Berlin, Heidelberg: Springer-Verlag; 2005. p. 685–93.

[23] Kumari S, Rath PK. Extraction and characterization of chitin and chitosan from (Labeo rohit) fish scales. Procedia Mater Sci 2014;6:482–9.

[24] Lozano VMP. Sustainable production of biofuel (bioethanol) from shellfish waste. Universidad de Alicante; 2015. p. 1–8.

[25] Inokuma K, Hasunuma T, Kondo A. Ethanol production from *N*-acetyl-d-glucosamine by *Scheffersomyces stipitis* strains. Amb Express 2016;6:83. https://doi.org/10.1186/s13568-016-0267-z.

Chapter 11

Phosphate solubilization by microorganisms: overview, mechanisms, applications and advances

Neha Prabhu[1], Sunita Borkar[2], Sandeep Garg[1]
[1]*Department of Microbiology, Goa University, Taleigao Plateau, Goa, India;* [2]*Department of Microbiology, P.E.S's R.S.N. College of Arts and Science, Goa, India*

11.1 Introduction

Phosphorus is the 11th most abundant element in earth's crust. The element phosphorus is highly reactive and occurs in combined form with other elements to form phosphates in nature [1]. Phosphate compounds present in soil can be classified into three groups, viz. soluble orthophosphate, insoluble inorganic phosphate, and insoluble organic phosphate. Orthophosphate is the only form that can be taken up by the plants. It reacts with numerous inorganic and organic constituents of soil and therefore becomes least mobile and unavailable for uptake by plants [2]. The complete cycle of phosphorus in the environment is presented in Fig. 11.1.

Phosphorus is classified as a major nutrient for plants as an adequate amount of phosphorus is required for optimum plant growth and yield [3]. The orthophosphate form of phosphorus enters the plant from water in the soil through root cells. In plants, phosphorus is essential for a number of physiological functions; its influence on plant growth is presented in Fig. 11.2.

The global cycling of insoluble inorganic and organic phosphate forms is attributed to microbes. Microorganisms that solubilize phosphorus-bearing insoluble inorganic and organic compounds are termed as phosphate-solubilizing microorganisms [4,5].

11.2 Phosphate-solubilizing microorganisms: an overview

The first report on microbial solubilization of insoluble inorganic phosphate was published by Pikovskaya in 1948 [6]. This report led to a significant

Advances in Biological Science Research. https://doi.org/10.1016/B978-0-12-817497-5.00011-2

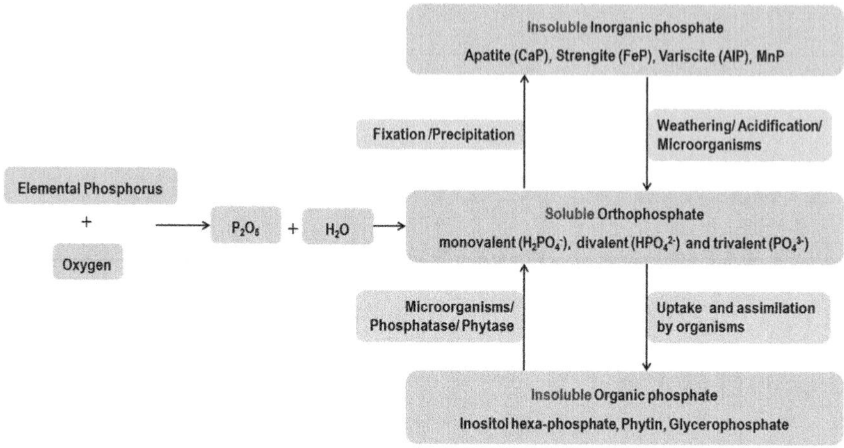

FIGURE 11.1 Phosphorus cycle in the environment.

increase in knowledge of phosphate solubilization. Phosphate-solubilizing microorganisms are ubiquitous in nature [7–9]. Numerous strains of bacteria, actinobacteria, and fungi have been reported and investigated for phosphate-solubilizing abilities [10–12]. The greatest number of species reported as phosphate solubilizers belong to the genus *Bacillus* followed by *Pseudomonas*. Among phosphate-solubilizing fungi, the genera *Aspergillus* and *Penicillium* are most common, followed by *Trichoderma* and *Rhizoctonia* [13–15]. Scientists have also reported novel genera and species of phosphate-solubilizing microorganisms [16,17].

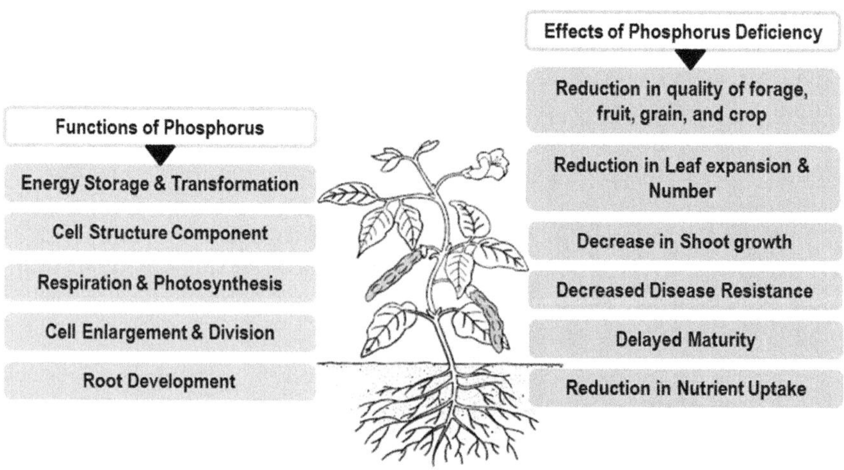

FIGURE 11.2 Functions of phosphorus in plants and effects during phosphorus deficiency.

In the past decade, researchers have proactively investigated phosphate solubilization in stress-tolerant and extremophilic microorganisms. Stress-tolerant phosphate-solubilizing microorganisms can be potential bioinoculants for stress-affected land for agriculture. It is evident from the literature that microorganisms help plants cope with abiotic stress. Stress-tolerant microorganisms modulate plant growth under stressed environment by altering the phytohormone ethylene. Ethylene is a stress phytohormone and affects seed germination, leaf abscission, leaf senescence, fruit ripening, flowering and flower wilting, root initiation, elongation, and branching [18]. 1-Aminocyclopropane-1-carboxylate (ACC) is the precursor of ethylene. Microorganisms that promote plant growth under stressed environment produce the enzyme ACC deaminase, which cleaves ACC. This leads to a decrease in or no ethylene production and indirect promotion of plant growth [19].

Bacteria have been reported to solubilize phosphate under the influence of abiotic stress such as drought, low or high pH, salinity, and temperature [20]. Cold-tolerant *Pseudomonas* sp., *Pantoea* sp., *Mycobacterium* sp., *Mycoplasma* sp., and *Acinetobacter* sp. have exhibited phosphate-solubilizing ability at low temperature from 4 to 16°C [21–24]. *Pseudomonas* sp. and *Azospirillum* sp. solubilized phosphate under the drought conditions [25,26]. *Aerococcus* sp., *Arthrobacter* sp., *Bacillus* sp., *Pantoea* sp., and *Pseudomonas* sp. solubilized phosphate under saline conditions (1%–10% NaCl) [27–30]. Isolates screened by Nakbanpote et al. demonstrated phosphate solubilization at 8% (w/v) salt concentration [31] *Bacillus* sp., *Pantoea* sp., and *Pseudomonas* sp. solubilized phosphate at alkaline pH (11.0) [32]. Phosphate-solubilizing alkaliphilic *Bacillus* sp. is reported to solubilize phosphate at pH 10.0 [33]. Nautiyal et al. have isolated phosphate solubilizing bacteria from alkaline soils, which solubilized phosphate at pH 12 and 10% NaCl [34]. Phosphate solubilization at two extreme conditions of high pH (10.0) and 25% crude salt concentration is reported in haloalkaliphilic bacterium *Chromohalobacter* sp. [35]. *Paenibacillus* sp., a drought, salt, and metal-resistant bacterium, was reported to confer abiotic factor resistance in *Arabidopsis* sp. [36].

11.2.1 Screening microorganisms for phosphate solubilization

Since the report of Pikovskaya (1948), several media for inorganic phosphate solubilization. including Pikovskaya's agar, such as bromophenol blue dye method, National Botanical Research Institute Phosphate medium, and Basal Sperber Medium, have been proposed [37–39]. Haloarchaea Phosphate Solubilization medium was recently devised by Yadav et al. for screening phosphate solubilization in haloarchaea [40]. Different sources of insoluble inorganic phosphate are used like aluminum phosphate, iron phosphate, tricalcium phosphate, and zinc phosphate. Microorganisms exhibiting inorganic phosphate solubilization produce a clear zone or halo around the colonies as seen in Fig. 11.3.

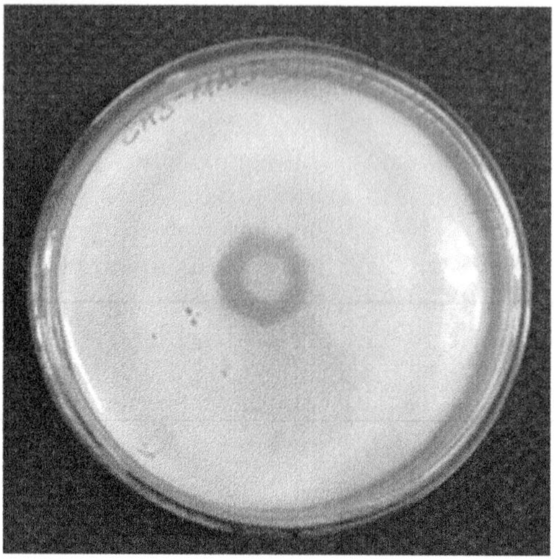

FIGURE 11.3 Tricalcium phosphate solubilization observed as a zone of clearance around the colony on Pikovskaya's agar.

The primary mechanism of inorganic phosphate solubilization is acidification due to the production of organic acids. Therefore, numerous researchers incorporate indicators in the medium to detect acid production. However, in a few studies, production of organic acids was not correlated with phosphate solubilization. Therefore, if an indicator is used in the medium, observation of zone of clearance is mandatory to report phosphate solubilization in addition to change in color of the medium around the colony.

Screening on agar medium is now considered not a suitable test for phosphate solubilization. Phosphate solubilization ability of microorganisms must be observed in liquid media. The decrease in the amount of inorganic phosphate must be demonstrated by gravimetric analysis to detect phosphate solubilization.

Phosphate solubilizing microorganisms improve plant phosphorus nutrition by mobilizing inorganic and organic phosphates. Several mechanisms of phosphate solubilization have been reported.

11.3 Phosphate solubilizing microorganisms: mechanisms

In microorganisms, a wide range of mechanisms to solubilize phosphate exist (Fig. 11.4).

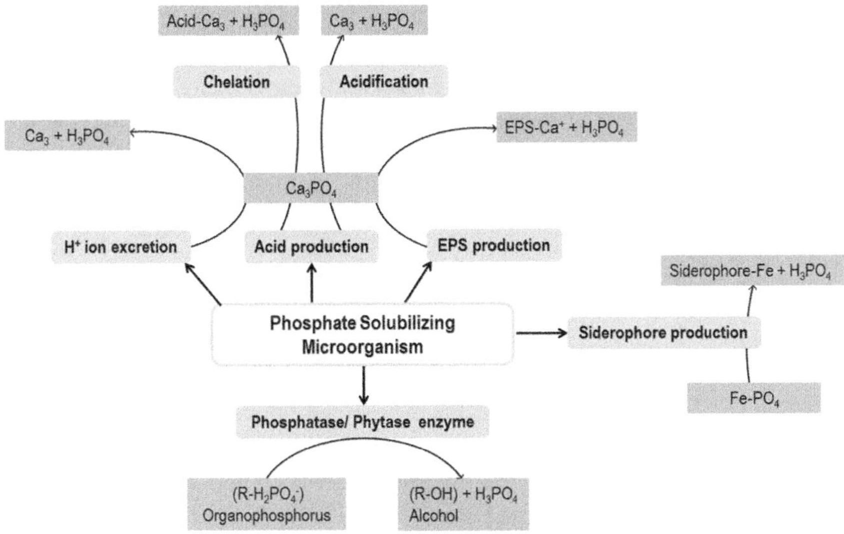

FIGURE 11.4 Mechanisms of inorganic and organic phosphate solubilization by microorganisms.

11.3.1 Inorganic phosphate-solubilization mechanisms

11.3.1.1 Organic acid production

The principal mechanism of inorganic phosphate solubilization is said to be organic acid production. The ability of organic acids to solubilize phosphate rock is due to acidification of the medium. A fall in pH due to the production of organic acids during the growth of phosphate-solubilizing microorganisms in a liquid medium has been reported by various authors [41–43]. Phosphate-solubilizing microorganisms are found to produce acids like acetic, formic (monocarboxylic acids); lactic, gluconic, glycolic (monocarboxylic hydroxy acids); 2-keto gluconic (monocarboxylic keto acid); oxalic, succinic (dicarboxylic acids); malic (dicarboxylic hydroxy acids); and citric (tricarboxylic hydroxy acids) in liquid media. Among the different acids produced by bacteria, gluconic acid has been regarded as the key acid for phosphate solubilization [44].

11.3.1.2 Chelation

Chelating substances also play a role in solubilization of phosphate. Acids like 2-keto-gluconic acid, humic acid, and fulvic acid are known as a powerful chelator of cations such as calcium, iron, and aluminum. They are effective in solubilization of inorganic phosphate complexed with calcium, iron, and aluminum [45,46]. Humic and fulvic acids are released by microorganisms during degradation of plant debris [47].

11.3.1.3 Inorganic acid production

Sulfuric acid, nitric acid, and carbonic acids are also reported to solubilize phosphate [48]. However, the effectiveness of inorganic acids' contribution in orthophosphate release appears to be less effective as compared to organic acids [49]. Inorganic acids are produced by nitrifying and sulfur-oxidizing bacteria during oxidation of nitrogenous or inorganic sulfur compounds. Inorganic acids produced react with insoluble phosphate compounds and convert them into soluble forms.

11.3.1.4 Proton extrusion

Solubilization of phosphates has also been reported to occur even in the absence of acid production. Acidification of the medium due to H^+ excretion is reported as an alternate mechanism of inorganic phosphate solubilization. H^+ excretion originates from $NH_4{}^+$ assimilation, respiratory H_2CO_3 production, and extrusion of organic acid anions [50,51]. Nitrogen supply influences the number of protons released into the external medium. The degree of pH reduction depends on the type of inorganic nitrogen source; NH_4 as the sole inorganic nitrogen source reduces the pH far more than NO_3 [52].

11.3.1.5 Exopolysaccharide production

Microbial exopolysaccharides are said to have an indirect effect on phosphate solubilization [53]. Exopolysaccharides are largely produced by microorganisms in response to stress. Studies on microbial exopolysaccharides have revealed their ability to bind with metals in soil. Therefore, they can influence solubility of metal phosphates in soil [54]. Microbial exopolysaccharide has exhibited tricalcium phosphate solubilization along with organic acid anions [55]. A positive correlation between the rate of phosphate solubilization and concentration of exopolysaccharide was observed. However, the role of high-molecular-weight exopolysaccharides in phosphate solubilization from soil constituents needs further investigation.

11.3.1.6 Siderophore production

Siderophores are substances secreted by microorganisms that show strong affinity to chelate iron [56]. Many phosphate-solubilizing microorganisms have exhibited production of siderophores [57]. Siderophores influence the solubility of iron phosphates in soil. However, a direct relationship between siderophore production and phosphate solubilization is not known.

11.3.2 Organic phosphate solubilization mechanisms

11.3.2.1 Enzyme production

Different examples of organic phosphate compounds in soil are nucleic acids, phospholipids, phosphonates, phytic acid, polyphosphonates, and sugar phosphates. Organic phosphate solubilization occurs in soil by the action of microbial enzymes like phosphatase or phosphohydrolase, phytase, phosphonatase, and C—P lyase [58,59].

Phosphatase enzymes render high-molecular-weight organic phosphate into low-molecular-weight compounds by the hydrolysis of ester phosphate bonds, leading to the release of phosphate ions. Based on their pH optima, these enzymes are classified as acid, neutral, and alkaline phosphatase [60]. Phytase enzyme hydrolyzes phytic acid or myo-inositolphosphate compounds. Phosphonatase and C—P lyase hydrolyze ester bonds of phosphonates (e.g., phosphoenol pyruvate, phosphonoacetate) and converts phosphonates into hydrocarbons and phosphate ions for assimilation [61].

11.4 Phosphate-solubilizing microorganisms: applications and advances

In the past years, phosphate-solubilizing microorganisms have only been used as bioinoculants. Recently their use in bioremediation and phytoremediation has been observed.

11.4.1 Biofertilizer

Biofertilizers are microbiologically active, low-cost products applied to soil for growth promotion of plants [62]. Formulation of biofertilizers has evolved over the years (Fig. 11.5).

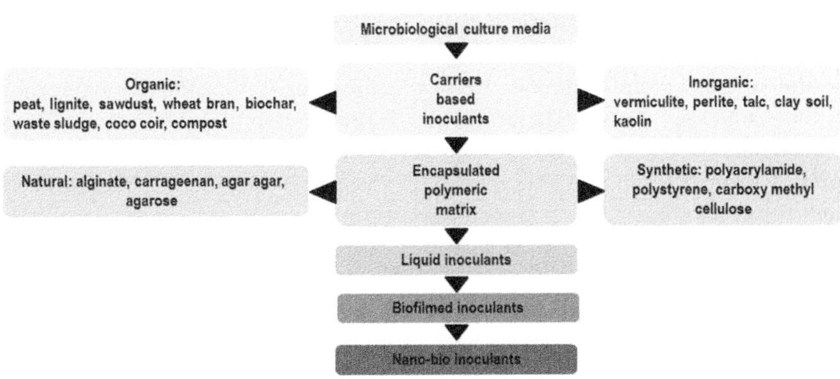

FIGURE 11.5 Advances in the formulation of phosphate-solubilizing biofertilizer.

They were earlier prepared as carrier-based inoculants in a form of fine powders (peat coal, clay, and soil), plant waste materials (compost, farmyard manure, soybean meal, byproducts of sugar industry, agricultural waste, spent mushroom compost), inert materials (vermiculite, perlite, rock phosphate, calcium sulfate), and plain lyophilized bacterial cultures [63,64]. Carrier-based formulation is the most widely tested formulation of biofertilizers. Inoculation of carrier-based phosphate-solubilizing bacteria has promoted growth and yield of *Cicer arietinum*, *Glycine max*, *Gossypium*, *Helianthus*, *Oryza sativa*, *Solanum lycopersicum*, *Vigna unguiculata*, and *Zea mays* plants [65–68]. Talc-based formulation of *Bacillus subtilis* and *Pseudomonas fluorescens* has significantly promoted the growth of *Cajanus cajan*, *Cicer arietinum*, *Gossypium*, *Oryza sativa*, *Saccharum officinarum*, *Solanum tuberosum*, and *Triticum* [69]. Two carrier-based formulations of *Bacillus cereus* and *Pseudomonas moraviensis* promoted growth in *Triticum* [70]. Ground maize straw and sugarcane husk were used as carriers. Biochar and fly ash–based formulations of *Bacillus* sp. strain A30 and *Burkholderia* sp. strain L2 promoted growth and yield of *Solanum lycopersicum* [71].

Liquid inoculants have overtaken carrier-based inoculants due to several advantages like better shelf life, easy handling, etc. To formulate optimum medium for liquid biofertilizers, scientists use statistical techniques like response surface methodology (RSM), Plackett–Burman Design, Simplex Search Technique, etc. [72]. RSM has been successfully applied in many areas of agricultural biotechnology. RSM has been applied to optimize variables for maximum phosphate solubilization in *Aspergillus japonicus* SA22P3406, *Pseudomonas putida* Rs-198, and *Acinetobacter calcoaceticus* TM8 [73–75]. Liquid inoculants now dominate the biofertilizer industry, followed by carrier-based inoculants (Table 11.1).

Two new techniques added to biofertilizer formulation technology are biofilm inoculants and nano-bio inoculants. In biofilm inoculants, two microorganisms are involved. One microorganism (mostly bacterium) colonizes over the second microorganism (either bacterium or fungus). The second microorganism acts as a biotic surface for the first microorganism, forming metabolically enhanced biofilms compared to single culture. This type of biofilm occurs naturally in soil. However, the density of naturally occurring biofilm is lower in soil. Therefore, the formulated biofilm is beneficial as bioinoculant for agricultural soil. The first in vitro biofilm was developed using nonmycorrhizal fungi and rhizobia [76]. Phosphate-solubilizing biofilms too have been developed using *Penicillium* spp., *Pleurotus ostreatus*, and *Xanthoparmelia mexicana*. Interestingly, it was observed that this consortium increased phosphate solubilization two times compared to single cultures [77]. As compared to biofilm formulations of nitrogen-fixing microorganisms, biofilm formulations using phosphate-solubilizing microorganisms is developing slowly.

TABLE 11.1 Phosphate-solubilizing biofertilizers available in the market.

Trade name	Culture	Formulation	Crops
Bio-Phospho	*Bacillus subtilis*	Liquid	NS
Phospho MAX	*Bacillus megaterium*	Liquid	Cereals
Phosphate-Solubilizing Biofertilizer	*Bacillus coagulans*	Liquid	NS
P-Sol	*Bacillus megaterium (NCIM 2087, ATCC 10778)*	Liquid	Cereals, pulses, oilseeds, cotton, jute, banana, turmeric
Bio Star	*Bacillus* sp., *Pseudomonas* sp.	Liquid	NS
Dr. Bacto's	*Bacillus* sp., *Pseudomonas* sp.	Carrier, liquid	NS
Flor-PSB	*Pseudomonas putida, Bacillus megaterium*	Carrier	NS
AG Phosbase	*Bacillus megaterium, Bacillus polymyxa*	Liquid	NS
Phosbac	*Bacillus megaterium*	Carrier	Cereals, pulses, oilseeds, cotton, and jute
Avtar PSB	*Bacillus megaterium* var. *phosphaticum*	Carrier and granular	NS

NS, nonspecific.

Nano-bio inoculants employ integration of whole phosphate-solubilizing microorganism with nanoparticles or nanostructures [78,79]. Different ways of using nanoparticles include encapsulating microbes with micronutrient nanoparticles, priming microbes using nanoparticles, and using nanoparticles as delivery vehicles. Formulation of biofertilizers using nanoparticles confers stability to biofertilizers from desiccation, heat, and UV inactivation. Nano-bio inoculants using gold nanoparticles have accelerated growth promotion ability of *P. fluorescens*, *B. subtilis*, *Paenibacillus elgii*, and *Pseudomonas putida* [80]. Besides gold, silver, lead, platinum, copper, and iron nanoparticles have also been used in nano-bio inoculants [81].

11.4.2 Phytoremediation

Phytoremediation is an eco-friendly, efficient, and economic method of bio-remediating metal-contaminated soil. However, increased concentration of metals causes decreased plant biomass and stunted growth. Phosphate-solubilizing microorganisms such as *Pseudomonas*, *Enterobacter*, *Klebsiella*, etc., have proven effective in bioremediation of metals and influencing phytoremediation by phytoextraction or by phytostabilization in metal-contaminated soil [82]. Various phosphate-solubilization mechanisms like organic acid production, H^+ excretion, siderophore production, and exopolymeric substance production contribute in bioremediation as presented in Table 11.2. Researchers believe in using a consortium of phosphate-solubilizing micro-organisms compared to axenic cultures for efficient bioremediation [93].

TABLE 11.2 Mechanisms of phosphate-solubilizing bacteria (PSB) that aid in phytoremediation of metal contaminated soil.

Metal chelated	PSB	Host plant	Phosphate-solubilization mechanism	Reference
Ni	*Rhodococcus globerulus X80619 Rhodococcus erythropolis X79289*	*Alyssum serpyllifolium*	N.R.	[83]
Cu	*Pseudomonas* sp.	*Maize and Sunflower*	Organic acid, Siderophore	[84]
Zn	*Rahnella* sp.	*Brassica napus*	Organic acid, Siderophore	[85]
Cr	*Pseudomonas aeruginosa*	*Cider arietinum*	Organic acid, EPS	[86]
Co, Pb, Zn	*Enterobacter ludwigii*	*Helianthus annuus*	Organic acids	[87]
Cd, Pb, Zn	*Enterobacter* sp., *Klebsiella* sp.	*Brassica napus*	Organic acids	[88]
Cd	*Rahnella* sp.	*Amaranthus* sp.	Siderophores, organic acids	[89]
Ni, Zn, Fe	*Psychrobacter* sp., *Pseudomonas*	*Brassica juncea, Ricinus communis*	Organic acids, Siderophore	[90]
As Pb	*Trichoderma*	*Helianthus annuus*	Siderophore, phosphatase	[91]
Cu	*Paenibacillus polymyxa*	*Wedelia trilobata*	N.R.	[92]

N.R, not reported.

11.5 Conclusion

Phosphorus is an indispensable nutrient for plant growth as its functions cannot be performed by any other nutrient. Ten decades of significant research have helped in the deeper understanding of the phenomenon of phosphate solubilization. Phosphate-solubilizing microorganisms exhibit multiple mechanisms toward solubilization of complex insoluble phosphates. They perform sustainably to enhance the productivity of crops in soil as biofertilizer agents. This chapter has revealed the dynamics of phosphate solubilization and the use of phosphate-solubilizing microorganisms in enhancing fertility and bioremediation of soil. Considering the immense potential of phosphate-solubilizing microorganisms, in future, a detailed study must be performed to understand the ecology of phosphate-solubilizing microorganisms, especially under abiotic stress. The outcome of such research will be potential phosphate-solubilizing microorganisms that can be developed as biofertilizers for soil affected by abiotic stress.

References

[1] Ehrlich HL, Newman DK, Kappler A, editors. Geomicrobiology. Boca Raton: Taylor & Francis Group; 2016.

[2] Behera BC, Singdevsachan SK, Mishra RR, Dutta SK, Thatoi HN. Diversity, mechanism and biotechnology of phosphate solubilising microorganisms in mangrove- a review. Biocatal Agric Biotechnol 2014;3:97–110.

[3] Ashley K, Cordell D, Mavinic D. A brief history of phosphorus: from the philosophers's stone to nutrient recovery and reuse. Chemosphere 2011;84(2011):737–46.

[4] Spohn M, Treichel NS, Cormann M, Schloter M, Fischer D. Distribution of phosphatase activity and various bacterial phyla in the rhizosphere of *Hordeum vulgare* L. depending on P availability. Soil Biol Biochem 2015;89:44–51.

[5] Yadav BK, Verma A. Phosphate solubilisation and mobilization in soil through microorganisms under arid ecosystems. In: Ali M, editor. The functioning of ecosystems. Shanghai: InTech; 2012. p. 93–108.

[6] Pikovskaya RI. Mobilization of phosphorus in soil in connection with vital capacity of source microbial species. Microbiologia 1948;17:362–70.

[7] Selvakumar G, Mohan M, Kundu S, Gupta AD, Joshi P, Nazim S. Cold resistance and plant growth promotion potential of *Serratia marcescens* strain SRM (MTCC8708) isolated from flowers of summer squash (*Cucurbita pepo*). Lett Appl Microbiol 2008;46:171–5.

[8] Rashid S, Charles TC, Click BR. Isolation and characterization of new plant growth promoting bacterial endophytes. Appl Soil Ecol 2012;61:217–24.

[9] Awais M, Tariq M, Ali A, Ali Q, Khan A, Tabassum B, et al. Isolation, characterization and inter-relationship of phosphate solubilising bacteria from the rhizosphere of sugarcane and rice. Biocatal Agric Biotechnol 2017;11:312–21.

[10] Whitelaw MA. Growth promotion of plants inoculated with phosphate solubilising fungi. Adv Agron 2000;69:99–151.

[11] Igual JM, Valverde A, Cervantes E, Velazquez E. Phosphate solubilising bacteria as inoculants for agriculture: use of updated molecular techniques in their study. Agronomie 2001;21:561–8.

[12] Zaidi A, Khan MS, Ahemad M, Oves M. Plant growth promotion by phosphate solubilising bacteria. Acta Microbiol Immunol Hung 2009;56(3):263–84.

[13] Kucey RM. Phosphate solubilising bacteria and fungi from various cultivated and virgin Alberta soils. Can J Soil Sci 1989;63:671–8.

[14] Altomare C, Norvell WA, Bjorkman T, Harman GE. Solubilisation of phosphates and micronutrients by the plant growth promoting and biocontrol fungus *Trichoderma harzianum* Rifai 1295-22. Appl Environ Microbiol 1999;65(7):2926–33.

[15] Jacobs H, Boswell GP, Ritz K, Davidson FA, Gadd GM. Solubilisation of calcium phosphate as a consequence of carbon translocation by *Rhizoctonia solani*. FEMS Microbiol Ecol 2002;40(1):65–71.

[16] Loganathan P, Nair S. *Swaminathania salitolerans* gen. nov., sp. nov., a salt-tolerant, nitrogen-fixing and phosphate-solubilising bacterium from wild rice (*Porteresia coarctata* Tateoka). Int J Syst Evol Microbiol 2004;54(4):1185–90.

[17] Madhaiyan M, Poonguzhali S, Senthilkumar M, Pragatheswari D, Lee JS, Lee KC. *Arachidicoccus rhizosphaerae* gen. nov., sp. nov., a plant-growth-promoting bacterium in the family Chitinophagaceae isolated from rhizosphere soil. Int J Syst Evol Microbiol 2015;65(2):578–86.

[18] Glick BR, Cheng Z, Czarny J, Duan J. Promotion of plant growth by ACC deaminase producing soil bacteria. In: Bakker PAHM, Raaijmakers JM, Bloemberg G, Hofte M, Lemanceau P, Cooke BM, editors. New perspectives and approaches in plant growth-promoting rhizobacteria research. Dordrecht: Springer; 2007. p. 329–39.

[19] Glick BR. Bacteria with ACC deaminase can promote plant growth and help to feed the world. Microbiol Res 2014;169(1):30–9.

[20] Vassilev N, Eichler-Lobermann B, Vassileva M. Stress-tolerant P-solubilising microorganisms. Appl Microbiol Biotechnol 2012;95(4):851–9.

[21] Egamberdiyeva D, Hoflich G. Influence of growth promoting bacteria on the growth of wheat in different soils and temperatures. Soil Biol Biochem 2003;35:973–8.

[22] Pandey A, Trivedi P, Kumar B, Palni LMS. Characterization of a phosphate solubilising and antagonistic strain of *Pseudomonas putida* (B0) isolated from a sub-alpine location in the Indian Central Himalaya. Curr Microbiol 2006;53(2):102–7.

[23] Trivedi P, Sa T. *Pseudomonas corrugata* (NRRI B-30409) mutants increased phosphate solubilisation, organic acid production and plant growth at lower temperatures. Curr Microbiol 2008;56(2):140–4.

[24] Gulati A, Vyas P, Rahi P, Kasana RC. Plant growth promoting and rhizosphere competent *Acinetobacter rhizosphaerae* strain BIHB 723 from the cold deserts of the Himalayas. Curr Microbiol 2009;58(4):371–7.

[25] Sandhya V, Ali SK, Grover M, Reddy G, Venkateswarlu B. Effect of plant growth promoting *Pseudomonas* spp. on compatible solutes, antioxidant status and plant growth of maize under drought stress. Plant Growth Regul 2010;62:21–30.

[26] Arzanesh MH, Alikhani HA, Khavazi K, Rahimian HA, Miransari M. Wheat (*Triticum aestivum* L.) growth enhancement by *Azospirillum* sp. under drought stress. World J Microbiol Biotechnol 2011;27(2):197–205.

[27] Srinivasan R, Yandigeri MS, Kashyap S, Alagwadi AR. Effect of salt on survival and p-solubilisation potential of phosphate solubilising microorganisms from salt-affected soils. Saudi J Biol Sci 2012;19(4):427–34.

[28] Son HJ, Park GT, Cha MS, Heo MS. Solubilisation of insoluble inorganic phosphates by a novel salt and pH tolerant *Pantoea agglomerans* R042 isolated from soybean rhizosphere. Bioresour Technol 2006;97(2):204–10.

[29] Johri JK, Surange S, Nautiyal CS. Occurrence of salt, pH, and temperature-tolerant phosphate solubilising bacteria in alkaline soils. Curr Microbiol 1999;39(2):89–93.

[30] Banerjee S, Palit R, Sengupta C, Standing D. Stress induced phosphate solubilisation by *Arthrobacter* sp. and *Bacillus* sp. isolated from tomato rhizosphere. Aust J Crop Sci 2010;4(6):78–383.

[31] Nakbanpote W, Panitlurtumpai N, Sangdee A, Sakulpone N, Sirisom P, Pimthong A. Salt-tolerant and plant growth promoting bacteria isolated from Zn/Cd contaminated soil: identification and effect on rice under saline conditions. J Plant Interact 2014;9(1):379–87.

[32] Mittal S, Meyer JM, Goel R. Isolation and characterization of aluminium and cooper resistant P solubilising alkalophilic bacteria. Indian J Biotechnol 2003;2:583–6.

[33] Prabhu N, Borkar S, Garg S. Phosphate solubilisation mechanisms in alkaliphilic bacterium *Bacillus marisflavi* FA7. Curr Sci 2018;114(4):845–53.

[34] Nautiyal CS, Bhadauria S, Kumar P, Lal H, Mondal R, Verma D. Stress-induced phosphate solubilisation in bacteria isolated from alkaline soils. FEMS Microbiol Lett 2000;182(2):291–6.

[35] Prabhu N, Borkar S, Garg S. Alkaliphilic and haloalkaliphilic phosphate solubilising bacteria from coastal ecosystems of Goa. Asian J Microbiol Biotechnol Environ Sci 2017;19(3):703–14.

[36] Sukweenadhi J, Kim YJ, Choi ES, Koh SC, Lee SW, Kim YJ, et al. *Paenibacillus yonginensis* DCY84 induces changes in *Arabidopsis thaliana* gene expression against aluminum, drought, and salt stress. Microbiol Res 2015;172:7–15.

[37] Gupta RR, Singal R, Shanker A, Kuhad RC, Saxena RK. A modified plate assay for screening phosphate solubilising microorganisms. J Gen Appl Microbiol 1994;40(3):255–60.

[38] Mehta S, Nautiyal CS. An efficient method for qualitative screening of phosphate solubilising bacteria. Curr Microbiol 2001;43(1):51–6.

[39] Nautiyal CS. An efficient microbiological growth medium for screening phosphate solubilising microorganisms. FEMS Microbiol Lett 1999;170(1):265–70.

[40] Yadav AN, Sharma D, Guati S, Singh S, Dey R, Pal KK, et al. Haloarchaea endowed with phosphorus solubilisation attribute implicated in phosphorus cycle. Sci Rep 2015:1–10. https://www.nature.com/articles/srep12293.

[41] Freitas JR, Banerjee MR, Germida JJ. Phosphate solubilising rhizobacteria enhance the growth and yield but not phosphorus uptake of canola (*Brassica napus* L.). Biol Fertil Soils 1997;24(4):358–64.

[42] Gulati A, Sharma N, Vyas P, Sood S, Rahi P, Pathania V. Organic acid production and plant growth promotion as a function of phosphate solubilisation by *Acinetobacter rhizosphaerae* strain BIHB 723 isolated from the cold desserts of the Trans-Himalayas. Arch Microbiol 2010;192(11):975–83.

[43] Collavino MM, Sansberro PA, Mroginski LA, Aguilar M. Comparison of in vitro solubilisation activity of diverse phosphate solubilising bacteria native to acid soil and their ability to promote *Phaseolus vulgaris* growth. Biol Fertil Soils 2010;46:727–38.

[44] Rashid M, Khalil S, Ayub N, Alam S, Farooq L. Organic acids production and phosphate solubilisation by phosphate solubilising microorganisms (PSM) under in vitro conditions. Pakistan J Biol Sci 2004;7(2):187–96.

[45] Sperber JI. Solution of apatite in soil microorganisms producing organic acids. Aust J Agric Res 1958;9(6):782–7.

[46] Katznelson H, Bose B. Metabolic activity and phosphate dissolving capability of bacterial isolates from wheat roots, rhizosphere, and non-rhizosphere soil. Can J Microbiol 1959;5(1):79–85.

[47] Stevenson FJ. Cycles of soil: carbon, nitrogen, phosphorus, sulfur, micronutrients. New York: John Wiley & Sons; 2005.

[48] Rodriguez H, Fraga R. Phosphate solubilising bacteria and their role in plant growth promotion. Biotechnol Adv 1999;17(4–5):319–39.

[49] Vazquez P, Holguin G, Puente ME, Lopez-Cortes A, Bashan Y. Phosphate solubilising microorganisms associated with the rhizosphere of mangroves in a semiarid coastal lagoon. Biol Fertil Soils 2000;30(5–6):460–8.

[50] Juriank JJ, Dudley LM, Allen MF, Knight WG. The role of calcium oxalate in the availability of phosphorus in soil of semiarid regions: therodynamic study. Soil Sci 1986;142:255–61.

[51] Arvieu JC, Leprince F, Plassard C. Release of oxalate and protons by ectomycorrhizal fungi in response to P-deficiency and calcium carbonate in nutrient solution. Ann For Sci 2003;60:815–21.

[52] Sharan A, Shikha, Darmwal NS. Efficient phosphorus solubilisation by mutant strain of *Xanthomonas campestris* using different carbon, nitrogen and phosphorus sources. World J Microbiol Biotechnol 2008;24:3087–90.

[53] Ionescu M, Belkin S. Overproduction of exopolysaccharides by an *Escherichia coli* K-12 rpoS mutant in response to osmotic stress. Appl Environ Microbiol 2009;75(2):483–92.

[54] Ochoa-Loza FJ, Artiola JF, Maier RM. Stability constants for the complexation of various metals with a rhamnolipid biosurfactant. J Environ Qual 2001;30(2):479–85.

[55] Yi Y, Huang W, Ge Y. Exopolysaccharide: a novel important factor in the microbial dissolution of tricalcium phosphate. World J Microbiol Biotechnol 2008;24(7):1059–65.

[56] Gaonkar T, Bhosle S. Effect of metals on a siderophore producing isolate and its implications on microbial assisted bioremediation of metal contaminated soils. Chemosphere 2013;93(9):1835–43.

[57] Vassilev N, Vassileva M, Niklaeval I. Simultaneous P-solubilising and biocontrol activity of microorganisms potential and future trends. Appl Microbiol Biotechnol 2006;71(2):137–44.

[58] Eivazi F, Tabatabai MA. Phosphatases in soils. Soil Biol Biochem 1977;9(3):167–72.

[59] Jorquera MA, Hernandez MT, Rengel Z, Marschner P, Mora MD. Isolation of culturable phosphobacteria with both phytate-mineralization and phosphate solubilisation activity from the rhizosphere of plants grown in a volcanic soil. Biol Fertil Soils 2008;44:1025–34.

[60] Cabugao KG, Timm CM, Carrell AA, Childs J, Lu TYS, Pelletier DA, et al. Root and rhizosphere bacterial phosphatase activity varies with tree species and soil phosphorus availability in Puerto Rico tropical forest. Front Plant Sci 2017;8(1834):1–14.

[61] Wanner BL. Phosphorus assimilation and control of the phosphate regulon. In: Neidhardt FC, Curtiss R, Ingraham JL, Lin ECC, Low KB, Magasanik B, et al., editors. *Escherichia coli* and *Salmonella*: cellular and molecular biology. Washington DC: ASM Press; 1996. p. 1357–81.

[62] Boraste A, Vamsi KK, Jhadav A, Khairnar Y, Gupta N, Trivedi S. Biofertilizers: a novel tool for agriculture. Int J Microbiol Res 2009;1(2):23–31.

[63] Malusa E, Sas-Paszt L, Ciesielska J. Technologies for beneficial microorganisms inocula used as biofertilizers. Sci World J 2012;2012:1–12.

[64] Mahdi SS, Hassan GI, Samoon SA, Rather HA, Dar SA, Zehra B. Bio-fertilizers in organic agriculture. J Phytol 2010;2(10):42–54.
[65] Sundara B, Natarajan V, Hari K. Influence of phosphorus solubilising bacteria on the change in soil available phosphorus and sugarcane and sugar yield. Field Crop Res 2002;77:43–9.
[66] Sahin F, Cakmakci R, Kantar F. Sugar beet and barley yields in relation to inoculation with N2-fixing and phosphate solubilising bacteria. Plant Soil 2004;265(1–2):123–9.
[67] Wu SC, Cao ZH, Li ZG, Cheung KC, Wong MH. Effects of biofertilizer containing N-fixer, P and K solubilizers and AM fungi on maize growth: a greenhouse trial. Geoderma 2005;125:155–66.
[68] Cakmakci R, Donmez MF, Erdogan U. The effect of plant growth promoting rhizobacteria on barley seedling growth, nutrient uptake, some soil properties, and bacterial counts. Turk J Agric For 2007;31(3):189–99.
[69] Arora NK. Plant microbe symbiosis: fundamentals and advances. New Delhi: Springer; 2013.
[70] Hassan T, Biofertilizer BA. A novel formulation for improving wheat growth, physiology and yield. Pakistan J Bot 2016;48(6):2233–41.
[71] Tripti A, Kumar A, Usmani Z, Kumar V, Anshumali R. Biochar and flyash inoculated with plant growth promoting rhizobacteria act as potential biofertilizer for luxuriant growth and yield of tomato plant. J Environ Manag 2017;190:20–7.
[72] Tabora JE, Domagalski N. Multivariable analysis and statistics in pharmaceutical process research and development. Annu Rev Chem Biomol Eng 2017;8:403–26.
[73] Nopparat C, Jatupornpipat M, Rittiboon A. Optimization of the phosphate solubilising fungus *Aspergillus japonicus* SA22P3406 in solid state cultivation by response surface methodology. Kasetsart J Nat Sci 2009;43:172–81.
[74] Peng Y, He Y, Wu Z, Lu J, Li C. Screening and optimization of low-cost medium for *Pseudomonas putida* Rs-198 culture using RSM. Braz J Microbiol 2014;45(4):1229–37.
[75] Mishra VK, Ali S, Gupta RK, Shoket H. TCP solubilisation by growth promotory endophytic *Acinetobacter calcoaceticus* TM8 from tomato. Curr Trends Biotechnol Pharm 2015;9(2):164–74.
[76] Taktek S, St-Arnaud M, Piche Y, Fortin JA, Antoun H. Igneous phosphate rock solubilisation by biofilm-forming mycorrhizobacteria and hyphobacteria associated with *Rhizoglomus irregulare* DAOM 197198. Mycorrhiza 2017;27(1):13–22.
[77] Babu VS, Triveni S, Reddy RS, Sathyanarayana J. Persistence of PSB-fungi biofilmed biofertilizer in the soils and its effect on growth and yield of maize. Int J Curr Microbiol Appl Sci 2017;6(12):1812–21.
[78] Singh DP, Singh HB, Prabha R, editors. Microbial inoculants in sustainable agricultural productivity. India: Springer; 2016.
[79] Gouda S, Kerry RG, Das G, Paramithiotis S, Shin HS, Patra JK. Revitalization of plant growth promoting rhizobacteria for sustainable development in agriculture. Microbiol Res 2018;206:131–40.
[80] Shukla SK, Kumar R, Mishra RK, Pandey A, Pathak A, Zaidi MG, et al. Prediction and validation of gold nanoparticles (GNPs) on plant growth promoting rhizobacteria (PGPR): a step toward development of nano-biofertilizers. Nanotechnol Rev 2015;4(5):439–48.
[81] Dikshit A, Shukla SK, Mishra RK, editors. Exploring nanomaterials with PGPR in current agricultural scenario. Deutschland: Lambert Academic Publishing; 2013.
[82] Ahemad M. Phosphate-solubilising bacteria-assisted phytoremediation of metalliferous soils: a review. 3 Biotech 2015;5(2):111–21.

[83] Becerra-Castro C, Prieto-Fernandez A, Alvarez-Lopez V, Monterroso C, Cabello-Conejo MI, Acea MJ, et al. Nickel solubilising capacity of rhizobacteria isolated from hyperaccumulating and non-hyperaccumulating subspecies of *Alyssum serpyllifolium*. Int J Phytoremediation 2011;13(Suppl. 1):229–44.

[84] Li K, Ramakrishna W. Effect of multiple metal resistant bacteria from contaminated lake sediments on metal accumulation and plant growth. J Hazard Mater 2011;189(1–2):531–9.

[85] He H, Ye Z, Yang D, Yan J, Xiao L, Zhong T, et al. Characterization of endophytic *Rahnella* sp. JN6 from *Polygonum pubescens* and its potential in promoting growth and Cd, Pb, Zn uptake by *Brassica napus*. Chemosphere 2013;90(6):1960–5.

[86] Oves M, Khan MS, Zaidi A. Chromium reducing and plant growth promoting novel strain *Pseudomonas aeruginosa* OSG41 enhance chickpea growth in chromium amended soils. Eur J Soil Biol 2013;56:72–83.

[87] Arunakumara KK, Walpola BC, Song JS, Shin M, Lee C, Yoon M. Phytoextraction of heavy metals induced by bioaugmentation of a phosphate solubilising bacterium. Korean J Environ Agric 2014;33(3):220–30.

[88] Jing YX, Yan JL, He HD, Yang DJ, Xiao L. Characterization of bacteria in the rhizosphere soils of *Polygonum pubescens* and their potential in promoting growth and cd, pb, zn uptake by *Brassica napus*. Int J Phytoremediation 2014;16(4):321–33.

[89] Yuan M, He H, Xiao L, Zhong T, Liu H, Li S, et al. Enhancement of Cd phytoextraction by two *Amaranthus* species with endophytic *Rahnella* sp. JN27. Chemosphere 2014;103:99–104.

[90] Ma Y, Rajkumar M, Rocha I, Oliveira R, Freitas H. Serpentine bacteria influence metal translocation and bioconcentration of *Brassica juncea* and *Ricinus communis* growing in multi-pollute soils. Front Plant Sci 2015;5(5):1–10.

[91] Govarthanan M, Mythili R, Selvakumar T, Kamala-Kannan S, Kim H. Myco-phytoremediation of arsenic- and lead-contaminated soils by *Helianthus annuus* and wood rot fungi, *Trichoderma* sp. isolated from decayed wood. Ecotoxicol Environ Saf 2018;151:279–84.

[92] Lin M, Jin M, Xu K, He L, Cheng D. Phosphate solubilising bacteria improve the phytoremediation efficiency of *Wedelia trilobata* for Cu-contaminated soil. Int J Phytoremediation 2018;20(8):813–22.

[93] Gupta P, Kumar V. Value added phytoremediation of metal stressed soils using phosphate solubilising microbial consortium. World J Microbiol Biotechnol 2017;33(9):1–15.

Chapter 12

Metagenomics a modern approach to reveal the secrets of unculturable microbes

Kashif Shamim[1], Sajiya Yusuf Mujawar[1], Milind Mutnale[2]
[1]*Laboratory of Bacterial Genetics and Environmental Biotechnology, Department of Microbiology, Goa University, Goa, India;* [2]*National Centre for Polar and Ocean Research (NCPOR), Vasco-da-Gama, Goa, India*

12.1 Introduction

The most diverse life forms that exist on planet Earth are microorganisms, but still, only 0.1%–1% have been cultivated so far, resulting in creating a great obstacle in understanding the world of microbes. This limitation nowadays has been overcome by the use of data being gathered using metagenomics and single-cell genomic techniques, which bypasses the need for cultivation of microbes in laboratory conditions [1].

The identification of bacterial and archaeal phyla using a small subunit ribosomal RNA database till date has resulted into 89 bacterial and 20 archaeal phyla, which is much less than the expected bacterial phyla of approximately 1500 [2–7]. The lack of data act as an indicator for the incapability of cultivating microorganisms, as the majority of what we have understood so far about microbial life form is based on cultivable organisms. Therefore, the phylogenic information available is being dominated mostly by the Proteobacteria, Actinobacteria, Firmicutes, and Bacteroidetes representing bacteria, with halotolerant and methanogens members of the Euryarchaeota in case of the archaea [3].

The realization of how diverse the untapped microbial community is in any particular environment came from the analysis of RNA (16S rRNA or SSU) genes sequencing directly from various environmental econiches [8]. The outcome of the study revealed that a single cultivable representative was present in less than half of the known microbial phyla [1]. Those phyla that contain exclusively uncultivable microbes are known as Candidate Phyla (CP). This CP is also referred to as microbial dark matter as these microbes account for a large portion of the Earth's biomass as well as biodiversity, but still not

Advances in Biological Science Research. https://doi.org/10.1016/B978-0-12-817497-5.00012-4

much is known about their metabolic and ecological properties. To tap these CP currently proves to be a great challenge to the scientific community, and until we solve the mystery of these CP our knowledge toward microbial communities will be minuscule [9].

Scientists during the last 7 to 8 years have addressed this issue of uncultivable microbes using advanced and newer technologies of genome sequencing. Nowadays the microbial genome from any environmental sample can be directly sequenced using metagenomics and single-cell genomics, but these techniques have their own pros and cons. Shotgun sequencing followed by its assembly in the case of metagenomic approach results in genome fragments from different organisms, which subsequently can be binned into separate genomes using shared features, viz., homology, tetranucleotides, and codon usage. However, the limitation of this technique is that sometimes binning does not reveal the identity of its strains, representing a different composite of genomic fragments from diverse populations [10]. In contrast to the metagenomic approach, single-cell genomics involves the separation of a single cell from the environment followed by its lysis, amplification, and sequencing of the genomic DNA. Although this approach overcomes the drawback of the metagenomic approach, due to limitations in amplification of the entire genome, it often results in fragments representing incomplete genome. These two techniques have contributed quite a lot in recent years in terms of new insights into discovering the hidden uncultivable microbes in any environment [1].

Every microorganism in any environment possesses a unique set of genes within its genome, and the combination of the genomes of all the microbial members in any particular environment is known as metagenome. The metagenome technology, i.e., metagenomics, has helped in accumulating the genomic DNA sequences possessing unique properties for various novel biotechnological applications [11]. The presence of an overwhelming majority of uncultivable microbes in any environment will always result in uncovering the hitherto unknown proteins and sequences pertaining to it, thus making this approach more advantageous over the culture-based traditional approach [12]. The importance of metagenomic approach can be assessed merely based on the fact that nearly one million novel open reading frames encoding microbial enzymes were successfully identified in a single sample of marine prokaryotic plankton obtained from the Sargasso Sea [13].

12.2 History of metagenomic approach

The term *metagenomics* was first coined by a group led by Jo Handelsman [14]. Metagenomics is also known as ecogenomics, community genomics, and environmental genomics [15]. The DNA cloning directly from the environmental sample was first proposed by Pace [16], and as early as 1991, the first such type of cloning was reported in phage vectors [17]. The construction of the metagenomic library from the mixture of organisms enriched on dried grasses was achieved in similar lines [18]. The library resulted in an expression of cellulolytic activity, which at that time was referred to as zoolibraries (the term was not much used in this field) [18]. The major outcome came from the team led by DeLong that constructed metagenomic libraries from prokaryotes in seawater samples [19]. The library constructed by DeLong resulted into 40 kb clone constituting a 16S rRNA gene, identifying this clone as archaeon that previously had never been cultivated.

When metagenomics became popularized, some aspects of this technology were patented for the first time by biotechnological companies, including Diverse (San Diego, California, USA) and TerraGen Discovery (Vancouver, Canada) in 1996 [13]. Cloning of PCR-generated microbial genes from soil DNA into the partial polyketide synthase gene cluster of recipient *Streptomyces* strain was shown for the first time by TerraGen Discovery company in order to explain the importance of cloning and expression of environmental genes [20,21].

Metagenomics became well known when two different but important research works using this technique were published in 2004. These research works clearly demonstrated the application of random whole-genome shotgun sequencing in understanding the microbial populations present in diverse habitats [22,23]. The research group headed by Tyson [22] used samples from the extreme environment that generated only 76 Mb DNA sequence data and resulted in the complete assembly of genomes of the dominant species along with their metabolic pathway. Whereas the research group of Craig Venter [23] analyzed environmental samples containing a large number of species resulting in a sequence database of 2 Gb. These two projects led to a flood of information in the area of metagenomics as nearly 200 projects with over 450 different environmental samples came into existence within a very short period according to the GOLD database [24].

12.3 Approach, strategies, and tools used in the metagenomic analysis

The metagenomic approach starts with the isolation of metagenomic DNA from an environmental sample followed by its cloning and transformation into a suitable cloning vector and host system (Fig. 12.1A and B). The resulted transformant is then screened for either phylogenetic markers/anchors, such as 16S rRNA gene, or it can be screened for any particular gene of interest using multiplex PCR or hybridization or enzyme substrate reaction or antibiotic production [22,23,25–34]. Each approach has its own strengths as well as weaknesses, but together these approaches have so far broadened our knowledge about the uncultivable microbes, which otherwise would have remained a mystery to us.

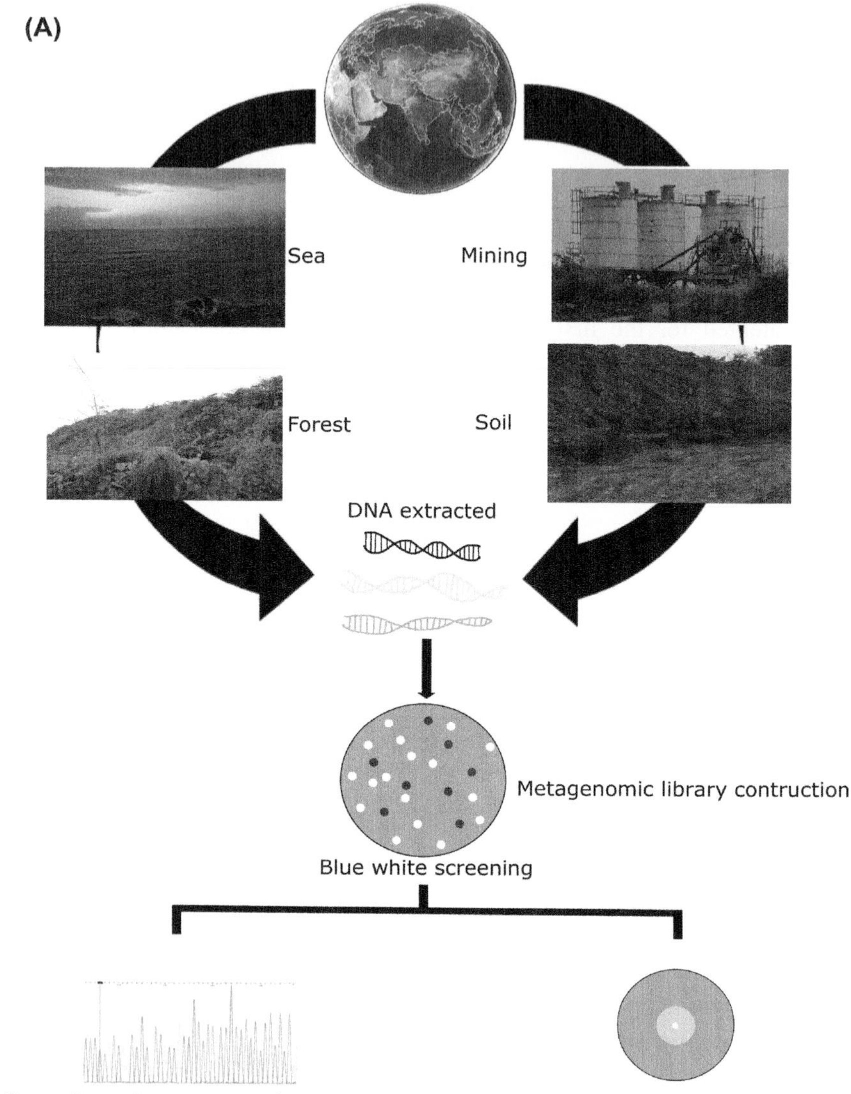

FIGURE 12.1 (A) Schematic representation of metagenomic approach. (B) Cloning and transformation of metagenomic DNA into suitable vector and host.

12.3.1 Isolation of metagenomic DNA

Metagenomic DNA recovery from any environmental sample suitable for metagenomic library construction is a challenging task as various contaminations, viz., humic, fulvic acids, DNases present in soil or sediment samples [35–37], cause major hindrances in downstream process of isolated DNA.

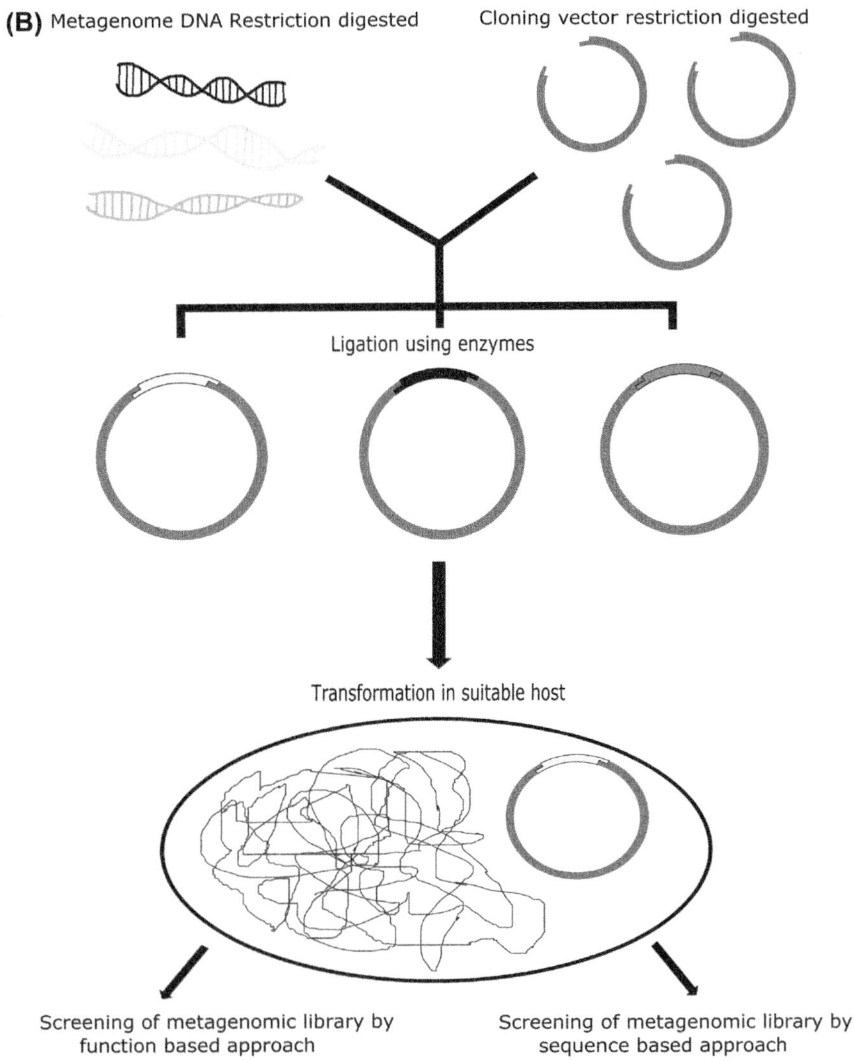

FIGURE 12.1 cont'd

The second drawback in case of isolation of intact high-molecular-weight DNA/metagenomic DNA is there are always chances of DNA getting degraded or sheared in the harsh process of extraction. To ensure the correct representation of any community metagenome, the quality as well as the quantity of extracted metagenomic DNA play an important role [29]. In order to overcome this limitation, several protocols have been devised that are fast, efficient, and yield pure metagenomic DNA from various environmental samples [38–43].

Apart from the above-mentioned protocols for extraction of metagenomic DNA, there are several commercially available kits in the market currently that claim to yield high as well as pure metagenomic DNA from various sources. These kits are easy to use but sometimes require sophisticated equipment in order to carry out the DNA extraction in an efficient manner. Some of the commercially available kits are PowerSoil, UltraClean, and RNA PowerSoil (Mo Bio Laboratories, California, USA); FastDNA Spin kit for soil and FastRNA Pro soil-direct kit (MP Biomedicals, Solon, Ohio, USA); and SoilMaster (Epicentre Biotechnologies, Madison, Wisconsin, USA).

12.3.2 Cloning vector and host

The vector system used in any metagenomic library construction depends upon many criteria such as the desired insert size of the library, screening method or strategy that will be used, quality of the isolated DNA, and vector copy number [44]. The construction of the metagenomic libraries can be classified into two groups: the first type with small insert libraries made in plasmid vector with <15 kbps insert size and the second type with large insert made in either cosmid or fosmid with 40 kbps insert size or in bacterial artificial chromosomes (BACs) with >40 kbps of insert size. Fosmids and BACs are the vectors that are most commonly used for metagenomic studies due to their capacity to clone larger inserts as well as maintaining the stability of prokaryotic as well as eukaryotic clones [21,45,46].

In most of the cases, the strains of *Escherichia coli* is preferred as a host in cloning and expression studies but there is a limitation in terms of expression of the genes in the *E. coli* host from an organism that is more distantly related to Enterobacteriaceae family. Therefore, in recent years, scientists have successfully tested other host organisms such as *Streptomyces lividans*, *Pseudomonas putida*, *bacillus subtilis*, and *Rhizobium leguminosarum*are for their studies [21,47,48].

12.3.3 Screening of metagenomic clones

Once the metagenomic library is constructed successfully, this library can be screened for either "functional" or "sequence-based" screening procedures [49,50].

Functional metagenomics is based on the expression of genes from the cloned metagenomic DNA in the respective host using specific substrates. This technique is advantageous to explore novel genes encoding various industrially important enzymes and other biomolecules, including antibiotics. There are several limiting factors in functional screening, viz., an expression of the gene(s) inserted in the foreign hosts, availability of substrates, and recognition of regulatory elements such as operators and promoters [51].

The sequence-based metagenomic approach relies on DNA sequencing of the metagenomic library followed by identification of the gene(s) encoding enzymes regulating different metabolic pathways as well as the microbial community structure that defines microbial diversity involved in specific metabolic/degradative pathways. The sequence-based approach requires the use of computers along with their respective software for genome assembly in order to find out the type of organisms as well as functions of different genes in specific samples [17,19,50].

12.3.4 Sequencing and bioinformatics analysis of the metagenomic clones

The complexity of microbiomes can be assessed in an example stated by Hess et al. [52] who estimated that approximately 1000 operational taxonomic units can be present in a single cow rumen, that's why sequencing technology needs to be very sensitive in order to capture the sequences of all the species within any microbiome sample. Nowadays, second- and third-generation sequencing, known as next-generation sequencing, has aided in the analysis of metagenome in a promising manner. The second-generation sequencing consists of Ion Torrent and Illumina, which are able to produce millions of short reads (200–400 bps), whereas third-generation sequencing procedure includes PacBio and ONT produces longer reads, i.e., 6–20 kbps, but they result in fewer reads per run.

The next thing after sequencing is its assembly or genome assembly from the smaller sequenced fragments. The assembling process of any genome from the smaller fragments of sequences is very tedious technique, due to the presence of repetitive elements within the genome. The de novo type of assembler is a reference-free type of strategy for constructing the contiguous sequences known as contigs. The de novo assembly software tools use either overlap layout consensus (OLC) or the de Bruijn graph approach. In the OLC approach, pairwise comparing of all reads are done to find out the regions with significant overlaps, whereas in de Bruijn graph approach a graph is constructed by reading the consecutive k-mers within each read [53]. The de Bruijn graph approach is less expensive than the OLC approach but is more sensitive to sequencing error than that of OLC approach. The list of other software used in the metagenomic analysis is tabulated in Table 12.1.

12.4 Application of the metagenomic approach

In the advancement of culture-independent approach for studying microbial community, we are now evaluating the microorganisms that inhabit each and every ecological niche from the deepest oceans to the gut of almost every organism. This technique has changed the way the bacterial phyla

TABLE 12.1 Commonly used software in metagenomic analysis.

S. No.	Software	Application	Links	References
1	PRINTSEQ	Quality control tool for sequence trimming.	http://prinseq.sourceforge.net/	[54]
2	FastQC	Quality control tool for high-throughput sequence data.	http://www.bioinformatics.babraham.ac.uk/projects/fastqc/	[55]
3	NGS QC toolkit	Tool for quality control analysis performed in parallel environment.	http://www.nipgr.res.in/ngsqctoolkit.html	[56]
4	Mothur	From reads quality analysis to taxonomic classification.	http://www.mothur.org/	[57]
5	Meta-QC-Chain	Parallel environment tool for quality control.	http://www.computationalbioenergy.org/qc-chain.html	[58]
6	QIIME	Quality pretreatment of raw reads, taxonomic annotation, calculus of diversity estimators, and comparison of metaprofiling or metagenomic data.	http://qiime.org/	[59]
7	PICRUSt	Predictor of metabolic potential from taxonomic information.	http://picrust.github.io/picrust/	[60]
8	CARMA	Phylogenetic classification of reads based on Pfam conserved domains.	http://omictools.com/carma-s1021.html	[61]
9	MOCAT	Pipeline that includes quality treatment of metagenomic reads and taxonomic annotation.	http://vm-lux.embl.de/~kultima/MOCAT2/index.html	[62]
10	Parallel-meta	Taxonomic annotation of ribosomal gene markers sequences obtained by metaprofiling or metagenomic reads.	http://www.computationalbioenergy.org/parallel-meta.html	[63]

11	TETRA	Taxonomic classification by comparison of tetranucleotide patterns.	http://omictools.com/tetra-s1030.html	[64]
12	MetaVelvet	De novo assembler of metagenomic short reads.	http://metavelvet.dna.bio.keio.ac.jp/	[65]
13	MetaORFA	Assembly of peptides.	Not available	[66]
14	ProViDE	Analysis of viral diversity in metagenomic samples.	http://metagenomics.atc.tcs.com/binning/ProViDE	[67]
15	MetagenomeSeq	Analysis of differential abundance of 16S rRNA gene in metaprofiling data.	http://bioconductor.org/packages/release/bioc/html/metagenomeSeq.html	[68]
16	MG-RAST	Taxonomic and functional annotation, comparative metagenomics.	http://metagenomics.anl.gov	[69]
17	IMG/M	Functional annotation, phylogenetic distribution of genes and comparative metagenomics.	https://img.jgi.doe.gov/cgi-bin/m/main.cgi	[70]
18	RayMeta	Assembler of de novo metagenomic reads and taxonomy profiler by Ray Communities.	http://denovoassembler.sourceforge.net/	[71]
19	IDBA-UD	Assembler de novo of metagenomic sequences with uneven depth.	http://i.cs.hku.hk/~alse/hkubrg/projects/idba_ud/	[72]

were defined previously by culture-dependent approach as the fact is that approximately 70% of all known bacterial phyla are totally devoid of any cultivable representative [73]. The great diversity among metagenomic 16S rRNA sequences can be easily detected using phylogenetic analysis (Fig. 12.2). In the case of pharmaceutical industries, bacterial strains are well known for the discovery of novel antibiotics to combat the problem of antibiotic-resistant pathogens [74]. To date, more than 10,000 biologically active molecules against serious human pathogens, viz., HIV, cancer, and inflammation have been isolated from known cultured actinomycetes and mostly from a single genus of *Streptomyces* [75]. Therefore, we can truly hypothesize the potential of discovering novel biomolecules from the uncultivable microbes as they are known to outnumber the cultivable microbes.

The industrial application of the metagenomic approach has resulted in the isolation of numerous industrially important enzymes (Table 12.2). In the field of other natural products, viz., drug discovery or metabolites, putative pathways have been reported from various environmental samples, i.e., from soil [93,94], tunicates [95,96], sponges [97] and from insects [98]. But still, the challenge to link these biosynthetic genes to a product has hindered discovery, whereas only in a few cases successful expression or in vitro characterization has been achieved [99,100]. The rate of discovery of newer genes has been accelerated with the advent of the metagenomic approach, and therefore, the application of the metagenomic approach to humankind seems to be endless.

12.5 Conclusion remarks

The viewpoint of microbiologists has been changed after the emergence of the metagenomic approach in recent years. Many concepts defined by cultural microbes have been altered drastically with the discovery of uncultivable microbes. In terms of its application to various biotechnological industries, metagenomics is limitless as it has led to some novel discoveries. This technique had shown a promising future with the advancement of sequencing methods and other bioinformatics analysis tools. There are still more extreme environments, such as thermal vents in the deep sea, the frozen Antarctic region, vertebrate gut microbiomes, cold soils, and plant rhizospheres, that needs to be explored with the help of this technique. Furthermore, this state of art technology, along with its improved methods for analysis, in near future will attract the scientific community of diverse fields. Metagenomics seems to have a very promising future in helping mankind to solve problems and develop a better understanding of the microbial community.

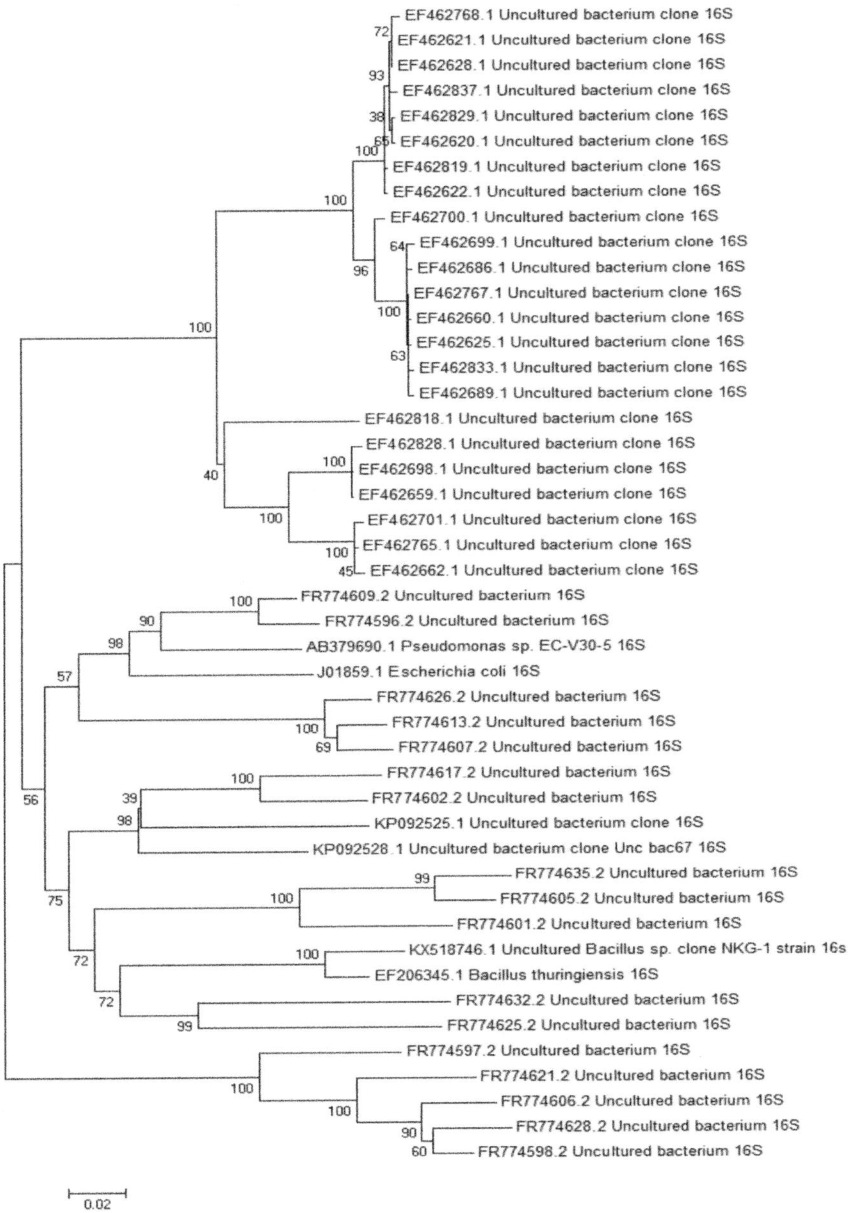

FIGURE 12.2 Phylogenetic analysis showing diversity among unculturable microbes using metagenomic approach.

TABLE 12.2 Various enzymes isolated using metagenomics approach.

S. No.	Source	Gene Name	Host/ Vector	Avg. Insert size (kb)	References
1	Soil	Esterase/lipase	*E. coli*, plasmid	6	[76]
2	Compost	Esterase, amylase, phosphatase, dioxygenase, protease	*E. coli*, plasmid	3.2	[77]
3	Sediment enrichment	Cellulase	*E. coli*, λ phage	6	[78]
4	Loam soil	Oxygenase	*E. coli*, plasmid	5.5	[79]
5	Soil	β-Lactamase	*E. coli*, plasmid	5	[80]
6	Activated sludge treating coke plant wastewater	Extradiol dioxygenase	*E. coli*, fosmid	33	[81]
7	Seawater	Chitinase	*E. coli*, λ phage	5	[82]
8	Glacial ice	DNA polymerase I	*E. coli*, plasmid	4	[83]
9	Soil/sediment enrichment	Dehydratase	*E. coli*, plasmid	4	[28]
10	Cow rumen	Mannanase/ glucanase/ xylanase	*E. coli*, phagemid	3	[84]
11	Sediment	Metalloprotease	*E. coli*	4	[85]

Sequence-based screening

S. No.	Source	Gene Name	Method	References
1	3-Chlorobenzoate enrichment	Benzoate 1,2-dioxygenase, chlorocatechol 1,2-dioxygenase	Degenerate PCR	[86]
2	Grassland soil	Nitrite reductase, nitrous oxide reductase	Probe hybridization	[87]
3	Sediment from hot spring	Pullulanase	Degenerate PCR	[88]

TABLE 12.2 Various enzymes isolated using metagenomics approach.—cont'd

Sequence-based screening				
S. No.	Source	Gene Name	Method	References
4	Bioreactors treating gold-bearing concentrates	Sulfur oxygenase reductase	Degenerate PCR	[89]
5	Deep-sea sediment	Alkane hydroxylase	Degenerate PCR	[90]
6	Marine sponge	Related polyketide synthesis	Degenerate PCR	[91,92]

Acknowledgments

We are thankful to Prof. Santosh Kumar Dubey for his assistance in preparing this chapter, and we are also very grateful to editors Dr. Milind Mohan Naik and Dr. Surya Nandan Meena for giving us the opportunity to contribute a chapter in their prestigious book.

References

[1] Solden L, Lloyd K, Wrighton K. The bright side of microbial dark matter: lessons learned from the uncultivated majority. Curr Opin Microbiol 2016;31:217–26.

[2] Castelle CJ, Wrighton KC, Thomas BC, Hug LA, Brown CT, Wilkins MJ, Frischkorn KR, Tringe SG, Singh A, Markillie LM, et al. Genomic expansion of domain archaea highlights roles for organisms from new phyla in anaerobic carbon cycling. Curr Biol 2015;25.

[3] Rinke C, Schwientek P, Sczyrba A, Ivanova NN, Anderson IJ, Cheng J-F, Darling A, et al. Insights into the phylogeny and coding potential of microbial dark matter. Nature 2013;499(7459):431–7.

[4] Brown CT, Hug LA, Thomas BC, Sharon I, Castelle CJ, Singh A, Wilkins MJ, Wrighton KC, Williams KH, Banfield JF. Unusual biology across a group comprising more than 15% of domain bacteria. Nature 2015;523(7559):208–11.

[5] Baker BJ, Lazar CS, Teske AP, Dick GJ. Genomic resolution of linkages in carbon, nitrogen, and sulfur cycling among widespread estuary sediment bacteria. Microbiome 2015;3:14.

[6] Hug LA, Baker BJ, Anantharaman K, Brown CT, Probst AJ, Castelle CJ, Butterfield CN, et al. A new view of the tree of life. Nat Microbiol 2016:16048.

[7] Yarza P, Yilmaz P, Pruesse E, Glockner FO, Ludwig W, Schleifer KH, Whitman WB, Euzeby J, Amann R, Rossello-Mora R. Uniting the classification of cultured and uncultured bacteria and archaea using 16S rRNA gene sequences. Nat Rev Microbiol 2014;12(9):635–45.

[8] Baker BJ. Omic approaches in microbial ecology: charting the unknown. Microbe 2013;8(9):353–9.

[9] Hedlund BP, Dodsworth JA, Murugapiran SK, Rinke C, Woyke T. Impact of single-cell genomics and metagenomics on the emerging view of extremophile "microbial dark matter". Extremophiles 2014;18(5):865–75.

[10] Sharon I, Banfield JF. Genomes from metagenomics. Science 2013;342:1057–8.

[11] Ferrer M, Martinez-Abarca F, Golyshin PN. Mining genomes and 'metagenomes' for novel catalysts. Curr Opin Biotechnol 2005;16:588–93.

[12] Rajendhran J, Gunasekaran P. Strategies for accessing soil metagenome for desired applications. Biotechnol Adv 2008;26(6):576–90.

[13] Lorenz P, Eck J. Metagenomics and industrial applications. Nat Rev Microbiol 2005;3(6):510.

[14] Handelsman J, Rondon MR, Brady SF, Clardy J, Goodman RM. Molecular biological access to the chemistry of unknown soil microbes: a new frontier for natural products. Chem Biol 1998;5(10):R245–9.

[15] Chistoserdova L. Recent progress and new challenges in metagenomics for biotechnology. Biotechnol Lett 2010;32(10):1351–9.

[16] Pace NR. Analyzing natural microbial populations by rRNA sequences. ASM News 1985;51:4–12.

[17] Schmidt TM, DeLong EF, Pace NR. Analysis of a marine picoplankton community by 16S rRNA gene cloning and sequencing. J Bacteriol 1991;173(14):4371–8.

[18] Healy FG, Ray RM, Aldrich HC, Wilkie AC, Ingram LO, Shanmugam KT. Direct isolation of functional genes encoding cellulases from the microbial consortia in a thermophilic, anaerobic digester maintained on lignocellulose. Appl Microbiol Biotechnol 1995;43(4):667–74.

[19] Stein JL, Marsh TL, Wu KY, Shizuya H, DeLong EF. Characterization of uncultivated prokaryotes: isolation and analysis of a 40-kilobase-pair genome fragment from a planktonic marine archaeon. J Bacteriol 1996;178(3):591–9.

[20] Seow KT, Meurer GU, Gerlitz MA, Wendt-Pienkowski EV, Hutchinson CR, Davies JU. A study of iterative type II polyketide synthases, using bacterial genes cloned from soil DNA: a means to access and use genes from uncultured microorganisms. J Bacteriol 1997;179(23):7360–8.

[21] Wang GY, Graziani E, Waters B, Pan W, Li X, McDermott J, Meurer G, Saxena G, Andersen RJ, Davies J. Novel natural products from soil DNA libraries in a *Streptomycete* host. Org Lett 2000;2(16):2401–4.

[22] Tyson GW, Chapman J, Hugenholtz P, Allen EE, Ram RJ, Richardson PM, Solovyev VV, Rubin EM, Rokhsar DS, Banfield JF. Community structure and metabolism through reconstruction of microbial genomes from the environment. Nature 2004;428(6978): 37–43.

[23] Venter JC, Remington K, Heidelberg JF, Halpern AL, Rusch D, Eisen JA, Wu D, Paulsen I, Nelson KE, Nelson W, Fouts DE. Environmental genome shotgun sequencing of the Sargasso Sea. Science 2004;304(5667):66–74.

[24] Liolios K, Chen IM, Mavromatis K, Tavernarakis N, Hugenholtz P, Markowitz VM, Kyrpides NC. The Genomes on Line Database (GOLD) in 2009: status of genomic and metagenomic projects and their associated metadata. Nucleic Acids Res 2009;38 (Suppl. 1):D346–54.

[25] Courtois S, Cappellano CM, Ball M, Francou FX, Normand P, Helynck G, Martinez A, Kolvek SJ, Hopke J, Osburne MS, August PR. Recombinant environmental libraries provide access to microbial diversity for drug discovery from natural products. Appl Environ Microbiol 2003;69(1):49–55.

[26] Diaz-Torres ML, McNab R, Spratt DA, Villedieu A, Hunt N, Wilson M, Mullany P. Novel tetracycline resistance determinant from the oral metagenome. Antimicrob Agents Chemother 2003;47(4):1430–2.

[27] Gillespie DE, Brady SF, Bettermann AD, Cianciotto NP, Liles MR, Rondon MR, Clardy J, Goodman RM, Handelsman J. Isolation of antibiotics turbomycin A and B from a metagenomic library of soil microbial DNA. Appl Environ Microbiol 2002;68(9):4301–6.

[28] Knietsch A, Bowien S, Whited G, Gottschalk G, Daniel R. Identification and characterization of coenzyme B12-dependent glycerol dehydratase-and diol dehydratase-encoding genes from metagenomic DNA libraries derived from enrichment cultures. Appl Environ Microbiol 2003;69(6):3048–60.

[29] Lorenz P, Liebeton K, Niehaus F, Eck J. Screening for novel enzymes for biocatalytic processes: accessing the metagenome as a resource of novel functional sequence space. Curr Opin Biotechnol 2002;13(6):572–7.

[30] Lorenz P, Schleper C. Metagenome-a challenging source of enzyme discovery. J Mol Catal B Enzym 2002;19:13–9.

[31] MacNeil IA, Tiong CL, Minor C, August PR, Grossman TH, Loiacono KA, Lynch BA, Phillips T, Narula S, Sundaramoorthi R, Tyler A. Expression and isolation of antimicrobial small molecules from soil DNA libraries. J Mol Microbiol Biotechnol 2001;3(2):301–8.

[32] Majernık A, Gottschalk G, Daniel R. Screening of environmental DNA libraries for the presence of genes conferring Na+ (Li+)/H+ antiporter activity on *Escherichia coli*: characterization of the recovered genes and the corresponding gene products. J Bacteriol 2001;183(22):6645–53.

[33] Rondon MR, August PR, Bettermann AD, Brady SF, Grossman TH, Liles MR, Loiacono KA, et al. Cloning the soil metagenome: a strategy for accessing the genetic and functional diversity of uncultured microorganisms. Appl Environ Microbiol 2000;66(6):2541–7.

[34] Schloss PD, Handelsman J. Biotechnological prospects from metagenomics. Curr Opin Biotechnol 2003;14(3):303–10.

[35] Tsai YL, Olson BH. Rapid method for separation of bacterial DNA from humic substances in sediments for polymerase chain reaction. Appl Environ Microbiol 1992;58(7):2292–5.

[36] Tebbe CC, Vahjen W. Interference of humic acids and DNA extracted directly from soil in detection and transformation of recombinant DNA from bacteria and a yeast. Appl Environ Microbiol 1993;59(8):2657–65.

[37] Alm EW, Zheng D, Raskin L. The presence of humic substances and DNA in RNA extracts affects hybridization results. Appl Environ Microbiol 2000;66(10):4547–54.

[38] Zhou J, Bruns MA, Tiedje JM. DNA recovery from soils of diverse composition. Appl Environ Microbiol 1996;62(2):316–22.

[39] Yeates C, Gillings MR, Davison AD, Altavilla N, Veal DA. Methods for microbial DNA extraction from soil for PCR amplification. Biol Proced Online 1998;1(1):40–7.

[40] Bürgmann H, Pesaro M, Widmer F, Zeyer J. A strategy for optimizing quality and quantity of DNA extracted from soil. J Microbiol Methods 2001;45(1):7–20.

[41] Amorim JH, Macena TN, Lacerda-Junior GV, Rezende RP, Dias JC, Brendel M, Cascardo JC. An improved extraction protocol for metagenomic DNA from a soil of the Brazilian Atlantic Rainforest. Genet Mol Res 2008;7(4):1226–32.

[42] Sagar K, Singh SP, Goutam KK, Konwar BK. Assessment of five soil DNA extraction methods and a rapid laboratory-developed method for quality soil DNA extraction for 16S rDNA-based amplification and library construction. J Microbiol Methods 2014;97:68–73.

[43] Shamim K, Sharma J, Dubey SK. Rapid and efficient method to extract metagenomic DNA from estuarine sediments. 3 Biotech 2017;7(3):182.

[44] Daniel R. The metagenomics of soil. Nat Rev Microbiol 2005;3(6):470.

[45] Bierman M, Logan R, O'brien K, Seno ET, Rao RN, Schoner BE. Plasmid cloning vectors for the conjugal transfer of DNA from *Escherichia coli* to *Streptomyces* spp. Gene 1992;116(1):43–9.

[46] Shizuya H, Birren B, Kim UJ, Mancino V, Slepak T, Tachiiri Y, Simon M. Cloning and stable maintenance of 300-kilobase-pair fragments of human DNA in *Escherichia coli* using an F-factor-based vector. Proc Natl Acad Sci Unit States Am 1992;89(18):8794–7.

[47] Martinez A, Kolvek SJ, Yip CL, Hopke J, Brown KA, MacNeil IA, Osburne MS. Genetically modified bacterial strains and novel bacterial artificial chromosome shuttle vectors for constructing environmental libraries and detecting heterologous natural products in multiple expression hosts. Appl Environ Microbiol 2004;70(4):2452–63.

[48] Wexler M, Bond PL, Richardson DJ, Johnston AW. A wide host-range metagenomic library from a wastewater treatment plant yields a novel alcohol/aldehyde dehydrogenase. Environ Microbiol 2005;7(12):1917–26.

[49] Handelsman J. Metagenomics: application of genomics to uncultured microorganisms. Microbiol Mol Biol Rev 2004;68(4):669–85.

[50] Cruz JM, Ortega MA, Cruz JC, Ondina P, Santiago R, Rios-Velazquez C. Unravelling activities by functional-based approaches using metagenomic libraries from dry and rain forest soils in Puerto Rico. In: Current research technology and education topics. Applied microbiology and microbial biotechnology. Badajoz, Spain: Formatex; 2010. p. 1471–8.

[51] Craig JW, Chang FY, Kim JH, Obiajulu SC, Brady SF. Expanding small-molecule functional metagenomics through parallel screening of broad-host-range cosmid environmental DNA libraries in diverse Proteobacteria. Appl Environ Microbiol 2010;76(5):1633–41.

[52] Hess M, Sczyrba A, Egan R, Kim TW, Chokhawala H, Schroth G, Luo S, Clark DS, Chen F, Zhang T, Mackie RI. Metagenomic discovery of biomass-degrading genes and genomes from cow rumen. Science 2011;331(6016):463–7.

[53] Pevzner PA, Tang H, Waterman MS. An Eulerian path approach to DNA fragment assembly. Proc Natl Acad Sci Unit States Am 2001;98(17):9748–53.

[54] Schmieder R, Edwards R. Quality control and preprocessing of metagenomic datasets. Bioinformatics 2011;27(6):863–4.

[55] Andrews S. Babraham bioinformatics-FastQC a quality control tool for high throughput sequence data. 2015.

[56] Patel RK, Jain M. NGS QC Toolkit: a toolkit for quality control of next generation sequencing data. PLoS One 2012;7(2). e30619.

[57] Schloss PD, Westcott SL, Ryabin T, Hall JR, Hartmann M, Hollister EB, Lesniewski RA, Oakley BB, Parks DH, Robinson CJ, Sahl JW. Introducing mothur: open-source, platform-independent, community-supported software for describing and comparing microbial communities. Appl Environ Microbiol 2009;75(23):7537–41.

[58] Zhou Q, Su X, Jing G, Ning K. Meta-QC-Chain: comprehensive and fast quality control method for metagenomic data. Genom Proteom Bioinform 2014;12(1):52–6.

[59] Caporaso JG, Kuczynski J, Stombaugh J, Bittinger K, Bushman FD, Costello EK, Fierer N, Pena AG, Goodrich JK, Gordon JI, Huttley GA. QIIME allows analysis of high-throughput community sequencing data. Nat Methods 2010;7(5):335.

[60] Langille MG, Zaneveld J, Caporaso JG, McDonald D, Knights D, Reyes JA, Clemente JC, Burkepile DE, Thurber RL, Knight R, Beiko RG. Predictive functional profiling of

microbial communities using 16S rRNA marker gene sequences. Nat Biotechnol 2013;31(9):814.

[61] Krause L, Diaz NN, Goesmann A, Kelley S, Nattkemper TW, Rohwer F, Edwards RA, Stoye J. Phylogenetic classification of short environmental DNA fragments. Nucleic Acids Res 2008;36(7):2230–9.

[62] Kultima JR, Sunagawa S, Li J, Chen W, Chen H, Mende DR, Arumugam M, Pan Q, Liu B, Qin J, Wang J. MOCAT: a metagenomics assembly and gene prediction toolkit. PLoS One 2012;7(10). e47656.

[63] Su X, Pan W, Song B, Xu J, Ning K. Parallel-META 2.0: enhanced metagenomic data analysis with functional annotation, high performance computing and advanced visualization. PLoS One 2014;9(3). e89323.

[64] Teeling H, Waldmann J, Lombardot T, Bauer M, Glöckner FO. TETRA: a web-service and a stand-alone program for the analysis and comparison of tetranucleotide usage patterns in DNA sequences. BMC Bioinf 2004;5(1):163.

[65] Namiki T, Hachiya T, Tanaka H, Sakakibara Y. MetaVelvet: an extension of Velvet assembler to de novo metagenome assembly from short sequence reads. Nucleic Acids Res 2012;40(20):e155.

[66] Ye Y, Tang H. An ORFome assembly approach to metagenomics sequences analysis. Comput Syst Bioinformatics Conf 2008;7:3–13.

[67] Ghosh TS, Mohammed MH, Komanduri D, Mande SS. ProViDE: a software tool for accurate estimation of viral diversity in metagenomic samples. Bioinformation 2011;6(2):91.

[68] Paulson JN, Stine OC, Bravo HC, Pop M. Differential abundance analysis for microbial marker-gene surveys. Nat Methods 2013;10(12):1200–2.

[69] Meyer F, Paarmann D, D'Souza M, Olson R, Glass EM, Kubal M, Paczian T, Rodriguez A, Stevens R, Wilke A, Wilkening J. The metagenomics RAST server—a public resource for the automatic phylogenetic and functional analysis of metagenomes. BMC Bioinf 2008;9(1):386.

[70] Markowitz VM, Chen IM, Chu K, Szeto E, Palaniappan K, Grechkin Y, Ratner A, Jacob B, Pati A, Huntemann M, Liolios K. IMG/M: the integrated metagenome data management and comparative analysis system. Nucleic Acids Res 2011;40(D1):D123–9.

[71] Boisvert S, Raymond F, Godzaridis É, Laviolette F, Corbeil J. Ray Meta: scalable de novo metagenome assembly and profiling. Genome Biol 2012;13(12):R122.

[72] Peng Y, Leung HC, Yiu SM, Chin FY. IDBA-UD: a de novo assembler for single-cell and metagenomic sequencing data with highly uneven depth. Bioinformatics 2012;28(11):1420–8.

[73] Achtman M, Wagner M. Microbial diversity and the genetic nature of microbial species. Nat Rev Microbiol 2008;6(6):431–40.

[74] Wright GD. Antibiotics: a new hope. Chem Biol 2012;19(1):3–10.

[75] Bérdy J. Bioactive microbial metabolites. J Antibiot 2005;58(1):1–26.

[76] Henne A, Schmitz RA, Bömeke M, Gottschalk G, Daniel R. Screening of environmental DNA libraries for the presence of genes conferring lipolytic activity on *Escherichia coli*. Appl Environ Microbiol 2000;66(7):3113–6.

[77] Lämmle K, Zipper H, Breuer M, Hauer B, Buta C, Brunner H, Rupp S. Identification of novel enzymes with different hydrolytic activities by metagenome expression cloning. J Biotechnol 2007;127(4):575–92.

[78] Rees HC, Grant S, Jones B, Grant WD, Heaphy S. Detecting cellulase and esterase enzyme activities encoded by novel genes present in environmental DNA libraries. Extremophiles 2003;7(5):415–21.

[79] Van Hellemond EW, Janssen DB, Fraaije MW. Discovery of a novel styrene monooxygenase originating from the metagenome. Appl Environ Microbiol 2007;73(18):5832–9.

[80] Gabor EM. Harvesting novel biocatalysts from the metagenome. University Library Groningen; 2004.

[81] Suenaga H, Ohnuki T, Miyazaki K. Functional screening of a metagenomic library for genes involved in microbial degradation of aromatic compounds. Environ Microbiol 2007;9(9):2289–97.

[82] Cottrell MT, Moore JA, Kirchman DL. Chitinases from uncultured marine microorganisms. Appl Environ Microbiol 1999;65(6):2553–7.

[83] Simon C, Herath J, Rockstroh S, Daniel R. Rapid identification of genes encoding DNA polymerases by function-based screening of metagenomic libraries derived from glacial ice. Appl Environ Microbiol 2009;75(9):2964–8.

[84] Palackal N, Lyon CS, Zaidi S, Luginbühl P, Dupree P, Goubet F, Macomber JL, Short JM, Hazlewood GP, Robertson DE, Steer BA. A multifunctional hybrid glycosyl hydrolase discovered in an uncultured microbial consortium from ruminant gut. Appl Microbiol Biotechnol 2007;74(1):113–24.

[85] Shamim K, Sharma J, Mutnale M, Mujawar S, Dubey SK. Characterization of a metagenomic serine metalloprotease and molecular docking studies. Process Biochem 2018;71:69–75.

[86] Morimoto S, Fujii T. A new approach to retrieve full lengths of functional genes from soil by PCR-DGGE and metagenome walking. Appl Microbiol Biotechnol 2009;83(2):389–96.

[87] Demanèche S, Philippot L, David MM, Navarro E, Vogel TM, Simonet P. Characterization of denitrification gene clusters of soil bacteria via a metagenomic approach. Appl Environ Microbiol 2009;75(2):534–7.

[88] Tang K, Kobayashi RS, Champreda V, Eurwilaichitr L, Tanapongpipat S. Isolation and characterization of a novel thermostable neopullulanase-like enzyme from a hot spring in Thailand. Biosci Biotechnol Biochem 2008;72(6):1448–56.

[89] Chen ZW, Liu YY, Wu JF, She Q, Jiang CY, Liu SJ. Novel bacterial sulfur oxygenase reductases from bioreactors treating gold-bearing concentrates. Appl Microbiol Biotechnol 2007;74(3):688–98.

[90] Xu M, Xiao X, Wang F. Isolation and characterization of alkane hydroxylases from a metagenomic library of Pacific deep-sea sediment. Extremophiles 2008;12(2):255–62.

[91] Fieseler L, Hentschel U, Grozdanov L, Schirmer A, Wen G, Platzer M, Hrvatin S, Butzke D, Zimmermann K, Piel J. Widespread occurrence and genomic context of unusually small polyketide synthase genes in microbial consortia associated with marine sponges. Appl Environ Microbiol 2007;73(7):2144–55.

[92] Fisch KM, Gurgui C, Heycke N, Van Der Sar SA, Anderson SA, Webb VL, Taudien S, Platzer M, Rubio BK, Robinson SJ, Crews P. Polyketide assembly lines of uncultivated sponge symbionts from structure-based gene targeting. Nat Chem Biol 2009;5(7):494.

[93] Chang FY, Brady SF. Cloning and characterization of an environmental DNA-derived gene cluster that encodes the biosynthesis of the antitumor substance BE-54017. J Am Chem Soc 2011;133(26):9996–9.

[94] Feng Z, Kallifidas D, Brady SF. Functional analysis of environmental DNA-derived type II polyketide synthases reveals structurally diverse secondary metabolites. Proc Natl Acad Sci Unit States Am 2011;108(31):12629–34.

[95] Rath CM, Janto B, Earl J, Ahmed A, Hu FZ, Hiller L, Dahlgren M, Kreft R, Yu F, Wolff JJ, Kweon HK. Meta-omic characterization of the marine invertebrate microbial consortium that produces the chemotherapeutic natural product ET-743. ACS Chem Biol 2011;6(11):1244–56.
[96] Schmidt EW, Donia MS, McIntosh JA, Fricke WF, Ravel J. Origin and variation of tunicate secondary metabolites. J Natural Prod 2012;75(2):295–304.
[97] Freeman MF, Gurgui C, Helf MJ, Morinaka BI, Uria AR, Oldham NJ, Sahl HG, Matsunaga S, Piel J. Metagenome mining reveals polytheonamides as posttranslationally modified ribosomal peptides. Science 2012;338:387–90.
[98] Piel J. A polyketide synthase-peptide synthetase gene cluster from an uncultured bacterial symbiont of *Paederus* beetles. Proc Natl Acad Sci Unit States Am 2002;99(22):14002–7.
[99] Brady SF, Chao CJ, Handelsman J, Clardy J. Cloning and heterologous expression of a natural product biosynthetic gene cluster from eDNA. Org Lett 2001;3(13):1981–4.
[100] Long PF, Dunlap WC, Battershill CN, Jaspars M. Shotgun cloning and heterologous expression of the patellamide gene cluster as a strategy to achieving sustained metabolite production. Chembiochem 2005;6(10):1760–5.

Chapter 13

Halophilic archaea as beacon for exobiology: recent advances and future challenges

Abhilash Sundarasami, Akshaya Sridhar, Kabilan Mani
Department of Biotechnology, PSG College of Technology, Coimbatore, India

13.1 Introduction

Exobiology, a term coined by Joshua Lederberg, a Noble Prize-winning molecular biologist, deals with the possibility of life on other planets or in space. The field emerged in the 1950s and 1960s because of the advent in rocket technology, a newfound interest in the search of exoplanets, origin-of-life studies, and the search for extraterrestrial intelligence. We can define exobiology as a subset of astrobiology, where astrobiology is a field that sits at the interface of a Venn diagram whose sets include biological sciences, chemistry, earth sciences, and space sciences. Recently it has gained new momentum due to new efforts in microbiology, geosciences, planetary sciences, and astronomy [1].

Exobiology sets itself to find answers to these questions: How did life originate and diversify? How does life co-evolve on a planet? What lies in the future for life? And finally, the most important question: Are we alone in the universe? Or are we the only type of life in the universe? Exobiology as a field collectively tries to answer all these questions in a quest to quench the thirst of human curiosity. Exobiology recognizes the fact that life on Earth cannot be understood completely without understanding its cosmic links. According to Panspermia theory, life on Earth could have originated from bacterial species transported by meteorites that were able to adapt and proliferate on our planet [2]. We usually start with questions like, how did life originate? Where did it start? Was it inevitable? What was the evidence for early life? All encompass the area of this research. Once life started spreading across Earth, then the question arises regarding the limits of life, such as to what extremities could life survive on Earth? Once we figure out the extremities in which life can exist, then we can start looking for planetary bodies with similar chemical and

Advances in Biological Science Research. https://doi.org/10.1016/B978-0-12-817497-5.00013-6

physical conditions to host life. In the rest of the chapter, we will look at some of the missions that contributed a significant amount of knowledge toward exobiology and about some saline environments in space, and further on why halophiles are an important class of model organisms for exobiological studies.

13.2 Missions with exobiological significance

With the advancement in rocket technology and the ability to put gigantic telescopes in space, the quest toward search for life elsewhere outside of our planet got a huge boost with the discovery of hundreds of exoplanets capable of hosting life. In this section we discuss the missions dedicated toward astrobiology and exobiology. A concise list of space exploratory missions is presented in Table 13.1.

13.2.1 1960–2000

In 1964, when **Mariner 4** was launched, it was the first spacecraft to do a successful flyby of Mars, and it transmitted the first-ever close-up images of Mars. It redefined the way astrobiologists looked at Mars, and it identified geological patterns showing that water could have once flowed on the surface of Mars. In 1976, **Viking 1 and 2** were launched, and they were among the first spacecraft to land on Mars and transmit images of the surface of Mars. The spacecraft landed at two different regions on Mars and conducted four different biological experiments, which included gas chromatography-mass spectrometry, gas exchange, labeled release, and pyrolytic release. While the labeled-release experiments indicated positive signals for exobiologists at both locations, the remaining three experiments turned out to be negative. In the labeled-release experiment, the Martian soil was inoculated with seven ^{14}C radiolabeled nutrients, which were Miller–Urey products. The air above the soil was monitored for any release of radioactive $^{14}CO_2$ as a result of metabolization of any of the nutrients due to microbial activity. To everyone's surprise, there was a steady release of radioactive $^{14}CO_2$, indicating the presence of life, while the control turned out to be negative [3]. However, when the experiments were repeated weeks later, they reported negative results. The result being ambiguous and since the other three experiments reported negative results, it was concluded that the presence of life on Mars was inconclusive. In 1989, **Galileo** was commissioned to study Jupiter. It aided in the identification of salty water on surface of Jupiter's moon Europa, eventually marking it as a hotspot for many further exobiological studies [4]. This discovery made scientists look for life on bodies outside of the habitable zone around the star, since Europa is outside the habitable zone of the solar system. In 1990, the most popular and one of the largest space telescopes, **Hubble space telescope,** was launched. The contributions of Hubble to

TABLE 13.1 Space missions carried out by various space exploratory agencies with exobiological significance.

Missions	Year	Significant findings	References
Mariner 4	1964	Identified geological patterns indicating water flowed on the surface of Mars.	[3]
Viking 1 & 2	1976	Transmitted images of the surface of Mars and presence of life on Mars were inconclusive.	[3]
Galileo	1989	Used as a hotspot for many astrobiological studies.	[4]
Hubble space telescope	1990	Discovered the survival of life in exoplanets.	[5]
Cassini Huygens	1997	Discovered the presence of saltwater and organic chemicals in Saturn and its moons, Titan and Enceladus.	[6]
Dawn	2004	Identified the molecules delivered to Earth from space, which helps us in understanding the origins of life on Earth.	[7]
Deep impact	2005	Data regarding the development of planets and origin of life were obtained through studies on comet.	[8]
Phoenix lander	2008	Discovered the salt concentration that act as nutrients for life in Mars.	[8]
Expose	2008	Evidences regarding origin, evolution, and distribution of life in the universe were collected through irradiation studies.	[9]
Kepler	2009	Identified several exoplanets to host life.	[10]
Oreos	2010	Studied the long-term survival, growth, and metabolic activity of microorganisms.	[10]
Juno	2011	Studied the origin and evolution of Jupiter.	[10]
MSL	2011	Discovered the environment capable of supporting microbial life.	[11]
OSIRIS-REx	2016	Identified the molecular precursors of life and oceans on Earth and determined the properties of asteroids.	[11]
Insight	2018	Identified the habitability conditions of Mars and provided data regarding the history of solar system.	[11]

astronomy and exobiology have been immense through the discovery of hundreds of exoplanets that may possibly host life and has further aided us in studying the planets in our solar system and the potential for life in our system. It has aided us in studying the origins of our universe by enabling us to observe the creation of planets from dust and debris around stars billions of light-years away [5]. In 1997, **Cassini–Huygens** Saturn Orbiter and Titan Probe was launched for examining Saturn and its moons. It arrived at Saturn in 2004 and discovered lakes and seas filled with hydrocarbons on the surface of Titan, Saturn's moon, making it an interesting place for search for life. Cassini discovered icy plumes erupting from Enceladus, another moon of Saturn. It did a flyby collecting samples from the erupting plumes and discovered the presence of saltwater and organic chemicals. This discovery opened our eyes toward new possibilities of habitable environments on the moons of giant planets around distant stars [6].

13.2.2 2000–10

In 2004, the **Dawn** spacecraft was launched by NASA to investigate the solar system's protoplanets Ceres and Vesta, remnants of the early solar system. They both could harbor water in the form of ice; the presence of water was further confirmed by the discovery of vapors by the European Space Agency's (ESA) Herschel telescope. Studying these bodies will help in identifying the molecules delivered to early Earth from space, aiding us in understanding the origins of life on Earth [7]. **Deep Impact** was launched in 2005 to analyze the interior of the surface of a comet known as Tempel 1. It conducted the first-ever analysis of the components of a comet and helped gather data regarding the development of planets and origins of life. It revealed that comets contain significant amounts of organic compounds and that interior parts of a comet are shielded from solar heat; hence the deeper parts of the comet would remain unchanged since the formation of the solar system. Various other missions, such as **EPOXI** and **Stardust**, repeated what the Deep Impact spacecraft did to study various other comets. **Phoenix Lander** was launched in 2008 with two primary objectives, one to search for a habitable zone where microbial life could exist, and the second to study the history of water on Mars. It landed in the polar region of Mars and discovered the presence of small concentrations of salt that could act as nutrients for life and also the presence of perchlorate salts that affect the physical and chemical properties of ice and soil. Further presence of calcium carbonate, which is formed as a result of the activity of liquid water, was also detected [8]. **EXPOSE** (2008) is a facility built by ESA and mounted to the International Space Station dedicated to studies related to exobiology. It studies the long-term effects of exposing biological specimens to outer space. EXPOSE-E, EXPOSE-R, and EXPOSE-R2 were the three experiments carried out by this facility between 2008 and 2015. These experiments studied the long-term effects of solar UV and cosmic radiation on organic compounds, such as amino acids and dipeptides. It also studied the potential for existence of microbial life on Mars with respect to the intensity of radiation. The effect of cosmic radiation on biological samples, such as plant seeds, bacterial spores, fungi, ferns, biological pigments, cellular components, and biofilms. These experiments showed a few promising results with some bacterial species and certain organic compounds were able to survive the ultraviolet radiation of spectrum similar to that of Mars [9]. These studies would help us understand the origin, evolution, and distribution of life in the universe. **Kepler** (2009) is a space-based telescope intended to study Earth-sized planets around distant stars and that are present within the habitable zone of the stars. Kepler has found thousands of exoplanets that could possibly host life. One of the most remarkable findings of Kepler was a planet named Kepler-186f, which is the first Earth-sized planet in the habitable zone of a star other than our sun. The findings of Kepler are directly related to studying the potential of life in our universe.

13.2.3 2010–18

O/OREOS (2010) Organic/Organism Exposure to Orbital Stresses **(OREOS)** is a small-scale satellite that conducted two exobiological experiments in the Earth's orbit at an altitude of 400 miles from the surface. The goal of this mission was to study the long-term survival, growth, and metabolic activity of two different microorganisms, *Bacillus subtilis* and *Halorubrum chaoviator*, in response to the stresses of the space environment.

Juno (2011) Juno is a spacecraft meant to study the origin and evolution of Jupiter. Jupiter, being the largest planet in our solar system, has played a vital role in the formation of our solar system. Hence, studying Jupiter will help exobiologists understand what conditions on Earth led to the formation of life. In addition, it will also help us to study the interaction between the moons of Jupiter, which are of major interest for exobiologists and Jupiter, in particular the icy world of Europa, thus enabling us to understand whether the oceans underneath the ice sheet of Europa could host life [10]. **MSL (2011)** Mars Science Laboratory's Curiosity rover is studying the climate and geology of Mars and whether it is capable of supporting microbial life at present or in the past. It discovered that water once flowed on Mars in a place called the Gale crater. It also made a discovery that a site named "Yellowknife bay" once had an environment capable of supporting microbial life [11]. **OSIRIS-REx (2016)** Origins, Spectral Interpretation, Resource Identification, Security, Regolith Explorer is a mission dedicated to collecting samples from an asteroid known as Bennu and sending them back to Earth. It is a carbon-rich asteroid that records the earliest history of our solar system and it may serve as the key in identifying the molecular precursors of life and oceans on Earth. The spacecraft will help us to determine the physical and chemical properties of the asteroid. **Insight (2018)** Interior Exploration Using Seismic Investigations,

Geodesy, and Heat Transport is a geophysical lander meant to study the interior of Mars deep beneath its surface. It will provide us with vital information regarding the formation of terrestrial planets; planets which are composed of silicon-based rocks or metals. Mercury, Venus, Earth, and Mars, are considered as terrestrial planets of the solar system. Hence, this mission will help us understand the habitability conditions of Mars in the past and also provide us some valuable data regarding the history of our solar system. **TESS (2018)** Transiting Exoplanet Survey Satellite is meant to survey and find new exoplanets, ranging from Earth-sized planets to gas giants. The primary objective of this explorer is to find small planets with bright host stars so that the detailed characterization of planets and its atmosphere can be made easily, since it is one of the important steps in identifying the habitability of an exoplanet.

13.3 Extremophiles—a general overview

Microorganisms growing in extreme conditions, where other life is deemed impossible, are considered as extremophiles (Table 13.2). These extremophiles include both Archaea and Bacteria. They are known to grow in temperatures ranging from −20°C up to 113°C [12,13], classified as psychrophiles (*Methanogenium frigidum*, *Methanococcoides burtonii*) isolated from Ace Lake, Antarctica [14,15]), thermophiles and hyperthermophiles (*Hyperthermus butylicus*) isolated from the island of Sao Miguel, Azores [13] and *Pyrococcus abyssi*, which was isolated from a deep-sea hydrothermal vent in North Fiji basin [16]). Thermophiles (*Thermococcus guaymasensis* and *Thermococcus aggregans*) isolated from Guaymas basin hydrothermal vent site [17]) are organisms that thrive at temperatures ranging from 41°C up to 121°C, and hyperthermophiles are a class of thermophiles that grow in extremely hot environments at temperatures above 80°C, such as in deep-sea hydrothermal vents where the temperature ranges from 60°C to 464°C. These organisms could hold the secrets of life from the primeval environment, helping us to better understand the origin and evolution of life. While psychrophiles inhabit Antarctic and Arctic regions, psychrotolerant microbes may remain docile under cold conditions and retain growth in warmer conditions. Thermophilic organisms are usually found in hot springs and in areas with volcanic activity. Some species of thermophilic bacteria produce spores, and bacterial spores can resist the harshest of conditions and can retain viability even for millions of years. Then there are acidophiles (*Acidianus infernus* and *Acidianus brierleyi*) isolated from solfatara fields and marine hydrothermal vents [18]) and alkaliphiles (*Natrialba hulunbeirensis* and *Natrialba chahannaoensis* isolated from soda lakes in China [19]), which can grow in the extremes of pH as the name suggests, the former being able to grow at lower pH (<2.0) and the latter in environments with high pH (8.5−11). Xerophiles (*Geodermatophilus arenarius*) isolated from the sands of Saharan desert in Chad [20]) are organisms that can tolerate dry conditions. Barophiles (*Thermococcus barophilus*) isolated from a

TABLE 13.2 General classification of organisms based on physical conditions with their example habitat.

Conditions	Classification	Organisms	Habitat	References
Above 80°C	Hyperthermophiles	*Hyperthermus butylicus, Pyrococcus abyssi*	Island of Sao Miguel, Azores, deep-sea hydrothermal vent in North Fiji basin	[13,16]
−20°C up to 113°C	Psychrophiles	*Methanogenium frigidum, Methanococcoides burtonii*	Ace Lake, Antarctica	[14,15]
41°C up to 121°C	Thermophiles	*Thermococcus guaymasensis Thermococcus aggregans*	Guaymas basin hydrothermal vent site	[17]
pH <2.0	Acidophiles	*Acidianus infernus Acidianus brierleyi*	Solfatara fields and marine hydrothermal vents	[18]
pH 8.5−11	Alkaliphiles	*Natrialba hulunbeirensis Natrialba chahannaoensis*	Soda lakes in China	[19]
Dry conditions	Xerophiles	*Geodermatophilus arenarius*	Sands of Saharan desert in Chad	[20]
High pressure (>0.1 Mpa)	Barophiles	*Thermococcus barophilus*	Hydrothermal vent at a depth of 3550 m in a mid-Atlantic ridge	[21]

hydrothermal vent at a depth of 3550 m in a mid-Atlantic ridge [21]) are organisms that can survive in conditions under high pressure (>0.1 MPa). Studying the extremophilic organisms, their habitats and their ability to survive in extreme environments, will help us in understanding the possibility of life in other planets, which can harbor similar conditions. In the following sections, we will take look in detail at halophiles an another group of extremophilic organisms and their extraordinary ability to survive in various extreme conditions apart from thriving in hypersaline conditions.

13.4 Halophiles in the universe

The search for extraterrestrial life has been set as the major goal of 21st century by most of the space agencies around the world. Hence, studying the model environments on Earth will give us some valuable insight about life elsewhere. The surfaces of moons and planets could be the potential habitats for microorganisms. With speculations of the presence of salt water ocean beneath the icy surface of Jupiter's moon Europa and the abundance of salts on Mars, this only makes it a priority to study the survival and adaptation of halophiles in extraterrestrial environment. Halophiles are a class of extremophiles that are known to survive in extremes of salinity, at levels usually higher than the salinity levels of seawater (Fig. 13.1). Halophilic life on Earth is extremely diverse and they are known to survive even at saturation levels of NaCl [22,23]. They have evolved different modes of energy generation, which includes oxygenic and anoxygenic photosynthesis, respiration using electron acceptors such as oxygen or sulfate or nitrogen, chemoautotrophy, and fermentation. The reason that they can survive in extremely hypersaline environments is through adaptation of various molecular machineries such as selective influx and accumulation of potassium ions within the cell to help maintain the osmotic balance; known as "salt-in" strategy. Another widely adapted method is the accumulation of organic solutes (polyols and derivatives, sugars and derivatives, amino acids and derivatives, betaines and ectoines) termed as osmotic solutes or compatible solutes for the maintenance of internal osmotic pressure. The former strategy is usually adapted by halophilic archaea and extremely halophilic bacteria, *Salinibacter ruber*, with the latter by moderately halophilic and halotolerant bacteria.

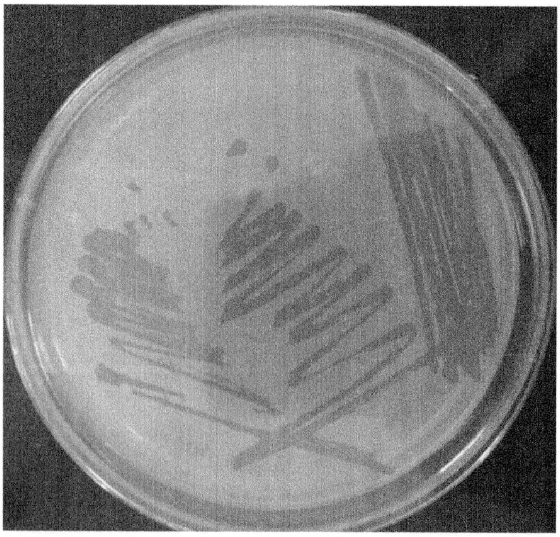

FIGURE 13.1 A pure halophilic archaeal isolate obtained from a solar saltern growing on NTYE (NaCl Tryptone Yeast Extract) medium containing 25% NaCl (wt/vol).

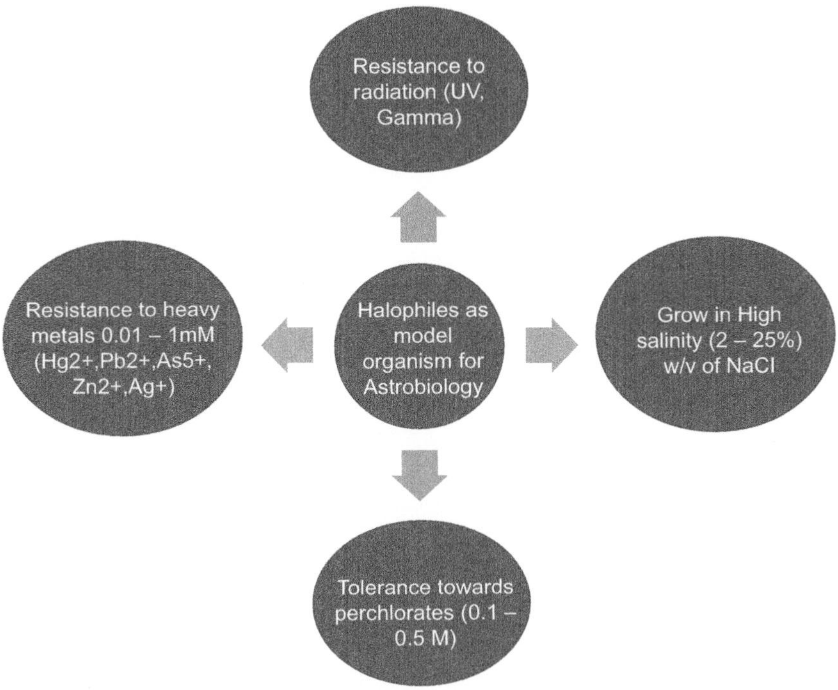

FIGURE 13.2 Specific characteristics of halophilic archaea that are important for their role as model organism in exobiology.

The members of halophilic archaea, classified under the class *Halobacteria*, can survive and grow in the highest salt concentrations (~30% NaCl w/v) and are adapted to life at extreme salinity levels, where other microbial communities cannot exist. The class consists of three orders and six families as of 2017 [24]. The members of this class are aerobic chemoheterotrophic microorganisms that obtain their energy by oxidation of simple organic compounds using oxygen as an electron acceptor for respiration [25]. What makes them good models for astrobiological studies on Mars, Europa, and Enceladus is their metabolic versatility and their unusual properties, which will be discussed in the following section (Fig. 13.2; Table 13.3).

13.5 Modes of energy generation in halophilic archaea

In conditions favorable to life, halophilic archaea are generally aerobic chemoheterotrophs. However, when exposed to certain conditions they exhibit different modes of energy generation and even anaerobic growth in the absence of oxygen. One such example is the use of light-driven proton pathway in place of chemical energy obtained through respiration with molecular oxygen as electron acceptor or through other molecules as electron

TABLE 13.3 Halophilic prokaryotes that have been demonstrated to exhibit significant exobiological properties.

Name of the organism	Condition tested	Findings	Reference
Halorubrum chaoviator	Exposed to high radiation for 2 weeks.	Cells retained viability (10^7 – 10^8cells).	[28]
Halococcus dombrowskii, *Halobacterium salinarum* NRC-1, *Haloarcula japonica*	Exposed to UV radiation with flux of (200–400 nm).	No loss of viability after exposing to 21 KJ/m^2.	[29]
Halococcus hamelinensis	Exposed to UV light.	Nucleotide excision repair.	[30]
Haloferax mediterranei, *Haloferax denitrificans*, *Haloferax gibbonsii*, *Haloarcula marismortui*, *Haloarcula vallismortis*	0.4 M of perchlorate salts in an NaCl medium.	Life could thrive on Mars.	[39]

acceptors in the absence of oxygen enabling anaerobic growth [26]. This light-driven proton pathway can be attributed to the production of purple-colored membrane-bound protein known as bacteriorhodopsin, which acts as a light-driven proton pump. The generation of the proton gradient through this pump is used for the generation of adenosine triphosphate (ATP). Halophilic archaea are also known to use other electron acceptors in the place of oxygen such as trimethylamine N-oxide, dimethyl sulfoxide, fumarate, and perchlorates and chlorates, which are widespread on Mars, making it a really good alternative, since oxygen is absent on Mars [25]. They also have the ability to reduce nitrate and grow by denitrification, which is a widespread technique employed by majority of halophilic archaea. Growth through means of fermentation is exhibited by few species of halophilic archaea. Members of the genera *Halobacterium* are able to grow anaerobically by fermentation of arginine, an amino acid. Interestingly, *Halorhabdus tiamatea*, a species of halophilic archaea isolated from the Red Sea, prefers an anaerobic lifestyle, growing very poorly in aerobic conditions [27].

13.6 Radiation resistance in halophilic archaea

The atmosphere of Mars got eroded away due to a weak magnetic field and solar winds. As a result, the cosmic radiation enters the atmosphere of Mars unfiltered whereas on Earth, the presence of ozone layer filters the harmful ultraviolet radiation. So, if any life is to exist on Mars or any other celestial body with an irradiated surface, it needs to be able to possess various repair mechanisms and should have the ability to resist lethal levels of radiation. Numerous studies have been made studying the response of halophilic archaea in response to radiation. A halophilic archaeon, *Halorubrum chaoviator*, was exposed to high radiation for 2 weeks in the lower orbit of Earth in space through the BIOPAN facility by the ESA. After exposure to ionizing radiation in high vacuum, a considerable number of cells were able to retain viability (about one out of 10^7-10^8 cells, when compared with the control) [28]. In another study, cells of *Halococcus dombrowskii*, *Halobacterium salinarum* NRC-1, and *Haloarcula japonica* were embedded in a laboratory-grown halite to simulate the natural environment and then exposed to UV radiation (200–400 nm), simulating the flux on the surface of Mars. The study reported that the cells showed no loss of viability after exposure to 21 kJ/m^2 of radiation (assessed using LIVE/DEAD kit dyes) and after exposure to a total of 148 kJ/m^2 radiation, the cells resumed growth after 12 days in liquid medium [29].

As discussed in the modes of energy generation, some halophilic archaea possess the ability to synthesize ATP through light-driven proton pumps. So generally, in saline environments these organisms tend to move toward the surface of the brine to get more exposure to the sun for energy production, eventually getting exposed to higher levels of UV radiation. In order to negate the effects of the radiation, halophilic archaea possess a variety of DNA damage repair mechanisms such as photoreactivation and various nucleotide excision repair mechanisms [30]. One specific study with *Halococcus hamelinensis* showed that, when exposed to UV light, there was an upregulation of genes such as *uvr A*, *uvr B*, and *uvr C*, which are responsible for nucleotide excision repair, and when incubated in light, the organism displayed a 20-fold increase in the *phr 2* gene, which is a photolyase enzyme, proving that this specific species uses light energy to repair the DNA damage. The same genes were found to take part in the nucleotide excision repair in another halophilic archaea, *Halobacterium salinarum* NRC-1 [31]. Halophilic archaea are also known to produce high levels of carotenoids, imparting shades of orange, pink, and red. These pigments belong to bacterioruberin (C_{50}) class of carotenoids protecting halophilic archaea from the UV damage from reactive oxygen species (ROS). Similarly the presence of high levels of intracellular salts is also known to prevent the damage caused by intense UV radiation by scavenging the ROS [32]. The methods mentioned above might explain the ability of numerous members of halophilic archaea to resist ionizing radiation.

13.7 Halophilic archaea from ancient halite crystals

During the Permian and Triassic periods, increasing temperatures caused the extremely saline seas to evaporate and led to the formation of large sediments of salt [33]. During this time, halite crystals were formed due to the evaporation of salt-saturated brines, and eventually halophilic archaea present in the brine inclusions became entrapped within these crystals. Many halophilic microorganisms have been isolated from these ancient salt deposits. Novel strains of halophilic archaea belonging to *Halococcus salfodinae* and *H. dombrowskii* were isolated from the permo-Triassic rocks found in the alpine region [34]. Scientists have isolated halophilic archaea from Death Valley, California, that were able to survive in subsurface halites for 34,000 years [35]. This study has greatly helped us to understand the ability of halophiles isolated from sediments to survive for prolonged periods in the state of dormancy. However, microscopic examination of the prokaryotic cells showed that they were smaller in cell size when compared to their modern counterparts grown in nutrient-rich media. The possible answer for their survival in the absence of organic carbon and energy sources lies in the fact that not only halophilic archaea become entrapped within the growing salt crystals but also the cells of the alga *Dunaliella,* which are present in salt-saturated brines along with the halophilic prokaryotes. *Dunaliella* cells produce glycerol as an osmoprotectant, thereby providing osmotic stabilization in saline environment. So, the organic material of *Dunaliella* may act as a carbon and energy source to support survival of halophilic archaea for prolonged periods [25].

Since the halophiles are able to survive for longer periods inside the halite crystals on Earth, it is possible that they could survive in the same way in extraterrestrial conditions. This fact is further bolstered by the discovery of halite on the SCN meteorite that fell in Shergotty (India), Chassigny (France), and Nakhla (Egypt), which stemmed from Mars [33]. This provides a strong contention to believe in the existence of similar halites on Mars, Europa, and Enceladus.

13.8 Adaptation of halophilic archaea to extreme temperatures and pH

Many members of halophilic archaea are known to be moderately thermophilic, with their optimum growth temperatures being 35–50°C, and the maximum temperature at which the growth was reported was 50–56°C [36]. Not many psychrophilic habitats of halophilic archaea have been explored or identified, hence the halophilic archaea's life at low temperatures is poorly understood. The best-studied environment that hosts halophilic archaea is the Deep Lake located in the Vestfold Hills region of Antarctica, which never freezes due to its hypersalinity. *Halorubrum lacusprofundi*, a halophilic archaeon, was isolated from this lake, which was able to grow slowly at 4°C, a temperature too low for the other halophilic archaea [37]. Many of the halophilic archaea have an optimum pH level of 7, and only some of them are alkalophilic, growing at pH 8.5 and above, such as *Halorubrum alkaliphilum*, *Halalkalicoccus tibentensis*, and *Halorubrum vacuolatum* [23]. A study was conducted in the acidophilic hypersaline lake Lake Magic located in West Australia, which has a pH of 1.6 and salinity as high as 32%, in terms of dissolved solids that were analyzed for prokaryotic composition. Various optical methods were employed in studying the microbial communities of Lake Magic, and researchers were able to identify several prokaryotes, eukaryotes, archaea, and several organic components within the halite crystals of Lake Magic. This yields an interesting result, as it is of high significance for studies about implications of halophilic life on Mars [38].

13.9 Growth of halophilic archaea in the presence of perchlorates

Ever since the discovery of perchlorates by the wet chemistry laboratory of the Phoenix Lander, many researchers have tried to study the implications of perchlorate on life on Mars, which accounts for up to 0.4–0.6 wt% of total soil composition. Perchlorates are strong oxidizing agents at higher temperatures, and due to their oxidizing properties, the organic compound availability will be affected, in turn affecting the microbial community. However, they can also be used as electron acceptors by microorganisms in the absence of oxygen in anaerobic conditions for respiration. Due to the hygroscopic nature of the perchlorate salts, they can lead to local formation of liquid droplets, which aids in the availability of local brines for the sustenance of the halophilic archaea [39].

A study showed that *Haloferax mediterranei*, *Haloferax denitrificans*, *Haloferax gibbonsii*, *Haloarcula marismortui*, and *Haloarcula vallismortis* were able to use perchlorates as electron acceptors and grow well at concentrations up to 0.4 M of perchlorate salts in an NaCl-based medium [39]. This shows that life could very well thrive on Mars. These results show that perchlorates, rather than being toxic to halophilic archaea, support their development in the absence of molecular oxygen, provided that a suitable electron donor and energy source is available. A more recent paper published in 2017 threw some light on the toxic nature of perchlorates in the presence of ionizing radiation. In this study, it was shown that perchlorates enhanced the bactericidal potential by tenfold when tested on a common soil-occurring bacteria, *Bacillus subtilis* [40]. However, the effect of irradiated perchlorates on halophilic archaea is yet to be studied.

13.10 Saline environments in space

The following section will throw some light on the most promising abodes for halophilic archaea in our solar system and conditions that are suitable for halophilic life to thrive on.

13.10.1 Mars

It has been shown that neither brine nor liquid water could exist in a stable form on the surface of Mars [41]. In certain cases, it could be temporarily present in some areas of the planet. Due to the evaporation into the extremely dry atmosphere and due to the extremely high temperature and pressure, liquid phase of pure water could not exist, and under such conditions water would neither freeze nor boil. In certain places, liquid brines could exist where sublimation couldn't be possible due to the presence of nearly saturated air [41]. Liquid brines could exist subsurface, because even a thin layer of soil would protect the water from sublimation, enabling brines to be accumulated over a period of time [41]. Mars is rich in salts such as $MgSO_4$, $NaSO_4$, NaCl, and other chlorides, and $MgCO_3$ and $CaCO_3$ [42]. Presence of these hygroscopic salts along with the recently discovered perchlorate salts of sodium and magnesium, which are highly hygroscopic, could lead to formation of water droplets at freezing temperatures, thus adding to the speculation of the presence of liquid brines saturated with salts on Mars.

13.10.2 Europa

Europa is the second largest moon of Jupiter. The Galileo spacecraft launched to study Jupiter and its moons discovered the presence of salty oceans beneath the icy surface of Europa. Discovery of the presence of hydrated sodium carbonates shed some light on the presence of alkaline oceans on Europa. The water activity a_w of saline oceans is 0.6—1 [43]. Usually halophilic archaea are known to survive at water activity levels as low as 0.75, which makes it a potential candidate for studying about halophilic life outside of our planet. Based on various models, the oceans on Europa are speculated to contain similar salts as those on Mars, such as NaCl, $MgSO_4$, MgCl, and KCl [44]. These speculations provide a compelling reason to study the ability of halophilic archaea to adapt and survive in the conditions prevalent on Europa.

13.10.3 Enceladus

Enceladus is one of Saturn's moons, which is known to often eject plumes of ice into space, which inturn contributes to the components of the E ring of Saturn, one of the rings of Saturn. These plumes are believed to contain salt-rich ice particles, which make up 99% of the ejected solids; however, diminishing in the population escaping into Saturn's E ring [45]. This leads to speculation for possibility of halophilic life on the Saturn's moon, hence making a strong case for study of the ability of halophilic microorganisms to survive in extreme conditions.

13.11 Methods for detecting halophilic archaea in saline econiches

Various Raman spectroscopy methods, such as FT-Raman spectroscopy, resonance Raman spectroscopy, and micro-Raman spectroscopy, are adopted to study the presence of halophilic archaea present in extraterrestrial space based on the detection of prominent peaks due to the dispersion by C50-carotenoid pigments [46,47]. In a study, nine different species of halophilic archaea were embedded in laboratory-grown halite and were subjected to laser excitation at 1064 nm in FT-Raman spectroscopy and 514.5 nm in micro-Raman spectroscopy [46]. The results indicated that the peaks were observed at 1507, 1152, and 1002 cm^{-1}, attributed to the presence of C_{50} carotenoid compounds [46]. Hence, these C_{50}-carotenoid molecules, especially bacterioruberin pigments, can be used as biomarkers to study the presence of halophilic archaea on any potential extraterrestrial habitat. The main advantage of this method is nonintrusiveness without the need for any chemical extractions [48]. Another method is the use of biomarkers and microarray chips like SOLID2, an antibody array-based life detector instrument containing 157 antibodies in protein microarrays, which has been tested in a simulated Mars drilling mission against ground core samples. This method is also useful due to the ability to detect antibodies in a very wide range of biomarkers simultaneously [49].

13.12 Conclusion

Considering the past decades of exploratory astronomy, finding hospitable places that support life outside planet Earth is closer than ever. Apart from the exoplanets, there has always been a constant interest in studying the primordial Earth conditions from which early life would have originated, and exobiology could lead us further in understanding the early earth conditions. Extremophiles provide a promising avenue for exobiology research due to the robustness in adjusting to the conditions otherwise deleterious for other organisms. Halophilic archaea in particular have proven promising for further research avenues in exobiology due to their multitude of properties like resistance to radiation, growth in high-saline conditions, tolerance of perchlorates, etc. Occurrence of saline habitats elsewhere other than Earth only strengthens this viewpoint. Further research focusing on the molecular and cellular processes behind the adaptability of halophilic archaea to exobiological conditions would give us valuable information regarding the adaptability of other life to spatial conditions. Apart from being a model organism in exobiology, halophilic archaea in the extraterrestrial habitats can be a potential source of novel β-carotene pigments, since these ecosystems are exposed to high UV radiation. Further, these isolates may also serve as a source for novel enzymes and biomolecules like halocins (antimicrobial peptide) and bioplastics (polyhydroxyalkanoates). The metabolites from halophilic archaea might have potential roles in harsh industrial conditions where metabolites from other organisms may fail. For an instance, halophilic enzymes may tolerate higher temperature, pressure, and extreme pH conditions and be stable in the presence of solvent. Similarly, β-carotene pigments isolated from halophilic archaea could have potential applications in neutraceutical and food processing industries. Therefore, bioprospecting of these organisms can lead us to the identification of novel bioproducts that can have potential bioapplications.

References

[1] Cockell CS. Astrobiology: understanding life in the Universe. 1st ed. New Jersey: John Wiley & Sons; 2015.

[2] Zagorski ZP. Question 2: relation of panspermia-hypothesis to astrobiology. Orig Life Evol Biosph 2007;37(4−5):351−5.

[3] Levin GV, Straat PA. Completion of the viking labeled release experiment on Mars. J Mol Evol 1979;14(1−3):167−83.

[4] McCord TB, Hansen GB, Fanale FP, Carlson RW, Matson DL, Johnson TV, Smythe WD, Crowley JK, Martin PD, Ocampo A, Hibbitts CA. Salts on Europa's surface detected by Galileo's near infrared mapping spectrometer. Science 1998;280(5367):1242−5.

[5] Garner R. Highlights of Hubble's exploration of the universe [Internet]. NASA; 2017. Available from: https://www.nasa.gov/content/goddard/2017/highlights-of-hubble-s-exploration-of-the-universe.

[6] Parkinson CD, Liang MC, Hartman H, Hansen CJ, Tinetti G, Meadows V, Kirschvink JL, Yung YL. Enceladus: Cassini observations and implications for the search for life. Astron Astrophys 2007;463(1):353−7.

[7] NASA Astrobiology [Internet]: NASA. Available from: https://astrobiology.nasa.gov/missions/dawn/.

[8] Hecht MH, Kounaves SP, Quinn RC, West SJ, Young SM, Ming DW, Catling DC, Clark BC, Boynton WV, Hoffman J, DeFlores LP. Detection of perchlorate and the soluble chemistry of martian soil at the Phoenix lander site. Science 2009;325(5936):64−7.

[9] Wassmann M, Moeller R, Rabbow E, Panitz C, Horneck G, Reitz G, Douki T, Cadet J, Stan-Lotter H, Cockell CS, Rettberg P. Survival of spores of the UV-resistant Bacillus subtilis strain MW01 after exposure to low-Earth orbit and simulated martian conditions: data from the space experiment ADAPT on EXPOSE-E. Astrobiology 2012;12(5):498−507.

[10] NASA Astrobiology [Internet]: NASA. Available from: https://astrobiology.nasa.gov/missions/juno/.

[11] NASA Astrobiology [Internet]: NASA. Available from: https://astrobiology.nasa.gov/missions/msl.

[12] Rivkina EM, Friedmann EI, McKay CP, Gilichinsky DA. Metabolic activity of permafrost bacteria below the freezing point. J Appl Environ Microbiol 2000;66(8):3230−3.

[13] Zillig W, Holz I, Janekovic D, Klenk HP, Imsel E, Trent J, Wunderl S, Forjaz VH, Coutinho R, Ferreira T. *Hyperthermus butylicus*, a hyperthermophilic sulfur-reducing archaebacterium that ferments peptides. J Bacteriol 1990;172(7):3959−65.

[14] Franzmann PD, Liu Y, Balkwill DL, Aldrich HC, De Macario EC, Boone DR. *Methanogenium frigidum* sp. nov., a psychrophilic, H_2-using methanogen from Ace Lake, Antarctica. Int J Syst Evol Microbiol 1997;47(4):1068−72.

[15] Franzmann PD, Springer N, Ludwig W, De Macario EC, Rohde M. A methanogenic archaeon from Ace Lake, Antarctica: *Methanococcoides burtonii* sp. nov. Syst Appl Microbiol 1992;15(4):573−81.

[16] Erauso G, Reysenbach AL, Godfroy A, Meunier JR, Crump B, Partensky F, Baross JA, Marteinsson V, Barbier G, Pace NR, Prieur D. *Pyrococcus abyssi* sp. nov., a new hyperthermophilic archaeon isolated from a deep-sea hydrothermal vent. Arch Microbiol 1993;160(5):338−49.

[17] Canganella F, Jones WJ, Gambacorta A, Antranikiang G. *Thermococcus guaymasensis* sp. nov. and *Thermococcus aggregans* sp. nov., two novel thermophilic archaea isolated from the Guaymas Basin hydrothermal vent site. Int J Syst Evol Microbiol 1998;48(4):1181–5.

[18] Segerer A, Neuner A, Kristjansson JK, Stetter KO. *Acidianus infernus* gen. nov., sp. nov., and *Acidianus brierleyi* comb. nov.: facultatively aerobic, extremely acidophilic thermophilic sulfur-metabolizing archaebacteria. Int J Syst Evol Microbiol 1986;36(4):559–64.

[19] Xu Y, Wang Z, Xue Y, Zhou P, Ma Y, Ventosa A, Grant WD. *Natrialba hulunbeirensis* sp. nov. and *Natrialba chahannaoensis* sp. nov., novel haloalkaliphilic archaea from soda lakes in Inner Mongolia Autonomous Region, China. Int J Syst Evol Microbiol 2001;51(5):1693–8.

[20] Montero-Calasanz MC, Göker M, Pötter G, Rohde M, Spröer C, Schumann P, Gorbushina AA, Klenk HP. *Geodermato philusarenarius* sp. nov., a xerophilic actinomycete isolated from Saharan desert sand in Chad. Extremophiles 2012;16(6):903–9.

[21] Marteinsson VT, Birrien JL, Reysenbach AL, Vernet M, Marie D, Gambacorta A, Messner P, Sleytr UB, Prieur D. *Thermococcus barophilus* sp. nov., a new barophilic and hyperthermophilic archaeon isolated under high hydrostatic pressure from a deep-sea hydrothermal vent. Int J Syst Evol Microbiol 1999;49(2):351–9.

[22] Oren A. Microbial life at high salt concentrations: phylogenetic and metabolic diversity. Saline Syst 2008;4(1):2.

[23] Bowers KJ, Wiegel J. Temperature and pH optima of extremely halophilic archaea: a minireview. Extremophiles 2011;15(2):119–28.

[24] Oren A, Ventosa A, Kamekura M. *Halobacteria:* Bergey's manual of systematics of archaea and bacteria. 1st ed. New Jersey: John Wiley & Sons; 2017.

[25] Oren A. Halophilic archaea on Earth and in space: growth and survival under extreme conditions. Phil Trans R Soc A 2014;372(2030):20140194.

[26] Hartmann R, Sickinger HD, Oesterhelt D. Anaerobic growth of halobacteria. Proc Natl Acad Sci Unit States Am 1980;77(7):3821–5.

[27] Antunes A, Taborda M, Huber R, Moissl C, Nobre MF, da Costa MS. *Halorhabdus tiamatea* sp. nov. a non-pigmented, extremely halophilic archaeon from a deep-sea, hypersaline anoxic basin of the Red Sea, and emended description of the genus Halorhabdus. Int J Syst Evol Microbiol 2008;58(1):215–20.

[28] Mancinelli RL, White MR, Rothschild LJ. Biopan-survival I: exposure of the osmophiles *Synechococcus* sp. (Nageli) and *Haloarcula* sp. to the space environment. Adv Space Res 1998;22(3):327–34.

[29] Fendrihan S, Bérces A, Lammer H, Musso M, Rontó G, Polacsek TK, Holzinger A, Kolb C, Stan-Lotter H. Investigating the effects of simulated Martian ultraviolet radiation on *Halococcus dombrowskii* and other extremely halophilic archaebacteria. Astrobiology 2009;9(1):104–12.

[30] Jones DL, Baxter BK. DNA repair and photoprotection: mechanisms of overcoming environmental ultraviolet radiation exposure in halophilic archaea. Front Microbiol 2017;8:1882.

[31] McCready S, Marcello L. Repair of UV damage in *Halobacterium salinarum*. Biochem Soc Trans 2003;31(3):694–8.

[32] Shahmohammadi HR, Asgarani E, Terato H, Saito T, Ohyama Y, Gekko K, Yamamoto O, Ide H. Protective roles of bacterioruberin and intracellular KCl in the resistance of *Halobacterium salinarium* against DNA-damaging agents. J Radiat Res 1998;39(4):251–62.

[33] Stan-Lotter H, Fendrihan S, Dorn-Mayr-Pfaffenhuemer M, Polacsek TK, Legat A, Grösbacher M, Weigl A. The likelihood of halophilic life in the Universe. In: Hanslmeier A, Kempe S, Seckbach J, editors. Life on earth and other planetary bodies. Dordrecht: Springer; 2012. p. 345–65.

[34] Stan-Lotter H, Radax C, Leuko S, Legat A, Gruber C, Pfaffenhuemer M, Wieland H, Weidler G. Viable halobacteria from ancient oceans—and in outer space? In: Seckbach J, Chlea-Flores J, Owen J, Raulin F, editors. Life in the Universe. Dordrecht: Springer; 2004. p. 207−10.

[35] Goh F, Leuko S, Allen MA, Bowman JP, Kamekura M, Neilan BA, Burns BP. *Halococcus hamelinensis* sp. nov., a novel halophilic archaeon isolated from stromatolites in Shark Bay, Australia. Int J Syst Evol Microbiol 2006;56(6):1323−9.

[36] Grant WD, Kamekura M, McGenity TJ, Class III VA. Halobacteria *class. nov*. In: Boone DR, Castenholz RW, Garrity GM, editors. Bergey's manual of systematic bacteriology. New York: Springer; 2001. p. 294−301.

[37] Franzmann PD, Stackebrandt E, Sanderson K, Volkman JK, Cameron DE, Stevenson PL, McMeekin TA, Burton HR. *Halobacterium lacusprofundi* sp. nov., a halophilic bacterium isolated from Deep Lake, Antarctica. Syst Appl Microbiol 1988;11(1):20−7.

[38] Conner AJ, Benison KC. Acidophilic halophilic microorganisms in fluid inclusions in halite from Lake Magic, Western Australia. Astrobiology 2013;13(9):850−60.

[39] Oren A, Bardavid RE, Mana L. Perchlorate and halophilic prokaryotes: implications for possible halophilic life on Mars. Extremophiles 2014;18(1):75−80.

[40] Wadsworth J, Cockell CS. Perchlorates on Mars enhance the bacteriocidal effects of UV light. Sci Rep 2017;7(1):4662.

[41] Martínez GM, Renno NO. Water and brines on Mars: current evidence and implications for MSL. Space Sci Rev 2013;175(1−4):29−51.

[42] Clark BC, Van Hart DC. The salts of Mars. Icarus 1981;45(2):370−8.

[43] Marion GM, Fritsen CH, Eicken H, Payne MC. The search for life on Europa: limiting environmental factors, potential habitats, and earth analogues. Astrobiology 2003;3(4):785−811.

[44] Brown ME, Hand KP. Salts and radiation products on the surface of Europa. Astron J 2013;145(4):110.

[45] Postberg F, Schmidt J, Hillier J, Kempf S, Srama R. A salt-water reservoir as the source of a compositionally stratified plume on Enceladus. Nature 2011;474(7353):620.

[46] Fendrihan S, Musso M, Stan Lotter H. Raman spectroscopy as a potential method for the detection of extremely halophilic archaea embedded in halite in terrestrial and possibly extraterrestrial samples. J Raman Spectrosc 2009;40(12):1996−2003.

[47] Marshall CP, Leuko S, Coyle CM, Walter MR, Burns BP, Neilan BA. Carotenoid analysis of halophilic archaea by resonance Raman spectroscopy. Astrobiology 2007;7(4):631−43.

[48] Villar SE, Edwards HG. Raman spectroscopy in astrobiology. Anal Bioanal Chem 2006;384(1):100−13.

[49] Fernández-Calvo P, Näke C, Rivas LA, García-Villadangos M, Gómez-Elvira J, Parro V. A multi-array competitive immunoassay for the detection of broad-range molecular size organic compounds relevant for astrobiology. Planet Space Sci 2006;54(15):1612−21.

Chapter 14

Bacterial probiotics over antibiotics: a boon to aquaculture

Samantha Fernandes, Savita Kerkar
Department of Biotechnology, Goa University, Taleigao Plateau, Goa, India

14.1 Introduction

Aquaculture is the active cultivation of aquatic organisms in a controlled environment to enhance their production [1]. A fast developing sector, aquaculture has now become an essential economic activity in many countries, with the total global production amounting to 80 million tonnes in 2016 (State of world fisheries and aquaculture, FAO, 2018). The report of World Aquaculture in 2012 stated that global production of farmed fish increased tremendously by 30% between 2006 and 2011, from 47.3 million tonnes to 63.6 million tonnes [2]. Over the last 2 decades, Asia has accounted for about 89% of world aquaculture production, with China contributing to around 61.5% followed by India (7.1%) (Food & Agricultural Organization, 2018). Aquaculture plays an important role in providing employment to thousands of skilled and unskilled workers, promoting economic income to farmers and catering to the increasing demand of seafood worldwide [3]. The United Nations FAO predicts that by 2020, 50% of the world's seafood requirement will be met by aquaculture, due to overexploitation of capture fisheries [4].

However, in large-scale production systems, where the farmed species are exposed to stressful situations, problems related to deterioration of environmental conditions and diseases often occur, resulting in serious economic losses. In several cases, intensive aquaculture has led to bacterial and viral outbreaks, which cause mass mortality in fish hatcheries and shrimp larviculture [5]. Some of the diseases include vibriosis (caused by *Vibrio anguillarum*, *Vibrio harveyi*, *Vibrio vulnificus*, *Vibrio parahaemolyticus*), aeromonasis (caused by *Aeromonas* spp.), edwardsiellosis (by *Edwardsiella anguillarum*, *Edwardsiella tarda*), pseudomonasis (caused by *Pseudomonas anguilliseptica*, *P. fluorescens*), streptococcosis (*Streptococcus agalactiae*,

Advances in Biological Science Research. https://doi.org/10.1016/B978-0-12-817497-5.00014-8

Streptococcus iniae), etc. (Food and Agricultural organization, 2016). In recent decades, the control and prevention of these pathogens have led to extensive use of antimicrobials and veterinary medicines [6]. Antibiotics have been the immediate treatment for bacterial infections, playing a critical role in modern medicine. Nevertheless, the efficacy of antibiotics as a precautionary measure has been questioned, as their exploitation can result in the emergence of resistant bacterial strains. This is seen mainly in shrimp farming where massive rise in production, overcrowding of animals and uncontrolled antibiotic usage has led to the development of several antibiotic-resistant bacteria and production crashes in many countries [7,8]. Antibiotic resistance can be easily transferred to other strains by horizontal gene transfer between cells or alteration to the existing genome [5]. Numerous bacterial pathogens can acquire plasmid-mediated resistance via natural transformation, transduction, or conjugation. Lateral exchange of plasmids carrying antibiotic-resistant genes has been found in many *Vibrio* species [6]. Wang et al., have reported the presence of antibiotics (fluoroquinolones, β-lactum, tetracyclines, sulfonamides, macrolides, phenicols) in finfish and shrimp sampled across Shanghai city and reported that 10% of the samples exceeded maximum residue limits [9]. Since these antibiotics are comparatively stable and nonbiodegradable, antibiotic residues can remain in fish and shellfish commercialized for consumption, thereby increasing the risk of transferring antibiotic-resistant genes from aquaculture to consumers [10]. These alarming drawbacks have provoked the aquaculture industry to identify and develop new strategies that are equally effective as antibiotics, sustainable, and eco-friendly [11,12]. One such remedy that is gaining importance within the aquaculture industry to control potential pathogens is the use of probiotic bacteria [1].

14.2 The probiotic approach

The word *probiotic* originates from Greek words *pro* and *bios* meaning "prolife" [2]. The work carried out by Elie Metchnikoff is regarded as the first study conducted on probiotics, describing them as "microbes ingested with the aim of promoting good health" [13]. Parker in 1974 defined probiotics as "organisms and substances that contribute to intestinal microbial balance" [14]. Later, in 1989 Fuller modified the definition as "live microbial feed supplement that benefits the host (human or animal) by improving the microbial balance of the body" [15]. To summarize, probiotics multiply a few beneficial microbes to compete with and suppress the growth of the harmful ones [1].

These microbial supplements used in aquaculture benefit their host by inhibiting pathogenic microorganisms, improving feed value, contributing enzymatically to digestion, secreting growth-promoting factors, stimulating the host immune response, and improving the pond water quality [16,17]. Probiotics include certain yeast, actinomycetes, and bacteria that are not harmful when continuously used for a long period of time [1].

14.3 Antimicrobial mechanism of probiotics

Understanding the underlying mechanism used by probiotics to compete with pathogens is an essential criterion for designing a protocol for probiotic selection [18]. In aquaculture, it is important to study the probiotic's ability to:

- Outcompete the pathogenic bacteria by production of antagonistic compounds;
- Compete with pathogens for energy, adhesion sites, space, nutrients in the gut of the host's body, attachment, and growth in mucus;
- Stimulate the immune system;
- Produce beneficiary compounds such as growth promoters and enzymes.

14.3.1 Production of antagonistic compounds

Microorganisms produce primary and secondary metabolites that may have bactericidal or bacteriostatic effects on other organisms. These metabolites include siderophores, hydrogen peroxide, lysozymes, bacteriocins, proteases, organic acids, antibiotics, etc [18]. Siderophores are iron-binding agents that chelate the iron from the surrounding niche suitable for the growth of microbes. Bacteria producing siderophores can survive in nutrient-limited conditions and deprive the pathogens of iron [19]. *Pseudomonas fluorescens* culture supernatant inhibited growth of *V. anguillarum* when grown in iron-limited conditions [20]. A study conducted by Lalloo et al. revealed *Bacillus cereus* produced siderophores directly at the start of batch culture, lowering the iron concentration to very minimal levels within the first 5 h, thus starving pathogen *Aeromonas hydrophila* of iron [21]. Some bacteria such as *Bacillus* and lactic acid bacteria produce proteins or protein complexes known as bacteriocins that could restrain the growth of other organisms [22]. A study conducted by Campos et al. reported three lactic acid bacteria, *Lactococcus lactis*, *Enterococcus faecium*, and *Enterococcus mundtii*, that inhibited the growth of *Listeria monocytogenes* and *Staphylococcus aureus* by production of heat-resistant bacteriocins [23]. Imada et al. isolated an *Alteromonas* sp. strain B-10-31 that produced an alkaline protease inhibitor called monastatin that exhibited inhibitory activity against fish pathogens *Aeromonas hydrophila* and *V. anguillarum* in vitro [24].

14.3.2 Competitive exclusion

Microbial interactions play an important role in regulating the equilibrium between competing beneficial and pathogenic microorganisms [6]. One credible mechanism to prevent the colonization by pathogens in the gut or other tissue surfaces is competition for adhesion sites [25]. Most pathogenic bacteria need to attach to the mucosal lining of the host's gastrointestinal tract during the initial stages of infection [25,26]. The ability of microorganisms to colonize the host's gastrointestinal tract, thereby inhibiting the attachment of pathogens, is an essential criterion for selection of probiotics [6,12]. Microbial manipulation by addition of beneficial bacteria in aquaculture is more effective if the probiotic is added at an early stage in larviculture indicating a prophylactic application of probionts [27]. *Lactobacilli* decreased the adhesion of *Carnobacterium piscicola*, *Yersinia ruckeri*, and *Aeromonas salmonicida* to the intestinal mucus of rainbow trout [6]. Studies have reported that bacterial strains associated with the skin and intestinal mucus of adult dab fish (*Limanda limanda*) and marine turbot (*Scophthalmus maximus*) halted the growth of the fish pathogen *V. anguillarum* [28].

The existence of any microbial population is dependent on its capacity to compete for chemicals, nutrients, and available energy with the other microorganisms in the same environment [29,30]. Verschuere et al. selected various strains with a positive effect on the growth and survival of *Artemia* juveniles. The filtrate experiments and in vitro antagonism tests demonstrated that no extracellular antagonistic compounds were involved in the protective role of these strains against *Vibrio proteolytics* CW8T2 pathogen. The study inferred that the selected bacteria protected *Artemia* by competing with the pathogenic *Vibrio* sp. for chemicals and available energy [30].

14.3.3 Immunomodulation

Fish larvae and shrimps have an underdeveloped immune system, depending mainly on their nonspecific immune response [31]. Accordingly to Fuller, probiotics may improve the immunity in three ways: (1) Enhancing the macrophage activity by increasing the ability to phagocytose microorganisms or certain carbon particles; (2) Increasing local antibodies at mucus surfaces, for instance, the gut wall; and (3) Stimulating the production of systematic antibodies, such as immunoglobulin [13,22].

A higher number of mucus-producing saccular cells have been detected in larvae exposed to increased bacterial concentration, suggesting that the augmented microflora stimulates the nonspecific defense mechanisms against invaders [32]. Several studies have reported that the usage of probiotics stimulated the cellular immune system by encapsulation, nodule formation, and phagocytosis; and the humoral immunity by production of agglutinins, anticoagulant proteins, phenol oxidase enzyme [33,34], antimicrobial peptides (defensins and chemokines), free radicals, bacteriocins, lysozymes, siderophores, monostatin, proteases, gramicidin, hydrogen peroxide, tyrotricidin, and organic acid [6]. A study by Gullian et al. assessed the immunostimulatory activity of a *Bacillus* sp. (P64) and *Vibrio* sp. (P62) using *Vibrio alginolyticus* as a positive control. They study inferred that P64 and *V. alginolyticus* were immunostimulants [35]. Taoka et al. analyzed Alchem Korea CO, Alchem Poseidon and Wonju Korea CO, containing a mixed consortium of bacteria (*Lactobacillus acidophilus, Bacillus subtilis,* and *Clostridium butyricum*) and yeast (*Saccharomyces cerevisiae*), stimulated the nonspecific immune parameters such as migration of neutrophils, lysozyme activity, and plasma bactericidal activity of tilapia (*Oreochromis niloticus*), resulting in resistance to *E. tarda* infection [36]. A study conducted by Standen et al. suggested that treatment with *Pediococcus acidilactici* caused the upregulation of proinflammatory cytokine TNF, a gene expression in probiotic-fed Nile tilapia (*O. niloticus*) [11].

14.3.4 Production of other beneficiary compounds

Probiotic organisms have been reported to improve the nutrition of the host by detoxifying harmful compounds in the feed, producing vitamins such as vitamin B12 and biotin, and denaturing indigestible ingredients in the diet by producing extracellular enzymes that aid in digestion [27]. Gupta et al., reported an increase in the digestive enzyme activity (protease, amylase, and lipase) of *Macrobrachium rosenbergii* when exposed to *Bacillus coagulans*, resulting in better feed absorption and growth of the host [37]. Studies have demonstrated that certain probiotics, particularly from *Bacteroides* and *Clostridium* sp., have the ability to secrete essential amino acids, fatty acids, and vitamins to the host [6,38]. Different bacteria isolated from the microflora of a variety of fish species, including channel catfish (*Ictalurus punctatus*), carp (*Cyprinus carpio)*, rainbow trout (*Oncorhynchus mykiss*), and tilapia species are known to produce vitamin B12 [18]. The benefit of including such bacteria in the fish diet needs to be investigated.

14.4 Screening and development of probiotics

The development of probiotics involves rigorous in vitro and in vivo screening of microorganisms from various sources that would benefit the host by conferring disease resistance, growth enhancement or improving its immediate surrounding environment.

Fig. 14.1 represents a flowchart of the protocol required to successfully develop a commercial probiotic for aquaculture.

14.4.1 In vitro screening for antimicrobial activity

Some of the major failures in probiotic research could be accredited to the selection of inappropriate microorganisms [5]. The primary criteria for selection of candidate probiotics involves methodically screening a vast number of potential microorganisms by in vitro tests. This step could be used to exclude the less-promising candidates, thus reducing the quantity of in vivo trials required to authenticate the effectiveness of the probiont [18]. One such test involves in vitro antagonism, in which pathogens are exposed directly to the candidate probionts or their extracellular metabolites in a solid medium,

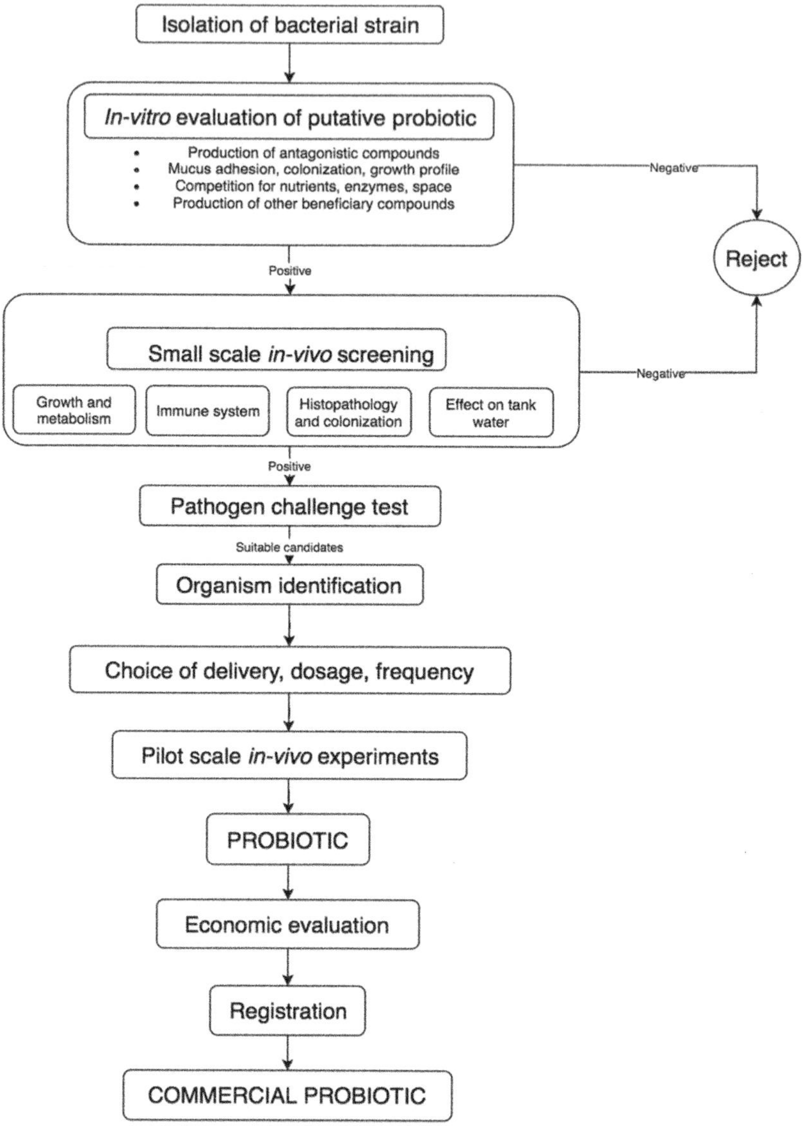

FIGURE 14.1 Research protocol to develop a probiotic for aquaculture.

such as well diffusion assay, or in liquid medium by tests such as coculture method [6,18,39]. However, the results of in vitro antagonism may not be a confirmed method to predict a promising in vivo effect [20]. A study conducted by Olsson et al., inferred that the same organism, when grown in two different media, produced varying quantities of inhibitory metabolites [28]. Similar observations were reported by Mayer-Harting et al., who suggested that the media composition might affect the amount of metabolite produced or the quantity released into the medium [40]. Therefore, besides screening for inhibitory compounds, further tests involving screening for siderophores, lysozymes, production of primary metabolites, competition for space, and nutrients needs to be performed.

14.4.2 Mucus adhesion, colonization, and growth profile

The ability of microorganisms to adhere to and colonize the gut or the external surface of the host, thereby preventing the colonization of pathogens, is an important criterion for selection of potential probiotics [18]. This involves the sustainability of the candidate probiotic within the host and within its culture environment for a significant time period. Some bacteria only secrete metabolites during the stationary phase of growth [41], which may not occur in vivo due to constant flushing in the gut. A common practice involves selecting bacteria that produce antimicrobial metabolites, however, determining the stage of growth for production of metabolites and their ability to compete for attachment sites is essential [42]. The incompetence to attach to the mucus of the gut wall indicates that these bacteria may not multiply adequately to compensate for being flushed during gut evacuation [18]. The growth profiles of the putative probiotics can be assessed in a standard microbiological media and compared to eliminate bacteria having similar growth and antagonistic abilities, thereby reducing the pool of probiotics to be tested further.

The tests carried out to study adhesion in vitro include measuring the adhesion of radioactively labeled bacteria, measuring the adhesion of fluorescent-tagged bacteria using a fluorometer and microscopic counting, detecting adhesion by crystal violet method and 4,6-diamidino-2-phenylindole [43]. For a probiotic to multiply in the gut of aquatic organisms, it is important to survive the bile salts present in the gut of the host [44]. Bile salts can affect the fatty acids and phospholipids present in microorganisms, thereby having a lethal effect. Some bacteria are capable of hydrolyzing bile salts using specific enzymes [45]. To effectively colonize the host gastrointestinal tract, the putative probiotic must be able to resist bile salts [44].

14.4.3 Pathogenicity test

Before a microorganism is used as a probiotic, it is mandatory to confirm that it is nontoxic and nonpathogenic to the host. Hence, the host organism should be challenged with the putative probiotic under both ordinary and stress conditions [25]. This can be validated by small-scale challenge experiments of the host species using temporary baths in the bacterial suspension, injection challenges, or direct addition of the putative probiotic to the tank water [6]. The pathogenicity of three bacterial strains toward Whiteleg shrimp,

Litopenaeus vannamei, was studied by adding the bacteria to nauplii cultures and monitoring for mortality for 4 days [46]. Similarly, Austin et al. injected a candidate probiotic suspension into Atlantic salmon intramuscularly and monitored it for 7 days, followed by examining the survivors for symptoms of disease [47]. Pathogenicity test can be combined with small-scale in vivo screening under monoxenic conditions.

The probiotic under study should preferably not contain any virulence or antibiotic resistance genes [19]. However, all 'pathogenic' bacteria are not necessarily harmful [6]. Some opportunistic bacteria such as *V. alginolyticus* has been effectively used as a probiotic in shrimp farming, fish aquaculture [47], and algal production [48]. Therefore, a probiont used for one species in aquaculture may turn out to be a pathogen for another species and vice versa.

14.4.4 Organism identification

Once the putative probiotic has been proven to be beneficial to the host by in vitro and small-scale in vivo tests, it should preferably be identified to strain level using 16S rRNA gene sequencing or fatty acid profile technique. The information about the identity of the organism would be useful in tracing the history of pathogenicity, duplicate organisms, culture requirements, and its suitability as a candidate probiont.

14.4.5 Route of delivery, dosage, and frequency

The delivery of probiotics should preferably begin in the early stages of larval development preceding exogenous feeding. Probiotics can be administered to the host or incorporated in its aquatic environment via live food, adding it to the culture water, injecting intramuscularly or intraperitoneally, addition to artificial diet or bathing [6,8,20,49,50]. The route of delivery should be determined based on the mode of action to ensure that the probiotic reaches the location where the effect is needed to take place [19].

Probiotics are more efficient if supplied on a regular basis or if they are able to colonize and persist in the aquatic host or in its ambient environment [19]. Vine et al. suggested that the probability of probiotics being ingested either directly or via live food is increased if the larvae are exposed to a probiotic concentration higher than that naturally occurring in the water [18]. This would result in better establishment of the probiotic in the mucus and gut epithelium, thereby suppressing the attachment of pathogenic bacteria. The probiotic dosage and frequency of exposure must be adequately determined so as to avoid a very low dose that could reduce the efficacy or an overdose that may result in high cost and low probiont efficiency [18]. Daily inoculation of

L. vannamei larval tanks with 10^5 cfu/mL probiotic bacteria prevented the colonization of bacterial pathogens during larval culture [6]. The frequency of the addition of probiotics usually depends on the probiotic species, culture conditions, stage of fish or shrimp development, diet, and selected probiotic concentration [18].

14.4.6 In vivo validation

A bacterium that exhibits antagonistic properties in the laboratory may not necessarily be inhibitory when associated with the host in vivo [19]. In vivo screening of candidate probiotics involves exposing the bacteria to the host under culture, followed by monitoring the growth, survival, immune, and physicochemical parameters for a defined time period. Vaseeharan and Ramasamy reported a decrease in the cumulative mortality of *Penaeus monodon* challenged with *V. harveyi* when treated with *B. subtilis* probiotic [51]. A researcher must also be able to re-isolate the organism from the gastrointestinal tract or the surface of the host in order to confirm resistance due to the putative probiotic. Zokaeifar et al. showed the colonization of *B. subtilis* strain L10 and G1 in the gut of *L. vannamei* after receiving a diet containing the two strains for 8 weeks. The study also reported a significant decrease in *Vibrio* spp. in the shrimp gastrointestinal tract [52].

If the microorganism has successively passed the screening stages discussed above, it can be termed as a "probiotic" [18]. However, for an industry to commercialize the probiotic, it is necessary to study the shelf life and viability of the organisms, storage conditions, and production cost of the probiotic.

14.4.7 Shelf life

Bacteria tend to lose their original properties of producing antagonistic compounds when cultured under artificial conditions, such as nutrient-rich media or without competition from other organisms. Hence it is necessary to culture and maintain the probiotic organisms on media that would be similar to the original environment of the bacteria of concern. The probiotic should be viable under normal storage as well as field conditions, and repeated subculturing of the organisms should be avoided [6,13,18,22]. Certain gram-positive bacteria such as *Bacillus* sp. produce endospores, making storage easier [18]. Commercial probiotics are generally available in liquid or lyophilized powder formulations or encapsulated in a colloidal matrix such as chitosan, alginate, carboxymethyl cellulose, etc. Encapsulation in a colloidal matrix protects the bacteria from the acidic pH and digestive enzymes of the host gastrointestinal tract [2]. The conditions such as temperature, osmolarity of the solution, and the degree of hydration for reconstituting the prepared probiotic are critical to ensure the viability and shelf life of the bacteria [53].

14.4.8 Economic evaluation

In order to commercialize the probiotic, it is necessary to evaluate the cost benefit of the probiotic product after successful accomplishment of in vivo trials. Investigating the economic feasibility of various product formulations, packaging possibilities, and dosage recommendations are mandatory [18]. The cost of the probiotic application could be estimated by mass producing the probiont and conducting comparative pilot-scale trials under hatchery or grow-out conditions in aquaculture farms. Similarly, effective legislation must be taken into account before beginning any commercial application [25].

14.5 Recent probiotics used in aquaculture

With an increase in demand for environmentally friendly aquaculture, many scientists are exploring the use of new probiotics for aquatic animals. The most frequently used bacterial probiotic species include genera *Lactobacillus*, strains of *Bacillus*, *Aeromonas*, *Bifidobacterium*, *Plesiomonas*, *Pseudomonas*, *Fusobacterium*, *Bacteroides*, *Agrobacterium*, *Eubacterium*, *Carnobacterium*, *Enterococcus*, *Bacteroides*, *Clostridium*, *Brevibacterium*, *Microbacterium*, and *Staphylococcus* spp. [6,29]. Table 14.1 represents the recent bacterial probiotics used to control pathogens in fish, crustacean, and mollusk aquaculture and their beneficial effects.

Besides bacteria, other probiotics such as yeast, algae and actinomycetes have also been used in many countries. The cell wall of *Saccharomyces cerevisiae* contains glucan, mannoproteins, and chitin, making it a suitable immunostimulating probiotic in mariculture [72]. Atlantic salmon fry fed with a diet comprising baker's yeast improved the survival compared to the control fish when challenged with *V. anguillarum*. Yeast *Debaryomyces hansenii* CBS 8339 demonstrated a probiotic effect in *Dicentrarchus labrax* (sea bass) larvae by promoting intestinal maturation and enhancing nutrient absorption, thus improving larval development [73]. You et al. reported a siderophore producing *Streptomyces* sp. that could influence the growth of pathogenic *Vibrio* sp. by competing with it for iron in the aquatic environment [74]. Actinomycetes *Rhodococcus* SM2 has been reported to improve the immunity and protection of rainbow trout against *V. anguillarum* [75].

14.6 Conclusion and future perspectives

Although numerous studies have reported the efficiency and mechanisms of probiotics, many questions are yet to be answered. Studies focusing on host-microbe interactions, proteome profiling of gut microbiota, interaction between gut microbes, gut immunity, antioxidant status, lipid level of hosts,

TABLE 14.1 Recent bacterial probiotics used as biocontrol agents in aquaculture.

Probiotic organism	Pathogenic strain	Host organism	Mode of action	Mode of application	Reference
Pseudoalteromonas sp. NC201	*Vibrio nigripulchritudo*	*Litopenaeus stylirostris*	Immunostimulatory, increased survival against *Vibrio parahaemolyticus nigripulchritudo*	Added to culture water	Louis et al. [54]
Clostridium butyricum	*Vibrio harveyi*	*Macrobrachium rosenbergii*	Antagonistic effect against *V. harveyi*	Incorporated in feed	Sumon et al. [55]
(*Bacillus subtilis* DCU, *Bacillus cereus* HL7, *Bacillus pumilus* BP	*Vibrio parahaemolyticus*	mud crab (*Scylla paramamosain*)	Immunostimulatory, increased survival against *V. parahaemolyticus*	Incorporated in feed	Wu et al. [56]
Lactobacillus plantarum	*Pseudomonas fluorescens*	Nile tilapia (*Oreochromis niloticus*)	Immunostimulatory, protection against *P. fluorescens*	Incorporated in feed	Iman et al. [57]
Alteromonas macleodii 0444, *Phaeobacter gallaeciensis*, *Neptunomonas* sp. 0536, *Pseudoalteromonas* sp. D41	*Vibrio coralliilyticus* and *V. pectenicida*, *V. splendidus*	Mollusk larviculture	Antagonistic effect against vibriosis	Added to culture water	Kesarcodi-Watson et al. [58]
Pseudomonas aeruginosa VSG-2	*Aeromonas hydrophila*	*Labeo rohita*	Immunostimulatory effect and increased survival	Dietary supplementation	Giri et al. [59]

Continued

TABLE 14.1 Recent bacterial probiotics used as biocontrol agents in aquaculture.—cont'd

Probiotic organism	Pathogenic strain	Host organism	Mode of action	Mode of application	Reference
Bacillus pumilus and *Bacillus clausii*	–	Grouper (*Epinephelus coioides*)	Immunostimulatory	Incorporated in feed	Yun-Zhang et al. [60]
Mixture of *B. subtilis, Lactobacillus plantarum* and *Saccharomyces cerevisiae*	–	Nile tilapia (*O. niloticus*)	Immunostimulant	Incorporated in feed	Magda et al. [61]
Enterobacter amnigenus	*Flavobacterium psychrophilum*	Rainbow trout (*Oncorhynchus mykiss*)	Resistance to rainbow trout fry syndrome	Incorporated in feed	Burbank et al. [62]
Alteromonas macleodii 0444 and *Neptunomonas* sp. 0536.	*Vibrio* sp. DO1 and *V. splendidus*	*Perna canaliculus* (Greenshell mussel)	Antagonistic, increased larval survival when challenged with pathogen	Added to tank water	Kesarcodi-Watson et al. [63]
Vibrio gazogenes	*Vibrio* sp.	*Litopenaeus vannamei*	Immunostimulatory effect and decrease in *Vibrio* count	Injection and incorporated in the diet	Thomson et al. [64]
Kocuria SM1	*Vibrio anguillarum* and *Vibrio ordalii*	Rainbow trout (*O. mykiss*)	Antagonistic to vibriosis	Incorporated in feed	Sharifuzzaman and Austin [65]
Bacillus cereus	*Aeromonas hydrophila*	Studied in vitro	Competitive exclusion by production of siderophores	Studied in vitro	Lalloo et al. [21]

Arthrobacter sp. and *Enterococcus* sp.	*Vibrio* sp.	Atlantic cod (*Gadus morhua*) larvae	Increase survival rate of larvae, decreased *Vibrio* count	Added to culture water	Lauzon et al. [66]
Bacillus circulans PB7	*A. hydrophila*	Catla catla	Immunostimulatory, increased survival against *A. hydrophila*	Incorporated in feed	Bandyopadhyay and Das Mohapatra [67]
Enterococcus faecium MC13 *Streptococcus phocae* PI80	*V. harveyi* and *V. parahaemolyticus*	*Penaeus monodon* (shrimp)	Increased survival against vibriosis	Incorporated in feed	Swain et al. [68]
Lactobacillus acidophilus and *B. subtilis*	*P. fluorescens* and *S. iniae*	Nile tilapia (*O. niloticus*)	Immunostimulatory effect	Incorporated in feed	Aly et al. [69]
B. subtilis	*A. hydrophila*	Indian major carp	Immunostimulation, increased growth and survival	Incorporated in feed	Kumar et al. [70]
Aeromonas media strain A199	*Saprolegnia parasitica*	*Anguilla australis* (eel)		Addition of live A199 cultures to the treatment tanks	Lategan et al. [71]
B. subtilis BT23	*V. harveyi*	*P. monodon*	90% reduction in shrimp mortality	Added to culture water	Vaseeharan and Ramasamy [51]

antagonistic activity or probable side effects of the probiotic still need to be carried out. Studies should also focus on production of viable probiotic formulations on a large scale at minimum operational cost. Besides this, an interactive approach and a combined effort to spread this scientific knowledge among scientists, academicians, producers and government organizations is essential to promote production, health and economic development of the aquaculture industry.

Acknowledgments

The authors are thankful to the Head, Department of Biotechnology, Goa University, for the facilities and UGC-MANF for providing the research fellowship (MANF-2017-18-GOA-87902).

References

[1] Sahu MK, Swarnakumar NS, Sivakumar K, Thangaradjou T, Kannan L. Probiotics in aquaculture: importance and future perspectives. Indian J Microbiol 2008;48:299–308.

[2] Cruz PM, Ibáñez AL, Hermosillo OAM, Saad HCR. Use of probiotics in aquaculture. ISRN Microbiol 2012;2012:1–13.

[3] Kumar V, Roy S, Meena DK, Sarkar UK. Application of probiotics in shrimp aquaculture: importance, mechanisms of action, and methods of administration. Rev Fish Sci Aquac 2016;24(4):342–68.

[4] Moriarty DJW. Control of luminous *Vibrio* species in penaeid aquaculture ponds. Aquaculture 1998;164:351–8.

[5] Gomez-Gil CB, Roque A, Turnbull JF. The use and selection of probiotic bacteria for use in the culture of larval aquatic organisms. Aquaculture 2000;191:259–70.

[6] Balcazar JL, Blas I, Ruiz-Zarzuela I, Cunningham D, Vendrell D, Muzquiz JL. The role of probiotics in aquaculture. Vet Microbiol 2006;114:173–86.

[7] Karunasagar I, Pai R, Malathi GR. Mass mortality of *Penaeus monodo*n larvae due to antibiotic resistant *Vibrio harveyi* infection. Aquaculture 1994;128:203–9.

[8] Moriarty DJW. Disease control in shrimp aquaculture with probiotic bacteria. In: Bell CR, Brylinsky M, Johnson-Green P, editors. Microbial biosystems: new frontiers: proceedings of the 8th international symposium on microbial ecology. Halifax, Canada: Atlantic Canada Society for Microbial Ecology; 1999.

[9] Wang H, Ren L, Yu X, Hu J, Chen Y, He G, Jiang Q. Antibiotic residues in meat, milk and aquatic products in Shanghai and human exposure assessment. Food Control 2017;80:217–25.

[10] Watts JEM, Schreier HJ, Lanska L, Hale MS. The rising tide of antimicrobial resistance in aquaculture: sources, sinks and solutions. Mar Drugs 2017;15:158.

[11] Standen BT, Rawling MD, Davies SJ, Castex M, Foey A, Gioacchini G. Probiotic *Pediococcus acidilactici* modulates both localised intestinal- and peripheral-immunity in tilapia (*Oreochromis niloticus*). Fish Shellfish Immunol 2013;35:1097–104.

[12] Lazado CC, Caipang CM, Estante EG. Prospects of host associated microorganisms in fish and penaeids as probiotics with immunomodulatory functions. Fish Shellfish Immunol 2015;45(1):2–12.

[13] Fuller R. History and development of probiotics. In: Fuller R, editor. Probiotics: the scientific basis. New York: Chapman & Hall; 1992. p. 1–8.

[14] Parker RB. Probiotics, the other half of the antibiotics story. Anim Nutr Health 1974;29:4–8.

[15] Fuller R. Probiotics in man and animals. J Appl Bacteriol 1989;66(5):365–78.

[16] Zhou X, Wang Y, Li W. Effect of probiotic on larvae shrimp (*Penaeus vannamei*) based on water quality, survival rate and digestive enzyme activities. Aquaculture 2009;287:349–53.

[17] Liu W, Bao Q, Jirimutu, Qing M, Siriguleng, Chen X, Sun T, Li M, Zhang J, Yu J, Bilige M, Sun T, Zhang H. Isolation and identification of lactic acid bacteria from Tarag in Eastern inner Mongolia of China by 16S rRNA sequences and DGGE analysis. Microbiol Res 2012;167:110–5.

[18] Vine NG, Leukes WD, Kaiser H. Probiotics in marine larviculture. FEMS Microbiol Rev 2006;30:404–27.

[19] Watson AK, Kaspar H, Lategan MJ, Gibson L. Probiotics in aquaculture: the need, principles and mechanisms of action and screening processes. Aquaculture 2008;274:1–14.

[20] Gram L, Melchiorsen J, Spanggaard B, Huber I, Nielsen TF. Inhibition of *Vibrio anguillarum* by *Pseudomonas fluorescens* AH2, a possible probiotic treatment of fish. Appl Environ Microbiol 1999;65(3):969–73.

[21] Lalloo R, Moonsamy G, Ramchuran S, Gorgens J, Gardiner N. Competitive exclusion as a mode of action of a novel *Bacillus cereus* aquaculture biological agent. Lett Appl Microbiol 2010;50:563–70.

[22] Farzanfar A. The use of probiotics in shrimp aquaculture. FEMS Immunol Med Microbiol 2006;48:149–58.

[23] Campos CA, Rodríguez O, Calo-Mata P, Prado M, Barros Velázquez J. Preliminary characterization of bacteriocins from *Lactococcus lactis, Enterococcus faecium* and *Enterococcus mundtii* strains isolated from turbot (*Psetta maxima*). Food Res Int 2006;39:356–64.

[24] Imada C, Simidu U, Taga N. Isolation and characterization of marine bacteria producing alkaline protease inhibitor. Bull Jpn Soc Sci Fish 1985;51:799–803.

[25] Verschuere L, Rombaut G, Sorgeloos P, Verstraete W. Probiotic bacteria as biological control agents in aquaculture. Microbiol Mol Biol Rev 2000;64(4):655–71.

[26] Adams CA. The probiotic paradox: live and dead cells are biological response modifiers. Nutr Res Rev 2010;23:37–46.

[27] Irianto A, Austin B. Use of probiotics to control furunculosis in rainbow trout, *Oncorhynchus mykiss* (Walbaum). J Fish Dis 2002;25:1–10.

[28] Olsson JC, Westerdahl A, Conway PL, Kjelleberg S. Intestinal colonization potential of turbot (*Scophthalmus maximus*) and dab (*Limanda limanda*) associated bacteria with inhibitory effects against *Vibrio anguillarum*. Appl Environ Microbiol 1992;58:551–6.

[29] Zorriehzahra MJ, Delshad ST, Adel M, Tiwari R, Karthik K, Dhama K, Lazado CC. Probiotics as beneficial microbes in aquaculture: an update on their multiple modes of action: a review. Vet Q 2016;36(4):228–41.

[30] Verschuere L, Heang H, Criel G, Dafnis S, Sorgeloos P, Verstraete W. Protection of *Artemia* against the pathogenic effects of *Vibrio proteolyticus* CW8T2 by selected bacterial strains. Appl Environ Microbiol 2000;66:1139–46.

[31] Vadstein O. The use of immunostimulation in marine larviculture: possibilities and challenges. Aquaculture 2000;155:401–17.

[32] Ottesen OH, Olafsen JA. Effects on survival and mucus cell proliferation of Atlantic halibut, *Hippoglossus hippoglossus* L, larvae following microflora manipulation. Aquaculture 2000;187:225–38.

[33] Lakshmi B, Viswanath B, Sai Gopal DVR. Probiotics as antiviral agents in shrimp aquaculture. J Pathog 2013;2013:1–13.
[34] Song SK, Beck BR, Kim D, Park J, Kim J, Kim HD, Ringo E. Prebiotics as immunostimulants in aquaculture: a review. Fish Shellfish Immunol 2014;40(1):40–8.
[35] Gullian M, Thompson F, Rodriguez J. Selection of probiotic bacteria and study of their immunostimulatory effect in *Penaeus vannamei*. Aquaculture 2004;233:1–14.
[36] Taoka Y, Maeda H, Jo JY, et al. Use of live and dead probiotic cells in tilapia *Oreochromis niloticus*. Fish Sci 2006;72(4):755–66.
[37] Gupta A, Dhawan A. Probiotic based diets for freshwater prawn *Macrobrachium rosenbergii* (de Man). Indian J Fish 2013;60:103–9.
[38] Tinh NTN, Dierckens K, Sorgeloos P, Bossier P. A review of the functionality of probiotics in the larviculture food chain. Mar Biotechnol 2007;10:1–12.
[39] Sotomayor MA, Balcazar JL. Inhibition of shrimp pathogenic vibrios by mixture of probiotic strain. Rev Aquat 2003;19:9–15.
[40] Mayer-Harting A, Hedges AJ, Berkeley RCW. Methods for studying bacteriocins. Methods Microbiol 1972;7:315–422.
[41] Monaghan RL, Gagliardi MM, Streicher SL. Culture preservation and inoculum development. In: Demain AL, Davies JE, Julian E, editors. Manual of industrial microbiology and Biotechnology. Washington: American Society for Microbiology; 1999. p. 29–48.
[42] Vanbelle M, Teller E, Focant M. Probiotics in animal nutrition: a review. Arch Anim Nutr 1990;40:543–67.
[43] Vesterlunda S, Palttab J, Karpb M, Ouwehanda AC. Measurement of bacterial adhesion – in vitro evaluation of different methods. J Microbiol Methods 2005;60(2):225–33.
[44] Vieira FN, Jatobá A, Mouriño ALP, Vieira EA, Soares M, Silva BC, Seiffert WQ, Martins ML, Vinatea LA. In vitro selection of bacteria with potential for use as probiotics in marine shrimp culture. Pesq agropec bras 2013;48(8):998–1004.
[45] Erkkilä S, Petäjä E. Screening of commercial meat starter cultures at low pH and in the presence of bile salts for potential probiotic use. Meat Sci 2000;55:297–300.
[46] Garriques D, Arevalo G. An evaluation of the production and use of a live bacterial isolate to manipulate the microbial flora in the commercial production of *Penaeus vannamei* postlarvae in Ecuador. In: Browdy CL, Hopkins JS, editors. Swimming through troubled water. Proceedings of the special session on Shrimp Farming, Aquaculture'95. Baton Rouge: World Aquaculture Society; 1995. p. 53–9.
[47] Austin B, Baudet E, Stobie M. Inhibition of bacterial fish pathogens by *Tetraselmis suecica*. J Fish Dis 1992;15:55–61.
[48] Gomez-Gil B, Roque A, Velasco-Blanco G. Culture of *Vibrio alginolyticus* C7b, a potential probiotic bacterium, with the microalga *Chaetoceros muelleri*. Aquaculture 2002;211:43–8.
[49] Austin B, Stuckey L, Robertson P, Effendi I, Griffith D. A probiotic strain of *Vibrio alginolyticus* effective in reducing diseases caused by *Aeromonas salmonicida, Vibrio anguillarum* and *Vibrio ordalii*. J Fish Dis 1995;18:93–6.
[50] Rengpipat S, Rukpratanporn S, Piyatiratitivorakul S, Menasaveta P. Immunity enhancement in black tiger shrimp *Penaeus monodon* by a probiont bacterium *Bacillus* S11. Aquaculture 2000;191:271–88.
[51] Vaseeharan B, Ramasamy P. Control of pathogenic *Vibrio* spp. by *Bacillus subtilis* BT23, a possible probiotic treatment for black tiger shrimp *Penaeus monodon*. Lett Appl Microbiol 2003;36:83–7.
[52] Zokaeifar H, Balcázar JL, Saad CR, Kamarudin MS, Sijam K, Arshad A, Nejat N. Effects of *Bacillus subtilis* on the growth performance, digestive enzymes, immune gene expression

and disease resistance of white shrimp, *Litopenaeus vannamei*. Fish Shellfish Immunol 2012;33:683–9.

[53] Muller JA, Ross RP, Fitzgeralk GF, Stanton C. Manufacture of probiotic bacteria. In: Charalampopoulos D, Rastall RA, editors. Prebiotics and probiotics science and technology. New York: Springer Verlag; 2009. p. 125–759.

[54] Louis S, Nelly W, Dominique A, Jean-Rene M, Marie P, Cong Z, Mathilde L, Viviane B, Dominique P. Survival improvement conferred by the *Pseudoalteromonas sp.* NC201 probiotic in *Litopenaeus stylirostris* exposed to *Vibrio nigripulchritudo* infection and salinity stress. Aquaculture 2018;495:888–98.

[55] Sumon MS, Ahmmed F, Khushi SS, Ahmmed MK, Rouf MA, Chisty MAH, Sarower MG. Growth performance, digestive enzyme activity and immune response of *Macrobrachium rosenbergii* fed with probiotic *Clostridium butyricum* incorporated diets. JKSUS 2018;30:21–8.

[56] Wu H, Sun L, Li C, Li Z, Zhang Z, Wen X, Hu Z, Zhang Y, Li S. Enhancement of the immune response and protection against *Vibrio parahaemolyticus* by indigenous probiotic *Bacillus* strains in mud crab (*Scylla paramamosain*). Fish Shellfish Immunol 2014;41:156–62.

[57] Iman MKA, Wafaa TA, Elham SA, Mohammad MNA, El-Shafei K, Osama MS, Gamal AI, Zeinab IS, El-Sayed HS. Evaluation of *Lactobacillus plantarum* as a probiotic in aquaculture: emphasis on growth performance and innate immunity. J Appl Sci Res 2013;9(1):572–82.

[58] Kesarcodi-Watson A, Miner P, Nicolas J, Robert R. Protective effect of four potential probiotics against pathogen-challenge of the larvae of three bivalves: Pacific oyster (*Crassostrea gigas*), flat oyster (*Ostrea edulis*) and scallop (*Pecten maximus*). Aquaculture 2012:344–9.

[59] Giri S, Sen SS, Sukumaran V. Effects of dietary supplementation of potential probiotic *Pseudomonas aeruginosa* VSG-2 on the innate immunity and disease resistance of tropical freshwater fish, *Labeo rohita*. Fish Shellfish Immunol 2012:1–6.

[60] Yun-Zhang S, Yang H, Ru-Long M, Lin W. Probiotic applications of two dominant gut Bacillus strains with antagonistic activity improved the growth performance and immune responses of grouper (*Epinephelus coioides*). Fish Shellfish Immunol 2010;29:803–9.

[61] Magda ME, Sabry SE, Mohamed AE, Santoch L, Said MD, Neven AE. The viability of probiotics as a factor influencing the immune response in the Nile tilapia (*Oreochromis niloticus*). Egypt J Aquat Biol Fish 2011;15(1):105–24.

[62] Burbank DR, Shah DH, La Patra SE, Fornshell G, Cain KD. Enhanced resistance to coldwater disease following feeding of probiotic bacterial strains to rainbow trout (*Oncorhynchus mykiss*). Aquaculture 2011;321:185–90.

[63] Kesarcodi-Watson A, Kaspar H, Lategan MJ, Gibson L. *Alteromonas macleodii* 0444 and *Neptunomonas* sp. 0536, two novel probiotics for hatchery-reared Greenshell (TM) mussel larvae, *Perna canaliculus*. Aquaculture 2010;309:49–55.

[64] Thompson J, Gregory S, Plummer S, Shields RJ, Rowley AF. An in vitro and in vivo assessment of the potential of *Vibrio* spp. as probiotics for the Pacific White shrimp, *Litopenaeus vannamei*. J Appl Microbiol 2010;109:1177–87.

[65] Sharifuzzaman SM, Austin B. Kocuria SM1 controls vibriosis in rainbow trout (*Oncorhynchus mykiss, Walbaum*). J Appl Microbiol 2010;108:2162–70.

[66] Lauzon HL, Gudmundsdottir S, Petursdottir SK, Reynisson E, Steinarsson A, Oddgeirsson M, Bjornsdottir R, Gudmundsdottir BK. Microbiota of Atlantic cod (*Gadus*

morhua L.) rearing systems at pre- and post-hatch stages and the effect of different treatments. J Appl Microbiol 2010;109:1775–89.

[67] Bandyopadhyay P, Das MPK. Effect of a probiotic bacterium *Bacillus circulans* PB7 in the formulated diets: on growth, nutritional quality and immunity of Catla catla (Ham.). Fish Physiol Biochem 2009;35:467–78.

[68] Swain SM, Singh C, Arul V. Inhibitory activity of probiotics *Streptococcus phocae* PI80 and *Enterococcus faecium* MC13 against vibriosis in shrimp *Penaeus monodon*. World J Microbiol Biotechnol 2009;25:697–703.

[69] Aly SM, Ahmed YAG, Ghareeb AAA, Mohamed MF. Studies on *Bacillus subtilis* and *Lactobacillus acidophilus* as potential probiotics, on the immune response and resistance of Tilapia nilotica (*Oreochromis niloticus*) to challenge infections. Fish Shellfish Immunol 2008;25:128–36.

[70] Kumar R, Mukherjee S, Prasad KP, Pal AK. Evaluation of *Bacillus subtilis* as a probiotic to Indian major carp *Labeo rohita* (Ham.). Aquacult Res 2006;37:1215–21.

[71] Lategan MJ, Gibson LF. Antagonistic activity of *Aeromonas media* strain A199 against *Saprolegnia* sp., an opportunistic pathogen of the eel, Anguilla australis Richardson. J Fish Dis 2003;26:147–53.

[72] Scholz U, Diaz GG, Ricque D, Suarez LEC, Albores FV, Latchford J. Enhancement of vibriosis resistance in juvenile *Penaeus vannamei* by supplementation of diets with different yeast products. Aquaculture 1999;176:271–83.

[73] Tovar-Ramirez D, Zambonino-Infante JL, Cahu C, Gatesoupe FJ, Vazquez-Juarez R. Influence of dietary live yeast on European sea bass (*Dicentrarchus labrax*) larval development. Aquaculture 2004;234:415–27.

[74] You J, Cao L, Liu G, Zhou S, Tan H, Lin Y. Isolation and characterization of actinomycetes antagonistic to pathogenic *Vibrio* spp. from near shore marine sediments. World J Microbiol Biotechnol 2005;21:679–82.

[75] Sharifuzzaman SM, Abbass A, Tinsley JW, Austin B. Subcellular components of probiotics *Kocuria* SM1 and *Rhodococcus* SM2 induce protective immunity in rainbow trout (*Oncorhynchus mykiss, Walbaum*) against *Vibrio anguillarum*. Fish Shellfish Immunol 2011;30:347–53.

Chapter 15

Recent advances in quorum quenching of plant pathogenic bacteria

Gauri A. Achari[1], R. Ramesh[2]
[1]*Department of Biological Sciences, Birla Institute of Technology and Science Pilani, KK Birla Goa Campus, Zuarinagar, India;* [2]*Crop Improvement and Protection Section, ICAR-Central Coastal Agricultural Research Institute, Old Goa, India*

15.1 Introduction

The term *quorum sensing* (QS) was coined by E. P. Greenberg and colleagues [1] and is defined as "bacterial cell-to-cell communication system" [2,3]. QS is regulated by diffusible, low-molecular-weight signal molecules called autoinducers (AIs), which increase in concentration as the cell population density increases [4,5]. QS regulates a plethora of biological activities such as bioluminescence, plasmid transfer, motility, expression of virulence, pigment production, siderophore production, epiphytic fitness, and biofilm formation. Interactions of plant-associated bacteria with the hosts, including colonization, control of tissue maceration, antibiotic production, toxin release, and horizontal gene transfer are governed by the QS mechanisms [6]. Bacterial QS molecules fall into two main categories: the short peptide and amino acids commonly produced by the gram-positive bacteria and the acyl homoserine lactones (AHL), which are the fatty acid derivatives produced by the gram-negative bacteria. Over the past decade many bacterial pathogens have been reported to produce diverse non-AHL AIs, and extensive research has shown that plant pathogens employ AHL- as well as non-AHL-based QS for the regulation of virulence. Deficiency in the QS leads to reduced virulence in plant pathogenic bacteria [7]. In simple terms, the term *quorum quenching* (QQ) can be defined as "interference in the QS system" as designated by Dong et al. [8]. Faure and Dessaux [9] defined QQ as a natural phenomenon or engineered procedures causing weakening of the expression of QS-regulated traits in bacteria. QQ strategies are nonlethal to bacteria and govern only the expression of virulence factors in pathogenic bacteria. Therefore, QQ does not exert selective pressure and this is of importance to curb the emergence of

Advances in Biological Science Research. https://doi.org/10.1016/B978-0-12-817497-5.00015-X

drug resistance in bacterial phytopathogens [10,11]. This chapter will focus on highlighting the recent research developments in QQ in bacterial phytopathogens proven useful in reducing the expression of virulence factors and controlling plant disease.

15.2 Overview of the different quorum sensing molecules of plant pathogenic bacteria

A majority of the phytopathogenic bacteria are gram negative and utilize the AHL-based QS systems for regulating virulence. AHL-based QS consists of a *Vibrio fischeri luxI* homologue encoding the AI synthase that synthesizes the AHL. It also has a cytoplasmic AI receptor/DNA-binding transcriptional activator protein, encoded by a *V. fischeri luxR* homologue [12]. In the cytoplasm, the AI forms a complex with its cognate receptor, which exhibits an increased affinity to the promoter regions of the genes controlled by QS [6]. AI synthases lactonize the methionine from *S*-adenosylmethionine (SAM) to fatty acyl chains on the acyl-acyl carrier proteins. The lactone ring of the AHL molecules is hydrophilic in nature, whereas the carbon chains (length varying from 4 to 18 carbons) are hydrophobic. Carbonyl or hydroxyl group substitutions can be present at the C3 atom of the AHL [13]. Substitutions present at the C3 atoms, the length of the acyl groups, and the degree of saturation are the prime determinants of the specificity of the AHL signals [14]. Some bacteria lacking the *luxI* gene homologue and having just the *luxR* gene homologue are called as Lux-R solos. These Lux-R solos can sense AI but cannot produce their own QS signal [15]. Table 15.1 provides an overview of the important bacterial phytopathogens and their respective QS molecules.

The plant pathogen *Ralstonia solanacearum* harbors AHL (N-hexanoyl homoserine lactone and N-octanoyl homoserine lactone) as well as non-AHL-based AI known as 3-hydroxy palmitic acid methyl ester (3OH-PAME). Recently, a novel signaling molecule, 3-hydroxy myristic acid methyl ester (R-3OH-MAME), has been reported in *R. solanacearum* [22]. Expression of virulence in *R. solanacearum* is chiefly governed by the 3OH-PAME and/or R-3OH-MAME QS system. SAM-dependent methyltransferase relocates the acyl carrier protein and links methyl group from SAM to 3-hydroxy palmitic acid to form 3OH-PAME. When the external concentration of 3OH-PAME increases above 5 nM, expression of QS-regulated virulence factors, mainly endoglucanase (Egl) and exopolysaccharides (EPS), starts, whereas when 3OH-PAME concentration is below 5 nM, cells are nonvirulent. Slight modifications in the acyl chain or the substitution in methyl group impedes QS activity in *R. solanacearum* [23]. The AHL QS system of *R. solanacearum* is not involved in the expression of virulence factors. *LuxI* and *luxR* homologues of *R. solanacearum* are named *solI* and *solR*, respectively. *SolI* encodes an enzyme that synthesizes N-hexanoyl-homoserine lactone and N-octanoyl-homoserine lactone, which in turn regulate the *aidA* gene. The 3OH-PAME

TABLE 15.1 Quorum sensing molecules in bacterial plant pathogens and quorum quenching mechanisms.

Major class QS molecules involved in regulation of virulence	Phytopathogens	Quorum quenching mechanisms	Quorum quenching organisms	Reference
Acyl homoserine lactones	*Pseudomonas syringae* *Agrobacterium tumefaciens* *Erwinia amylovora* *Pectobacterium carotovorum* *Pectobacterium atrosepticum*	Acyl homoserine lactone acylases, lactonases, and oxidoreductases; blocking of signal sensing and synthesis; structural analogues of AHL	*Rhodococcus* sp. *Pseudomonas* sp. *Klebsiella* sp. *Ralstonia* sp. Several plant extracts	[3,9,16,17]
Diffusible signal factor	*Xanthomonas oryzae* *Xanthomonas campestris* *Xanthomonas axonopodis* *Xylella fastidiosa*	Degradation and modification	*Paenibacillus*, *Microbacterium*, *Staphylococcus*, and *Pseudomonas* sp.	[18,19]
3-Hydroxy palmitic acid methyl ester (3OH-PAME), 3-hydroxy myristic acid methyl ester (3OH-MAME)	*Ralstonia solanacearum*	Degradation of 3OH-PAME, degradation of 3OH-MAME is not reported	*Ideonella* sp., *Acinetobacter* sp., *Stenotrophomonas maltophilia*, *Pseudomonas aeruginosa* and *Rhodococcus corynebacterioides*	[20,21]

system has a quorum of 10^7 cells/mL, whereas the SolRI system requires the cell density to exceed 10^8 cells/mL [4,23].

Another type of non-AHL QS molecules are the diffusible signal factors (DSFs). Chemically, the DSFs are *cis*-2-unsaturated fatty acids. *Cis*-11-methyl-2-dodecenoic acid was the first DSF identified in the *Xanthomonas campestris* pv. *campestris*, which basically regulates the genes named as regulation of pathogenicity factors (*rpf*), and also regulates the expression of extracellular enzymes (Egl and protease), xanthan, and other virulence factors. DSF is also synthesized by *Burkholderia cenocepacia* and *Pseudomonas aeruginosa*, which are however distantly related to the Xanthomonads. DSF synthesis in *X. campestris* pv. *campestris* is dependent on *rpfF*, which encodes a crotonase family enzyme acting on fatty acyl carrier protein substrates and a fatty acyl CoA ligase RpfB. Sensor RpfC and regulator RpfG form a two-component system for DSF sensing and signal transduction. DSF signals are released from the cell by a yet-unknown transport mechanism [24]. Other examples of DSF signals include *cis*-2-decenoic acid in *P. aeruginosa* and in *B, cenocepacia* (known as Burkholderia-DSF); *cis*-2-tetradecenoic acid (XfDSF1) and 2-*cis*-hexadecanoic acid (XfDSF2) in *Xylella fastidiosa*; and *cis*, *cis*-11 methyldodeca-2,5-dienoic acid in *Xanthomonas oryzae* [25,26]. In *Dickeya* sp. (formerly classified as *Erwinia chrysanthemi*), along with a classic AHL system known as the Exp system, there exists a newly discovered system comprising a virulence factor modulating molecule, with an unknown chemical structure [27,28]. However, QQ in this type of QS has not yet been elucidated.

15.3 Mechanisms of quorum quenching

QQ can occur by four major mechanisms as described in the following subsections.

15.3.1 Inhibition of synthesis of quorum sensing signal

Analogues of SAM, namely L-S-adenosyl homocysteine and sinefungin (an SAM-like antibiotic), inhibit the synthesis of AHL. Examples of other inhibitors of AHL synthesis include the triclosan aiming at the enoyl-ACP reductase activity [9]. Research needs to be focused toward determining the mechanisms for inhibition of 3OH-PAME or R-3OH-MAME synthesis in *R. solanacearum*, since there are no reports on this to date.

15.3.2 Inhibition of sensing of quorum sensing signal

Halogenated furanones produced by the algae *Delisea pulchra* are efficient QS inhibitors since they can link to the LuxR receptor, displacing the bound AHLs, thereby disrupting signal sensing. *Chlamydomonas reinhardtii* produces an unidentified AHL analogue having potential as QQ agent. Extracts of pea, grape, strawberry, soybean, vanilla, geranium, lily, garlic, clover, lotus, yam beans, and pepper inhibit AHL QS in several diverse bacterial species. These plant extracts exhibit QQ activity, plausibly due to the presence of an active lactonase in the extracts [16]. Fungal compounds patulin and penicillic acid are lactones and therefore act as bacterial AHL signal analogues. It is interesting to note that patulin occurs in apple, pear, peach, apricot, banana, and pineapple, which makes these fruits potential anti-QS phyto resources. Anti-QS activity is reported in structural analogues of AHLs such as phenyl-AHL and chlorophenyl-AHL [9]. However, structural analogues of 3OH-PAME or R-3OH-MAME of *R. solanacearum* useful in QQ are unreported to date.

15.3.3 Degradation of quorum sensing molecules

15.3.3.1 Acyl homoserine lactone degradation

AHLs are sensitive to elevated temperatures and under alkaline pH, a reversible lactonolysis can occur. AHLs can spontaneously convert to antibacterial tetramic acids that do not function in QS. Biological degradation of AHL was first observed in *Variovorax* sp., and subsequently in *Bacillus* sp., which occurs mainly via the action of AI degrading and modifying enzymes [29,30]. Degradation of AHL produced by several members of phyla Proteobacteria, Actinobacteria, and Firmicutes are reported to occur by the action of AHL acylase and AHL lactonase enzymes [9]. In addition, modification of AHL can actively occur by the action of bacterial oxidoreductases such as those produced by *Rhodococcus* sp [17].

AHL lactonase hydrolyzes the lactone ring of AHL molecule and reduces its effectiveness as QS molecule. This type of degradation is called lactone hydrolysis, and it works similarly to the lactonolysis occurring at alkaline pH. Two Zn^{2+}-dependent metalloproteins function as AHL lactonases in bacteria, namely, AiiA lactonase and QsdA lactonase. AiiA-type lactonases are present in *Bacillus* sp., *Agrobacterium* sp., *Rhodococcus* sp., *Pseudomonas* sp., and *Klebsiella* sp. QsdA-type lactonase of *Rhodococcus erythropolis* belongs to the phosphotriesterase family and is effective in QQ [3,17]. Another AHL degrading enzyme that is AHL acylase functions by irreversibly hydrolyzing the amide bond between the acyl chain and homoserine, thereby releasing homoserine lactone and a corresponding fatty acid, both of which fail to act as QS molecules. This reaction of AHL degradation is also known as amidohydrolysis. AHL acylases are present in *Ralstonia* (AiiD), *Streptomyces* sp. (AhlM), *P. aeruginosa* PAO1 (PvdQ and QuiP), and *Anabaena* sp. PCC7120 (AiiC). Additional AHL acylase producers include *Comamonas* sp., *Shewanella* sp., and *Variovorax* sp. [3,17]. Interestingly, the *Arabidopsis thaliana* fatty acid amide hydrolase can catalyze AHL amidolysis to form L-homoserine, which in turn upregulates several pathways involved in plant growth [31].

The effect of L-homoserine released due to degradation by plant-mediated amidolysis on plant growth is dependent on its concentration and the length of AHL acyl side chain [31]. The third class of enzymes is the AHL oxidoreductases that are involved in the modification of the acyl side chain of the AHL by oxidative or reductive reactions. The oxidoreductases act on the AHLs independently of the length or the type of the fatty acid side chain. P-450/NADPH-P450 reductase from *B. megaterium* modifies AHL, whereas an oxidoreductase from *R. erythropolis* reduces the keto group of 3-oxo-AHLs (C8 to C14 acyl chain) to the corresponding 3-hydroxy product [3,17].

15.3.3.2 3-Hydroxy palmitic acid methyl ester hydrolase

Enzymatic degradation of 3OH-PAME occurs by hydrolysis of the ester bond between the methyl group and the 3-hydroxy fatty acid molecule by an esterase produced by several genera of bacteria, including *Ideonella* sp., *Stenotrophomonas maltophilia*, *P. aeruginosa*, and *Rhodococcus corynebacterioides* [20,21].

15.3.3.3 Degradation of the diffusible signal factor

RpfB a fatty acid CoA ligase, which is involved in signal synthesis in *Xanthomonas* sp., plays an important role in the DSF degradation as suggested by the recent literature [26]. Orthologues of RpfB are prevalent in plant-associated bacterial species, mainly *Bacillus*, *Paenibacillus*, *Microbacterium*, *Staphylococcus*, and *Pseudomonas* [18,19]. The DSF signal quenching occurs due to degradation of DSF or its modification due to addition of a sugar moiety from uridine diphosphate (UDP) sugars such as UDP-glucose or UDP-galactose via the action of enzyme UDP sugar transferase. In QQ bacterial species these UDP sugars are produced by the activity of a carbamoylphosphate synthetase encoded by *carA* and *carB* genes, indicating an important role of this enzyme in QQ [18,19]. Fig. 15.1 provides an overview of different mechanisms of enzymatic degradation of QS molecules.

15.3.3.4 Other mechanisms for quorum quenching

Epiphytic *Pseudomonas* strain 114 inhibited QS in phytopathogenic *P. syringae* by producing a siderophore that sequesters Co^{2+} and Fe^{2+}, thereby limiting the availability of these metal ions for maximal QS expression [32]. Although this QS inhibition is not mediated by enzymes, the exact mechanism of the metal-dependent QS inhibition has not been elucidated. Recently, a novel strategy termed as "pathogen confusion," which involves disruption of DSF QS signal balance in *X. fastidiosa* by DSF overproduction in plants, was reported [33]. Overexpression of DSF causes decreased mobility of *X. fastidiosa*, which further leads to a reduction in disease symptoms in DSF-producing transgenic plants [33]. A compound called bismerthiazol (1,3,4-thiadiazole molecule) reduced DSF-regulated virulence in the bacterial leaf

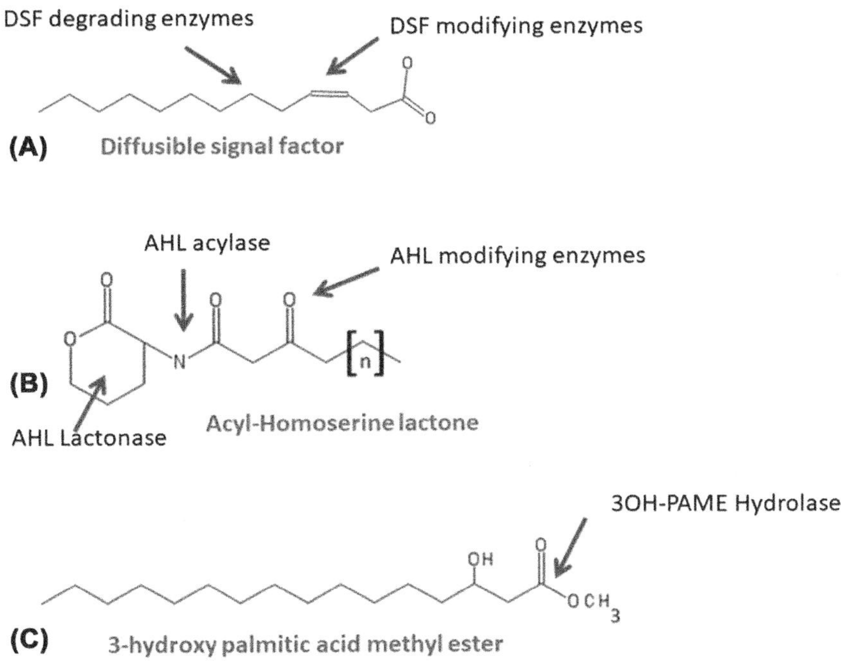

FIGURE 15.1 Figure illustrating mode of action of various enzymes involved in quorumquenching in phytopathogenic bacteria. (A) Degradation or modification of DSF degradingenzymes, (B) Degradation of acyl-homoserine lactones by the action of lactonases and acylases, and modification by the action of modifying enzymes, (C) degradation of 3-hydroxy palmitic acidmethyl ester by esterase enzymes.

blight pathogen *Xanthomonas oryzae* in rice by inhibiting the histidine utilization pathway important for expressing QS [34].

15.4 Quorum quenching against plant pathogens

Two main approaches for identifying QQ bacteria include screening of the isolated bacteria for inhibition of QS and another is a widely followed method that involves enrichment of QS degraders using QS molecules as sole source of carbon and/or nitrogen [9,29]. *Ideonella* sp. isolated from tomato rhizosphere degraded 3OH-PAME and reduced expression of virulence factors, mainly EPS in *R. solanacearum* [20]. Rhizosphere and endophytic tissue–colonizing QQ strains also exhibited biocontrol activity against bacterial wilt in eggplant when tested under greenhouse conditions [21]. *Bacillus* sp. with high activity toward AHL degradation was isolated by Dong et al. [8]. Dong et al. [2] have reported that *in planta*, AHL lactonase-producing strain of *B. thuringiensis* decreased the incidence of *Erwinia carotovora* infection and development of soft rot symptoms in potato. Biofilm forming ability of *P. aeruginosa* PAO1 and PAO1-JP2 was curbed by QQ strains of *Bacillus firmus* and *Enterobacter* sp. [35]. Phyllosphere-inhabiting strains *Pseudomonas* 114 and 120 sequestered Fe^{2+} from leaves and reduced QS-regulated traits in *P. syringae* pv. *syringae*, suggesting that the Fe competition on the leaves is an important parameter to control QS-regulated virulence in pathogens. However, in *P. syringae* pv. *syringae*, QS blocking may cause a rise in hyper-swarmers, which can invade the leaf faster and more frequently to cause disease [32]. DSF-degrading *Pseudomonas* and *Bacillus* sp. controlled the citrus canker symptoms in *Citrus sinensis* since they reduced biofilm formation and altered attachment patterns of the citrus canker pathogen on leaves [19]. Along with virulence, deficiency in QS inhibits colonization ability of the black leg pathogen *Pectobacterium carotovorum* sp. *brasiliense* in vascular tissues of potato [36]. Using structural analogues of AHL, growth of *R. erythropolis* was stimulated in the rhizosphere, so that it could efficiently degrade AHL of *Pectobacterium atrosepticum* [37–39]. Coinoculation of *X. campestris* pv. *campestris* with DSF-degrading bacteria into mustard and cabbage leaves and in grape stems reduced disease severity and disease incidence [18]. AHL lactonase-producing *Acinetobacter* sp., *Klebsiella* sp., and *Burkhoderia* sp. enriched from ginger rhizosphere effectively quenched QS and prevented *Erwinia carotovora* infection in potato plants [40]. Strains of *Pseudomonas* sp., *Variovorax* sp., *Comamonas* sp., and *Rhodococcus* sp. isolated from tobacco rhizosphere exhibited QQ activities [41]. Zwittermicin-producing strain of *Bacillus cereus* was genetically modified to express AHL lactonase and was found to reduce the incidence of *E. carotovora* infections [42]. *Burkolderia* sp., an endophyte from *Oryza sativa* (rice), was engineered to produce QQ lactonase for biocontrol applications [43]. Gene-encoding AiiA (AI inactivation-A) from *Bacillus* sp., when cloned in pathogenic *E. carotovora*, affected the release of its own AHL. In addition, decreased extracellular pectolytic enzymes and reduced virulence of *E. carotovora* on potato, tobacco, eggplant, cabbage, cauliflower, carrot, and celery were described [30].

15.5 Transgenic plants expressing quorum quenching molecules

Transgenic tobacco and potato plants expressing AHL lactonase were generated by Dong et al. [8] and Fray [44]. These transgenic plants quenched QS signaling and showed reduced tissue maceration and enhanced resistance to *E. carotovora* infection. Interestingly, transgenic tobacco plants expressing *E. carotovora* gene for AHL biosynthesis (*expl* gene) also showed increased resistance to *E. carotovora*. It is reported that plants expressing AHL prematurely trigger plant cell wall–degrading enzymes in *E. carotovora* during early stages of infection and induce plant defenses leading to enhanced resistance [17]. The expression of *X. fastidiosa* RpfF protein in grape and citrus reduced the virulence of *X. fastidiosa* and *Xanthomonas citri*, respectively, although the underlying mechanisms are not fully understood [33,45]. Also, the expression of DSF signals in by plants can activate the premature production of virulence factors in plant pathogens *in planta* that leads to triggered plant defenses that can overwhelm the smaller number of bacteria actually infecting the plant under experimental conditions [26]. DSF by itself can stimulate induction of innate immunity by callose deposition, induction of the pathogenesis-related protein 1 gene and hypersensitive reactions in leaves of *A. thalianais* and *Nicotiana benthamiana* and roots of *O. sativa* [46]. Root application of *Serratia marcescens* heightened the induced systemic resistance response against pathogen *Pectobacterium carotovorum* subsp. *carotovorum* and also against *Pseudomonas syringae* pv. *tabaci*, in transgenic plants that expressed AHL, whereas the induced systemic resistance decreased in QQ plants expressing *AiiA* [47]. Similarly, DSF is reported to prime plant responses toward the microbe-associated molecular patterns such as flagellin and lipopolysaccharides of *X. campestris* [26]. *Amorphophallus konjac* expressing a QQ lactonase (*AiiA*) from *B. thuringiensis* exhibited enhanced resistance to *Pectobacterium carotovorum* infection (scaled based on the lesion size on leaves) when compared to the control plants [48].

Newer niches such as marine environments and even marine organisms are potential resources for the isolation of novel QQ bacteria [49]. In addition, several newer in silico approaches using bioinformatics tools have been reported useful in detecting *luxR* regulators in plant pathogenic Actinobacteria, and can be applied to detect QS and QQ molecules in other important phytopathogens as well [50].

15.6 Summary and future research needs

The literature suggests that QQ strategies are emerging as new antivirulence approaches to prevent the occurrence of plant diseases. Reports on the existence of newer QS molecules with unknown structures indicate a need for exhaustive future research to delineate the complex and multiple QS processes that exist in bacteria for regulation of their virulence to plants. Bioinformatics is emerging as an interesting and useful approach to screen strains for the presence of QS homologues and QQ enzymes. There is also a need for better and efficient QS indicators and bioassays for rapid screening of QQ bacterial strains. Since a majority of the QQ molecules are active enzymes, research must be focused on studying their stability and efficacy in soil environment when secreted out from bacterial cells and determining their resistance to environmental extremes, which is important for their application to control plant diseases. A potential approach is also to screen novel bioactive molecules from terrestrial and aquatic flora for muting QS in plants, essentially the compounds that block the signal sensing.

Acknowledgments

The library facilities at Birla Institute of Technology and Science Pilani KK Birla Goa Campus, Goa, India are greatly acknowledged for their assistance. GAA graciously thanks the Science and Engineering Research Board (SERB), Department of Science and Technology (DST), Govt. of India, for financial support (Grant No. PDF/2016/001893).

References

[1] Fuqua C, Parsek MR, Greenberg EP. Regulation of gene expression by cell-to-cell communication: acyl-homoserine lactone quorum sensing. Annu Rev Genet 2001;35:439–68.

[2] Dong YH, Zhang XF, Xu JL, Zhang LH. Insecticidal *Bacillus thuringiensis* silences *Erwinia carotovora* virulence by a new form of microbial antagonism, signal interference. Appl Environ Microbiol 2004;70(2):954–60.

[3] Chen F, Gao Y, Chen X, Yu Z, Li X. Quorum quenching enzymes and their application in degrading signal molecules to block quorum sensing-dependent infection. Int J Mol Sci 2013;14:17477–500. https://doi.org/10.3390/ijms140917477.

[4] Whitehead NA, Barnard AML, Slater H, Simpson NJL, Salmond GPC. Quorum- sensing in Gram-negative bacteria. FEMS Microbiol Rev 2001;25:365–404.

[5] Zhang LH, Dong YH. Quorum sensing and signal interference: diverse Implications. Mol Microbiol 2004;53(6):1563–71.

[6] von Bodman SB, Bauer WD, Coplin DL. Quorum sensing in plant pathogenic bacteria. Annu Rev Phytopathol 2003;41:455–82.

[7] Soto MJ, Sanjuan J, Olivares J. Rhizobia and plant-pathogenic bacteria: common infection weapons. Microbiol 2006;152:3167–74.

[8] Dong YH, Wang LH, Xu JL, Zhang HB, Zhang XF, Zhang LH. Quenching quorum-sensing dependent bacterial infection by an N-acyl homoserine lactonase. Nature 2001;411:813–7.

[9] Faure D, Dessaux Y. Quorum sensing as a target for developing control strategies for the plant pathogen *Pectobacterium*. Eur J Plant Pathol 2007;119:353–65. https://doi.org/10.1007/s10658-007-9149-1.

[10] Tay SB, Yew WS. Development of quorum-based anti-virulence therapeutics targeting Gram-negative bacterial pathogens. Int J Mol Sci 2013;14:16570–99. https://doi.org/10.3390/ijms140816570.

[11] Defoirdt T. Quorum-sensing systems as targets for antivirulence therapy. Trends Microbiol 2017. https://doi.org/10.1016/j.tim.2017.10.005.

[12] Waters CM, Bassler BL. Quorum sensing: cell-to-cell communication in bacteria. Annu Rev Cell Dev Biol 2005;21:319–46.

[13] Schikora A, Schenk ST, Hartmann A. Beneficial effects of bacteria-plant communication based on quorum sensing molecules of the N-acyl homoserine lactone group. Plant Mol Biol 2016;90(6):605–12. https://doi.org/10.1007/s11103-016-0457-8.

[14] Kalia VC, Purohit HJ. Quenching the quorum sensing system: potential antibacterial drug targets. Crit Rev Microbiol 2011;37(2):121–40.

[15] Patel HK, Suárez-Moreno ZR, Degrassi G, Subramoni S, González JF, Venturi V. Bacterial LuxR solos have evolved to respond to different molecules including signals from plants. Front Plant Sci 2013;4:447. https://doi.org/10.3389/fpls.2013.00447.

[16] Helman Y, Chernin L. Silencing the mob: disrupting quorum sensing as a means to fight plant disease. Mol Plant Pathol 2015;16(3):316–29.

[17] Uroz S, Dessaux Y, Oger P. Quorum sensing and quorum quenching: the yin and yang of bacterial communication. Chembiochem 2009;10:205–16.

[18] Newman KL, Chatterjee S, Ho KA, Lindow SE. Virulence of plant pathogenic bacteria attenuated by degradation of fatty acid cell-to-cell signaling factors. Mol Plant-Microbe Interact 2008;21(3):326–34.

[19] Caicedo JC, Villamizar S, Ferro MI, Kupper KC, Ferro JA. Bacteria from the citrus phylloplane can disrupt cell–cell signalling in *Xanthomonas citri* and reduce citrus canker disease severity. Plant Pathol 2016;65(5):782–91.

[20] Shinohara M, Nakajima N, Uehara Y. Purification and characterization of a novel esterase (β-hydroxypalmitate methyl ester hydrolase) and prevention of the expression of virulence by *Ralstonia solanacearum*. J Appl Microbiol 2007;103:152–62.

[21] Achari GA, Ramesh R. Characterization of bacteria degrading 3-hydroxy palmitic acid methyl ester (3OH-PAME), a quorum sensing molecule of *Ralstonia solanacearum*. Lett Appl Microbiol 2015;60(5):447–55. https://doi.org/10.1111/lam.12389.

[22] Kai K, Ohnishi H, Shimatani M, Ishikawa S, Mori Y, Kiba A, Ohnishi K, Tabuchi M, Hikichi Y. Methyl 3-hydroxymyristate, a diffusible signal mediating *phc* quorum sensing in *Ralstonia solanacearum*. Chembiochem 2015;16(16):2309–18.

[23] Schell MA. Control of virulence and pathogenicity genes of *Ralstonia solanacearum* by an elaborate sensory network. Annu Rev Phytopathol 2000;38:263–92.

[24] Ryan RP, An SQ, Allan JH, McCarthy Y, Dow JM. The DSF family of cell–cell signals: an expanding class of bacterial virulence regulators. PLoS Pathog 2015;11(7). e1004986.

[25] Ionescu M, Yokota K, Antonova E, Garcia A, Beaulieu E, Hayes T, Iavarone AT, Lindow SE. Promiscuous diffusible signal factor production and responsiveness of the *Xylella fastidiosa* Rpf system. mBio September 7, 2016;7(4). e01054-16.

[26] Dow JM. Diffusible signal factor-dependent quorum sensing in pathogenic bacteria and its exploitation for disease control. J Appl Microbiol 2017;122(1):2–11.

[27] Potrykus M, Hugouvieux-Cotte-Pattat N, Lojkowska E. Interplay of classic Exp and specific Vfm quorum sensing systems on the phenotypic features of *Dickeya solani* strains exhibiting different virulence levels. Mol Plant Pathol 2018;19(5):1238–51.

[28] Nasser W, Dorel C, Wawrzyniak J, Van Gijsegem F, Groleau MC, Déziel E, Reverchon S. Vfm a new quorum sensing system controls the virulence of *Dickeya dadantii*. Environ Microbiol 2013;15(3):865–80.

[29] Leadbetter JR, Greenberg EP. Metabolism of acyl-homoserine lactone quorum sensing signals by *Variovorax paradoxus*. J Bacteriol 2000;182(24):6921–6.

[30] Dong YH, Xu JL, Li XZ, Zhang LH. *AiiA*, an enzyme that inactivates the acylhomoserine lactone quorum-sensing signal and attenuates the virulence of *Erwinia carotovora*. Proc Natl Acad Sci USA 2000;97(7):3526–31.

[31] Palmer AG, Senechal AC, Mukherjee A, Ane JM, Blackwell HE. Plant responses to bacterial N-acyl L-homoserine lactones are dependent on enzymatic degradation to L-homoserine. ACS Chem Biol 2014;9(8):1834–45.

[32] Dulla GF, Krasileva KV, Lindow SE. Interference of quorum sensing in *Pseudomonas syringae* by bacterial epiphytes that limit iron availability. Environ Microbiol 2010;12(6):1762–74.

[33] Lindow S, Newman K, Chatterjee S, Baccari C, Iavarone AT, Ionescu M. Production of *Xylella fastidiosa* diffusible signal factor in transgenic grape causes pathogen confusion and reduction in severity of Pierce's disease. Mol Plant-Microbe Interact 2014;27(3):244–54.

[34] Liang X, Yu X, Pan X, Wu J, Duan Y, Wang J, Zhou M. A thiadiazole reduces the virulence of *Xanthomonas oryzae* pv. *oryzae* by inhibiting the histidine utilization pathway and quorum sensing. Mol Plant Pathol 2018;19(1):116–28.

[35] Rajesh PS, Ravishankar RV. Quorum quenching activity in cell-free lysate of endophytic bacteria isolated from *Pterocarpus santalinus* Linn., and its effect on quorum sensing regulated biofilm in *Pseudomonas aeruginosa* PAO1. Microbiol Res 2013. https://doi.org/10.1016/j.micres.2013.10.005.

[36] Moleleki LN, Pretorius RG, Tanui CK, Mosina G, Theron J. A quorum sensing-defective mutant of *Pectobacterium carotovorum* ssp. *brasiliense* 1692 is attenuated in virulence and unable to occlude xylem tissue of susceptible potato plant stems. Mol Plant Pathol 2017;18(1):32−44.

[37] Crépin A, Barbey C, Cirou A, Tannières M, Orange N, Feuilloley M, Dessaux Y, Burini JF, Faure D, Latour X. Biological control of pathogen communication in the rhizosphere: a novel approach applied to potato soft rot due to *Pectobacterium atrosepticum*. Plant Soil 2012;358:27−37.

[38] Cirou A, Mondy S, An S, Charrier A, Sarrazin A, Thoison O, DuBow M, Faure D. Efficient biostimulation of native and introduced quorum-quenching *Rhodococcus erythropolis* populations is revealed by a combination of analytical chemistry, microbiology, and pyrosequencing. Appl Environ Microbiol 2012;78(2):481−92. https://doi.org/10.1128/AEM.06159-11.

[39] Barbey C, Crépin A, Bergeau D, Ouchiha A, Mijouin L, et al. *In planta* biocontrol of *Pectobacterium atrosepticum* by *Rhodococcus erythropolis* involves silencing of pathogen communication by the Rhodococcal gamma-lactone catabolic pathway. PLoS One 2013;8(6). https://doi.org/10.1371/journal.-pone.0066642. e66642.

[40] Chan KG, Atkinson S, Mathee K, Sam CK, Chhabra SR, Cámara M, Koh CL, Williams P. Characterization of N-acylhomoserine lactone degrading bacteria associated with the *Zingiber officinale* (ginger) rhizosphere: Co-existence of quorum quenching and quorum sensing in *Acinetobacter* and *Burkholderia*. BMC Microbiol 2011;11:51. http://www.biomedcentral.com/1471-2180/11/51.

[41] Uroz S, D'Angelo-Picard C, Carlier A, Elasri M, Sicot C, Petit A, Oger P, Faure D, Dessaux Y. Novel bacteria degrading N-acylhomoserine lactones and their use as quenchers of quorum-sensing-regulated functions of plant-pathogenic bacteria. Microbiol 2003;149:1981−9.

[42] Zhao C, Zeng H, Yu Z, Sun M. *N*-Acyl homoserine lactonase promotes prevention of *Erwinia* virulence with zwittermicin A-producing strain *Bacillus cereus*. Biotechnol Bioeng 2008;100:599−603.

[43] Cho HS, Park SY, Ryu CM, Kim JF, Kim JG, Park SH. Interference of quorum sensing and virulence of the rice pathogen *Burkholderia glumae* by an engineered endophytic bacterium. FEMS Microbiol Ecol 2007;60:14−23.

[44] Fray RG. Altering plant-microbe interactions through artificially manipulating bacteria quorum sensing. Ann Bot 2002;89:245−53.

[45] Caserta R, Picchi SC, Takita MA, Tomaz JP, Pereira WE, Machado MA, Ionescu M, Lindow S, De Souza AA. Expression of *Xylella fastidiosa* RpfF in citrus disrupts signaling in *Xanthomonas citri* subsp. *citri* and thereby its virulence. Mol Plant Microbe Interact 2014;27(11):1241−52.

[46] Kakkar A, Nizampatnam NR, Kondreddy A, Pradhan BB, Chatterjee S. *Xanthomonas campestris* cell−cell signalling molecule DSF (diffusible signal factor) elicits innate immunity in plants and is suppressed by the exopolysaccharide xanthan. J Exp Bot 2015;66(21):6697−714.

[47] Ryu CM, Choi HK, Lee CH, Murphy JF, Lee JK, Kloepper JW. Modulation of quorum sensing in acylhomoserine lactone-producing or-degrading tobacco plants leads to alteration of induced systemic resistance elicited by the rhizobacterium *Serratia marcescens* 90−166. Plant Pathol J 2013;29(2):182.
[48] Ban H, Chai X, Lin Y, Zhou Y, Peng D, Zhou Y, Zou Y, Yu Z, Sun M. Transgenic *Amorphophallus konjac* expressing synthesized acyl-homoserine lactonase (aiiA) gene exhibit enhanced resistance to soft rot disease. Plant Cell Rep 2009;28(12):1847−55.
[49] Ma Z, Song Y, Cai Z, Lin Z, Lin G, Wang Y, Zhou J. Anti-quorum sensing activities of selected coral symbiotic bacterial extracts from the South China Sea. Front Cell Infect Microbiol 2018;8:144.
[50] Santos CL, Correia-Neves M, Moradas-Ferreira P, Mendes MV. A walk into the *LuxR* regulators of Actinobacteria: phylogenomic distribution and functional diversity. PLoS One 2012;7(10). e46758.

Chapter 16

Trends in production and fuel properties of biodiesel from heterotrophic microbes

Gouri Raut[1], Srijay Kamat[2], Ameeta RaviKumar[3]
[1]*Bioenergy division, Agharkar Research Institute, Pune, India;* [2]*Department of Biotechnology, Goa University, Taleigao Plateau, Goa, India;* [3]*Institute of Bioinformatics and Biotechnology (IBB), Savitribai Phule Pune University, Pune, India*

16.1 Introduction

Biodiesel, a renewable energy source, is rapidly emerging as an alternative fuel for conventional petrodiesel. It has been estimated that approximately 22% of the total greenhouse emissions arise due to the use of fossil transport fuels [1]. One-fifth of the total CO_2 released is due to the use of transport fuels, and hence the use of alternative, renewable fuels like biodiesel is recommended. Biodiesel offers many advantages like miscibility with petrodiesel, reduction in exhaust emissions, higher flash point, inherent lubricity, high energy return, low content of sulfur and phosphorus, a high cetane number, and ease in biodegradability [2].

In the transport sector, global biofuel production increased by 2.5% in 2017 compared to the previous year. The United States registered a 1.6% increase in biodiesel production with 6 billion liters and was the largest producer of this renewable fuel. Brazil was the second largest producer with a total production of 4.3 billion liters, a 13% increase. Germany, which produced 3.5 billion liters, was the largest biodiesel producer in Europe. Argentina produced 3.3 billion liters, followed by Indonesia with a production of 2.5 billion liters. Biodiesel commands a share of 29% in the liquid biofuel market. The total production of biodiesel was 36.6 billion liters in 2017, which was a 1% increase compared to 2016 [3]. In India, 150 million liters of biodiesel was produced in 2017 and another 10 million liters would be added in 2018 [4]. The European Union (EU) has specified that every member country should derive 20% of its energy requirement from biofuels and 10% of its transport fuel should be obtained from renewable sources [5]. Biodiesel production is expected to increase every year by 4.5% to reach 41 billion liters in 2022 [6].

Advances in Biological Science Research. https://doi.org/10.1016/B978-0-12-817497-5.00016-1

Chemically, biodiesel is defined as fatty acid methyl esters (FAMEs) of vegetable and microbial oils or animal fats. The quality of biodiesel varies according to the source of plant/microbial oil or animal fats, i.e., on the type of fatty acids present in them. Amongst these, microbes are the most preferred due to their faster and shorter life cycle; they do not require arable land or aquatic bodies for growth unlike plants and autotrophic algae. Also, microbes can be easily scaled up from shake flask to bioreactor level, with their cultivation being independent of climate and soil conditions. Apart from these advantages, heterotrophic microbes like bacteria, fungi, and microalgae are renewable in nature and hence are emerging as the feedstocks of choice for future commercial biodiesel production facilities. However, their production costs at industrial scale would depend on the carbon source used and still need to be ascertained [7].

This chapter mainly focuses on various downstream processes that are required for the microbial production of biodiesel. There is special emphasis on screening, harvesting, cell lysis, lipid extraction, and transesterification of oleaginous microbes along with fuel properties of FAMEs, which form the crux of the biodiesel production process.

16.2 Growth of different sources of biodiesel on various substrates

Microbial sources like bacteria, microalgae, yeasts, and fungi are cultivated on carbon-rich and nitrogen-limited media to yield lipid-rich biomass. Fig. 16.1 illustrates the advantages and disadvantages of different types of biodiesel. As compared to autotrophic sources like plants and algae, heterotrophic microbes are at a disadvantage since they need an external source of carbon. But this disadvantage could be used for their growth on waste renewable, cheap carbon sources (agro-residues, wastewaters, waste oils, etc.), which cuts down the production cost since the carbon source contributes to the total cost by 80% [8].

Table 16.1 illustrates the most recent studies where various microbes (bacteria, microalgae, yeasts, and fungi) have been grown on waste substrates. After the biomass of the microbe grown on the substrate is harvested, the intracellular lipid is extracted by lysing the cells. The extracted lipid is then converted to fatty acid methyl esters (FAMEs), which chemically is biodiesel.

16.2.1 Screening of lipid-producing microorganisms

Screening is an essential part of any research program and often requires a method that can be executed rapidly to efficiently identify, in this case, lipid-accumulating microorganisms.

Sudan dyes, a class of nonfluorescent diazo solvent dyes, have been used to stain lipids since 1896 and comprises mainly Sudan I-IV, Oil red O, and Sudan

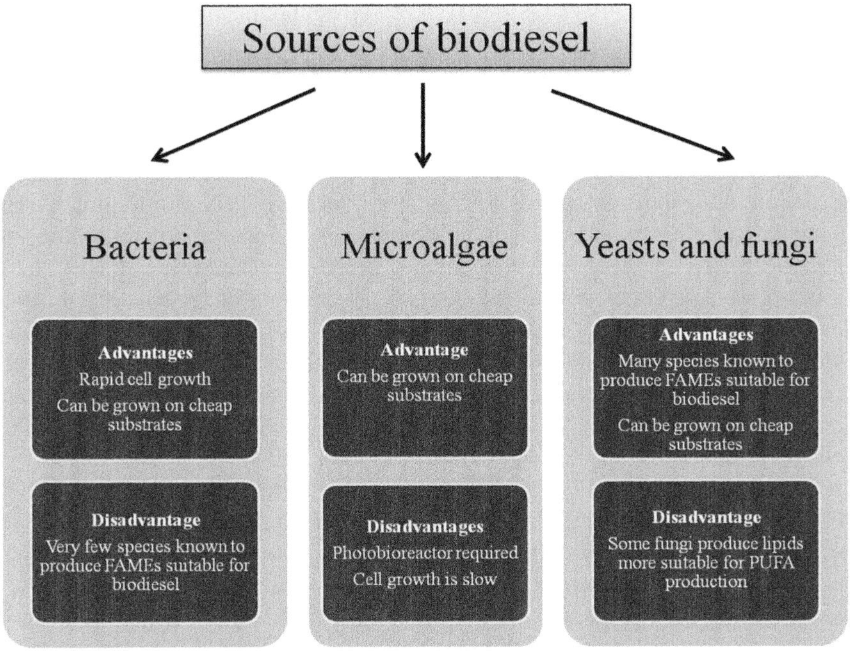

FIGURE 16.1 Advantages and disadvantages of different sources of biodiesel.

TABLE 16.1 Growth of heterotrophic microbes on various renewable substrates.

Microbe	Name of organism	Substrate	Reference
Bacteria	*Rhodococcus opacus* PD630	Fishery waste	[9]
	R. opacus DSM 43205	Biomass gasification waste water (BGWW)	[10]
Microalgae	*Chlorella minutissima*	Stagnant nonpotable pond water	[11]
	Chlorella sp. C2	Waste from biomass power plants	[12]
Yeast	*Trichosporon oleaginosus*	Crude glycerol	[13]
	Cryptococcus curvatus MTCC 2698	Vegetable waste hydrolysate	[14]
Fungi	*Fusarium verticillioides*	Waste cooking oils (WCOs)	[15]
	Mortierella isabellina NRRL 1757	Orange peel extract (OPE), Ricotta cheese whey (RCW)	[16]

Black B (SBB), [17]. SBB was used to screen potential oleaginous microbes by direct qualitative microscopic observation of lipid granules [18]. The researchers also found a change in color of SBB from black to blue with increasing degree of saturation of the lipids. SBB (saturated solution in 70% ethanol) can be used to stain all classes of lipids after a pre-exposure to 2.5% aqueous bromine [19]. The only class of lipids that does not stain with SBB is crystalline fatty acids and their esters. Convenience of SBB lipid staining made it popular for diagnostic applications, but the lack of information regarding its carcinogenicity is a cause for concern and needs to be determined [20].

A protocol has been developed using the fluorescent probe Nile red for the detection of intracellular neutral lipid by fluorescence microscopy and flow cytofluorometry in smooth muscle cells and cultured peritoneal macrophages [21]. This method has also been adapted for the estimation of intracellular neutral lipid in oleaginous fungi and yeasts [22]. Briefly, appropriately diluted cells were stained with 0.24–0.47 μg/ml of Nile red for 5 min, followed by excitation at 488 nm and scanning for an emission spectrum between 400 and 700 nm. The fluorescence intensity is generally used to quantitate the neutral lipid content (concentration range of 2–5000 μg lipid/ml of broth) in a variety of microorganisms. Though Nile red has been successfully applied for screening bacteria, fungi, and yeasts [23–25], it is less suitable for oleaginous algae due to the interference of chlorophyll fluorescence in the same range (650–750 nm) [26].

Hence, for microalgal lipid detection such as in *Chlorella vulgaris*, *Dunaliella primolecta*, and *Chaetoceros calcitrans*, BODIPY 505/515, a lipophilic green fluorophore, was found to be more suitable [27]. This dye could be used at a lower staining concentration (0.067 μg/ml), had a minimum staining time (2 min), and maintained fluorescence efficacy up to 30 min. BODIPY is not affected by chlorophyll fluorescence as it has an emission wavelength of 510 nm and stains a wider range of lipids like fatty acids, phospholipids, cholesterol, cholesteryl esters, and ceramides compared to Nile red [28].

A new method was developed for noninvasive and label-free determination of lipids, proteins, sugars, water, and whole intact microbial cells using Fourier transform infrared spectroscopy (FTIR) [29]. Infrared spectroscopy utilizes electromagnetic radiation between visible light and radio waves, especially the spectrum in the middle infrared (mid-IR; 4000–400/cm) and near infrared (near-IR; 12,500––4000/cm) [30]. The mid-IR region, together with the Fourier transform signal processing, has been used not only for qualitative and quantitative analysis in analytical chemistry but also for the identification and characterization of microorganisms [31]. FTIR spectroscopy needs to be combined with multiple linear regression, least-squares method, classical least-squares method, inverse least-squares model, principal component regression, partial least squares, and artificial neural networks for the analysis of complex mixtures [29].

A rapid-screening technique based on the principles of FTIR microspectroscopy and multivariate analysis corroboration was developed to analyze and monitor lipid accumulation in intact oleaginous yeasts, *Cryptococcus curvatus* and *Rhodosporidium toruloides* [32]. The FTIR analysis was also verified using classical techniques such as microscopy, flow cytometry, and gas chromatography (GC). The researchers found that the new method can discriminate among different classes of fatty acids in addition to detection of cell wall modifications during fatty acid accumulation.

Lipid and carbohydrate compositions of the microalgae *Isochrysis zhangjiangensis* and *Tetraselmis subcordiformis* have also been studied using FTIR [33]. *I. zhangjiangensis* was found to have the highest lipid content, whereas *T. subcordiformis* produced the highest carbohydrate content, under nitrogen limitation, as determined by traditional methods. FTIR spectroscopy was also able to show a similar diversion of carbon allocation from protein to neutral lipid and carbohydrate under nitrogen limitation conditions for *I. zhangjiangensis*.

FTIR spectroscopy has also been used to monitor the efficiency of lipid extraction from filamentous fungi *Mucor circinelloides* and *Mortierella alpina* [31]. Solvent extractions of the filamentous fungal biomass were carried out according to protocols described by Folch et al., [34], Bligh and Dyer [35], and Lewis et al. [36] to recover the lipid. The extracted lipids were derivatized as FAMEs and quantified using GC to validate the method. The method was found to successfully detect the lipid fraction in the intact fungal cells as well as at the different stages of all three extraction processes. In addition, the method was also able to visualize the efficiency of lipid extraction after acid digestion and bead-beating pretreatments [31].

The recent developments in the use of FTIR spectroscopy for the simultaneous study of a number of biomolecules, including lipids, in intact whole cells is very effective. Infrared radiation is highly absorbed by water, and the selectivity and specificity of FTIR spectra can interfere with the absorption by numerous functional groups generally present in molecules of complex mixtures [29]. Reduction in the water content of samples is hence essential and achieved by using low volumes of aqueous suspensions followed by extensive drying [31,32]. The use of multivariate calibration techniques has been successfully implemented to solve problems arising due to overlapping spectra of multiple functional groups. Normalization of data for accurate comparison between various samples is generally done by dividing all the absorbance values in a spectrum by the largest absorbance value like the amide I band [31,33]. When applied to intact microbial cells, the FT-IR technique provides spectral fingerprints of the complex biological structures such as proteins, lipids, nucleic acids, and polysaccharides under investigation. Microbial samples suitable for FT-IR investigations can be obtained by taking a smear from solid agar plates. Such samples can then be measured as dried films, or in cases where KBr pellets are used for obtaining

FTIR spectra, standardization of pellet weights and microbial cell to KBr proportion is employed [33]. FT-IR microspectrometry, which combines infrared spectroscopy and microscopy. can be applied to colonies of microorganisms or single cells, does not require additional reagents or stains, and can be performed without chemical sample modifications. For the generation of spatially resolved chemical IR maps by FT-IR microspectrometry, a high signal-to-noise ratio is required, together with a highly sensitive detector. These recent developments have resulted in the use of FTIR spectroscopy for the simultaneous study and quantification of a number of biomolecules, including lipids, in intact whole cells.

16.3 Harvesting of cellular biomass from fermentation broth

This step involves the separation of cellular biomass from the bulk liquid phase after fermentation. Centrifugation, flocculation, and filtration are techniques routinely employed for recovery of cellular biomass of yeast, fungi, bacteria, and microalgae [37]. In case of microalgae, specialized techniques of harvesting such as flotation, magnetic-based separation, microstraining, electrolysis, and electrophoresis have been used [38].

Centrifugation and filtration are quick methods to recover all types of heterotrophic biomass with recovery efficiencies ranging from 95% to 100% in laboratory-scale experiments. However, high capital costs and energy requirements limit its use to process fermentation broth of oleaginous microorganisms at an industrial scale. Given the high cell densities achievable in both microbial cell culture processes, primary recovery can be a significant bottleneck in commercial manufacturing. Both microfiltration and centrifugation coupled with depth filtration have been employed successfully as primary recovery processing steps. Advances in the design and application of membrane technology for microfiltration and dead-end filtration have contributed to significant improvements in process performance and integration, in some cases allowing for a combination of multiple unit operations in a given step [39].

Flocculation is a cost-effective method routinely used for the recovery of microalgal biomass grown autotrophically in raceway ponds or heterotrophically at high cell densities [40]. There are different types of flocculation based on nature of the flocculent and its mechanism of action. Chemical flocculation uses multivalent metal ions (alum or iron salts) or polysaccharides (chitosan or cationic starch) [41]. Autoflocculation is observed when the pH of the culture medium is increased, and bioflocculation is observed when normal microfloral bacteria produce biofilms that help to flocculate the biomass [42].

Filtration is another major technique that has been used to recover cellular biomass of yeast fungi and microalgae from spent fermentation medium. This technique allows for the recovery of shear-sensitive microorganisms with 70%–89% efficiency [37]. Dead-end filtration is more appropriate for the recovery of large cells (diameter >70 μm), whereas smaller cells (diameter <70 μm) can be recovered by tangential flow filtration [42].

Pellet formation is a known phenomenon that occurs during the submerged cultivation of filamentous fungi [43]. The application of cell pelletization to facilitate biomass harvesting was first reported by Xia et al. [44], where *Mucor circinelloides* pellets were harvested following pH adjustment. It was also demonstrated in this study that cell pelletization does not have a significant effect on growth and lipid accumulation. In the following year, the same group reported that in a co-culture of *Aspergillus niger* and *Chlorella vulgaris*, the filamentous fungal pellets could be used to harvest the microalgae from the culture medium [45]. A more recent study tested the flocculation efficiencies of filamentous fungi for harvesting a freshwater microalga, *Chlorella protothecoides*, and a marine microalga, *Tetraselmis suecica*, and found many benefits [46]. In addition to the efficient and cost-effective harvesting of both microalga, co-culture of fungus and algae showed an enhancement of total oil production as well as optimization of oil composition. This co-culture method was also found to improve nutrient cycling of primary nutrients like nitrogen, phosphorus, and other microelements.

Large-scale operation of centrifugation and filtration involve high capital expenditure and continuous maintenance. Although high-capacity centrifuges are available that can operate in continuous mode [38], the cost-effective methods of flocculation and pelletization can be explored at the industrial level.

16.4 Cell lysis

The efficient extraction of lipid from harvested biomass depends largely on lipid accessibility, mass transfer of lipids to the solvent phase, and emulsion formation of lipids during extraction [39]. Cell disruption methods are classified into four categories based on the approach used [47]:

- Mechanical
- Physical
- Chemical
- Biological

Mechanical methods involve techniques like bead milling, ultrasound-assisted cell disruption, and high-pressure homogenization. Bead milling is achieved by grinding the biomass with zirconium oxide, zircon-silica, or titanium carbide beads and has been used extensively for bacteria [48], yeast [49], fungi [24], and microalgae [50]. This method, however, is ideally suited for laboratory-scale experiments and is not cost efficient on the industrial scale [39]. Ultrasound-assisted cell lysis is largely used for microalgal biomass [51], although a recent report has developed the technique for extraction of lipids from yeast biomass [52]. High-pressure homogenization is another promising mechanical technique used for cell disruption of mainly single-celled microbes for extraction of high-value proteins [53]. The performance of this technique at industrial level for biodiesel production still needs to be studied [39].

Physical methods used for cell disruption include microwave-assisted lysis and pulsed electric field/electroporation. Microwave-assisted extraction was developed for the extraction of different classes of compounds from varied sources [54]. Samples were ground in appropriate solvent followed by irradiation for 30 s without allowing the sample to boil. The microwaves (300 MHz−300 GHz) heat the free water molecules in the sample, leading to nonuniform temperature changes that result in spontaneous cell rupture [37]. Pulsed electric field/electroporation is another physical method used for cell disruption of yeast and microalgal cell suspensions [55]. An electric field is applied to the cell suspension to induce pore formation and leaching of intracellular contents [56]. Electroporation is more suitable for the extraction of water-soluble polar compounds such as carbohydrates and proteins rather than nonpolar high-molecular-weight compounds such as lipids [57]. Both microwave-assisted lysis and electroporation are energy-intensive processes, and their utilization on the industrial scale would increase biodiesel cost [39].

Chemical cell disruption methods are a promising and inexpensive alternative to many of the above-mentioned methods. These include acid digestion, ionic liquid disruption, surfactant-mediated cell lysis, and pressurized fluids [39,47]. A study by Yu et al. [58] compared the efficiencies of autoclaving, bead-beating, microwaving, sonication, acid digestion (HCl), and Soxhlet extraction on the yeast *Cryptococcus curvatus*, the fungus *Mortierella isabellina*, and the microalga *Chlorella sorokiniana*. Acid digestion was found to be the best method for the extraction of lipids from *C. curvatus* and the second best method for *M. isabellina* and *C. sorokiniana*. The use of mineral acid for cell disruption is cost-effective, but the extreme pH involved often leads to corrosion of the reactors, leading to increased maintenance costs [59].

Ionic liquids (ILs) are a new class of solvents that have negligible vapor pressure and whose polarity and hydrophobicity can be manipulated through choice of anion and/or cation structure [60]. ILs like 1-ethyl-3-methylimidazolium ethylsulfate [61], 1-ethyl-3-methyl-imidazolium acetate [62], 1-butyl-3-methylimidazolium trifluoromethanesulfonate, 1-butyl-3-methylimidazolium methyl sulfate, and 1-ethyl-3-methylimidazolium methyl sulfate [63] have been used for cell lysis of yeasts and microalgae. ILs offer an environmentally safe alternative for cell disruption and lipid recovery as an alternative to organic solvent extraction. However, the high cost, especially of imidazolium- and pyridinium-based ILs, and high viscosities limit their use at the industrial scale [64].

Detergent-based cell lysis is a new and useful technique for lipid recovery from cell biomass. Cationic detergents, like cetrimonium bromide, myristyltrimethyl-ammonium bromide, and 3-(decyldimethylammonio)-propanesulfonate inner salt, have been used to improve lipid recovery from negatively charged microalgal cells [51]. An anionic detergent (N-lauroyl sarcosine)-based lipid extraction method was developed to reduce the time taken to extract lipids from *Yarrowia lipolytica* [65]. Compared to the 12 h required for lipid extraction by the conventional method, the detergent-based method successfully extracted 95.43% of the lipid in 8.8 min. Nonionic detergents, like Triton X-100, Brij 58, IGEPAL CA-630, and Tergitol NP-40, have been found to induce free fatty acid release in *Aspergillus oryzae* [66]. Detergent-based lysis is a promising alternative for cell disruption although a subsequent step of detergent removal from the biodiesel is required.

Pressurized fluids (CO_2 and H_2O) have been used for the lysis of all types of microorganisms [67], but the main drawback of the technology is the high equipment and operational costs that limit its use in the industrial-scale biodiesel production [51].

Biological extraction of lipids involves the use of a cocktail of cell wall–degrading enzymes, such as xylanase, cellulase, amylase, papain, and pectinase, for cell disruption to release intracellular lipid bodies [68]. Although this method has a low operating cost and energy requirement, the cost of developing an enzyme cocktail to specifically degrade the cell wall of an oleaginous microbe is high [47].

A number of cell disruption methods have been developed for the lysis of a wide variety of heterotrophic microorganisms, each with their specific advantages and drawbacks. A number of these methods have also been combined to increase their extraction efficiency and have been tabulated in Table 16.2.

16.5 Lipid extraction

The most cited method for lipid extraction is the Folch method, which was initially developed to extract lipids from animal tissues [34]. In this method, the cell biomass was homogenized in chloroform/methanol (2:1, v/v) to extract the lipid followed by water or salt solution washes to remove unwanted substances. The second most popular method is the Bligh and Dyer method, where a similar homogenization of biomass in chloroform/methanol (2:1, v/v) is followed by a sequential addition of chloroform and water [35]. Acidification of the salt solution in the Bligh and Dyer method with acetic acid [84] and hydrochloric acid [85] has been found to improve the recovery of acidic lipids. The Folch method was adapted for lipid extraction from yeasts by incorporating sequential washes with 0.034% $MgCl_2$, 2N KCl/methanol (4:1, v/v) and artificial upper phase: chloroform/methanol/water (3:48:47; per vol) after initial solvent extraction [86].

The use of methyl-tert-butyl ether instead of chloroform was found to improve recovery of all classes of lipids from a variety of microbial cells [87]. Although a number of solvents have been used for the extraction of lipids from oleaginous microbes, nonpolar, water-immiscible solvents such as hexane, chloroform and chloroform with low-evaporation energies are preferred [39].

TABLE 16.2 Combination of different methods to design new approaches for cell disruption to release lipid feedstock for biodiesel production.

Cell disruption technique	Cell type	Microorganism name	Lipid recovery (%)	Reference
Mano-thermo-sonication	Yeast	*Rhodosporidium toruloides*	26.94	[69]
Ultrasonication-microwave treatment	Yeast	*Rhodosporidium kratochvilovae*	70.86	[70]
Enzyme cocktail–mediated lysis	Microalgae	*Chlorella zofingiensis* ATCC 30412 *Chlorella vulgaris* UTEX 395 *Chlorella pyrenoidosa* FACHB-9	62.49 65.27 59.43	[71]
Amino solvent and ionic liquid-mediated lysis	Microalgae	Commercial DHA strain	90.00	[72]
Biodiesel-mediated cell disruption	Microalgae	*Chlorella* sp.	68.00	[73]
Microwave-assisted chemical hydrolysis	Bacteria	*Methylomicrobium buryatense*	95.00	[74]
Phage-mediated cell disruption	Bacteria	*Rhodococcus opacus* PD630	36.11	[75]
Cryo-pulverization (glass beads)	Filamentous fungi	*Absidia glauca* CCM 451 *Mortierella alpina* ATCC 32222 *Cunninghamella blakesleeana* CCM F-705	NA	[76]

The steps discussed so far comprise the conventional biodiesel production process, which consists of screening, harvesting, cell disruption, and solvent extraction and finally transesterification. This process comprises a number of steps required for biodiesel production and only serves to increase the price of the final product. Considerable attention is now being given to direct in situ transesterification to reduce biodiesel cost and reduce the use of solvents and expensive capital investment [37].

16.6 Transesterification/FAME preparation—conventional two-step, one-step, use of lipases

16.6.1 Transesterification process

The extracted lipid reacts with excess methanol in the presence of catalyst in the transesterification reaction. The catalyst could be homogeneous or heterogeneous.

16.6.1.1 Homogeneous catalyzed transesterification

Acid or alkali is generally used for transesterifying the lipid to yield FAMEs. Acid-catalyzed reaction to yield biodiesel using 2% (v/v) H_2SO_4 was used for the bacterium *Rhodoccus opacus* PD630 [77]. The microalgae *Chlorella* sp. BDUG 91771 was transesterified using 3% (v/v) H_2SO_4 [88]. Whereas for fungi, FAMEs were obtained using H_2SO_4 as a catalyst for transesterification [89].

FAME production by alkali-catalyzed transesterification for the bacterium *Serratia* sp. ISTD04 resulted in a biodiesel yield of 95% [90]. Biodiesel production using NaOH-catalyzed reaction has been previously done for the marine microalga *Tetraselmis* sp [91]. The lipids extracted from microalgae are known to be rich in free fatty acids, which could result in soap formation and hence alkali catalysis is generally not used for microalgae [92]. In another study, the fungus *Aspergillus terreus* IBB M1 was transesterified using NaOH as a catalyst to yield FAMEs suitable for biodiesel production [24]. The total lipid extracted from the oleaginous yeast *Y. lipolytica* NCIM 3589 was transesterified using NaOH to yield FAMEs that were suitable for biodiesel [25].

16.6.1.2 Heterogeneous catalysts for transesterification

Different heterogeneous catalysts have been used for the transesterification reaction. Heterogeneous catalysis when used has some advantages like easy recovery, reuse, and regeneration [93]. Unlike bacteria and fungi, many reports exist wherein heterogeneous catalysis has been used for microalgae. The lipid extracted from the microalga *Scenedesmus obliqus* sp. was transesterified using a tungstated zirconia catalyst, resulting in 94.58% conversion to FAMEs [94]. When CaO derived by calcinating waste shells of Angel Wing shell

(AWS), a mollusk, was used as a renewable catalyst for biodiesel production from *Nannochloropsis oculata*, a yield of 84.11% FAME was obtained [93]. *S. obliqus* was grown in a raceway pond and the lipids extracted from its biomass were transesterified using a heterogeneous solid acid catalyst, chromium-aluminium mixed oxide, a FAME conversion of 98.28% was obtained [95].

16.6.1.3 Direct or in situ transesterification

Direct or in situ methods for transesterification have been developed wherein all the reaction components are added together and the reaction is carried out in a single step requiring less solvent and catalyst while simultaneously making the process convenient and less time-consuming.

The in situ transesterification of the dry biomass of microalga *Acutodesmus obliqus* was done using a heterogeneous nanoCaO catalyst derived from egg shell waste and it was found to yield 86.41% biodiesel [96]. The wet biomass of *Nannochloropsis* sp. was converted to biodiesel using 1-butyl-3-methylimidazoliumhydrogen sulfate ([Bmim][HSO_4]), which acted as a catalyst, giving a biodiesel yield of 95.2% [97]. In case of fungi, the dried biomass of the fungus *Aspergillus candidus* IBB G4 yielded 68 mg/g of FAME [98]. The wet biomass of *Mucor circinelloides* URM 4182 was converted to fatty acid ethyl esters (FAEE) using ethanol and catalyzed by heteropolyacid impregnated on alumina (H_3PMo/Al_2O_3), to achieve a yield of >97% [99]. A vigorously stirred tank reactor was used for the dry and wet biomass of *Y. lipolytica* TISTR 5151 with 2% (v/v) H_2SO_4 catalyst to obtain yields of 94.99% and 80.91%, respectively [100]. An in situ transesterification using 0.1 M H_2SO_4 was performed using biomass of *Y. lipolytica* NCIM 3589 and yielded 0.88 g FAME [81].

Noncatalytic in situ transesterification of dried biomass of *Spirulina platensis* using supercritical methanol gave a high yield of 99.32% of biodiesel [101]. Another study using a novel vortex fluidic device–assisted in situ transesterification for the wet biomass of *Chloroparva pannonica* resulted in a 96% efficiency of fatty acid to FAME conversion [102] (Table 16.3).

16.6.1.4 Lipase-catalyzed transesterification

Alkali catalysis could lead to soap formation if the microbial oil is rich in water and free fatty acids [103]. Acid catalysis is a good option, but these chemicals need a large amount of water during purification and hence enzyme-catalyzed reactions are fast gaining importance due to their environmentally friendly nature and ease in purification [104]. In this context, the yields of alkali-catalyzed and lipase-catalyzed reactions were compared for the microalga *Tetraselmis* sp. and it was found that the yield of the reaction catalyzed by lipase from *Candida rugosa* was seven times higher [91]. Direct transesterification of the dried biomass of *Botyrococcus* sp. catalyzed using

TABLE 16.3 In situ transesterification for biodiesel production from various bacteria, microalgae, yeasts, and fungi.

Microbe	Organism name	Catalyst	Additional technique	Reference
Bacteria	*Rhodococcus opacus* PD630	2% H_2SO_4	–	[77]
Microalgae	*Scenedesmus* sp.	Tungstated zirconia (WO_3/ZrO_2)	Microwave and sonication	[78]
	Nannochloropsis gaditana	2% H_2SO_4	–	[79]
Yeasts	*Cryptococcus curvatus*	KOH	Microwave	[80]
	Yarrowia lipolytica NCIM 3589	0.1 M H_2SO_4 (0.1%)	–	[81]
Fungi	*Mucor circinelloides* URM 4182	12-Molybdo phosphoric acid supported on	Sonication	[82]
	Aspergillus candidus	0.1 M H_2SO_4		[83]

immobilized *Candida antarctica* lipase B (Novozym CAL-B) carried out along with ultrasound treatment led to a 88% methyl ester yield [105].

The lipid extracted from the fungus *M. circinelloides* URM 4182 yielded 93% FAEE in the presence of *C. antarctica* (Novozym 435) immobilized lipase [106]. Another study reported the use of intracellular lipase of *M. circinelloides* IBT-83 and conversion of its lipids to FAMEs using the endogenous lipase with a yield of 80% in 1 h [107].

16.6.1.5 Other methods of transesterification

Other techniques have been used for transesterifying microbial oil to FAMEs.

A novel ultrasound-assisted in situ method for transesterification of *Chlorella* sp. using response surface methodology and artificial neural network to determine FAME yield and exergy efficiency was developed [108]. It was found that under optimized conditions, a FAME content of 81.2% and exergy efficiency of 79.8% was achieved. Wet torrefaction was used as a pretreatment method to convert biomass of *Chlorella vulgaris* EPS-31 and it was found that the higher heating value (HHV) of the torrefied microalga increased by 21% [109]. A combination of wet in situ transesterification and hydrothermal liquefaction was used for the non-catalytic FAEEs production from *Nannochloropsis gaditana* with a yield of 76.54% [110]. Conventional hydrothermal liquefaction involves the use of water at very high temperature (above 350°C) and pressure (16 bars). The modified method used wet microalgal biomass and solvents to achieve the desired yield at a lower temperature (185°C).

The lipids from *Nannochloropsis salina* were converted to FAMEs and triacetin using a supercritical methyl acetate technology. The triglycerides of microalgae react with methyl acetate to yield diacetylglyceride and FAME. The resulting diacetylglyceride reacts with methyl acetate to form methyl ester and monoacetylglyceride, which finally yields methyl ester and triacetin in a supercritical reactor. Instead of the conventional glycerol, the triacetin produced as the second product finds an additional application as a biodiesel fuel [111]. A novel ionic liquid, deep eutectic solvent was used for the conventional two-step and in situ transesterification of wet biomass of *Chlorella* sp. and it was found that the FAME content in the in situ method improved by 30% compared to the conventional method [112]. In another study, an ohmic-heating strategy involving an alternating current with frequency 5 Hz was applied for 2 min at a temperature of 70°C in order to pretreat the suspension of *Chlorella* sp. TISTR 8990, resulting in increase in the rate of transesterification by almost twofold [113].

The various stages involved in biodiesel production from heterotrophic microbes are illustrated in Fig. 16.2. The FAMEs produced after transesterification are evaluated for their fuel properties.

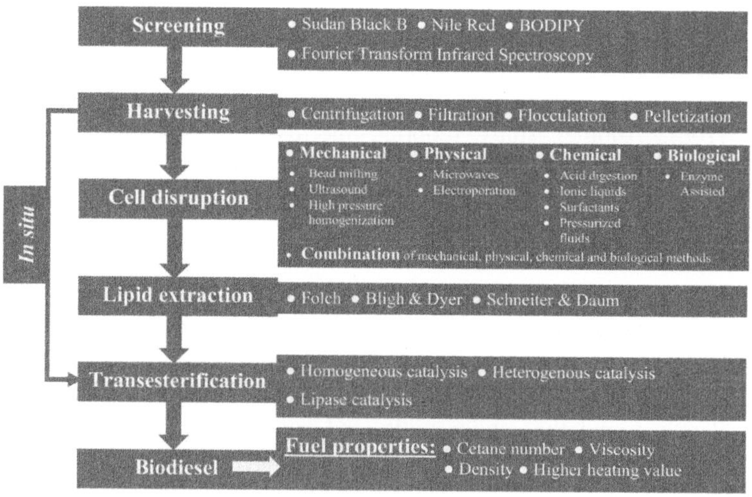

FIGURE 16.2 Different steps in biodiesel production from heterotrophic microbes.

16.7 Determination of fuel properties of heterotrophic microbes

The different fuel properties of biodiesel, viz., cetane number (CN), viscosity, density, saponification number, iodine value, and HHV need to be evaluated in order to determine the suitability as fuel.

16.7.1 Cetane number

This is the most important fuel property, which indicates the ignition delay time. Higher cetane number is desired since it means that the ignition delay time is shorter [114]. The CN limits have been set by the US Standard, ASTM 6751 (minimum 47), and the European standard, EN 14214 (minimum 51). CN depends on the fatty acid profile of the microbial feedstock since biodiesel essentially is a mixture of FAMEs [115]. Various mathematical equations can be used to predict the CN of biodiesel.

A CN of 51.77 was obtained for the biodiesel produced from the microalga *S. obliqus* using the formula as below:

$$CN = -0.1209 \times DU + 65.0958 \tag{16.1}$$

Where, DU is the degree of unsaturation calculated as given in the equation [116]. In another study, the microalga *N. gaditana* biodiesel obtained after in situ transesterification was evaluated for CN using the standard ASTM 976 and was determined to possess a value of 48 [92].

The CN of *Rhodotorula mucilaginosa*, *A. oryzae*, and *Mucor plumbeus* was found to be 73.3, 61.8, and 61.9 [117], respectively, which was calculated using the mathematical equation [118]:

$$\varnothing_i = -7.8 + 0.302 \times M_i - 20 \times N \tag{16.2}$$

Where, $\varnothing_i$ indicates the CN of the *i*th FAME, molecular weight of the *i*th FAME is indicated by M_i, and N is the number of double bonds. The CN of *A. terreus* IBB M1 was found to be 50.41 [119].

The CN of a mutant YlE1 of the yeast *Y. lipolytica* NCIM 3589 was predicted using mathematical equations and was found to be 68.6 [120]. The CN of biodiesel produced *Candida* sp. LEB-M3 was 67.38 and was calculated using the prediction model:

$$CN = \sum \frac{CN_{me}\%ME}{100} \tag{16.3}$$

Where, *me* denotes each individual methyl ester, *% ME* denotes the percentage of individual methyl ester present in the mixture of FAMEs [121].

16.7.2 Viscosity

Viscosity, another important fuel property. affects the atomization of fuel and, like CN, depends on the fatty acid profile of the biodiesel feedstock. ASTM 6751 and the EN 14214 norms have specified that the viscosity of a biodiesel fuel should be between 1.9–6.0 and 3.5–5.0 mm^2/s, respectively at 40°C. Accordingly, the kinematic viscosity of biodiesel from *Rhodococcus opacus* PD630 was found to be 4.1 mm^2/s at 40°C [77]. The experimental value of the kinematic viscosity of biodiesel produced by the microalga *N. gaditana* was found to be 3.9 mm^2/s at 40°C, as determined using the standard method ASTM 445 [92].

M. circinelloides URM 4182 biodiesel produced by in situ transesterification of the fungal biomass yielded a viscosity of 4.45 mm^2/s at 40°C, as determined experimentally using an LVDVII Brookfield viscosimeter according to the ASTM 445 standard method [82]. The viscosity of *A. terreus* was determined as 4.5 mm^2/s using the equation as below:

$$\ln \eta_i = -12.503 + 2.496 \times \ln(Mi) - 0.178 \times N \qquad (16.4)$$

Where, the kinematic viscosity of the *i*th FAME is denoted by η_i, Mi is the molecular weight of the *i*th FAME, and N indicates the number of double bonds [122]. The viscosity of *A. terreus* IBB M1 determined using prediction model was 3.32 mm^2/s [123].

The biodiesel obtained from *Candida* sp. LEB-M3 demonstrated a kinematic viscosity of 4.28 mm^2/s, which was calculated according to the formula:

$$\ln \nu = \sum \frac{\ln(\nu me)\%ME}{100} \qquad (16.5)$$

Where, ν indicates kinematic viscosity of the sample, *me* indicates individual methyl ester, and *%ME* indicates the percentage present of each methyl ester in the mixture of biodiesel [121]. The viscosity of biodiesel obtained after in situ transesterification of biomass of *Y. lipolytica* NCIM 3589 was 4.65 mm^2/s [81].

16.7.3 Density

Density of biodiesel fuel is of paramount importance since injection equipment for fuel follows a certain volume metering system and a fixed volume needs to be delivered for optimum combustion [124]. If the density is high, higher mass of fuel would be injected into the engine. The ASTM 6751 and the EN 14214 norms define the density of a biodiesel fuel to be between 0.860 and 0.900 g/cm^3. The density of biodiesel from *R. opacus* PD630 was found to be 0.895 g/cm^3 [77]. The density of the FAME produced by *N. gaditana* was evaluated experimentally using the EN ISO 3675 official methods and was found to be 0.885 g/cm^3 [92]. The experimental determination of density of

Chlorella sp. BDUG 91771 was done using standard methods and was found to be 0.895 g/cm^3 [88].

The biodiesel produced by *M. isabellina* was found to have a density of 0.874 g/cm^3 using the predictive equation as below:

$$\rho_i = 0.8463 + \frac{4.9}{Mi} + 0.0118 \times N \tag{16.6}$$

Where, the density of the *i*th FAME is denoted by ρ_i, Mi is the molecular weight of the *i*th FAME, and N indicates the number of double bonds [122]. The density of the biodiesel produced by the fungus *M. circinelloides* URM 4182 was determined experimentally at 20°C using a DMA 35 N EX digital densimeter according to the ASTM D 4052 method and was found to be 0.878 mm^2/s [82]. The density of *Y. lipolytica* NCIM 3589 produced by in situ transesterification was calculated as 0.880 mm^2/s [81] using the following equation:

$$\rho = \sum c_i \rho_i \tag{16.7}$$

Here, c_i indicates the concentration, ρ_i, the density of the *i*th FAME obtained from database.

16.7.4 Higher heating value

The actual energy content of the fuel is indicated by the HHV, which is the energy released when 1 g of fuel is combusted completely to water and CO_2. The HHV of *R. mucilaginosa*, *A. oryzae*, and *M. plumbeus* was found to be 40.64, 40.11, and 39.10 MJ/kg, respectively [117], which was calculated using the equation as follows:

$$\delta_i = 46.19 + \frac{1794}{Mi} + 0.21 \times N \tag{16.8}$$

Here, δ_i indicates the HHV in MJ kg^{-1} of the *i*th FAME, M_i denotes the molecular weight of individual FAME. Carvalho et al. [82] investigated the biodiesel characteristics of FAMEs from *M. circinelloides* URM 4182 using a software Biodiesel analyzer ver 2.2, which calculated the fuel properties based on Eq. (16.8) as above and found that the HHV of biodiesel was 39.8 MJ/kg.

The empirical Eq. (16.8) was also used to calculate the HHV of biodiesel produced by the microalga *Chlorella sorokiniana*, which was found to be 39.7 41 MJ/kg [125]. The HHV of *Scenedesmus incrassatulus* CLHE-Si01 was found to be 41 MJ/kg and was calculated according to the prediction equation based on its density value as:

$$HHV = 0.4625\nu + 39.45 \tag{16.9}$$

Where, ν indicates the density of the microalgal oil [126].

The HHV of biodiesel obtained from *Y. lipolytica* NCIM 3589 biomass after two-step transesterification was found to be 36.77 MJ/kg [25].

16.8 Conclusions and future perspectives

The biodiesel obtained from heterotrophic microbes and algae as well as from mixotrophic algae that use light for organic carbon assimilation are set to gradually replace petrodiesel. Cost-effectiveness, relative simplicity in operations, and easy maintenance are the main attractions of the heterotrophic growth approach, and the need for supply of the carbon source can be fulfilled by the use of renewable substrates for growth of these heterotrophic microbes, which drastically cuts down production costs. Moreover, heterotrophic cultivation can be performed in any fermenter without illumination; hence, there is no requirement of photobioreactor, which reduces the overall production cost. Another important bottleneck in biodiesel production from oleaginous heterotrophic microbes is the selection of suitable candidate by high-throughput methods, which now has become easy owing to the advent of techniques like FTIR, wherein the lipid content of the intact microbial cell can be estimated. Compared to the conventional two-step transesterification, the in situ reaction, wherein all reactants are added in a single pot, is fast, convenient, and economical. The assessment of fuel properties—CN, kinematic viscosity, density, and HHV—using predictive and experimental methods has been used to determine the suitability of the FAMEs as fuel.

The production of biodiesel from these heterotrophic microbes is not yet commercialized owing to many problems: economic feasibility, low product-to-substrate ratios, continuous availability of cheap substrates for their growth, and overall, lack of an existing microbial technology for industrial scale-up for biodiesel production. The use of random mutagenesis of selected strains to achieve higher yields could solve problems of low yield. A further exploration of cheap, continuously available substrates that could result in optimized and increased yields of the chosen microbe and thus, simultaneously take care of the issues related to disposal of such wastes in the environment could pave the way for future technologies with reduced costs.

Acknowledgments

GR is a recipient of the DBT-Research Associateship and would like to thank the Department of Biotechnology (DBT), Government of India for the financial assistance. SK is a recipient of the Dr. D.S. Kothari Postdoctoral Fellowship (No.F.4-2/2006 (BSR)/BL/17-18/0121) and would like to thank the University Grants Commission, Government of India for financial support.

References

[1] Mahmudul HM, Hagos FY, Mamat R, Adam AA, Ishak WFW, Alenezi R. Production, characterization and performance of biodiesel as an alternative fuel in diesel engines – a review. Renew Sustain Energy Rev 2017;72:497–509. https://doi.org/10.1016/j.rser.2017.01.001.

[2] Mardhiah HH, Ong HC, Masjuki HH, Lim S, Lee HV. A review on latest developments and future prospects of heterogeneous catalyst in biodiesel production from non-edible oils. Renew Sustain Energy Rev 2017;67:1225–36. https://doi.org/10.1016/j.rser.2016.09.036.

[3] Renewables REN21. Global status report. 2018, ISBN 978-3-9818911-3-3.

[4] GAIN report. India Biofuels Annual. 2017.

[5] Avinash A, Subramaniam D, Murugesan A. Bio-diesel – a global scenario. Renew Sustain Energy Rev 2014;29:517–27. https://doi.org/10.1016/j.rser.2013.09.007.

[6] Monteiro MR, Kugelmeier CL, Pinheiro RS, Batalha MO, da Silva César A. Glycerol from biodiesel production: technological paths for sustainability. Renew Sustain Energy Rev 2018;88:109–22. https://doi.org/10.1016/j.rser.2018.02.019.

[7] Li Q, Du W, Liu D. Perspectives of microbial oils for biodiesel production. Appl Microbiol Biotechnol 2008;80:749–56. https://doi.org/10.1007/s00253-008-1625-9.

[8] Tsigie YA, Wang C, Truong C, Ju Y. Lipid production from *Yarrowia lipolytica* Po1g grown in sugarcane bagasse hydrolysate. Bioresour Technol 2011;102:9216–22. https://doi.org/10.1016/j.biortech.2011.06.047.

[9] Palmer JD, Brigham CJ. Feasibility of triacylglycerol production for biodiesel, utilizing *Rhodococcus opacus* as a biocatalyst and fishery waste as feedstock. Renew Sustain Energy Rev 2016;56:922–8. https://doi.org/10.1016/j.rser.2015.12.002.

[10] Goswami L, Tejas Namboodiri MM, Vinoth Kumar R, Pakshirajan K, Pugazhenthi G. Biodiesel production potential of oleaginous *Rhodococcus opacus* grown on biomass gasification wastewater. Renew Energy 2017;105:400–6. https://doi.org/10.1016/j.renene.2016.12.044.

[11] Arora N, Patel A, Pruthi PA, Poluri KM, Pruthi V. Utilization of stagnant non-potable pond water for cultivating oleaginous microalga *Chlorella minutissima* for biodiesel production. Renew Energy 2018;126:30–7. https://doi.org/10.1016/j.renene.2018.03.033.

[12] Chen H, Wang J, Zheng Y, Zhan J, He C, Wang Q. Algal biofuel production coupled bioremediation of biomass power plant wastes based on *Chlorella* sp. C2 cultivation. Appl Energy 2018;211:296–305. https://doi.org/10.1016/j.apenergy.2017.11.058.

[13] Chen J, Yan S, Zhang X, Tyagi RD, Surampalli RY, Valéro JR. Chemical and biological conversion of crude glycerol derived from waste cooking oil to biodiesel. Waste Manag 2018;71:164–75. https://doi.org/10.1016/j.wasman.2017.10.044.

[14] Chatterjee S, Mohan SV. Microbial lipid production by *Cryptococcus curvatus* from vegetable waste hydrolysate. Bioresour Technol 2018;254:284–9. https://doi.org/10.1016/j.biortech.2018.01.079.

[15] Kamoun O, Ayadi I, Guerfali M, Belghith H, Gargouri A, Trigui-Lahiani H. *Fusarium verticillioides* as a single-cell oil source for biodiesel production and dietary supplements. Process Saf Environ Protect 2018;118:68–78. https://doi.org/10.1016/j.psep.2018.06.027.

[16] Carota E, Crognale S, D'Annibale A, Petruccioli M. Bioconversion of agro-industrial waste into microbial oils by filamentous fungi. Process Saf Environ Protect 2018;117:143–51. https://doi.org/10.1016/j.psep.2018.04.022.

[17] Hartman TL. The use of Sudan black B as a bacterial fat stain. Biotech Histochem 1940;15:23–8. https://doi.org/10.3109/10520294009110328.

[18] Thakur MS, Prapulla SG, Karanth NG. Microscopic observation of Sudan Black B staining to monitor lipid production by microbes. J Chem Technol Biotechnol 1988;42:129–34. https://doi.org/10.1002/jctb.280420206.

[19] Bayliss OB, Adams CWM. Bromine-Sudan Black: a general stain for lipids including free cholesterol. Histochem J 1972;4:505–15. https://doi.org/10.1007/BF01011130.

[20] Fonovich TM. Sudan dyes: are they dangerous for human health. Drug Chem Toxicol 2013;36:343–52. https://doi.org/10.3109/01480545.2012.710626.

[21] Greenspan P, Mayer EP, Fowler SD. Nile red: a selective fluorescent stain for intracellular lipid droplets. J Cell Biol 1985;100:965–73. https://doi.org/10.1021/cr9001879.

[22] Kimura K, Yamaoka M, Kamisaka Y. Rapid estimation of lipids in oleaginous fungi and yeasts using Nile red fluorescence. J Microbiol Methods 2004;56:331–8. https://doi.org/10.1016/j.mimet.2003.10.018.

[23] Röttig A, Hauschild P, Madkour MH, Al-Ansari AM, Almakishah NH, Steinbüchel A. Analysis and optimization of triacylglycerol synthesis in novel oleaginous *Rhodococcus* and *Streptomyces* strains isolated from desert soil. J Biotechnol 2016;225:48–56. https://doi.org/10.1016/j.jbiotec.2016.03.040.

[24] Khot M, Kamat S, Zinjarde S, Pant A, Chopade B, RaviKumar A. Single cell oil of oleaginous fungi from the tropical mangrove wetlands as a potential feedstock for biodiesel. Microb Cell Factories 2012;11:71. https://doi.org/10.1186/1475-2859-11-71.

[25] Katre G, Joshi C, Khot M, Zinjarde S, Ravikumar A. Evaluation of single cell oil (SCO) from a tropical marine yeast *Yarrowia lipolytica* NCIM 3589 as a potential feedstock for biodiesel. AMB Express 2012;2:36. https://doi.org/10.1186/2191-0855-2-36.

[26] Hounslow E, Noirel J, Gilmour DJ, Wright PC. Lipid quantification techniques for screening oleaginous species of microalgae for biofuel production. Eur J Lipid Sci Technol 2017;119:1–24. https://doi.org/10.1002/ejlt.201500469.

[27] Govender T, Ramanna L, Rawat I, Bux F. BODIPY staining, an alternative to the Nile Red fluorescence method for the evaluation of intracellular lipids in microalgae. Bioresour Technol 2012;114:507–11. https://doi.org/10.1016/j.biortech.2012.03.024.

[28] Rumin J, Bonnefond H, Saint-Jean B, Rouxel C, Sciandra A, Bernard O, et al. The use of fluorescent Nile red and BODIPY for lipid measurement in microalgae. Biotechnol Biofuels 2015;8:1–16. https://doi.org/10.1186/s13068-015-0220-4.

[29] Cadet F, Garrigues S, de la Guardia M. Quantitative analysis, infrared. In: Meyers RA, editor. Encyclopedia of Analytical Chemistry. John Wiley & Sons, Ltd.; 2006. p. 16504. https://doi.org/10.1002/9780470027318.a5610.pub2.

[30] Zhang WB. Review on analysis of biodiesel with infrared spectroscopy. Renew Sustain Energy Rev 2012;16:6048–58. https://doi.org/10.1016/j.rser.2012.07.003.

[31] Forfang K, Zimmermann B, Kosa G, Kohler A, Shapaval V. FTIR spectroscopy for evaluation and monitoring of lipid extraction efficiency for oleaginous fungi. PLoS One 2017;12:1–17. https://doi.org/10.1371/journal.pone.0170611.

[32] Ami D, Posteri R, Mereghetti P, Porro D, Doglia SM, Branduardi P. Fourier transform infrared spectroscopy as a method to study lipid accumulation in oleaginous yeasts. Biotechnol Biofuels 2014;7:12. https://doi.org/10.1186/1754-6834-7-12.

[33] Meng Y, Yao C, Xue S, Yang H. Application of fourier transform infrared (FT-IR) spectroscopy in determination of microalgal compositions. Bioresour Technol 2014;151:347–54. https://doi.org/10.1016/j.biortech.2013.10.064.

[34] Folch J, Lees M, Sloane Stanley GH. A simple method for the isolation and purification of total lipids from animal tissues. J Biol Chem 1957;226:511.

[35] Bligh E, Dyer W. A rapid method of total lipid extraction and purification. JCP Biochem Phys 1959;37:911–7. https://doi.org/10.1139/cjm2014-0700.

[36] Lewis T, Nichols PD, Mcmeekin TA. Evaluation of extraction methods for recovery of fatty acids from lipid-producing microheterotrophs. J Microbiol Methods 2000;43:107–16.

[37] Yellapu SK, Bharti KR, Kumar LR, Tiwari B, Zhang X, et al. Recent developments of downstream processing for microbial lipids and conversion to biodiesel. Bioresour Technol 2018;256:515–28. https://doi.org/10.1016/j.biortech.2018.01.129.

[38] Enamala MK, Enamala S, Chavali M, Donepudi J, Yadavalli R, Kolapalli B, et al. Production of biofuels from microalgae – a review on cultivation, harvesting, lipid extraction, and numerous applications of microalgae. Renew Sustain Energy Rev 2018;94:49–68. https://doi.org/10.1016/j.rser.2018.05.012.

[39] Roush DJ, Lu Y. Advances in primary recovery: centrifugation and membrane technology. Biotechnol Prog 2008;24(3):488–95.

[40] Schlesinger A, Eisenstadt D, Bar-Gil A, Carmely H, Einbinder S, Gressel J. Inexpensive non-toxic flocculation of microalgae contradicts theories; overcoming a major hurdle to bulk algal production. Biotechnol Adv 2012;30:1023–30. https://doi.org/10.1016/j.biotechadv.2012.01.011.

[41] Vandamme D, Foubert I, Muylaert K. Flocculation as a low-cost method for harvesting microalgae for bulk biomass production. Trends Biotechnol 2013;31:233–9. https://doi.org/10.1016/j.tibtech.2012.12.005.

[42] Barros AI, Gonçalves AL, Simões M, Pires JCM. Harvesting techniques applied to microalgae: a review. Renew Sustain Energy Rev 2015;41:1489–500.

[43] Metz B, Kossen NWF. The growth of molds in the form of pellets—a literature review. Biotechnol Bioeng 1977;19:781–99. https://doi.org/10.1002/bit.260190602.

[44] Xia C, Zhang J, Zhang W, Hu B. A new cultivation method for microbial oil production: cell pelletization and lipid accumulation by *Mucor circinelloides*. Biotechnol Biofuels 2011;4:1–10. https://doi.org/10.1186/1754-6834-4-15.

[45] Zhang J, Hu B. A novel method to harvest microalgae via co-culture of filamentous fungi to form cell pellets. Bioresour Technol 2012;114:529–35. https://doi.org/10.1016/j.biortech.2012.03.054.

[46] Muradov N, Taha M, Miranda AF, Wrede D, Kadali K, Gujar A, et al. Fungal-assisted algal flocculation: application in wastewater treatment and biofuel production. Biotechnol Biofuels 2015;8. https://doi.org/10.1186/s13068-015-0210-6.

[47] Yousuf A, Ethiraj B, Khan MR, Pirozzi D. Fungal biorefinery for the production of single cell oils as advanced biofuels. In: Kumar S, Dheeran P, Taherzadeh M, Khanal S, editors. Fungal biorefineries. Cham: Springer International Publishing; 2018. p. 185–213. https://doi.org/10.1007/978-3-319-90379-8_9.

[48] Shields-Menard SA, AmirSadeghi M, Green M, Womack E, Sparks DL, Blake J, et al. The effects of model aromatic lignin compounds on growth and lipid accumulation of *Rhodococcus rhodochrous*. Int Biodeterior Biodegrad 2017;121:79–90. https://doi.org/10.1016/j.ibiod.2017.03.023.

[49] Vasconcelos B, Teixeira JC, Dragone G, Teixeira JA. Optimization of lipid extraction from the oleaginous yeasts *Rhodotorula glutinis* and *Lipomyces kononenkoae*. AMB Express 2018:1–20. https://doi.org/10.1186/s13568-018-0658-4.

[50] Ghasemi NF, González González LM, Chan W, Schenk PM. Progress on lipid extraction from wet algal biomass for biodiesel production. Microb Biotechnol 2016;9:718–26. https://doi.org/10.1111/1751-7915.12360.

[51] Jeevan Kumar SP, Vijay Kumar G, Dash A, Scholz P, Banerjee R. Sustainable green solvents and techniques for lipid extraction from microalgae: a review. Algal Res 2017;21:138–47. https://doi.org/10.1016/j.algal.2016.11.014.

[52] Patel A, Pruthi V, Singh RP, Pruthi PA. Synergistic effect of fermentable and non-fermentable carbon sources enhances TAG accumulation in oleaginous yeast *Rhodosporidium kratochvilovae* HIMPA1. Bioresour Technol 2015;188:136–44. https://doi.org/10.1016/j.biortech.2015.02.062.

[53] Middelberg APJ. Process-scale disruption of microorganisms. Biotechnol Adv 1995;13:491–551. https://doi.org/10.1016/0734-9750(95)02007-.

[54] Ganzler K, Salgó A, Valkó K. Microwave extraction: a novel sample preparation method for chromatography. J Chromatogr 1986;371:299–306.

[55] Vorobiev E, Lebovka N. Selective extraction of molecules from biomaterials by pulsed electric field treatment. In: Miklavcic D, editor. Handbook of Electroporation. Springer International Publishing; 2016. p. 1–16. https://doi.org/10.1007/978-3-319-26779-1_163-1.

[56] Ochsenreither K, Glück C, Stressler T, Fischer L, Syldatk C. Production strategies and applications of microbial single cell oils. Front Microbiol 2016;7. https://doi.org/10.3389/fmicb.2016.01539.

[57] Eing C, Goettel M, Straessner R, Gusbeth C, Frey W. Pulsed electric field treatment of microalgae – benefits for microalgae biomass processing. IEEE Trans Plasma Sci 2013;41:2901–7. https://doi.org/10.1109/TPS.2013.2274805.

[58] Yu X, Dong T, Zheng Y, Miao C, Chen S. Investigation of cell disruption methods for lipid extraction from oleaginous microorganisms. Eur J Lipid Sci Technol 2015;117:730–7. https://doi.org/10.1002/ejlt.201400195.

[59] Kadir WNA, Lam MK, Uemura Y, Lim JW, Lee KT. Harvesting and pre-treatment of microalgae cultivated in wastewater for biodiesel production: a review. Energy Convers Manag 2018;171:1416–29. https://doi.org/10.1016/j.enconman.2018.06.074.

[60] Cooney MJ, Benjamin K. Ionic liquids in lipid extraction and recovery. AOCS Press; 2016. https://doi.org/10.1016/B978-1-63067-047-4.00009-X.

[61] Ward VCA, Munch G, Cicek N, Rehmann L. Direct conversion of the oleaginous yeast *Rhodosporidium diobovatum* to biodiesel using the ionic liquid [C2mim][EtSO4]. ACS Sustainable Chem Eng 2017;5:5562–70. https://doi.org/10.1021/acssuschemeng.7b00976.

[62] Huang Q, Wang Q, Gong Z, Jin G, Shen H, Xiao S, et al. Effects of selected ionic liquids on lipid production by the oleaginous yeast *Rhodosporidium toruloides*. Bioresour Technol 2013;130:339–44. https://doi.org/10.1016/j.biortech.2012.12.022.

[63] Gong M, Hu Y, Yedahalli S, Bassi A. Oil extraction processes in microalgae. Recent Adv Renew Energy 2017;1:377–411.

[64] Zhao H, Baker GA. Ionic liquids and deep eutectic solvents for biodiesel synthesis: a review. J Chem Technol Biotechnol 2013;88:3–12. https://doi.org/10.1002/jctb.3935.

[65] Yellapu SK, Bezawada J, Kaur R, Kuttiraja M, Tyagi RD. Detergent assisted lipid extraction from wet yeast biomass for biodiesel: a response surface methodology approach. Bioresour Technol 2016;218:667–73. https://doi.org/10.1016/j.biortech.2016.07.011.

[66] Tamano K, Miura A, Koike H, Kamisaka Y, Umemura M, Machida M. High-efficiency extracellular release of free fatty acids from *Aspergillus oryzae* using non-ionic surfactants. J Biotechnol 2017;248:9–14. https://doi.org/10.1016/j.jbiotec.2017.03.002.

[67] Howlader MS, Rai N, Todd French W. Improving the lipid recovery from wet oleaginous microorganisms using different pretreatment techniques. Bioresour Technol 2018;267:743–55. https://doi.org/10.1016/j.biortech.2018.07.092.

[68] Patel A, Mikes F, Matsakas L. An overview of current pretreatment methods used to improve lipid extraction from oleaginous microorganisms. Molecules 2018;23:1562. https://doi.org/10.3390/molecules23071562.

[69] Meullemiestre A, Breil C, Abert-Vian M, Chemat F. Manothermosonication as a useful tool for lipid extraction from oleaginous microorganisms. Ultrason Sonochem 2017;37:216–21. https://doi.org/10.1016/j.ultsonch.2017.01.014.

[70] Patel A, Arora N, Pruthi V, Pruthi PA. A novel rapid ultrasonication-microwave treatment for total lipid extraction from wet oleaginous yeast biomass for sustainable biodiesel production. Ultrason Sonochem 2018. https://doi.org/10.1016/j.ultsonch.2018.05.002.

[71] He Y, Wu T, Wang X, Chen B, Chen F. Cost-effective biodiesel production from wet microalgal biomass by a novel two-step enzymatic process. Bioresour Technol 2018;268:583–91. https://doi.org/10.1016/j.biortech.2018.08.038.

[72] Yang H, Lu W, Chen Y, Chen K, Teng J, Wan H. New algal lipid extraction procedure using an amphiphilic amine solvent and ionic liquid. Biomass Bioenergy 2017;100:108–15. https://doi.org/10.1016/j.biombioe.2017.03.018.

[73] Huang W, Park CW, Kim J. A novel microalgal lipid extraction method using biodiesel (fatty acid methyl esters) as an extractant. Bioresour Technol 2016. https://doi.org/10.1016/j.biortech.2016.12.013.

[74] Dong T, Fei Q, Genelot M, Smith H, Laurens LML, Watson MJ, et al. A novel integrated biorefinery process for diesel fuel blendstock production using lipids from the methanotroph, *Methylomicrobium buryatense*. Energy Convers Manag 2017;140:62–70. https://doi.org/10.1016/j.enconman.2017.02.075.

[75] Gill JJ, Wang B, Sestak E, Young R, Chu K. Characterization of a novel *Tectivirus* phage toil and its potential as an agent for biolipid extraction. Sci Rep 2018:1–11. https://doi.org/10.1038/s41598-018-19455-2.

[76] Kosa G, Zimmermann B, Kohler A, Ekeberg D, Afseth NK, Mounier J, et al. High - throughput screening of Mucoromycota fungi for production of low – and high – value lipids. Biotechnol Biofuels 2018;11:66. https://doi.org/10.1186/s13068-018-1070-7.

[77] Wahlen BD, Morgan MR, McCurdy AT, Willis RM, Morgan MD, Dye DJ, et al. Biodiesel from microalgae, yeast, and bacteria: engine performance and exhaust emissions. Energy Fuels 2013;27:220–8. https://doi.org/10.1021/ef3012382.

[78] Guldhe A, Singh B, Rawat I, Bux F. Synthesis of biodiesel from *Scenedesmus* sp. by microwave and ultrasound assisted in situ transesterification using tungstated zirconia as a solid acid catalyst. Chem Eng Res Des 2014;92:1503–11. https://doi.org/10.1016/j.cherd.2014.05.012.

[79] Macías-Sánchez MD, Robles-Medina A, Jiménez-Callejón MJ, Hita-Peña E, Estéban-Cerdán L, González-Moreno PA, et al. Optimization of biodiesel production from wet microalgal biomass by direct transesterification using the surface response methodology. Renew Energy 2018;129:141–9. https://doi.org/10.1016/j.renene.2018.06.001.

[80] Cui Y, Liang Y. Direct transesterification of wet *Cryptococcus curvatus* cells to biodiesel through use of microwave irradiation. Appl Energy 2014;119:438–44. https://doi.org/10.1016/j.apenergy.2014.01.016.

[81] Katre G, Raskar S, Zinjarde S, Kumar VR, Kulkarni BD, Ravikumar A. Optimization of the in situ transesterification step for biodiesel production using biomass of *Yarrowia lipolytica* NCIM 3589 grown on waste cooking oil. Energy 2018;142:944–52. https://doi.org/10.1016/j.energy.2017.10.082.

[82] Carvalho AKF, Bento HBS, Rivaldi JD, de Castro HF. Direct transesterification of *Mucor circinelloides* biomass for biodiesel production: effect of carbon sources on the accumulation of fungal lipids and biofuel properties. Fuel 2018;234:789–96. https://doi.org/10.1016/j.fuel.2018.07.029.

[83] Kakkad H, Khot M, Zinjarde S, Ravikumar A, Ravi Kumar V, Kulkarni BD. Conversion of dried *Aspergillus candidus* mycelia grown on waste whey to biodiesel by in situ acid transesterification. Bioresour Technol 2015;197:502–7. https://doi.org/10.1016/j.biortech.2015.07.118.

[84] Weerheim AM, Kolb AM, Sturk A, Nieuwland R. Phospholipid composition of cell-derived microparticles determined by one-dimensional high-performance thin-layer chromatography. Anal Biochem 2002;302:191–8. https://doi.org/10.1006/abio.2001.5552.

[85] Jensen SK. Improved Bligh and dyer extraction procedure. Lipid Technol 2008;20:280–1. https://doi.org/10.1002/lite.200800074.

[86] Schneiter R, Daum G. Extraction of yeast lipids. In: Wei X, editor. Yeast Protocol. 2nd ed., vol. 313; 2006. p. 41–5.

[87] Matyash V, Liebisch G, Kurzchalia TV, Shevchenko A, Schwudke D. Lipid extraction by methyl- *tert* -butyl ether for high-throughput lipidomics. J Lipid Res 2008;49:1137–46. https://doi.org/10.1194/jlr.D700041-JLR200.

[88] Mathimani T, Uma L, Prabaharan D. Homogeneous acid catalysed transesterification of marine microalga *Chlorella* sp. BDUG 91771 lipid - an efficient biodiesel yield and its characterization. Renew Energy 2015;81:523–33. https://doi.org/10.1016/j.renene.2015.03.059.

[89] Vicente G, Bautista LF, Rodríguez R, Gutiérrez FJ, Sádaba I, Ruiz-Vázquez RM, et al. Biodiesel production from biomass of an oleaginous fungus. Biochem Eng J 2009;48:22–7. https://doi.org/10.1016/j.bej.2009.07.014.

[90] Bharti RK, Srivastava S, Thakur IS. Production and characterization of biodiesel from carbon dioxide concentrating chemolithotrophic bacteria, *Serratia* sp. ISTD04. Bioresour Technol 2014;153:189–97. https://doi.org/10.1016/j.biortech.2013.11.075.

[91] Teo CL, Jamaluddin H, Zain NAM, Idris A. Biodiesel production via lipase catalysed transesterification of microalgae lipids from *Tetraselmis* sp. Renew Energy 2014;68:1–5. https://doi.org/10.1016/j.renene.2014.01.027.

[92] Vicente G, Carrero A, Rodríguez R, del Peso GL. Heterogeneous-catalysed direct transformation of microalga biomass into biodiesel-grade FAMEs. Fuel 2017;200:590–8. https://doi.org/10.1016/j.fuel.2017.04.006.

[93] Nur Syazwani O, Rashid U, Taufiq Yap YH. Low-cost solid catalyst derived from waste *Cyrtopleura costata* (Angel Wing Shell) for biodiesel production using microalgae oil. Energy Convers Manag 2015;101:749–56. https://doi.org/10.1016/j.enconman.2015.05.075.

[94] Guldhe A, Singh P, Ansari FA, Singh B, Bux F. Biodiesel synthesis from microalgal lipids using tungstated zirconia as a heterogeneous acid catalyst and its comparison with homogeneous acid and enzyme catalysts. Fuel 2017;187:180–8. https://doi.org/10.1016/j.fuel.2016.09.053.

[95] Guldhe A, Moura CVR, Singh P, Rawat I, Moura EM, Sharma Y, et al. Conversion of microalgal lipids to biodiesel using chromium-aluminum mixed oxide as a heterogeneous solid acid catalyst. Renew Energy 2017;105:175–82. https://doi.org/10.1016/j.renene.2016.12.053.

[96] Pandit PR, Fulekar MH. Egg shell waste as heterogeneous nanocatalyst for biodiesel production: optimized by response surface methodology. J Environ Manag 2017;198:319–29. https://doi.org/10.1016/j.jenvman.2017.04.100.

[97] Sun Y, Cooke P, Reddy HK, Muppaneni T, Wang J, Zeng Z, et al. 1-Butyl-3-methylimidazolium hydrogen sulfate catalyzed in-situ transesterification of *Nannochloropsis* to fatty acid methyl esters. Energy Convers Manag 2017;132:213–20. https://doi.org/10.1016/j.enconman.2016.10.071.

[98] Kakkad H, Khot M, Zinjarde S, RaviKumar A. Biodiesel Production by Direct in situ transesterification of an oleaginous tropical mangrove fungus grown on untreated agro-residues and evaluation of its fuel properties. Bioenergy Res 2015;8:1788–99. https://doi.org/10.1007/s12155-015-9626-x.

[99] Carvalho AKF, da Conceição LRV, Silva JPV, Perez VH, de Castro HF. Biodiesel production from *Mucor circinelloides* using ethanol and heteropolyacid in one and two-step transesterification. Fuel 2017;202:503–11. https://doi.org/10.1016/j.fuel.2017.04.063.

[100] Louhasakul Y, Cheirsilp B, Maneerat S, Prasertsan P. Direct transesterification of oleaginous yeast lipids into biodiesel: development of vigorously stirred tank reactor and process optimization. Biochem Eng J 2018;137:232–8. https://doi.org/10.1016/j.bej.2018.06.009.

[101] Shirazi H, Karimi-Sabet J, Ghotbi C. Biodiesel production from *Spirulina* microalgae feedstock using direct transesterification near supercritical methanol condition. Bioresour Technol 2017;239:378–86. https://doi.org/10.1016/j.biortech.2017.04.073.

[102] Sitepu EK, Corbin K, Luo X, Pye SJ, Tang Y, Leterme SC, et al. Vortex fluidic mediated direct transesterification of wet microalgae biomass to biodiesel. Bioresour Technol 2018;266:488–97. https://doi.org/10.1016/j.biortech.2018.06.103.

[103] Tran D, Chen C, Chang J. Effect of solvents and oil content on direct transesterification of wet oil-bearing microalgal biomass of *Chlorella vulgaris* ESP-31 for biodiesel synthesis using immobilized lipase as the biocatalyst. Bioresour Technol 2013;135:213–21. https://doi.org/10.1016/j.biortech.2012.09.101.

[104] López EN, Medina AR, Moreno PAG, Cerdán LE, Valverde LM, Grima EM. Biodiesel production from *Nannochloropsis gaditana* lipids through transesterification catalyzed by *Rhizopus oryzae* lipase. Bioresour Technol 2016;203:236–44. https://doi.org/10.1016/j.biortech.2015.12.036.

[105] Sivaramakrishnan R, Incharoensakdi A. Direct transesterification of *Botryococcus* sp. catalysed by immobilized lipase: ultrasound treatment can reduce reaction time with high yield of methyl ester. Fuel 2017;191:363–70. https://doi.org/10.1016/j.fuel.2016.11.085.

[106] Carvalho AKF, Rivaldi JD, Barbosa JC, de Castro HF. Biosynthesis, characterization and enzymatic transesterification of single cell oil of *Mucor circinelloides* - a sustainable pathway for biofuel production. Bioresour Technol 2015;181:47–53. https://doi.org/10.1016/j.biortech.2014.12.110.

[107] Szczesna-Antczak M, Struszczyk-Świta K, Rzyska M, Szeląg J, Stańczyk Ł, Antczak T. Oil accumulation and in situ trans/esterification by lipolytic fungal biomass. Bioresour Technol 2018;265:110–8. https://doi.org/10.1016/j.biortech.2018.05.094.

[108] Karimi M. Exergy-based optimization of direct conversion of microalgae biomass to biodiesel. J Clean Prod 2017;141:50–5. https://doi.org/10.1016/j.jclepro.2016.09.032.

[109] Bach QV, Chen WH, Lin SC, Sheen HK, Chang JS. Wet torrefaction of microalga *Chlorella vulgaris* ESP-31 with microwave-assisted heating. Energy Convers Manag 2017;141:163–70. https://doi.org/10.1016/j.enconman.2016.07.035.

[110] Kim B, Park J, Son J, Lee JW. Catalyst-free production of alkyl esters from microalgae via combined wet in situ transesterification and hydrothermal liquefaction (iTHL). Bioresour Technol 2017;244:423–32. https://doi.org/10.1016/j.biortech.2017.07.129.

[111] Patil PD, Reddy H, Muppaneni T, Deng S. Biodiesel fuel production from algal lipids using supercritical methyl acetate (glycerin-free) technology. Fuel 2017;195:201–7. https://doi.org/10.1016/j.fuel.2016.12.060.

[112] Pan Y, Alam MA, Wang Z, Huang D, Hu K, Chen H, et al. One-step production of biodiesel from wet and unbroken microalgae biomass using deep eutectic solvent. Bioresour Technol 2017;238:157–63. https://doi.org/10.1016/j.biortech.2017.04.038.

[113] Yodsuwan N, Kamonpatana P, Chisti Y, Sirisansaneeyakul S. Ohmic heating pretreatment of algal slurry for production of biodiesel. J Biotechnol 2018;267:71–8. https://doi.org/10.1016/j.jbiotec.2017.12.022.

[114] Knothe G. Dependence of biodiesel fuel properties on the structure of fatty acid alkyl esters. Fuel Process Technol 2005;86:1059–70. https://doi.org/10.1016/j.fuproc.2004.11.002.

[115] Ramos MJ, Fernández CM, Casas A, Rodríguez L, Pérez Á. Influence of fatty acid composition of raw materials on biodiesel properties. Bioresour Technol 2009;100:261–8. https://doi.org/10.1016/j.biortech.2008.06.039.

[116] Guldhe A, Singh B, Rawat I, Permaul K, Bux F. Biocatalytic conversion of lipids from microalgae *Scenedesmus obliquus* to biodiesel using *Pseudomonas fluorescens* lipase. Fuel 2015;147:117–24. https://doi.org/10.1016/j.fuel.2015.01.049.

[117] Ahmad FB, Zhang Z, Doherty WOS, Hara IMO, Crops T. Evaluation of oil production from oil palm empty fruit bunch by oleaginous micro-organisms. Biofuels, Bioprod Biorefining 2016;10:378–92. https://doi.org/10.1002/bbb.1645.

[118] Ramírez-Verduzco LF, Rodríguez-Rodríguez JE, Jaramillo-Jacob ADR. Predicting cetane number, kinematic viscosity, density and higher heating value of biodiesel from its fatty acid methyl ester composition. Fuel 2012;91:102–11. https://doi.org/10.1016/j.fuel.2011.06.070.

[119] Kamat S, Khot M, Zinjarde S, Ravikumar A, Namdeo W. Coupled production of single cell oil as biodiesel feedstock, xylitol and xylanase from sugarcane bagasse in a biorefinery concept using fungi from the tropical mangrove wetlands. Bioresour Technol 2013;135:246–53. https://doi.org/10.1016/j.biortech.2012.11.059.

[120] Katre G, Ajmera N, Zinjarde S, RaviKumar A. Mutants of *Yarrowia lipolytica* NCIM 3589 grown on waste cooking oil as a biofactory for biodiesel production. Microb Cell Factories 2017;16:176. https://doi.org/10.1186/s12934-017-0790-x.

[121] Duarte SH, Ansolin M, Maugeri F. Cultivation of *Candida* sp. LEB-M3 in glycerol: lipid accumulation and prediction of biodiesel quality parameters. Bioresour Technol 2014;161:416–22. https://doi.org/10.1016/j.biortech.2014.03.096.

[122] Zheng Y, Yu X, Zeng J, Chen S. Feasibility of filamentous fungi for biofuel production using hydrolysate from dilute sulfuric acid pretreatment of wheat straw. Biotechnol Biofuels 2012;5:50. https://doi.org/10.1186/1754-6834-5-50.

[123] Khot M, Gupta R, Barve K, Zinjarde S, Govindwar S, RaviKumar A. Fungal production of single cell oil using untreated copra cake and evaluation of its fuel properties for biodiesel. J Microbiol Biotechnol 2015;25:459–63. https://doi.org/10.4014/jmb.1407.07074.

[124] Khot M, Katre G, Zinjarde S, Ravikumar A. Single Cell Oils (SCOs) of oleaginous filamentous fungi as a renewable feedstock: a biodiesel biorefinery approach. Microb Cell Factories 2018:145–83. https://doi.org/10.1186/s12934-017-0790-x.

[125] Zheng Y, Li T, Yu X, Bates PD, Dong T, Chen S. High-density fed-batch culture of a thermotolerant microalga *Chlorella sorokiniana* for biofuel production. Appl Energy 2013;108:281–7. https://doi.org/10.1016/j.apenergy.2013.02.059.

[126] Arias-Peñaranda MT, Cristiani-Urbina E, Montes-Horcasitas C, Esparza-Garćıa F, Torzillo G, Cañizares-Villanueva RO. *Scenedesmus incrassatulus* CLHE-Si01: a potential source of renewable lipid for high quality biodiesel production. Bioresour Technol 2013;140:158–64. https://doi.org/10.1016/j.biortech.2013.04.080.

Chapter 17

Advances and microbial techniques for phosphorus recovery in sustainable wastewater management

Meghanath Shambhu Prabhu[1,2], Srikanth Mutnuri[2]

[1]*Porter School of the Environment and Earth Sciences, Tel Aviv University, Tel Aviv, Israel;* [2]*Applied and Environmental Biotechnology Laboratory, Department of Biological Sciences, Birla Institute of Technology and Science, Pilani, K K Birla Goa Campus, Zuarinagar, Goa, India*

17.1 Introduction

Nitrogen (N) and phosphorus (P) are life-essential nutrient elements that are most abundantly used in agriculture. Their consumption along with price are increasing day by day all over the world due to the increasing human population. In the agricultural sector, consumption of phosphorus fertilizer in the EU-15 is $>$ 1.2 million tons per year [1]. At present, the synthetic N and P fertilizers are produced through processes that require tremendous energy inputs and use nonrenewable resources [2]. A sedimentary rock phosphate deposit is the main source of phosphorus on earth. Due to the decline in rock phosphate deposits, more than half of the P deposits on earth are likely to be gone in coming decades [3,4]. This makes phosphorus recovery and recycling of significant importance to face the emerging global challenges of food security [5].

Prior to the sanitation revolution of the 19th and 20th centuries, both human and animal manure were returned to agricultural fields to recycle the organic phosphorus. Due to the installation of centralized wastewater treatment systems in urban areas for better sanitation systems and for prevention of the spread of disease outbreaks, the valuable nutrients found their way into waterbodies bypassing the land application [6,7]. Urine is a major source of nitrogen and phosphorous in domestic wastewater treatment plant (WWTPs). Use of synthetic detergents further adds to the phosphorus content in sewage. Nitrogen concentration is important for the optimum functioning of WWTPs and for disposal of effluent on land [7]. Generally, nitrogen content in the

Advances in Biological Science Research. https://doi.org/10.1016/B978-0-12-817497-5.00017-3

untreated sewage is observed to be in the range of 20–50 mg/L, measured as total Kjeldahl nitrogen [2]. The concentration of phosphorus in domestic sewage is generally adequate to support aerobic biological WW treatment. The concentration of PO_4 in raw sewage is generally observed in the range of 5–10 mg/L [2]. The phosphate concentration in sludge digester liquors can be as high as up to 85–95 g P/m^3 [8]. Inadequate treatment of WW in WWTPs results in high amount of N and P in the effluent. Effluents from traditional wastewater treatment processes have total phosphorus concentrations ranging from 10 to 25 mg/L [7]. The excess P fed to dairy and beef cattle is also excreted into the environment through manure and urine. All of this P is ultimately transported into nearby water bodies and infiltrates into the groundwater. Discharge of such effluents can cause serious soil and water pollution, including eutrophication of water bodies [9]. To avoid the accumulation of nutrients such as N and P in the environment, they must be removed before entering into the WWTPs using suitable sustainable technologies. Currently, there are over 152 WWTPs in Indian cities providing WW collection and treatment capacity to around 51% of total WW generated [10]. WWTPs are vital for the protection of human health and environment. However, these treatment plants use significant amounts of energy and materials to take out valuable resources and to comply with discharge standards. These mechanized WW treatment systems turn out to be rather expensive in terms of both the installation as well as operation and maintenance costs, and hardly produce any resources from the treatment process, thus making them not sustainable. In order to address the issues of water, food, and energy demand, domestic WW is now being looked at more as a resource (a resource for water, for energy, and fertilizing nutrients for agriculture) than as a waste. Useful nutrient resources from WW can be recovered for secondary uses if treated properly [7]. It is estimated that through recovery and recycling of phosphorus from waste streams, 15%–20% of the world's phosphorus dependence on rock phosphate can be reduced [11].

The conventional WW treatment has only focused on eliminating these nutrients or making them biologically unavailable. The nutrients (nitrogen, phosphorus, and potassium) present in WW promote the growth of microbes in waterways (rivers, lakes, and coastal areas), leading to the problem of toxic algal blooms. Ammonia can kill fish and stimulate algal blooms, creating oxygen-depleted waters that impart the threat of eutrophication in other aquatic life. The recovery of nitrogen and phosphorus not only prevents eutrophication but also conserves limited natural resources. One the other hand, these same nutrients are major components of agricultural fertilizers and are critical for plant growth [12]. With WW being increasingly recognized as a valued source of multiple renewable resources such as nutrients, energy, organics, and clean water, the Environmental Protection Agencies are urging WWTPs to be viewed as waste resource recovery facilities (WRRFs) [2,7]. For instance, in the United States, this view is backed by the not-for-profit organization, the Water Environment Federation, which believes that "WWTPs are not waste disposal facilities, but rather water resource recovery facilities that produce clean water, and recover nutrients" [13]. For most of us, WW is irrelevant in our life, and thus we give it the least significant importance. But it is only recently that researchers have been developing the technology to recover the valuable resources from WW. It is a classic case of seeing a problem as an opportunity. WRRFs that would benefit from these new technologies need a scientifically sound view of the new technologies that can be adopted as part of their standard processes.

The global decline in soil fertility and hence crop productivity along with depletion in the resource deposits from the earth has stimulated interest in recovering nutrients from waste and converting them into plant fertilizers [7,14]. The recovery of valuable nutrients, from waste sources, allows minerals and carbon-based materials to be recycled and reused. Many researchers have used WW, WW sludge, and anaerobically digested fluids and other waste streams as a source of useful nutrients. As the nutrients in these waste streams represent a renewable resource, recovery of nutrients into a useable form from waste streams has emerged as a key component of sustainable approaches to managing global and regional nutrient use [15]. Nutrient (N, P, etc.) recovery from WW as fertilizers can be handled to offset the environmental loads associated with producing the equivalent amount of fertilizers from fossil fuels [7]. The challenge is to develop cost-effective means for recovery and purification of these valuable components.

17.2 Technologies for phosphorus recovery

Recovery of phosphorus in various forms such as calcium phosphate and aluminum phosphate has been tried using lime as a source of calcium and alum as a source of aluminum, respectively [15]. Ferric chloride is used to remove phosphorus through chemical sedimentation [16]. However, the chemicals used to remove phosphorus from wastewater comprise heavy metals, salts, and materials that are prohibited for use in soil amendments [17,18]. The adsorption-precipitation-filtration (reactive filtration) process for near-complete phosphorus removal (effluent phosphorus 0.011 mg/L) and the zeolite ammonia removal process to achieve near-complete ammonia removal appear to represent attractive and technically economical approaches (effluent ammonia-N less than 1.5 mg/l) [18].

17.2.1 The process of struvite crystallization

Struvite ($MgNH_4PO_4 \cdot 6H_2O$) is a white crystalline mineral substance containing equimolar amount (1:1:1) of magnesium ammonium and phosphate ions (also called as MAP) and is a good source of phosphorus (13% P), along with nitrogen (6% N) and magnesium and (10% Mg). Struvite is a slow-release inorganic fertilizer with solubility value 0.018 g/100 ml neutral water at 25°C [9,19]. It is very stable up to temperature of 40–50°C [14]. It is equivalent or superior to other chemical agricultural fertilizers [14,20]. Struvite was identified in WWTPs as early as 1939. Formation of struvite occurs when concentrations of soluble magnesium (Mg^{+2}), ammonium $\left(NH_4^{+}\right)$, and orthophosphate $\left(PO_4^{-3}\right)$ exceed levels that promote the formation of crystals (equimolar amount, 1:1:1), referred to as supersaturation [21–23]. Controlled struvite crystallization is a way of recycling nutrients and resource recovery from WW and sludge digester liquors because of high concentrations of phosphorus, ammonium, and sometimes, magnesium in them. The method imposes benefits such as phosphorus and nitrogen recovery, thereby cutting on eutrophication [24].

17.2.2 Recovery of struvite from wastes

Struvite precipitation occurs when the concentrations of magnesium ammonium and phosphate ions in wastewater exceed the supersaturation limit of >0.2 g/L [22,25]. The general formula for struvite precipitation is given in reversible reaction Eq. (17.1) (n = 0, 1, or 2) [26]:

$$Mg^{+2} + NH_4^{+} + H_nPO_4^{n-3} + 6H_2O \rightarrow MgNH_4PO_4 \cdot 6H_2O \downarrow + nH^{+} \quad (17.1)$$

Another condition for struvite formation is pH, which has to be between 8.5 and 9.5 as the solubility of struvite is pH dependent [23,26,27]. It is in a soluble form at low-pH levels and precipitates into solid forms at high-pH levels [26]. Wastewater pH levels are often too low for struvite formation; therefore pH levels must be raised [25,28]. Within a crystallizer reactor, pH values can be raised by the addition of sodium hydroxide [26,28]. Reports have shown that increase in pH achieved in such manner results into poor quality struvite formation. An increase in pH due to the accumulation of ammonium ions favors the formation of better quality struvite [14,26]. Mixing intensities and reactor types are other factors responsible for efficient struvite crystallization apart from molar ratio and pH [22]. Few factors can negatively affect struvite precipitation. Studies conducted using synthetic wastewater suggested that coexisting ion, SO_4^{2-} (1300 ppm), remarkably reduced the capability of a reactor to remove phosphate from solution. Few ions, such as NO_3^{-} and CH_3COO^{-} and F^{-}, reduced the crystallization ratio for struvite [23].

17.2.3 Source of magnesium for struvite formation

Often Mg^{+2} is a limiting nutrient in wastewaters and must be supplied externally for struvite crystal formation [26]. Although the sources of the Mg^{+2} used are diverse and expensive [29], inexpensive sources of magnesium such as MgO containing by-products [30], seawater [26,31] brine (reliable and inexpensive at seashore vicinities) [14], magnesite [32,33], and wood ash [34]) have also been used in some cases. Haiming et al. [35] showed a saving of 37% of the total cost for production of struvite by using bittern as the magnesium source instead of pure chemicals.

17.3 Struvite crystallization technologies

17.3.1 Lab-scale studies

To date, various kinds of reactor systems have been developed for struvite crystallization using processes such as chemical precipitation, crystallization, tertiary filtration, ion exchange methods, and biological crystallization [17,35]. Controlled struvite recovery from wastewater can be achieved through several approaches: chemical addition, carbon dioxide stripping, or electrolysis [36]. Continuously stirred tank reactor set-up was used for struvite precipitation with stable operating conditions and high phosphate removal efficiencies up to 98% [35]. Enhanced struvite recovery from wastewater was achieved using a novel cone-inserted fluidized bed reactor (FBR). Cones were sandwiched in FBR to reduce unwanted crystal loss up to 67.5%. Phosphorus removal efficiencies were more than 90% under the optimal operating conditions in such FBR reactor [35]. FBRs are most commonly used for crystallized struvite recovery because the design creates an abundance of the reactive surface area and solution turbulence. These conditions enable rapid molecular growth at the crystal surface interface, shortening the time required for crystal formation [37,38]. Lab-scale airlift reactor was used to recover struvite from fish canning industrial wastewater using the seawater as magnesium source [26,39]. Pastor et al. [40] reported struvite formation in a stirred reactor from the supernatants of an anaerobic digestion plant. As high pH is essential for struvite formation, they used air for increasing pH and thus struvite precipitation. This seems to be a better way as aeration also cleaned struvite crystals from suspended solids. Struvite has been recovered from various waste sources, using various sources of Mg^{+2}, and crystallized using the different type of reactors that are summarized in Table 17.1.

17.3.2 Biological struvite precipitation

Struvite can be crystallized from wastewaters having very low phosphorus concentration using bacterial machinery. The advantage of the biological method includes less chemical consumption, thus making the process of struvite crystallization a more economical one. Various soil and freshwater bacteria (e.g., *Staphylococcus aureus*, *Proteus* spp., *Escherichia coli*, *Myxococcus* sp., *Arthrobacter* sp., *Pseudomonas* sp.) have been shown to be capable of precipitating struvite in laboratory experiments [41]. Some authors have explored the potential of a metallophilic bacterium, *Enterobacter* sp. EMB19, for the recovery of phosphorus as phosphate-rich mineral; but the

TABLE 17.1 Various waste and Mg^{+2} sources used for struvite recovery process using different reactor types for precipitation.

Sr. no	Waste source	Reactor type	Process scale	Mg^{+2} source	P Removal (%)	Reference
1	Sewage treatment effluent	Stirred reactor	Lab scale	$MgCl_2$	80	[69]
2	Semiconductor wastewater	Jar test apparatus	Lab scale and pilot scale	$MgCl_2$	80	[20]
3	Aerobically and anaerobically treated sewage sludge and wastewater	NA*	Pilot scale	NA	90	[19]
4	Sewage sludge ash	Fluidized bed reactors or air-agitated columns	Pilot scale	NA	80	[40]
5	Human urine	NA	Lab scale	MgO, zeolite	95–100	[39]
6	Animal wastewater	NA	Bench scale	MgO, $MgCl_2$.	NA	[29]
7	Cow urine	Jar test apparatus	Lab scale	Brine, $MgCl_2$	NA	[14]
8	Swine wastewater	Activated sludge system	Lab scale	NA	97	[20]
9	Swine wastewater	Single chamber MFC	Lab scale	$MgCl_2$	NA	[70]
10	Swine wastewater	Continuous anaerobic sequencing batch reactor	Pilot scale	MgOH	NA	[49]
11	Piggery wastewater	NA	Lab scale	Struvite pyrolysate	96	[71]

*Note: *NA*- Values not given or mentioned by the author in the literature.

efficiency is about only 20% [42]. Enhanced biological phosphorus removal has been discussed in detail as an important technology for phosphorus recovery [11]. A group of specific bacteria known as phosphate accumulating organisms (PAOs), which are capable of accumulating excess phosphorus as intracellular polyphosphate granules (15%–20% dry cell weight), were used in effluents containing low level of phosphorus. (These PAOs belong to subclass 2 of *Betaproteobacteria, Actinobacteria.*) Generally, a carbon-to-phosphorus ratio of 30 was required for the efficient removal of phosphorus in this process. The sludge resulting due to the growth of PAOs was separated, stabilized, and used effectively for agricultural application. Further, the obtained sludge can be processed to produce the concentrated soluble phosphorus that can be recovered as struvite [11]. At low external ionic concentration of phosphate, bacterial cells, for instance, *Enterobacter* sp. EMB19, secrete ions and attain the supersaturation. This organism showed the crystallization of struvite within 3 to 4 days after inoculation in nutrient broth medium. It was observed that the crystals formed were of typical prismatic crystal habit and highly homogenous in their morphology. The secreted cellular proteins and cell fragments are used to form the struvite crystals in the surrounding medium [42]. Sinha et al. [42] also made a very interesting observation. The pH of the medium increased due to accumulation of ammonia from metabolic hydrolysis of nitrogenous substrates, which is favorable for struvite formation. The study was carried out to substantiate bacterial ability present in fixed-film bioreactor (FFB) to precipitate crystals of phosphate from domestic wastewaters [43]. The study was done in natural as well as artificial media. Precipitation observed in artificial media was more rapid and occurred after the third day of inoculation. Twelve platable heterotrophic bacteria were obtained from FFB, namely, *Pseudoxanthobacter*, two strains of *Rhodobacter*, *Escherichia*, *Paracoccus*, *Roseobacter*, two strains of *Sphingomonas*, two strains of *Agromyces*, *Ochrobactrum*, and *Alcaligenes*. Precipitation of phosphate minerals was observed near the bacterial colonies and in bacterial mass. These strains of bacteria precipitated minerals such as bobierrite, struvite, and baricite.

Microbial fuel cell (MFC) enables struvite recovery from digested sewage sludge along with energy produced by metabolic activity of *E. coli*. The iron phosphate ($FePO_4$) present in digested sewage sludge was precipitated with the help of MFC as the power source delivering electrons. Up to 82% of the orthophosphate was recovered in the form of struvite [44]. The technology fits well as sustainable decentralized process for phosphorus recovery and recycling. The electrochemical struvite precipitation from digestate with a fluidized bed cathode with a two-chamber microbial electrolysis cell (MEC) was used to precipitate struvite from anaerobic digester effluent, with up to 85% recovery simultaneously converting wastewater organics to hydrogen. In a two-chamber MEC, a cation exchange membrane is used to separate the inner cathode chamber from the outer anode chamber. This creates an ionic connection between chambers but prevents convective liquid flow. The organic

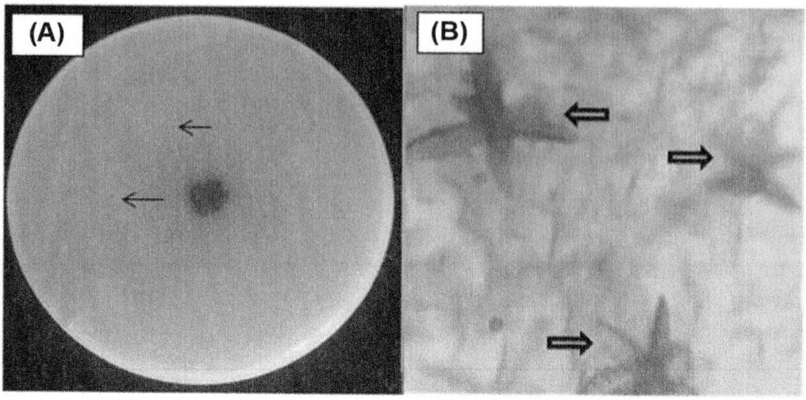

FIGURE 17.1 (A) *Klebsiella* sp. inoculated at the central well on solid SS1 agar, showing struvite crystal formed as white dots (pointed by an arrow). (B) Star-shaped struvite crystals were seen when the plate was observed under light microscope with 10 × magnification (arrow indicates crystals).

matter oxidation activity is carried out by mixed culture of anaerobic bacteria [37]. This can serve as a promising method to recover energy and nutrients at the same time from wastewater.

Bacteria such as *Staphylococcus* sp., *Pseudomonas* sp., *Bacillus* sp., *Escherichia* sp., *Yesinia* sp., *Klebsiella* sp., *Chromobacterium* sp., and *Proteus* sp. growing on synthetic solid medium (SS1) were shown to precipitate struvite in their surrounding growth area. *Proteus* sp. and *Klebsiella* sp. showed struvite formation when grown in industrial wastewater. The crystals formed were in different shapes and sizes (Fig. 17.1) [45].

17.3.3 Struvite formation within wastewater treatment plants: pilot-scale studies

Struvite crystallization has high nutrient recovery rates, especially for recovering precious phosphate resources. It is also economically feasible. It is assumed that a WWTP can recover 1 kg of struvite from 100 m^3 of WW, which will minimize sludge handling and disposal and reduce the operating cost [19]. It has been estimated that a WWTP with an average influent flow rate of 393 $m^{-3}\ d^{-1}$ has the ability to produce struvite worth £8400–20,000 per year [8]. Although unintentional struvite formation in sewage treatment plants can block valves, pipes, centrifuge bowls, and pumps, leading to decreased flow capacity and eventual equipment failure [21,24,46], struvite has great potential to be used as phosphate fertilizer.

The potential rate of struvite recovery varies with technology and local conditions and practices. Reports have shown that phosphorus recovery methods using struvite precipitation processes are very promising and likewise successful on large scale [47–49]. There are three full-scale facilities currently in operation in the United States utilizing struvite crystallization technologies provided by Ostara Nutrient Recovery Technologies, Inc. [50]. Several more sites are under construction. The University of British Columbia in Canada started a pilot project at the Lulu Island WWTP in 2006 to predict struvite precipitation levels through saturation index calculations and to recognize factors affecting struvite crystallization in fluidized bed reactors [38]. This pilot-scale reactor removed 75%–85% phosphate in the form of struvite from incoming rejected water flows. A pilot-scale study of the Unitika Ltd. Phosnix process at the Oxley Creek WWTP in Australia recovered almost 94% of the phosphate through struvite precipitation process [24]. The community-scale production of struvite and its use to grow the crops is being successfully practiced in Siddhipur in Nepal. There is a strong presence and acceptance of the urine-diverting dry toilets needed to collect urine separately at the source [47]. A low-cost, efficient, and reliable struvite reactor was built in Nepal with locally available materials with almost 90% of phosphorus recovery from urine. Decentralized technology at each household level to recover struvite from source-separated urine is emerging in Sweden. The struvite generated in this system can be made commercially available through existing technology [39].

Most of the current pilot-scale studies for struvite recovery are in Australia, Canada, and Spain [24,40]. The technology has not been widely applied in other countries. However, only Japan has implemented complete P removal and recovery as struvite from anaerobically digested sludge liquors with capacities ranging from 100 to 500 kg/day of struvite [24,51]. The resulting product is sold to fertilizer companies. A study has shown that it is more environmentally feasible to separate the urine at a source and produce struvite by addition of MgO as compared with centralized wastewater treatment system [30].

17.4 Use of struvite as fertilizer and its potential market

17.4.1 Use of struvite to increase soil fertility

Growing population has led to the increased demand for food that has led to the intensification and industrialization of agricultural production, and resulted in the increased use of energy and inorganic chemical fertilizers [52]. Human interventions have resulted in contamination of water bodies, and a large area of fertile soil is getting depleted year by year. This expression of soil degradation is very extensive and is not isolated to one region or even one continent—it is a worldwide problem [53]. According to the Food and Agriculture Organization of the United Nations, in their report "Status of the World's Soil Resources– 2015" climate change and land use dynamics are the major drivers of soil degradation [54]. Synthetic chemical fertilizers are usually applied to increase or restore the soil fertility, which further deteriorates the soil quality [55]. Various other factors that have contributed toward the decline in the soil fertility over the centuries are overproduction of crops, intensive animal grazing, volatilization, losses from leaching, losses due to crop (or product) removal, losses from various forms of erosion, and excessive use of artificial fertilizers [52]. In some parts of Asia, for example, up to six tons of chemical nutrient and hundreds of tons of organic fertilizers are applied per hectare each year in order to achieve high-yielding multiple cropping of vegetables. Between 50% and 60% of the nutrient inputs remain in the croplands after harvest. When these nutrients are later mobilized, they become a major source of pollution to waters via eutrophication. Use of toxic pesticides and herbicides along with heavy metals present in minerals further worsens the soil productivity [56].

The decline in soil fertility and productivity and loss of soil organic matter has stimulated interest in improving overall soil quality [57]. In developing countries, the increasing prices of chemical fertilizers coupled with growing concerns for sustaining soil productivity have led to renewed interest in the use of organic manures to restore soil fertility [58]. This can be achieved by measures of sustainable management by targeted additions of mineral to supply the nutrient needs of high-yielding crops [54]. Struvite is a slow-release fertilizer when applied to the soil. Studies have shown that the recovered struvite can be directly applied as fertilizer to grow crops such as *Vigna radiata* [14] ryegrass and fescue [59], Chinese cabbage [20], and wheat [60]. In all the studies, results clearly showed that the growth of the crops was promoted when the struvite deposit was used rather than compost fertilizers and other complex fertilizers. According to T. Karak et al. [61], struvite can be added to the compost to increase the nutrient value of compost by increasing the NPK content. Struvite formation during composting normally increases the electrical conductivity (EC) of the compost, thus limiting its usage. Man et al. [62] amended the compost containing struvite with 10% zeolite, which effectively reduced the EC and improved compost maturity. Furthermore, heavy metals and micropollutants in struvite recovered from various wastewaters are well below the regulatory limit for fertilizer usage. This encourages the use of struvite as commercial soil fertilizer and its use on long-term application basis [27]. Ryan et al. [63] used struvite as a fertilizer source for growing two microalgal strains, namely *Nannochloropsis salina* and *Phaeodactylum tricornutum*. The biomass and pigment productivities were found to be as high as when media formulations were used.

17.4.2 World and India's fertilizer requirements

World total fertilizer use increased steadily over the past 3 decades and is likely to continue to grow at slower rates. Global fertilizer consumption currently is in the range of 159–187 million tons (Mt). The consumption of N,

P_2O_5, and K_2O was 11.08, 4.12, and 1.60 Mt, respectively [54]. World fertilizer consumption in 2013–14 was seen as growing by 3.1% year on year, to 184 Mt nutrients [64]. Overall total nutrient consumption increased to 25.6 Mt during the full year of 2014–15 [65].

The global potential nitrogen fertilizer requirement is increasing and is expected to steadily rise during the forecast period, from 5.4% in 2015 to 6.9% in 2016, then 8.8% in 2017, and further reach 9.5% in 2018 [54]. World demand for the phosphorous is expected to rise from 2,700,000 tons in 2014 to 3,700,000 tons in 2018. Today, India is one of the largest producers and consumers of fertilizers in the world. By 2009–10, total fertilizer consumption in the country was 26.49 million nutrient tons. Currently, India produces in large amounts, a number of grades of NP/NPK complex and other fertilizers [66]. The total indigenous capacity of N and P_2O_5 increased from 17,000 to 21,000 tons in 1950–51 to 12,276 Mt and 5547 Mt in 2004–05 [54]. N and P_2O_5 production were at 12.43 Mt and 4.09 Mt during 2014–15 [65]. India imports mainly urea, DAP and potassium chloride (MOP). During 2014–15, import of urea was 8.75 Mt as compared to 7.09 Mt in 2013–14. Import of DAP was marginally increased up to 3.82 Mt in 2014–15 compared to 3.26 t in 2013–14. Import of MOP increased considerably significantly from 3.18 Mt in 2013–14 to 4.18 Mt in 2014–15 [65]. Struvite has the potential to replace a significant amount of the world's synthetic chemical fertilizers.

17.5 Economic feasibility of struvite recovery process

Struvite precipitation is a well-established and promising physicochemical treatment method for removing and recovering excess nitrogen and phosphorus from wastewater due to its higher efficiency. For instance, in India, Prabhu and Mutnuri [14] showed that a total of 12,176 tons struvite could be made from cow urine alone. K. Yetilmezsoy et al. [67] addressed the feasibility analysis of struvite recovery process for a full-scale fertilizer production industry with a 500 m^3/day capacity [67]. They determined that when the struvite sale price is raised to 560 €/ton, the facility will obtain a net profit of €445.62/day, with a payback period of approximately 6 years.

17.6 Conclusion

Phosphorus in wastewater represents a significant renewable resource that must be removed and recycled. Although it cannot fulfill the entirety of global P demands, it will reduce our reliance on phosphate rock [67]. Recycling of phosphorous from wastewater should be followed using appropriate technologies. Struvite precipitation using microbial technologies is emerging as less chemicals are needed making the process more economically viable. Various bacteria have shown promising results in terms of struvite precipitation from various waste sources. Fluidized bed cathode MEC is a dual promising method of sustainable electrochemical struvite and energy recovery method from nutrient-rich wastewaters [37]. The government seems to be assisting farmers with 50%–75% subsidy on use of chemical fertilizers containing phosphorus and magnesium [68], a practice that is ultimately unsustainable as it has adverse environmental repercussions [52]. Instead, assistance can be provided to buy technologies to recycle phosphorus. Removal of phosphorus in the form of struvite has many advantages over other forms of phosphorus recovery. When applied to crop soil it acts as a slow-release fertilizer, supplies P and N along with Mg to the crop plants. Hence struvite provides potential environmental benefits over conventional mineral P fertilizers, which are readily soluble and are responsible for eutrophication in water bodies. Recovery and use of struvite in agriculture will become more widespread and will surely take us one step further in sustainability in wastewater management and agriculture [67].

References

[1] Adam C, Peplinski B, Michaelis M, Kley G, Simon FG. Thermochemical treatment of sewage sludge ashes for phosphorus recovery. Waste Manag 2009;29:1122–8.

[2] Moss LH, Donovan JF, Carr S, Stone L, Christine PE, Black P, et al. Enabling the future: advancing resource recovery from biosolids. 2013.

[3] Kataki S, West H, Clarke M, Baruah DC. Phosphorus recovery as struvite: recent concerns for use of seed, alternative Mg source, nitrogen conservation and fertilizer potential. Resour Conserv Recycl 2016;107:142–56.

[4] Amanullah MM, Sekar S, Vincent S. Plant growth substances in crop production: a review. Asian J Plant Sci 2010;9:215–22.

[5] Cordell D, Rosemarin A, Schröder JJ, Smit AL. Towards global phosphorus security: a systems framework for phosphorus recovery and reuse options. Chemosphere 2011;84: 747–58.

[6] Bird AR. Evaluation of the feasibility of struvite precipitation from domestic wastewater as an alternative phosphorus fertilizer resource. Master's Thesis. University of San Francisco; 2015.

[7] Fricke K, Santen H, Wallmann R, Hüttner A, Dichtl N. Operating problems in anaerobic digestion plants resulting from nitrogen in MSW. Waste Manag 2007;27:30–43.

[8] Jaffer Y, Clark T, Pearce P, Parsons S. Potential phosphorus recovery by struvite formation. Water Res 2002;36:1834–42.

[9] Rahman MM, Salleh MAM, Rashid U, Ahsan A, Hossain MM, Ra CS. Production of slow release crystal fertilizer from wastewaters through struvite crystallization – a review. Arab J Chem 2014;7:139–55.

[10] Meghanath P. Resource recovery from wastewaters for sustainable development. Shodhganga; 2016.

[11] Yuan Z, Pratt S, Batstone DJ. Phosphorus recovery from wastewater through microbial processes. Curr Opin Biotechnol 2012;23:878–83.

[12] Yang J, Yamazaki A, Criddle C, Cantwell B, Wolak F. The Challenge: development of cost-effective technology for recovery of clean water, energy, and materials from wastewater. Wastewater as a Valuable Resource; 2010.

[13] Capuco CW. Waste not want not: the rise of resource recovery – WaterWorld. Water Wastewater Int; 2015. https://www.waterworld.com/articles/wwi/print/volume-28/issue-5/regional-spotlight-us-caribbean/waste-not-want-not-the-rise-of-resource-recovery.html.

[14] Prabhu M, Mutnuri S. Cow urine as a potential source for struvite production. Int J Recycl Org Waste Agric 2014;3:49.

[15] Zacharof MP, RWL. The use of mixed effluent liquid waste as a source of valuable nutrients. In: Popov V, Itoh H, Brebbia CA, editors. Waste Management and the Environment VI; 2010.

[16] Elliott HA, Taylor M. Phosphorus partitioning in co-dewatering biosolids and water treatment residuals. Water Sci Technol 2014;70:422.

[17] de-Bashan LE, Bashan Y. Recent advances in removing phosphorus from wastewater and its future use as fertilizer (1997–2003). Water Res 2004;38:4222–46.

[18] Sonune NA, Mungal N, Kamble S. Study of physico-chemical characteristics of domestic wastewater in Vishnupuri, Nanded, India. Int J Curr Microbiol App Sci 2015;4:533–6. https://doi.org/10.1016/J.CHEMOSPHERE.2011.03.001.

[19] Shu L, Schneider P, Jegatheesan V, Johnson J. An economic evaluation of phosphorus recovery as struvite from digester supernatant. Bioresour Technol 2006;97:2211–6.

[20] Ryu HD, Kim D, Lee SI. Application of struvite precipitation in treating ammonium nitrogen from semiconductor wastewater. J Hazard Mater 2008;156:163–9.

[21] Doyle JD, Parsons SA. Struvite formation, control and recovery. Water Res 2002;36: 3925–40. https://doi.org/10.1016/S0043-1354(02)00126-4.

[22] Peng L, Dai H, Wu Y, Peng Y, Lu X. A comprehensive review of phosphorus recovery from wastewater by crystallization processes. Chemosphere 2018;197:768–81.

[23] Shih YJ, Abarca RRM, de Luna MDG, Huang YH, Lu MC. Recovery of phosphorus from synthetic wastewaters by struvite crystallization in a fluidized-bed reactor: effects of pH, phosphate concentration and coexisting ions. Chemosphere 2017;173:466–73.

[24] Münch EV, Barr K. Controlled struvite crystallisation for removing phosphorus from anaerobic digester sidestreams. Water Res 2001;35:151–9.

[25] Barak P, Stafford A. Struvite: a recovered and recycled phosphorus fertilizer, vol. 45; 2006. p. 199.

[26] Sanchez A, Barros S, Mendez R, Garrido JM. Phosphorus removal from an industrial wastewater by struvite crystallization into an airlift reactor. In: International conferance of nutrition recovery from wastewater streams May 10–13, 2009, Westin Bayshore Hotel resort, Vancouver, Br. Columbia, Canada. IWA Publishing; 2009. p. 830.

[27] Uysal A, Yilmazel YD, Demirer GN. The determination of fertilizer quality of the formed struvite from effluent of a sewage sludge anaerobic digester. J Hazard Mater 2010;181:248–54.

[28] Rahaman MS, Mavinic DS, Meikleham A, Ellis N. Modeling phosphorus removal and recovery from anaerobic digester supernatant through struvite crystallization in a fluidized bed reactor. Water Res 2014;51:1–10.

[29] Burns RT, Moody LB. Phosphorus recovery from animal manures using optimized struvite precipitation. 2002.

[30] Ishii SKL, Boyer TH. Life cycle comparison of centralized wastewater treatment and urine source separation with struvite precipitation: focus on urine nutrient management. Water Res 2015;79:88–103.

[31] Lahav O, Telzhensky M, Zewuhn A, Gendel Y, Gerth J, Calmano W, et al. Struvite recovery from municipal-wastewater sludge centrifuge supernatant using seawater NF concentrate as a cheap Mg (II) source. Separ Purif Technol 2013;108:103–10.

[32] Gunay A, Karadag D, Tosun I, Ozturk M. Use of magnesit as a magnesium source for ammonium removal from leachate. J Hazard Mater 2008;156:619–23.

[33] Gunay A, Karadag D, Tosun I, Ozturk M. Use of magnesit as a magnesium source for ammonium removal from leachate. J Hazard Mater 2008;156:619–23.

[34] Sakthivel SR, Tilley E, Udert KM. Wood ash as a magnesium source for phosphorus recovery from source-separated urine. Sci Total Environ 2012;419:68–75.

[35] Huang H, Yang J, Li D. Recovery and removal of ammonia-nitrogen and phosphate from swine wastewater by internal recycling of struvite chlorination product. Bioresour Technol 2014;172:253–9.

[36] Cusick RD, Logan BE. Phosphate recovery as struvite within a single chamber microbial electrolysis cell. Bioresour Technol 2012;107:110–5.

[37] Cusick RD, Ullery ML, Dempsey BA, Logan BE. Electrochemical struvite precipitation from digestate with a fluidized bed cathode microbial electrolysis cell. Water Res 2014;54:297–306.

[38] Bhuiyan MIH, Mavinic DS, Koch FA. Thermal decomposition of struvite and its phase transition. Chemosphere 2008;70:1347–56.

[39] Ashley K. Energy efficient nutrient recovery from household wastewater using struvite precipitation and zeolite adsorption techniques. A pilot plant study in Sweden. Building 2009:511–20.

[40] Pastor L, Mangin D, Ferrer J, Seco A. Struvite formation from the supernatants of an anaerobic digestion pilot plant. Bioresour Technol 2010;101:118–25.

[41] Beavon J, Heatley NG. The occurrence of struvite (magnesium ammonium phosphate hexahydrate) in microbial cultures. J Gen Microbiol 1963;31:167–9.

[42] Sinha A, Singh A, Kumar S, Khare SK, Ramanan A. Microbial mineralization of struvite: a promising process to overcome phosphate sequestering crisis. Water Res 2014;54:33–43.

[43] Rivadeneyra A, Gonzalez-Martinez A, Gonzalez-Lopez J, Martin-Ramos D, Martinez-Toledo MV, Rivadeneyra MA. Precipitation of phosphate minerals by microorganisms isolated from a fixed-biofilm reactor used for the treatment of domestic wastewater. Int J Environ Res Publ Health 2014;11:3689–704.

[44] Fischer F, Bastian C, Happe M, Mabillard E, Schmidt N. Microbial fuel cell enables phosphate recovery from digested sewage sludge as struvite. Bioresour Technol 2011;102:5824–30.

[45] Rego O. Nutrient recovery from wastewaters using sustainable environmental technology. Goa University; 2017.

[46] Le Corre KS, Valsami-Jones E, Hobbs P, Parsons SA. Phosphorus recovery from wastewater by struvite crystallization. Review 2009;39.

[47] Etter B, Tilley E, Khadka R, Udert KM. Low-cost struvite production using source-separated urine in Nepal. Water Res 2011;45:852–62.

[48] Pstor L, Mngin D, BaratT R, Seco A. A pilot-scale study of struvite precipitation in a stirred tank reactor: conditions influencing the process. Bioresour Technol 2008;99:6285–91.

[49] Miles A, Ellis TG. Struvite precipitation potential for nutrient recovery from anaerobically treated wastes. Water Sci Technol 2001;43:259–66.

[50] Ostara Ostara Nutrient Recovery Technologies Inc. Http://OstaraCom/2018. http://ostara.com/(accessed August 6, 2018).

[51] Ueno Y, Fujii M. Three years experience of operating and selling recovered struvite from full-scale plant. Environ Technol 2001;22:1373–81.

[52] Campbell LC. Managing soil fertility decline. J Crop Prod 1998;1:29–52.

[53] DeLong C, Cruse R, Wiener J. The soil degradation paradox: compromising our resources when we need them the most. Sustainability 2015;7:866–79.

[54] FAO and ITPS. Status of the world's soil resources. 2015. Rome.

[55] Factura H, Bettendorf T, Buzie C, Pieplow H, Reckin J, Otterpohl R. Terra Preta sanitation: re-discovered from an ancient Amazonian civilisation – integrating sanitation, bio-waste management and agriculture. Water Sci Technol 2010;61:2673.

[56] Otterpohl R. Introduction: synergistic systems for living soil, water, food and energy. In: Terra preta sanitation. Deutsche Bundesstiftung Umwelt; 2015. p. 10–5.

[57] Suthar S. Potential of domestic biogas digester slurry in vermitechnology. Bioresour Technol 2010;101:5419–25.

[58] Ayala S, Rao EVSP. Perspectives of soil fertility management with a focus on fertilizer use for crop productivity. Curr Sci 2002;82:797–807.

[59] Achat DL, Sperandio M, Daumer ML, Santellani AC, Prud'Homme L, Akhtar M, et al. Plant-availability of phosphorus recycled from pig manures and dairy effluents as assessed by isotopic labeling techniques. Geoderma 2014;232–234:24–33.

[60] Talboys PJ, Heppell J, Roose T, Healey JR, Jones DL, Withers PJA. Struvite: a slow-release fertiliser for sustainable phosphorus management? Plant Soil 2016;401:109–23.

[61] Karak T, Sonar I, Nath JR, Paul RK, Das S, Boruah RK, et al. Struvite for composting of agricultural wastes with termite mound: utilizing the unutilized. Bioresour Technol 2015;187:49–59.

[62] Chan MT, Selvam A, Wong JWC. Reducing nitrogen loss and salinity during "struvite" food waste composting by zeolite amendment. Bioresour Technol 2016;200:838–44.

[63] Davis RW, Siccardi AJ, Huysman ND, Wyatt NB, Hewson JC, Lane TW. Growth of mono- and mixed cultures of *Nannochloropsis salina* and *Phaeodactylum tricornutum* on struvite as a nutrient source. Bioresour Technol 2015;198:577–85.

[64] Prudhomme PH and M. Fertilizer outlook 2014–2018. 2014.

[65] Lochan. 60th annual report 2014–15. 2015.

[66] DFMCF G of I. Indian fertilizer scenario 2010. 2010. New Delhi.

[67] Yetilmezsoy K, Ilhan F, Kocak E, Akbin HM. Feasibility of struvite recovery process for fertilizer industry: a study of financial and economic analysis. J Clean Prod 2017;152: 88–102.

[68] Govenrment of Goa D of A. A farmer friendly handbook. 2015. Panaji.

[69] Booker NA, Priestley AJ, Fraser IH. struvite formation in wastewater treatment plants: opportunities for nutrient recovery. Environ Technol 1999;20:777–82.

[70] Ichihashi O, Hirooka K. Removal and recovery of phosphorus as struvite from swine wastewater using microbial fuel cell. Bioresour Technol 2012;114:303–7.

[71] Huang H, Xu C, Zhang W. Removal of nutrients from piggery wastewater using struvite precipitation and pyrogenation technology. Bioresour Technol 2011;102:2523–8.

Chapter 18

Genotoxicity assays: the micronucleus test and the single-cell gel electrophoresis assay

Avelyno D'Costa, M.K. Praveen Kumar, S.K. Shyama
Department of Zoology, Goa University, Taleigao Plateau, Goa, India

18.1 Introduction

All organisms are constantly exposed to various factors that can cause damage not only at the cellular level but also at the level of DNA. These factors may either be chemical substances, radiation, dusts, biological toxins, etc., which may affect DNA directly or indirectly thereby resulting in genotoxic effects. Genetic toxicology is a branch of toxicology that deals with the studies on the genotoxic effects of substances, especially the induction of mutation by chemical means. It investigates the interaction of chemical and physical agents with genetic material, in relation to subsequent adverse effects, such as cancer (in case of alterations in somatic cells) or genetic disease in future generations (in case of alterations in germ cells). In other words, genetic toxicology is a branch of toxicology that identifies and analyzes the action of agents with toxicity directed toward the hereditary components of living systems. It can be widely implemented for identifying the genotoxic agents found in the environment, whose presence may alter the integrity of the gene pool of a wide range of organisms including human beings. It can also be used for the detection and mechanistic understanding of the action of carcinogenic agents. In vitro genotoxicity tests are usually used for screening the genotoxic potential of different substances such as pesticides, nanomaterials, pharmaceuticals, extracts, etc. and can be used as initial toxicity data prior to in vivo testing. If in vitro test results are positive, then in vivo genotoxicity testing can be done to ascertain if the substance causes toxicity in a living animal model. This is of great significance because in living systems, substances can get metabolized into more-harmful or even less-harmful forms and can get

Advances in Biological Science Research. https://doi.org/10.1016/B978-0-12-817497-5.00018-5

excreted out of the body with or without exerting toxic effects. The genotoxicity of substances may be due to their direct effects on the genome where the pollutant molecules directly interact with the nucleic acids or indirectly by interacting with other molecules such as water to give rise to highly reactive molecules. These molecules are called reactive oxygen species (ROS) and are implicated in the cellular oxidative stress within organisms. ROS include the hydroxyl radical ($^{\cdot}OH$), superoxide anion radical $\left(O_2^{\cdot -}\right)$, and hydrogen peroxide (H_2O_2). These chemical species are highly reactive and react with DNA, lipids, proteins, and carbohydrates, often leading to destructive effects [1]. These ROS can be produced under natural conditions by the normal basal metabolism or by the influence of environmental factors. Naturally produced ROS can easily be reduced by antioxidants that are produced by the body. However, ROS production mediated by environmental stressors build up to much higher levels and they can overcome the antioxidant defenses. Oxidative stress is therefore considered to be an imbalance between oxidants and antioxidants at the cellular level. Oxidative damage as a result of such an imbalance may cause oxidative modification of cellular macromolecules, such as the induction of cell death by apoptosis or necrosis, as well as structural tissue damage. As a result, these ROS can affect various cellular processes and organelles. DNA is one of the key cellular components that are highly susceptible to the action of ROS. More than 100 different products of DNA damage are known to be formed as a result of attack by ROS [2]. Hydroxyl radical attack on the sugar–phosphate backbone of DNA can cause different lesions, such as apurinic sites where the base has been removed, fragmentation of deoxyribose causing single-strand breaks, and oxidation of the sugar moiety [3–5].

In this chapter, we focus on the two commonly used genotoxicity tests in in vitro and in vivo laboratory testing: the micronucleus test and the comet assay (single-cell gel electrophoresis). These parameters are also routinely used as biomarkers for monitoring environmental pollution by genotoxic contaminants and can be combined with other physiological and biochemical biomarkers to fully assess the pollution status of various areas affected by pollution stress.

18.1.1 Micronucleus test

The micronucleus assay (MN assay) is a simple and sensitive assay for in vivo/in vitro evaluation of genotoxic properties of various agents. It is included in the Organization for Economic Co-Operation and Development (OECD) guidelines as an officially approved test along with the chromosomal aberration test, sister chromatid exchange assay, and bacterial reverse mutation test [6]. Schmid et al. [7] initially studied the formation of micronuclei (MNi) in bone marrow cells of Chinese hamsters. Since then, it is considered as one of the most rapid methods for screening genotoxic agents in various mammalian models. This test has also been adopted to study genotoxicity in aquatic organisms, such as fish, mollusks, and crustaceans, as an aspect of ecotoxicology.

Micronuclei are the smaller nuclei compared to the main nuclei of the cells. They arise in the mitotic cells from chromosomal fragments or chromosomes that lag behind in anaphase and are not integrated into the daughter nuclei. Small micronuclei are produced by the acentric chromosomal fragments, whereas the whole lagged chromosomes give rise to larger micronuclei. Micronuclei harboring chromosomal fragments result from direct DNA breakage, replication on a damaged DNA template, and inhibition of DNA synthesis. MNi harboring whole chromosomes is primarily formed from the failure of the mitotic spindle, kinetochore or other parts of the mitotic apparatus, or by damage to chromosomal substructures, alterations in cellular physiology, and mechanical disruption. Thus, an increase in the frequency of micronucleated cells is a biomarker of genotoxic effects that can reflect exposure to agents with clastogenic (chromosome breaking; DNA as target) or aneugenic (aneuploidogenic; effect on chromosome number; mostly non-DNA target) modes of action (Fig. 18.1).

In a study conducted by Borkotoky et al. [8], the micronucleus test was used to check the toxicity of a well-known drug, nimesulide, in Wistar rats in vivo. Different doses of the drug were selected and administered orally to the rats for a period of 14 days. A control group of rats administered with

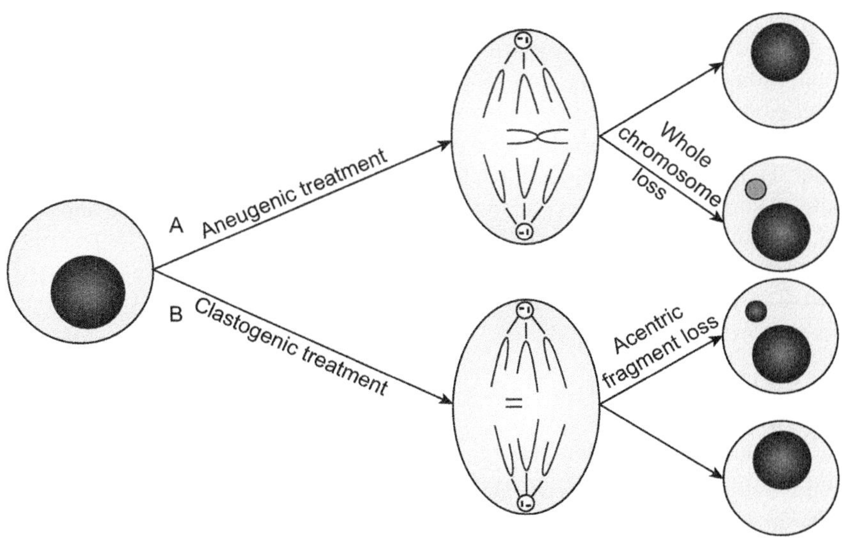

FIGURE 18.1 Mechanism of formation of micronuclei. *Source: Terradas et al. Genetic activities in micronuclei: is the DNA entrapped in micronuclei lost for the cell? Mutation Res 2010; 705:60–7.*

saline was also maintained in parallel in order to compare the potential toxicity of nimesulide. After the treatment period, the rats were sacrificed and bone marrow was withdrawn from the femur bones since the bone marrow cells divide at a rapid rate and can be used to screen micronucleated cells. It was observed that the rats treated with nimesulide exhibited a significant increase of micronucleated erythrocytes compared to the control group. Therefore, although widely used as a drug, nimesulide can potentially induce micronuclei in rats.

The micronucleus test is also extensively used in in vitro studies with cultured mammalian cells and human cells. In case of human cells, the cytokinesis-block micronucleus test is used because the data obtained are not confounded by altered cell division kinetics caused by cytotoxicity of agents tested or suboptimal cell culture conditions [9]. This technique involves the use of a cytokinesis inhibitor such as cytochalasin-B, which inhibits the assembly of the microfilaments. A number of studies are available on the toxicity of various substances in cultured mammalian cells using the in vitro micronucleus test (for review [10,11]). The in vitro micronucleus test can also be coupled with flow cytometry to allow for better precision in counting micronucleated cells as well as data analyses [12]. In case of in vivo studies in humans, minimally invasive procedures such as a sample of blood [13] or exfoliated buccal cells (for review [14]) may provide evidence for the correlation between micronucleated cells and the incidence of particular types of cancers.

In the mouse micronucleus test, the target cells are the bone marrow erythroblasts. Chemically induced micronuclei in the erythroblasts are retained in the erythrocytes after the extrusion of the main nuclei from the cells during maturation and can be scored in polychromatic erythrocytes (PCEs: young erythrocytes) and normochromatic erythrocytes (NCEs: mature erythrocytes). An increase in micronuclei in these erythrocytes (PCE + NCE) indicates genotoxicity of the test agent. Toxicity is also monitored by studies on bone marrow suppression. Bone marrow suppression is measured by the decrease of the ratio of PCE to NCE or to total erythrocytes/RBCs (PCE + NCE) in the bone marrow, which is commonly referred to as the PCE/NCE or PCE/RBC ratio. The PCE/NCE ratio is normally 1:1 in untreated condition. If there is an increase in the NCE population, it is a signal of cytotoxicity of a drug or chemical [15].

Bone marrow smears are prepared on glass micro slides, stained with May-Grunwald-Geimsa or acridine orange and scored for micronucleated erythrocytes. As an additional advantage for the micronucleus test, young erythrocytes (PCEs) stain differently from older forms (NCEs). For the duration of their adolescence, lasting approximately 24 h, they do not stain reddish pink as NCEs but bluish due to high RNA content. The micronuclei stain deep purple and are clearly distinguishable from the surrounding lightly stained cytoplasm.

The micronucleus test is also a very important biomarker for environmental pollution studies. A number of biomonitoring studies employ the micronucleus test to assess or quantify DNA damage as a result of pollutants in the environment. In case of organisms such as fish, mollusks, crustaceans, and birds, micronuclei can easily be identified in the red blood cells from peripheral blood, which makes this test ideal for ecotoxicological assessment of probable polluted areas. A small sample of blood is required, which is minimally invasive and can be obtained from a peripheral vein in case of fish and birds. In case of mollusks and crustaceans, the test can be conducted using the hemolymph or using the gill cells, which require the organisms to be sacrificed. A number of studies are available on the ecotoxicological significance of the micronucleus test in aquatic organisms [16–18] and birds [19,20].

18.1.2 Comet assay (single-cell gel electrophoresis)

The comet assay has a wide range of applications in clinical sciences as well as in biomonitoring studies. The comet assay is able to detect repairable DNA damage such as single- or double-strand DNA breakages, whereas the MN test detects more-persistent DNA damage, which is more difficult to repair [21,22]. Studies have shown that exposure to genotoxic agent leads to loss of DNA integrity through DNA strand breakage (SB), which can be used as a sensitive indicator of genotoxicity. Therefore techniques that can detect DNA damage in individual cells are needed. The single-cell gel electrophoresis or comet assay is a simple, rapid, noninvasive, visual, and sensitive technique for analyzing and quantifying DNA damage (induced by genotoxic agents even at low concentrations) in any eukaryotic cell, reflected as strand breaks under alkaline conditions. A damaged cell in the comet assay has the appearance of a "comet," with a brightly fluorescent intensity, and is related to the number of strand breaks present. Cells with undamaged DNA will appear as intact comet heads without tails after specific electrophoresis time (Fig. 18.2).

DNA damage as measured by the comet assay has been linked to a wide spectrum of genotoxic and cytotoxic compounds, such as PAHs and trace metals [23]. The amount of the DNA in the tail region (tail DNA) of a comet is commonly used for quantifying DNA strand breakage and represents the most reliable parameter [24,25].

The comet assay has been developed from the method of Rydberg and Johanson [26] who were the first to perform a quantification of DNA damage in single cells. Subsequently, Ostling and Johanson [27] improved the assay by developing an electrophoretic microgel technique under neutral conditions and stained the DNA with acridine orange. The image obtained looked like a "comet" with a distinct head, comprising of intact DNA and a tail, consisting of damaged or broken pieces of DNA, hence the name was given as "Comet" assay. The more versatile alkaline method of the comet assay was developed by Singh et al. [28]. This method was developed to measure low levels of

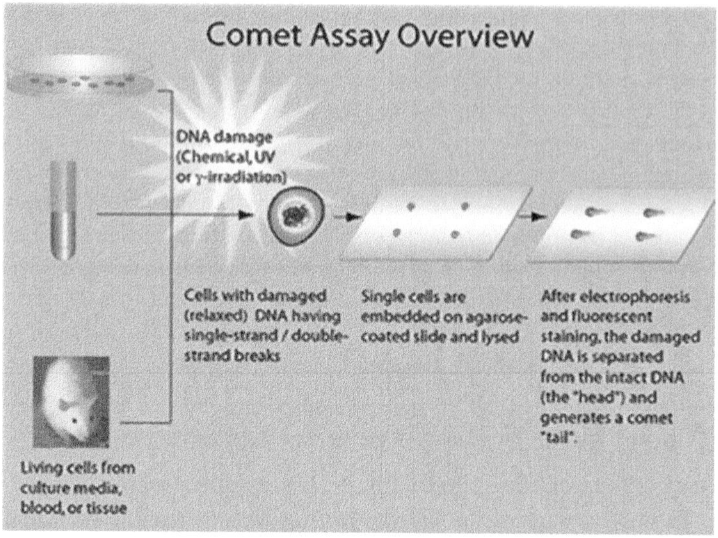

FIGURE 18.2 Overview of the comet assay. *Source: https://www.sigmaaldrich.com/life-science/cell-biology/cancer-research/learning-center/cancer-research-protocols/comet-assay.html.*

strand breaks with high sensitivity. This method is being widely used for measuring DNA damage in somatic cells in different organisms.

In comet assay, the cells are typically embedded in a thin agarose gel on a microscope slide. The cells are then subject to lysis by exposure to detergents and strong salt solutions to remove all cellular proteins, leaving behind the supercoiled DNA. The DNA is then allowed to unwind under alkaline/neutral conditions. Following unwinding, the DNA is electrophoresed, and broken DNA fragments (damaged DNA) or relaxed chromatin migrate away from the nucleus toward the positive electrode or anode and gives the appearance of a "comet." These comets can then be visualized using DNA-binding dyes such as ethidium bromide or SYBR green and observed under a fluorescence microscope. The extent of DNA liberated from the head of the comet is directly proportional to the DNA damage. DNA damage can then be analyzed by either a manual scoring method or with the aid of computer software. In the manual scoring method, comets can be analyzed by visual scoring by assigning a score ranging from 0 to 4, with 0 having no tail, 1 having a tail less than the diameter of the head, 2 having a tail almost equal to the diameter of the head, 3 having a tail more than the diameter of the head, and 4 having a tail more than twice the diameter of the head. In the computer software analyses, photographs of the comets have to be taken and loaded into the software such as CASP (casplabs.com) [29]. The software allows the quantification of various parameters besides the length of the tail such as % tail DNA, tail moment, and olive tail moment. These parameters are preferred over the manual scoring method due to ease of interpretation of data during statistical analyses.

Tail length: this is the distance of DNA migration from the body of the nuclear core and it is used to evaluate the extent of DNA damage.

Olive tail moment: this is defined as the product of the tail length and the fraction of total DNA in the tail. Tail moment incorporates a measure of both the smallest detectable size of migrating DNA (reflected in the comet tail length) and the number of relaxed/broken pieces (represented by the intensity of DNA in the tail):

$$Olive\ Tail\ Moment = \frac{Tail\ Mean - Head\ Mean}{100} \times \%\ Tail\ DNA$$

% Head DNA

$$= \frac{Head\ Optical\ Intensity}{Head\ Optical\ Density + Tail\ Optical\ Density} \times 100$$

% Tail DNA

$$= 100 - \%\ \text{Head DNA}$$

In a study conducted by Balasubramanium et al. [30], Wistar rats were used to check the toxicity of aluminum oxide nanoparticles in vivo. Rats were administered with different concentrations of these nanoparticles along with a control group. Peripheral blood was then taken from these rats and was used for the comet assay. They concluded that aluminum oxide nanoparticles indeed induced significant DNA strand breaks compared to the control.

Many studies employ the comet assay along with the micronucleus test in environmental monitoring. The procedure for obtaining the sample is the same as in the micronucleus test, from the peripheral blood or from other body tissues. The comet assay has been applied to a wide range of organisms from invertebrates to vertebrate systems [31]. In a majority of the ecotoxicological studies available, fish are the most employed organisms due to their interaction with the aquatic environment and can be used to determine the overall health of the environment. Fishes play a very important role as consumers in an aquatic ecosystem. Further, they exhibit the intake and accumulation of many of the pollutants of the aquatic environment and thereby contribute to their bioaccumulation/biomagnification through the food chain. In aquatic ecosystems, pollutants may accumulate in sediment as well as the organisms at different trophic levels in this food chain, including the benthic and pelagic animals. Fishes are the most predominant pelagic animals in an aquatic environment, representing the primary/secondary/tertiary consumers. Fishes being at the higher tropic level of aquatic food chain can readily accumulate a variety of contaminants by ingestion of smaller species, as well as through water intake. They also respond to mutagens at fairly low concentrations and are more sensitive to the induction of genetic damage. Man, being at the apex of this food chain, consumes varieties of fishes and bivalves as seafood and becomes the final recipient of these pollutants. Fishes are therefore the most popular animals used as biological detectors of genotoxic agents in the aquatic environment as they are sensitive indicator organisms, economically important food sources, higher order predators, capable of concentrating xenobiotics, and a major part of aquatic communities [32]. They act as "sentinel" organisms for indicating the potential for exposure of human populations to genotoxic chemicals in drinking water.

The comet assay is a highly sensitive method to detect low levels of DNA damage in vitro or in vivo. Cultured cells exposed to chemical agents can be subject to the comet assay to detect DNA strand breaks. As mentioned earlier, this may partly reflect the toxicity condition in an in vivo system.

The comet assay has also been modified over time to yield data that can give us information on the specific DNA lesions occurring in cells. The use of endonucleases such as formamidopyrimidine DNA glycosylase can detect lesions in the DNA, such as formamidopyrimidines, oxidized purines, and ring-opened N7 guanine adducts produced by alkylating agents, whereas and 8-oxo-guanine (8-oxoG) DNA glycosylase (OGG1) can detect oxidized purines and formamidopyrimidines [33–35]. The comet assay can also be combined with fluorescence in situ hybridization to assess the DNA repair process in defined sequences of DNA treated with genotoxic substances [36]. More modern usage of the comet assay has improved upon the time-consuming process, limited reproducibility, and error-prone analysis steps. Computational tools, such as HiComet, facilitate the analysis of high-throughput comet assay data, which can detect normal and damaged cells in a fully automated process and exclude the analysis of debris and artifacts that may confound data interpretation [37]. This method can therefore screen a large number of comets within a very small amount of time and can therefore be beneficial in clinical studies or ecotoxicological studies. Another high-throughput method, the EpiComet-Chip, detects DNA damage by methylation alterations and can be used to assess damage by epigenetic modes of action [38].

18.2 Conclusion

In conclusion, the micronucleus test and comet assay are simple, rapid, and powerful tests that can be used to test the toxicological profile of any substance or can be used as biomarkers for genotoxic agents in the environment. However, with rising ethical concerns over using animal models such as rats and mice in laboratories, toxicological testing is mostly being carried out in vitro using cell lines. It may also be noted that the behavior of potential genotoxicants may differ under in vivo conditions compared to those that are exposed in vitro, and thus in vitro experiments may not fully reflect the natural genotoxic potential of these substances. With regard to environmental monitoring, these two tests can be used in a noninvasive manner using peripheral blood from organisms exposed to environmental toxicants. This gives a real-time, baseline data of the "well-being" of the environment and can therefore enable researchers to conduct further studies of the potential stressors in the environment.

References

[1] Storey KB. Oxidative stress: animal adaptations in nature. Braz J Med Biol Res 1996;29:1715–33.

[2] Dizdaroglu M. Chemistry of free radical damage to DNA and nucleoproteins. In: Halliwell B, Aruoma OI, editors. DNA and free radicals. Chichester: Ellis Horwood, Ltd.; 1993. p. 19–39.

[3] Dizdaroglu M, Schulte-Frohlinde D, von Sonntag C. Strand breaks and sugar release by g-irradiation of DNA in aqueous solution. J Am Chem Soc 1975;97:2277–8.

[4] Breen AP, Murphy JA. Reactions of oxyl radicals with DNA. Free Radical Biol Med 1995;18:1033–77.

[5] Valavanidis A, Vlahogianni T, Dassenakis M, Scoullos M. Molecular biomarkers of oxidative stress in aquatic organisms in relation to toxic environmental pollutants. Ecotoxicol Environ Saf 2006;64:178–89.

[6] OECD. Test No. 474: mammalian erythrocyte micronucleus test. Paris: OECD Publishing; 2014. https://doi.org/10.1787/9789264224292-en.

[7] Schmid W. The Micronucleus test. Mutat Res 1975;31:9–15.

[8] Borkotoky D, Panda SK, Sahoo GR, Parija SC. Genotoxicity of nimesulide in Wistar rats. Drug Chem Toxicol 2014;37:178–83.

[9] Fenech M. Cytokinesis-block micronucleus cytome assay. Nat Protoc 2007;2:1084–104.

[10] Annangi B, Bonassi S, Marcos R, Hernández A. Biomonitoring of humans exposed to arsenic, chromium, nickel, vanadium, and complex mixtures of metals by using the micronucleus test in lymphocytes. Mutat Res-Rev Mutat 2016;770:140–61.

[11] Bolognesi C, Bruzzone M, Ceppi M, Kirsch-Volders M. The lymphocyte cytokinesis block micronucleus test in human populations occupationally exposed to vinyl chloride: a systematic review and meta-analysis. Mutat Res Rev Mutat Res 2017;774:1–11.

[12] Bemis JC, Bryce SM, Nern M, Raschke M, Sutter A. Best practices for application of attachment cells to in vitro micronucleus assessment by flow cytometry. Mutat Res Genet Toxicol Environ Mutagen 2016;795:51–9.

[13] Kadam SB, Shyama SK, Praveen Kumar MK, D'costa A, Almeida VG. Cytogenetic analysis on the yields of chromosomal aberrations induced by the scattered doses of γ-radiation. Nucl Med Radiat Ther 2016;7:1–8.

[14] Bolognesi C, Bonassi S, Knasmueller S, Fenech M, Bruzzone M, Lando C, Ceppi M. Clinical application of micronucleus test in exfoliated buccal cells: a systematic review and metanalysis. Mutat Res Rev Mutat Res 2015;766:20–31.

[15] Adler ID. Cytogenetic tests in mammals. In: Venitt S, Parry JM, editors. Mutagenicity testing- A practical approach. Oxford: IRL Press Ltd.; 1984. p. 275–306.

[16] Hussain B, Sultana T, Sultana S, Masoud MS, Ahmed Z, Mahboob S. Fish eco-genotoxicology: comet and micronucleus assay in fish erythrocytes as in situ biomarker of freshwater pollution. Saudi J Biol Sci 2018;25:393–8.

[17] Butrimavičienė L, Baršienė J, Greiciūnaitė J, et al. Environmental genotoxicity and risk assessment in the Gulf of Riga (Baltic Sea) using fish, bivalves, and crustaceans. Environ Sci Pollut Res 2018;25:24818–28.

[18] Bolognesi C, Hayashi M. Review: micronucleus assay in aquatic animals. Mutagenesis 2011;26:205–13.

[19] Quero AAM, Ferré DM, Zarco A, et al. Erythrocyte micronucleus cytome assay of 17 wild bird species from the central Monte desert, Argentina. Environ Sci Pollut Res 2016;23:25224–31.

[20] Souto HN, de Campos Júnior EO, Campos CF, et al. Biomonitoring birds: the use of a micronuclei test as a tool to assess environmental pollutants on coffee farms in southeast Brazil. Environ Sci Pollut Res 2018;25:24084–92.

[21] Hartmann A, Elhajouji A, Kiskinis E, Poetter F, Martus HJ, Fjällman A, Frieauff W, Suter W. Use of the alkaline assay for industrial genotoxicity screening: comparative investigation with the micronucleus test. Food Chem Toxicol 2001;39:843–58.

[22] Klobucar GI, Pavlica M, Erben R, Papes D. Application of the micronucleus test and comet assay to mussel *Dreissena polymorpha* haemocytes for genotoxicity monitoring of freshwater environments. Aquat Toxicol 2003;64:15–23.

[23] Lee RF, Steinert S. Use of the single cell gel electrophoresis/comet assay for detecting DNA damage in aquatic (marine and freshwater) animals. Mutat Res 2003;544:43–64.

[24] Mitchelmore CL, Chipman JK. DNA strand breakage in aquatic organisms and the potential value of the comet assay in environmental monitoring. Mutat Res 1998;399:135–47.

[25] Kumaravel TS, Jha AN. Reliable Comet Assay measurements for detecting DNA damage induced by ionising radiation and chemicals. Mutat Res Genet Toxicol 2006;605:7–16.

[26] Rydberg B, Johanson KJ. Estimation of DNA strand breaks in single mammalian cells. In: Hanawalt PC, Friedberg EC, Fox CF, editors. DNA repair mechanisms. New York: Academic Press; 1978. p. 465–8.

[27] Ostling O, Johanson KJ. Microelectrophoretic study of radiation –induced DNA damages in individual mammalian cells. BiochemBiophys Res Commun 1984;123:291–8.

[28] Singh NP, McCoy MT, Tice RR, Schneider EL. A simple technique for quantitation of low levels of DNA damage in individual cells. Exp Cell Res 1988;175:184–91.

[29] Konca K, Lankoff A, Banasik A, Lisowska H, Kuszewski T, Góźdz S, Koza Z, Wojcik A. A cross-platform public domain PC image-analysis program for the comet assay. Mutat Res-Gen Tox En 2003;534:15–20.

[30] Balasubramanyam A, Sailaja N, Mahboob M, Rahman MF, Hussain SM, Grover P. In vivo genotoxicity assessment of aluminium oxide nanomaterials in rat peripheral blood cells using the comet assay and micronucleus test. Mutagenesis May 2009;24:245–51.

[31] de Lapuente J, Lourenço J, Mendo SA, Borràs M, Martins MG, Costa PM, Pacheco M. The Comet Assay and its applications in the field of ecotoxicology: a mature tool that continues to expand its perspectives. Front Genet 2015;6:180.

[32] Kligerman AD. Fishes as biological detectors of the effects of genotoxic agents. In: Heddle JA, editor. Mutagenicity: new horizons in genetic toxicology. New York: Academic Press; 1982. p. 435–53.

[33] Collins AR. The comet assay for DNA damage and repair: principles, applications, and limitations. Mol Biotechnol 2004;26:249.

[34] Piperakis SM. Comet assay: a brief history. Cell Biol Toxicol 2009;25:1.

[35] Gunasekarana V, Raj GV, Chand P. A comprehensive review on clinical applications of comet assay. J Clin Diagn Res 2015;9:GE01–5.

[36] Spivak G, Cox RA, Hanawalt PC. New applications of the comet assay: comet-FISH and transcription-coupled DNA repair. Mutat Res 2009;681:44–50.

[37] Lee T, Lee S, Sim WY, Jung YM, Han S, Won J, Min H, Yoon S. HiComet: a high-throughput comet analysis tool for large-scale DNA damage assessment. BMC Bioinf 2018;19:44.

[38] Townsend TA, Parrish MC, Engelward BP, Manjanatha MG. The development and validation of EpiComet-Chip, a modified high-throughput comet assay for the assessment of DNA methylation status. Environ Mol Mutagen 2017;58:508–21.

Chapter 19

Advances in methods and practices of ectomycorrhizal research

Lakshangy S. Charya, Sandeep Garg
Department of Microbiology, Goa University, Taleigao Plateau, Goa, India

19.1 Introduction

Ectomycorrhizae (ECM), also termed as ectotrophic mycorrhizae, are the second most predominant type of mycorrhiza found in nature. A typical ectomycorrhizal root shows formation of characteristic features such as "mantle" and "hartig net." The fungal mantle, also known as the "sheath," is the clustering of fungal mycelia on the surface of the host root and the hartig net is the network of fungal hyphae formed between the epidermal or cortical cells of the host root without penetrating the root cells. From the mantle surface arise "rhizomorphs" that are hyphal strands interwoven to form bundles and these structures spread in the surrounding soil. In tropical forests, rhizomorph development can be prolific, sometimes traveling several meters away from the host root [1,2].

The ECM occur on about 6000 plant species that mainly include woody plants such as gymnosperms, angiosperms, and certain lower land plants such as hornworts and liverworts [3]. ECM association is found in most of the coniferous trees, including the Pinaceae family, in which all the species essentially form ECM. Other plant families commonly associating with ECM fungi include Betulaceae, Dipterocarpaceae, Fagaceae, Juglandaceae, Myrtaceae, and Salicaceae [2,4–6]. Over 20,000 fungal species are known to form ectomycorrhizae. The majority of ECM synthesizing fungi belong to classes Basidiomycetes and Ascomycetes that form fruiting bodies, like mushrooms, puffballs, coral fungi, toadstools, truffles, etc. [2,3]. The ECM fungi are classified based on their host range and the stage of the plant on which they appear. Some fungi have a narrow plant host range, such as *Boletus betulicola* forms mycorrhizae only with *Betula* species. Some fungi have a broad range, such as *Pisolithus tinctorius* that forms mycorrhiza with

Advances in Biological Science Research. https://doi.org/10.1016/B978-0-12-817497-5.00019-7

more than 46 tree species belonging to 20 genera [7]. Broadly, ECM fungi are grouped as early stage and late stage according to their colonization on young roots (plants of age <5 years) and older root systems (plants of age >5 years), respectively [8]. The carbon requirement of late-stage fungi is higher than the early stage ECM fungi [9]. Early stage fungi are very important for reclamation and forestry purposes as young seedlings considered for reclamation and forestry applications can be easily established on a large scale in the nursery with the inoculation of the early stage ECM fungi. In older plantations, with the change in the canopy and nutritional status of soil, the late stage fungi take over [9].

19.2 Benefits of ECM association

The vital success of this symbiosis is the exchange of nutrients between the symbionts. The plant provides carbon and essential vitamins to the fungus and in return the fungus provides inorganic nutrients to the host, thus creating an essential link between the two partners [1]. ECM particularly improves the uptake of nutrients present with low mobility in the soil, e.g., phosphorus [10]. ECM fungi assist plants through augmenting hydraulic conductivity, tolerance to drought, and resistance to soil-borne pathogens [1,11−15]. Extrametrical hyphae of ECM fungi form an extensive hyphal network, which liberates various chemicals in the soil. This holds the soil particles together and helps to improve the quality of soil [16].

Further benefit comes to the host tree by increased tolerances to extreme conditions of soil temperature, pH, and high concentrations of heavy metals [17−24]. Thus, an ECM plant grows and survives in low fertility or denuded soil. ECM fungi are also known to produce various growth-promoting substances, organic acids, antibiotics, and fatty acids [25,26]. Ectomycorrhizae can enhance the resistance of a plant to soil-borne pathogens by exhibiting different protective measures [1,27−29]. Through these different mechanisms, a mycorrhizal plant is benefitted by increased growth, improved fitness, and better survival.

Plants live in communities and their roots are connected by common mycelia of mycorrhizal fungi. Fungal hyphae can be connected to more than one plant of the same or different species. Plants can host different mycorrhizal fungi of the same or different species [30]. This helps the host plant and fungi to share carbon, nitrogen, phosphorus, and other resources within the community (Fig. 19.1). Ecologically, ECM fungi play an important role in the stability of forest ecosystems.

Besides being beneficial to plants, many ECM fungi are a source of food for humans and animals, e.g., *Agaricus bisporus*, *Boletus edulis*, *Lactarius deliciosus*, *Pleurotus* spp., *Tuber* spp., thus contributing to the economy of many human communities. Edible mushrooms are a high-value, nontimber

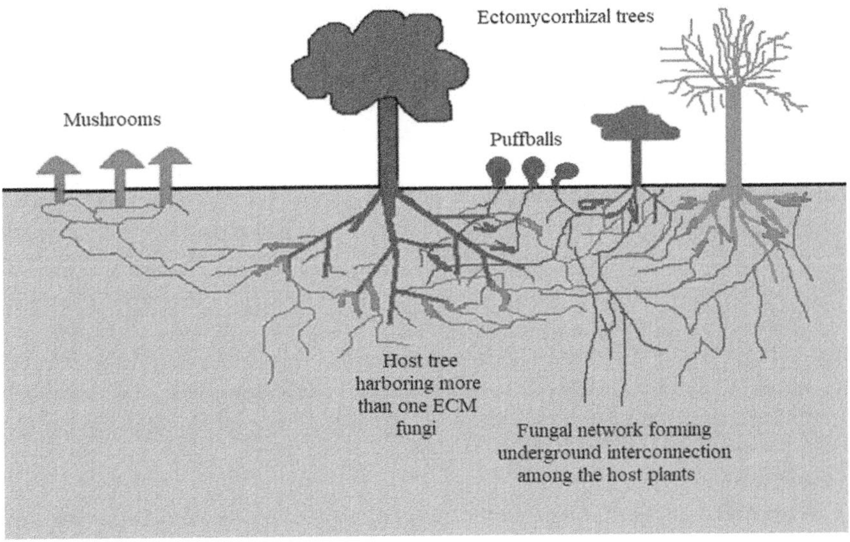

FIGURE 19.1 The importance of belowground part of ECM association. *Source: Chari L. Studies on* Pisolithus *sp.: stress response, pigment production and mycorrhization with forestry trees of Goa [Ph.D. thesis]. Goa: Goa University; 2012.*

forest product of increasing commercial importance in international markets. Some mushrooms and polypores are used for making dyes, e.g., *P. tinctorius* yields tan to gold dye [31].

19.3 Cultivation and physiology of ECM fungi

Unlike endomycorrhizal fungi, ECM fungi are neither intracellular nor require the presence of a host for cultivation in the laboratory [3,31]. These features make it possible to use it on a large scale. The requirement is to properly understand the physiology of the fungus so that the cultivation procedures can be manipulated. A detailed study is thus required to optimize the culture parameters for the promising ECM fungi.

19.3.1 Cultivation media for ECM fungi

Axenic culture of ECM fungi can be established in vitro from the spores, sporocarp tissues, or mycorrhizae. Usually, Hagems (Modess), modified Melin Norkrans (MMN), modified Pridham–Gottlieb, and potato dextrose agar media were routinely used by the researchers to establish in vitro culture of these fungi or investigate the physiology [4,32–34]. All the media contain one or more complex organic component. Rossi and Oliveira [35] reported modified Pridham–Gottlieb medium was suitable for accumulation of maximum biomass of *Pisolithus microcarpus* (Cooke and Massee) G. Cunn when they carried out a factorial assay using glucose, peptone, and yeast extract as variable components in the medium used for the production of biomass; the results obtained suggest that to achieve higher biomass production, glucose concentration can go up to 40% [35]. Moreover, yeast extract can be completely omitted and peptone could serve as the sole source of nitrogen. Hence, MMN could serve as a very good medium for isolation and maintenance of ECM culture. In the past, MMN was the most commonly used media for ECM research, and the majority of in vitro physiological studies on ECM fungi have been conducted using MMN. However, it is not possible to examine physiological characteristics using the complex organic medium as biochemical details are not revealed completely. Garg and team [36] modified the mineral medium using 2% glucose as sole carbon source. The medium supported a rapid luxuriant growth of ECM fungi and is very much suitable for physiological studies [37]. The compositions of all the different media are summarized in Table 19.1.

A relatively large number of studies on physiological growth parameters of ECM fungi using MMN medium are reported. The growth response of a diverse array of ECM fungal species has clearly indicated inter- and intraspecific variation. The effects of temperature [39]; pH [40]; saline condition [41–43]; heavy metal tolerance for metal species such as aluminum (Al), cadmium (Cd), chromium (Cr), copper (Cu), iron (Fe), lead (Pb), manganese (Mn), mercury (Hg), nickel (Ni), and zinc (Zn) [18,42,44]; and water potential induced using potassium chloride, sodium chloride, sucrose, or polyethylene glycol [15,45–49] were evaluated in vitro. Though largely the studies have been carried out in MMN medium, an organic medium, the ideal medium for such studies would be GMSM, a defined medium, that supported maximal growth of ECM fungi [36,37].

19.3.2 Isolation methods of ECM fungi

The fruiting body is the common source for isolation of the ECM fungus. Basidiocarps in mycorrhizal association with the host plant are collected and washed with sterile water. Sporocarps are then surface sterilized with sodium hypochlorite solution containing 2% (v/v) of active chlorine, cut open aseptically using a sterile scalpel. Inner tissue of mushroom is picked using sterile forceps and placed onto the modified Melin Norkran's agar medium (MMN, pH 6.5) and plates incubated at room temperature (RT). It is advisable not to pass the scalpel through entire basidiocarp, rather just give a sufficient cut and then pull apart. This would minimize contamination. Pure fungal mycelium obtained is maintained by periodic transfer on MMN and glucose mineral salt medium (GMSM, pH 6.0). The culture is maintained at RT and subcultured after every 45 days. Though the ECM fungi do wonders in harsh environmental conditions,

TABLE 19.1 Media used to cultivate ECM fungi.

Medium component	MMN [11]	Hagem (modess) [34]	MPG [32]	KHO [38]	GMSM [36]
Macro-nutrient (g L^{-1})					
Glucose		5.00	30.00	20.00	20.00
Sucrose	10.00			5.00	
Casein hydrolysate					
Peptone			10.00		
Malt extract	3.00	5.00			
Yeast extract			2.00		
NH_4Cl		0.50		0.50	1.00
$(NH_4)_2HPO_4$	0.25				
$(NH_4)NO_3$			3.00		
$(NH_4)_2C_4H_4O_6$					
KH_2PO_4	0.50	0.50	2.38	0.50	1.00
K_2HPO_4			5.65		
KCl					0.10
$MgCl_2 \cdot 4H_2O$					0.30
$MgSO_4 \cdot 7H_2O$	0.15	0.50	1.00	0.50	
$CaCl_2$	0.05				0.132
NaCl	0.025				0.10
Na_2EDTA					0.03752
$FeCl_3$ (1% soln.)	1.2 mL	1.2 mL			
$FeSO_4 \cdot 7H_2O$			0.0011		0.01053
$Fe_2(SO_4)_3$				0.05	
$FeC_6H_5O_6 \cdot 3H_2O$					
$Fe(NO_3)_3 \cdot 9H_2O$					
Micro-nutrient (μg L^{-1})					
$CuSO_4 \cdot 5H_2O$			6400		97
H_3BO_3					2784
$MnCl_2 \cdot 4H_2O$			1900		

Continued

TABLE 19.1 Media used to cultivate ECM fungi.—cont'd

Medium component	MMN [11]	Hagem (modess) [34]	MPG [32]	KHO [38]	GMSM [36]
$MnSO_4 \cdot H_2O$					3380
$Na_2MoO_4 \cdot 2H_2O$					338.46
$ZnSO_4 \cdot 7H_2O$			1500		201
Acid (μL L^{-1})					
H_2SO_4					240
Vitamins (μg L^{-1})					
Thiamine. HCl	100	50			100
Biotin					40
pH	5.50	4.60–4.80	4.00/ 6.00	5.50	6.0

Source: Garg S. Production of ectomycorrhizal fungal inoculum by submerged fermentation [Ph.D. thesis]. New Delhi: University of Delhi; 1999; Chari L. Studies on *Pisolithus* sp.: stress response, pigment production and mycorrhization with forestry trees of Goa [Ph.D. thesis]. Goa: Goa University; 2012.

the vegetative mycelium is fragile and very sensitive. Hence, mycelia plugs are used to transfer the live culture from one medium to another.

During isolation of the ECM fungi, contaminant bacteria, yeast, or fungi can often show up as coculture on the isolation medium that could serve as a biological stimulant or spore activator of the ECM fungi in the soil [50]. Fungus *Tritirachium roseum* and bacteria such as *Micrococcus roseus*, *Pseudomonas stutzeri*, and *Corynebacterium* spp. obtained from sporophores of *Hebeloma crustuliniforme* were seen to stimulate germination of the spore of that fungus.

19.4 Identification methods of ECM fungi

19.4.1 Conventional methods

The conventional method of ECM fungal identification involves noting the morphological characteristics of mushrooms such as their size, color, presence or absence of volva, stipe, ring, scales, reticulum, zonation, striation, warts, cap, areolae, and gills. Transverse sections of the sporocarps are prepared. Mycelial growth and morphology of dried spores are examined microscopically. Fungal hyphae are stained with lactophenol cotton blue solution and observed under microscope and images captured (Olympus BX51 compound microscope attached with Olympus DP71 digital camera). Basidiospore samples can be prepared for scanning electron microscopy using the method described by De Melo and Faull [51] to reveal the spore size and morphology.

19.4.2 Case study

The following description is of sporocarps of *Pisolithus* sp. collected from the vicinity of *Acacia mangium* from iron ore mine at Codli, Goa [37]. Sporocarps of *Pisolithus* sp. were of 1.0–11.5 cm in diameter, rounded or club-shaped, yellowish smooth shiny surface, with deeply rooted fibrous base (Fig 19.2A). A transverse section reveals yellowish brown spore sacs (peridioles) developing in a black gelatinous matrix (Fig 19.2B). Mature dry spores were of 7.5–8.5 μm in diameter, cinnamon brown, globose, and bearing triangular-shaped flattened curved spines (platelet spines) that were seen under a scanning electron microscope (Fig 19.2C). Vegetative growth of *Pisolithus* sp. PT1 isolate consists of conspicuous aerial mycelium emerging from a golden yellowish cottony growth with a dull yellow appearance on the leading edge (Fig 19.2D). After the isolate grew maximally, the colony changed to a tan color. Mycelia exuded brown pigment that colored the medium from yellowish (GMSM) to brownish (MMN). Moreover, PT1 colonies seldom exuded brown pigmented droplets from the aerial hyphae (Fig 19.2F). The hyphae grew interlaced and numerous clamp connections occurred throughout the 20-day-old culture (Fig. 19.2E).

FIGURE 19.2 Morphological and colonial characteristics of *Pisolithus tinctorius* PT1. (A) Fruiting bodies, (B) Transverse section of sporocarp, (C) Scanning electron micrograph of basidiospores, (D) Vegetative mycelium growing on MMN medium, (E) Hyphae with clamp connections, and (F) Pigment exuded from aerial hyphae on GMSM.

19.4.3 Challenges in the identification of ECM

Staining the vegetative mycelium of ECM fungi with lactophenol cotton blue was challenging as the fungus was very sensitive and soon formed a clump when we tried to pick it with a nichrome loop. So we autoclaved the Petri plates with the glass slide stuck to the lid with autoclavable tape. After the medium was poured, the culture was grown and when the aerial filamentous growth touched the slide, the slide with fungal hyphae was taken out, stained, and observed under a compound microscope. So with this method, the individual hypha could be easily stained without carrying out teasing exercise and thus microscopic details such as the presence of septum, the formation of clamp connection, etc. could be easily revealed.

19.4.4 Advances in identification of ECM

Though the morphological identification of ECM fungi is fairly good enough, molecular identification is equally essential. For example *Pisolithus* sp. can be identified by above-mentioned morphological features, however, the fungi exhibit conspecificity, for which identification has to be confirmed by molecular methods [7,51]. Molecular identification of the ECM fungi is carried out using sequence data of the ITS region of the nuclear ribosomal DNA [3]. The methods can be modified and optimized to get better results. The ITS region of rDNA was amplified using the primer pair ITS 1 and ITS 4 [52], and compared with the sequences available in the public nucleotide databases at the National Center for Biotechnology Information (NCBI) using its world wide web site (http://www.ncbi.nlm.nih.gov/entrez), and the BLAST (basic local alignment search tool) algorithm. Sequence alignment can be done using the multiple sequence alignment software ClustalX2 and a neighbor-joining (NJ) tree drawn using MEGA4. The sequence obtained is ideally deposited in GenBank.

Arai et al. [53] examined ECM fungal communities associated with *Quercus dentata* in a coastal broadleaf forest. The results obtained showed an association of five ECM fungi taxa (*Tomentella* sp., *Russula* spp., *Tricholoma* sp., *Hebeloma* sp., and Boletales sp.) identified by morphological characteristics and DNA sequencing of the root tips.

19.5 Assessment and quantification of ECM

A variety of methods have been used to examine ECM associations. Primarily root systems of target plant species are excavated, taking care that fine roots are well represented in samples and to exclude entangled roots of other species. Once separated from the soil, the root samples are cleaned with water over a fine sieve to ensure that finest laterals are not lost. Fresh roots are assessed with dissecting microscope for external features. Unstained ECM roots can usually be distinguished from nonmycorrhizal roots by differences in their color, thickness, texture, and branching patterns [54]. The characteristics of ECM associations are typically governed by mycorrhizal fungus and the host plant. *Pisolithus* sp. forms a prominent golden yellowish mycorrhizal association with the host plant, which is easy to identify by visual inspection. *Pinus radiata* roots show dichotomously branched root tips with several ECM fungi.

19.5.1 Conventional methods of assessment and quantification of ECM

Internal features of the root such as the presence of mantle and hartig net confirm the presence of ECM association. Sectioning and staining of the ECM roots are necessary to visualize the Hartig net in the root cortex [55,56]. Fresh roots are chopped into 2–4 cm long segments. Root fragments are cleared by heating in 10% KOH and stained with Trypan Blue (0.05%) or Chlorazol Black E (CBE-0.03%) in lactoglycerol. Structural details of the stained specimen are observed using compound microscopy and three-dimensional details are revealed using electron microscopy [54].

Trees harboring ECM exhibit modification in the branching pattern of laterate roots. This pattern shows the presence of short mycorrhizal lateral roots (short roots) along with the network of long roots, and this is known as heterorhizy. The mycorrhizal short roots are morphologically distinct from the other lateral roots in being thicker due to the presence of the mantle and hartig net. To exemplify, the ectomycorrhizal pine short roots could be unbranched, dichotomously branched, sparsely branched, or show a dense cluster of branches, whereas in eucalyptus mycorrhizal roots could show pinnate branching or even tuberculate mycorrhiza. The intensity of the ECM association and their numbers should be expressed relative to root length and soil volume. The root length of a sample can be measured with the gridline intersect method [54]. The ratio of the mycorrhizal roots and total roots give the percentage of root length colonized by mycorrhizal fungi. For certain mycorrhizae, the intensity of branching within a mycorrhizal cluster varies considerably and can be quantified with a branching density index. Branching density intensity is calculated as the ratio of the number of root tips and the total root length [57]. ECM associations with minimal changes to the host roots, e.g., *Eucalyptus*, can be recognized with practice but require more careful examination of roots [54].

ECM extrametrical mycelium plays a very important role in scavenging nutrients and water from the soil. The extent of extrametrical mycelium provides physiological competitiveness to the ECM. The ECM extrametrical biomass estimates in soils are usually obtained by measuring total hyphal length or by measuring the amount of fungal-specific biomarkers such as ergosterol and phospholipid fatty acids (18:26,9) [58–60]. These methods have one major drawback; although it is possible to estimate total fungal biomass present in the soil, they cannot differentiate the species. The estimates of particular EM fungus can be determined in the control conditions, i.e., only in the laboratory [60,61].

19.5.2 Molecular tools of assessment and quantification of ECM

For a long period the knowledge of the microbial populations in the rhizosphere has been limited, because they have always been studied by traditional culture-based techniques. These methods, which only allow the study of cultured microorganisms, do not allow the characterization of most organisms existing in nature. The introduction in the last few years of methodologies that are independent of culture techniques has bypassed this limitation. This, together with the development of high-throughput molecular tools, has given new insights into the biology, evolution, and biodiversity of mycorrhizal associations, as well as the molecular dialogue between plants and fungi. The genomes of many mycorrhizal fungal species have been sequenced so far, allowing to better understanding the lifestyle of these fungi, their sexual reproduction modalities, and metabolic functions [62].

19.5.2.1 Nucleic acid–based molecular methods

Landeweert et al. [63] demonstrated the possible use of molecular methods such as denaturing gradient gel electrophoresis (DGGE), restriction fragment length polymorphism (RFLP), competitive polymerase chain reaction (PCR), real-time quantitative PCR, and cloning-sequencing to identify and quantify EM mycelia from the soil. Such molecular techniques enable the use of genes as biomarkers and facilitate identification of particular fungi directly from a mixed population environment as in soil. The three molecular methods, i.e., DGGE, a clone, and real-time quantitative PCR, showed consistent results and enabled identification and relative quantification of two ECM fungi, *Suillus bovines* and *Paxillus involutus*, in mixed-species environment [63]. These methods clearly revealed the relative changes in fungal biomass of the two ECM species over a period. The other promising molecular techniques employed for the study of ECM ecology include temperature gradient gel electrophoresis, terminal restriction fragment length polymorphism, ribosomal intergenic spacer analysis (RISA), and automated RISA [64]. Recently, use of stable isotope labeling technique and RNA-sequencing approaches have been used to reveal nutrient channeling between the ECM partners, and this has been exemplified in three *Pisolithus* isolates and their host plant *Eucalyptus grandis* [65]. Recently, De La Varga et al. [66] carried out quantification of extraradical mycelia and ECM of *Bovis edulis* associated with *Pinus sylvestris*, by using real-time polymerase chain reaction using specific primers and a TaqMan probe to correlate the sporocarp productivity with the abundance of the ECM and soil mycelium.

19.5.2.2 Transcriptome analysis

The recent research on ECM focuses mostly on the molecular basis of the association. Transcriptome analysis is one of the important molecular tools in understanding ECM development [67]. In ECM association the two partners interact and undergo various developmental transitions that include metabolic changes and a substantial transcriptome reprogramming correlating the differential gene expressions. Daguerre et al. [67] carried out genome-wide expression profiling and functional analysis of the transcription factors to reveal their role in ECM development. It was evident from the results obtained that the ECM fungus *Laccaria bicolor* associated with poplar and Douglas fir differentially expressed ECM-specific transcription factors.

The first attempt was made by Heller et al. [68] to characterize the transcriptome of the plant partner in ECM symbiotic association. The model system used was the *P. sylvestri–L. bicolor*. The results highlighted the importance of different molecular events during the development of a fully functional symbiosis in a gymnosperm host. In a study evaluating the transcriptomic responses of Masson pine under different levels of drought conditions gave insight into the molecular mechanism of adaptation of the host to the drought stress. The results clearly indicated that several transcription factors had a key role in overcoming the drought stress [69]. The transcript profiling also revealed the possible role of ECM symbiont *Cenococcum geophilum* in water and nutrient transport. The results showed significant upregulation of membrane transporters (sugar transporters and aquaporin water channels), and mycorrhiza-induced small secreted proteins in ECM as that of free-living mycelium [70].

19.5.2.3 Proteomic analysis

In a recent study, the proteomic analysis revealed the different proteins involved with symbiotic germination of *Gastrodia elata*, a mycoheterotrophic orchid. The results could conclusively state that the metabolic change and defensive reaction would probably disrupt the balance between *Mycena*, a mycorrhizal fungus, and *G. elata* during mycorrhizal symbiotic germination [71].

19.6 Stress response and pigments/phenolics in ECM fungi

Mycorrhizal fungi are exposed to all or many of the environmental stresses that other fungi may experience. These include extremes of temperature and pH, anoxia, water stress, physical fragmentation, toxic metals, and other pollutants, as well as anthropogenic stresses arising from applications of fertilizers, lime, and wood ash. Hence, stress response in ECM fungi has attracted the attention of various researchers. Fungi can respond to these stresses by altering their morphology, modifying their external environment, or adapting their internal metabolism [72]. When any stress is applied, the most prominent physiological reactions are the production of a set of novel proteins or an increase in the number of certain types of existing proteins. Stress proteins or heat shock proteins (sHSPs) synthesized could be the high-molecular size (69–120 kDa), medium-molecular size (39–68 kDa), or low-molecular size (<38 kDa). The sHSPs ranging from 12 to 42 KDa are synthesized ubiquitously in eukaryotic and prokaryotic cells in response to heat and other stresses [23]. Synthesis and accumulation of protective organic compounds, such as sucrose, glycogen, and trehalose, are one of the tolerance mechanisms found in fungi to live in adverse environmental conditions [73]. Ferreira et al. [24] have demonstrated an accumulation of intracellular trehalose and an increase in trehalase activity in the mycelium of *Pisolithus* sp. RV82 in response to heat shock at 42°C. The methodology employed is protein profile of ECM fungus in optimum and stressed conditions that are carried by sodium dodecyl sulfate - polyacrylamide gel electrophoresis (SDS-PAGE) followed by protein staining by silver staining. Better resolution can be achieved by carrying out 2D electrophoresis; Isoelectric focusing (IEF) followed by SDS-PAGE. Recently, Ramesh et al. [74], characterized two metallothionine (MTs) genes, Hc*MT1* and Hc*MT2*, from the ECM fungus *Hebeloma cylindrosporum* under metal stress by performing competitive reverse transcription-PCR analysis. They observed that ECM fungi encode different MTs with particular expression pattern that could help the host plant to survive and grow in heavy metals contaminated ecosystems. The authors have identified a heat-shock protein gene from *L. bicolor* (Lbhsp) that could probably have a significant role in the establishment of functional ECM. The methodology included characterizing this gene and gene product is cDNA cloning, sequencing, and phylogenetic analysis, Northern and Southern analysis. The results suggest that Lbhsp mRNA is about 611 nucleotides long and codes for ~17 kDa protein and could play a key role in Ras-mediated mycorrhizal signaling pathway [73]. Metabolome profiling reveals the formation of stress-related metabolites like amino acids [75].

ECM fungi produce various pigmented compounds. Pigments in higher fungi are well documented [76,77]. The fruiting bodies of some macrofungi elaborate hues ranging from beautiful shades of pink, blue, yellow, red, and brown (e.g., *Boletus*, *Cortinarius*, *Hypholoma*, *Hydnellum*, *Hygrocybe*, spp.) [78]. Several varieties of *Dermocybe* species yield pink, salmon, and red dyes, *Gymnopilus spectabilis* yields butter yellow dye, and *Omphalotus olivascens* yields gray, green, and purple dyes [31].

The chromophores of mushroom dyes contain a variety of fascinating organic compounds. Their pigmentation may vary with the age and some undergo distinctive color changes on bruising. Many of the pigments of higher fungi are quinones or similar conjugated structures. Recently, the presence of naphthoquinone, a naphthalenoid pulvinic acid (pisoquinone), has been demonstrated in a white-skinned variant of the ectomycorrhizal gastromycete *Pisolithus arhizus* [79]. Phenolic compounds produced by ECM fungi are antimicrobial and involved in the protection against invasion by pathogens and competitive organisms. Antimicrobial phenolic compounds are synthesized in fruiting bodies, soil hyphae, and the mantle in ECM [80]. This probably explains host root protection that has been observed in ECM symbiosis. Experimentation is usually carried out by testing the antagonistic activity of cell-free culture broth or pigments extracted from fruiting bodies against phytopathogenic cultures by disc diffusion or bore well method [29].

19.7 Application in forestry: ECM fungi as bioinoculants

The biological requirement of many species of forest trees for ECM associations is mainly for afforestation and reforestation program for utilizing the cutover lands and other treeless areas. Moreover, soils deficient in native ECM fungi or having low species diversity, and the attempts of introducing exotic tree species in such soils, require essentially ECM association [81].

Although ectomycorrhizal fungi are naturally present in various types of soils, there is a lot of variation in their ability to colonize and benefit plants [82]. It is therefore advantageous to inoculate the seedling with specific ECM fungus that would be beneficial within the specific environment. This so-called "mycorrhization control" is usually done by planting seedlings that have been previously inoculated in the nursery with the chosen fungal strain. This practice improves not only the survival of the seedlings upon transplanting but also their subsequent growth [82,83]. The economic benefits of this practice, in terms of increased productivity, have been demonstrated in plantations in the United States and in France [82,84]. However, in order for mycorrhizal inoculation to become a routine practice in nurseries, it is necessary to establish methods for inoculant production at industrial scale. Efforts have been made to develop various inoculum formulations for the application. The selection of ectomycorrhizal fungal isolate is primarily based on the efficiency and the compatibility of the fungus. In the nursery, these studies can be performed in periods up to 6 months. However, under field conditions, it is necessary to follow plant development for several years [84].

The ideal ECM inoculant must contain sufficient numbers of infective propagules, remain viable during storage, transport and maintain its infectivity for several months after its production. Furthermore, the formulated inoculant must be easy to apply, free of contamination and production process should be cost-efficient [85]. Moreover, the inoculants should preferably contain a cocktail of different fungal species adapted to different soil properties and be compatible with other ingredients (i.e., rhizobacteria, biostimulants, surfactants, and other organics) that may be part of the product formulation [81]. Table 19.2 summarizes the commercial inoculum of ECM fungi [37].

TABLE 19.2 Commercial availability of ectomycorrhizal fungi inoculants.

Commercial product	Type/process	Species	Company
BioGrow Blend	Spores	–	Terra Tech, LLC
MycoApply-ecto	Spores	–	Mycorrhizal Applications Inc.
Mycorise Pro Reclaim	Propagules ecto + endo	–	Symbio Technologies Inc.
Myke Pro LF3	Propagules	–	Premier Tech Biotechnologies
MycorTree	Spores	–	Plant Health Care, inc.
MycoRhiz	Mycelium/SSF	*Pisolithus tinctorius*	Abbot Laboratories
Somycel PV	Mycelium/SSF	*Hebeloma crustuliniforme* *Laccaria laccata* *Paxillus involutus*	INRA-Somycel S.A.
Ectomycorrhizal Spawn	Mycelium/SSF	*Laccaria laccata* *Pisolithus tinctorius*	Sylvan Spawn Laboratories, Inc.
–	Mycelium/ Submerged	*Hebeloma crustuliniforme*	Rhone Poulene-INRA
Mycobead	Mycelium/ Submerged	Eleven ECM strains	Biosynthetica Pvt. Ltd.
Ectovit	–	–	Symbio-m Ltd.

Source: Chari L. Studies on *Pisolithus* sp.: stress response, pigment production and mycorrhization with forestry trees of Goa [Ph.D. thesis]. Goa: Goa University; 2012.

19.7.1 Types of ectomycorrhizal inoculants

Three main types of ectomycorrhizal inoculants commonly used include soil, fungal spores, and vegetative mycelia.

For soil inoculum, a thin layer of soil obtained from natural forests, old nurseries, or established plantations is spread on the top of nursery bed and mixed with the soil or planting substrate [86]. This method is still used in many parts of the world, particularly in developing countries. The use of forest soil inoculums has major disadvantages. Species composition of ECM fungi in the inoculums is usually not known, and the inoculums may also contain harmful microorganisms and noxious weeds. Moreover, there is no precise information about the fungal species that are being introduced and their infection potential [87]. Despite these disadvantages, soil inoculant is recommended if no other type of inoculant is available, with the premise that any ECM is better than none. For spore inoculum, fungal spores, obtained from fruiting bodies harvested in natural forests, old nurseries, or established plantations, are used in many parts of the world [88]. This type of inoculant is limited to those fungal species able to produce large numbers of spores and fruiting bodies. Root colonization by this type of inoculant is slower than that presented by vegetative inoculant of the same fungal isolate [89], yet abundantly used. Vegetative inoculant raised from pure cultures of ectomycorrhizal fungi that is the vegetative mycelia of the fungus is one of the best methods for inoculation of the seedlings. Pure cultures are generally obtained from fruiting bodies (mushrooms/sporocarps) or from mycorrhizas and maintained under laboratory conditions. For commercial production of vegetative inoculant, mycelium has to be grown in the solid substrate or in liquid culture.

19.7.1.1 Solid-state fermentation

Solid-state fermentation is carried out by growing the culture on a suitable substrate such as a mixture of peat and vermiculite, supplemented with a nutritive solution [83]. Vermiculite is a low-price material and absorbs nutrient solutions well. It also provides a bed structure that enables good aeration of the substrate. The mixture is generally distributed in glass flasks or plastic bags, being inoculated with mycelium plugs or a mycelium suspension, obtained from a previous culture in solid or liquid medium, respectively. It then takes 2–4 months of incubation for the final product to be ready to inoculate seedlings [90]. Inoculation of plants with vegetative inoculants produced in solid substrates has been done in the United States, France, Mexico, Brazil, etc. [84,87]. Vegetative inoculants of *P. tinctorius, Hebeloma crustuliniforme, Laccaria laccata, Suillus luteus, Cenococcum geophillum*, and *Thelephora terrestris* have also been tested with plant hosts of economic interest [91].

19.7.1.2 Submerged cultivation

Submerged cultivation of ectomycorrhizal fungi is achieved by growing the culture in a liquid medium in bioreactor followed by immobilization of mycelia in calcium alginate gel or other types of polymeric matrices [35,36,92–95]. Nursery studies have shown that this type of inoculant is more efficient than those produced by solid-state fermentation, probably due to better survival of the fungus inside the gel when applied to the planting substrate [96]. Despite being a very promising technique, submerged fermentation still has limited application as poor growth of ECM fungus is obtained in most cultivation systems and frequent contamination [95].

19.7.2 Ectomycorrhizal inoculants in field applications

Over the past few decades, a tremendous amount of research has been done on the practical use of mycorrhizal fungi demonstrating their effectiveness in improving survival and health of many plant species in diverse landscapes. Considering this aspect, one cannot overlook the reforestation success in the United States of the Ohio Division of Mine Land Reclamation – Abandoned Mined Lands program that used Pt-inoculated pine and oak seedlings to provide a low-cost, low-maintenance reclamation method for abandoned mine lands [97]. Another example is the reclamation of Utah Copper Mine Site using Pt and VAM fungi [80]. Reddy and Natarajan [98] have demonstrated the promising application of *Laccaria fraterna, L. laccata, P. tinctorius, Rhizopogon luteolus, Scleroderma citrinum*, and *T. terrestris* with several species of *Pinus*, in reforestation of the Nilgiri and Palni hills of Tamilnadu, South India. Sesa Goa, a mining company, has used *Pt-Acacia mangium* combination for reclamation of reject dumps.

Pisolithus, a Gasteromycete, has widespread global distribution and is often found in adverse environments. Sites where this fungus exists are commonly characterized by high soil temperature, extreme acidity, draughtiness, low fertility, and high levels of toxic metals. This, combined with the ability of *Pisolithus* to form extensive mycorrhizal relations with numerous tree species, makes it quite popular in reclamation efforts. Moreover, *P. tinctorius* is reported to have a wide spectrum of antibiotic activity. *P. tinctorius* yields a beautiful spectrum of gold to chestnut brown color dye that is used to dye yarn. *P. tinctorius* is thus a well-known economically and ecologically important ECM fungus. Our study focused on preparation of inoculum formulations of *P. tinctorius* PT1 in an attempt to reclaim mine reject dump soil. We inoculated 26 native Goan plant species transplanted on the mine reject dump with soil-based spore inoculums. The procedure used for preparing spore inoculum of *P. tinctorius* PT1 is shown in Fig. 19.3. The inoculation of transplanted seedlings with spore-based PT inoculum proved to be very promising. The positive results in terms of plant growth in inoculated seedlings as compared to control uninoculated plants are shown in Fig. 19.4.

19.8 Conclusion

The ECM association plays a key role in terrestrial environments, affecting soil structure and functioning of the ecosystem. Owing to the benefits that are incurred through the symbiotic interaction, interest in ECM research has grown over the years. For quite some time, ECM was studied by traditional culture-based techniques, and by this method the characterization of the association was limited. In recent years, the introduction of culture-independent methods and molecular tools gave insight into biochemistry, molecular signaling, biodiversity, and evolution of ECM associations. This clearly

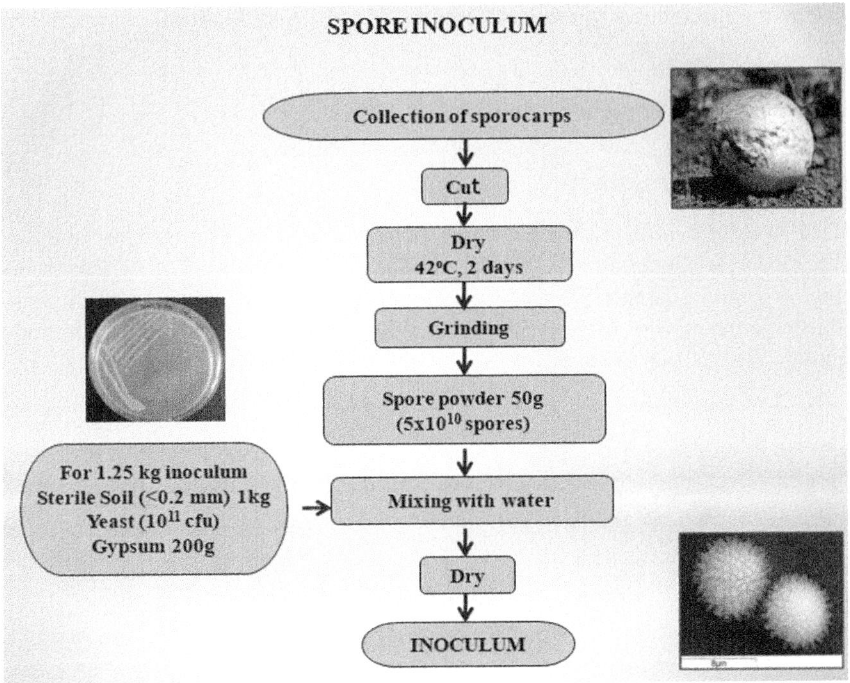

FIGURE 19.3 Procedure for preparation of soil-based spore inoculums using *Pisolithus tinctorius* PT1.

FIGURE 19.4 Application of soil-based spore inoculum for reclamation of mine reject dump soil of Codli mines, Goa. (Photos of plant response to inoculum, taken after 5 years of inoculation).

reveals the interaction of the two partners and also other soil microorganisms. Hence, the traditional methods along with newer molecular tools would enable biologists and ecologists to tap the ECM fungi and its products for ecological and industrial utilization.

19.9 Future prospects

ECM fungi are well known for their ecological and economical importance. The recent advancements in ECM research will probably unveil more and more potential characteristics of ECM fungi and their interaction with the host plants. There is a lot of scope for manipulation of the association at the molecular level as per the need of the people and their environment.

Acknowledgments

The authors are grateful to Sesa Goa Limited (Vedanta Limited) Panaji, Goa, India, for their financial assistance and experimental plot and labour.

References

[1] Sylvia DM, Fuhrmann JJ, Hartel PG, Zubere DA. Overview of mycorrhizal symbiosis. In: Principles and applications of soil microbiology. 2nd ed. Upper Saddle River, New Jersey: Prentice Hall Publishers; 2004.

[2] Futai K, Taniguchi T, Kataoka R. Ectomycorrhizae and their importance in forest ecosystems. In: Mycorrhizae: sustainable agriculture and forestry. Dordrecht: Springer; 2008. p. 241–85.

[3] Van Der Heijden MG, Martin FM, Selosse MA, Sanders IR. Mycorrhizal ecology and evolution: the past, the present, and the future. New Phytol 2015;205(4):1406–23.

[4] Marx DH. Tree host range and world distribution of the ectomycorrhizal fungus Pisolithus tinctoriu. Can J Microbiol 1977;23(3):217–23.

[5] Trappe JM. Selection of fungi for ectomycorrhizal inoculation in nurseries. Annu Rev Phytopathol 1977;15(1):203–22.

[6] Natarajan K, Senthilarasu G, Kumaresan V, Riviere T. Diversity in ectomycorrhizal fungi of a dipterocarp forest in Western Ghats. Curr Sci 2005;88(12):1893–5.

[7] Cairney JW, Chambers SM. Interactions between *Pisolithus tinctorius* and its hosts: a review of current knowledge. Mycorrhiza 1997;7(3):117–31.

[8] Rao CS, Sharma GD, Shukla AK. Distribution of ectomycorrhizal fungi in pure stands of different age groups of *Pinus kesiya*. Can J Microbiol 1997;43(1):85–91.

[9] Mason PA. Aseptic synthesis of sheathing (ecto-) mycorrhizas. 1980. p. 173–8.

[10] Bolan NS. A critical review on the role of mycorrhizal fungi in the uptake of phosphorus by plants. Plant Soil 1991;134(2):189–207.

[11] Marx DH. The influence of ectotrophic mycorrhizal fungi on the resistance in pine roots to pathogenic infections I. Antagonisms of mycorrhizal fungi to root pathogenic fungi and soil bacteria. Phytopathology 1969;59:53–163.

[12] Augé RM. Water relations, drought and vesicular-arbuscular mycorrhizal symbiosis. Mycorrhiza 2001;11(1):3–42.

[13] Porcel R, Ruiz-Lozano JM. Arbuscular mycorrhizal influence on leaf water potential, solute accumulation, and oxidative stress in soybean plants subjected to drought stress. J Exp Bot 2004;55(403):1743–50.

[14] Han XL, Jia GX, Niu Y. Research of the mechanism of ectomycorrhizae to drought resistance. Res Soil Water Conserv 2006;5:42–4.

[15] Zhang HH, Tang M, Yang Y. The response of ectomycorrhizal (ECM) fungi under water stress induced by polyethylene glycol (PEG) 6000. Afr J Microbiol Res 2011;5(4):365–73.

[16] Bogeat-Triboulot MB, Bartoli F, Garbaye J, Marmeisse R, Tagu D. Fungal ectomycorrhizal community and drought affect root hydraulic properties and soil adherence to roots of *Pinus pinaster* seedlings. Plant Soil 2004;267(1–2):213–23.

[17] Hung LL, Trappe JM. Growth variation between and within species of ectomycorrhizal fungi in response to pH in vitro. Mycologia 1983;75(2):234–41.

[18] Colpaert JV, Vandenkoornhuyse P, Adriaensen K, Vangronsveld J. Genetic variation and heavy metal tolerance in the ectomycorrhizal basidiomycete *Suillus luteus*. New Phytol 2000;147(2):367–79.

[19] Jentschke G, Godbold DL. Metal toxicity and ectomycorrhizas. Physiol plant 2000;109(2):107–16.

[20] Meharg AA. The mechanistic basis of interactions between mycorrhizal associations and toxic metal cations. Mycol Res 2003;107(11):1253–65.

[21] Tibbett M, Sanders F, Cairney J. Low-temperature-induced changes in trehalose, mannitol and arabitol associated with enhanced tolerance to freezing in ectomycorrhizal basidiomycetes (*Hebeloma* spp.). Mycorrhiza 2002;12(5):249–55.

[22] Collin-Hansen C, Pedersen SA, Andersen RA, Steinnes E. First report of phytochelatins in a mushroom: induction of phytochelatins by metal exposure in *Boletus edulis*. Mycologia 2007;99(2):161–74.

[23] Ferreira AS, Tótola MR, Kasuya MC, Araújo EF, Borges AC. Small heat shock proteins in the development of thermotolerance in *Pisolithus* sp. J Therm Biol 2005;30(8):595–602.

[24] Ferreira AS, Tótola MR, Borges AC. Physiological implications of trehalose in the ectomycorrhizal fungus *Pisolithus* sp. under thermal stress. J Therm Biol 2007;32(1):34–41.

[25] Ho I. Comparison of eight *Pisolithus tinctorius* isolates for growth rate, enzyme activity, and phytohormone production. Can J For Res 1987;17(1):31–5.

[26] Satyanarayana T, Gupta V, Garg S. Problems and prospects in inoculum production technology of ectomycorrhizal fungi in: fungi in Biotechnology. New Delhi: CBS Publishers and Distributors; 1996. p. 45–50.

[27] Suh HW, Crawford DL, Korus RA, Shetty K. Production of antifungal metabolites by the ectomycorrhizal fungus *Pisolithus tinctorius* strain SMF. J Ind Microbiol 1991;8(1):29–35.

[28] Yamaji K, Ishimoto H, Usui N, Mori S. Organic acids and water-soluble phenolics produced by *Paxillus* sp. 60/92 together show antifungal activity against Pythium vexans under acidic culture conditions. Mycorrhiza 2005;15(1):17–23.

[29] Chari LJ, Bhosle S, Garg S. Antagonistic activity of phenolics produced by *Pisolithus* sp. PT1. J Pure Appl Microbiol 2012;6(1):289–96.

[30] Carlisle MJ, Watkinson SC, Gooday GW. The fungi. 2nd ed. San Diego: Academic Press; 2001. p. 394–405.

[31] Dustin. Shades of Pisolithus. In: Mushroom dyes. Creative by nature; 2011. www.studiobluedesigns.net/blog/?p=845.

[32] Litchfield JH, Lawhon Jr WT. Aerobic submerged fermentation of sporulating, ectomycorrhizal fungi. inventors; Battelle Development Corp, assignee. 1982. United States patent US 4,327,181.

[33] Mason PA. Aseptic synthesis of sheathing (ecto-) mycorrhizas. In: Ingram DS, Hegeson JP, editors. Tissue culture methods for plant pathologists; 1980. p. 173–8.
[34] Modess O. Zur kenntnis der mykorrhizabildner von kiefer und fichte. Symb Bot Ups 1941;5:1–47.
[35] Rossi MJ, Oliveira VL. Growth of the ectomycorrhizal fungus *Pisolithus microcarpus* in different nutritional conditions. Braz J Microbiol 2011;42(2):624–32.
[36] Garg S. Production of ectomycorrhizal fungal inoculum by submerged fermentation [Ph.D. Thesis]. New Delhi: University of Delhi; 1999.
[37] Chari L. Studies on *Pisolithus* sp.: stress response, pigment production and mycorrhization with forestry trees of Goa [Ph.D. Thesis]. Goa: Goa University; 2012.
[38] Cudlin P, Krropacek K, Sarek V, mejtrik U, Neeraj VA. Production and application of ectomycorrhizal alginate inoculums. Indian J Microbiol 1992;32:371–81.
[39] Daza A, Manjón JL, Camacho M, De La Osa LR, Aguilar A, Santamaría C. Effect of carbon and nitrogen sources, pH and temperature on in vitro culture of several isolates of *Amanita caesarea* (Scop.: Fr.) Pers. Mycorrhiza 2006;16(2):133.
[40] Yamanaka T. The effect of pH on the growth of saprotrophic and ectomycorrhizal ammonia fungi in vitro. Mycologia 2003;95(4):584–9.
[41] Bois G, Bertrand A, Piché Y, Fung M, Khasa DP. Growth, compatible solute and salt accumulation of five mycorrhizal fungal species grown over a range of NaCl concentrations. Mycorrhiza 2006;16(2):99–109.
[42] Matsuda Y, Sugiyama F, Nakanishi II S-. Effects of sodium chloride on growth of ectomycorrhizal fungal isolates in culture. Mycoscience 2006;47:212–7.
[43] Madsen T, Mulligan D. Effect of salt on the growth of ectomycorrhizal isolates and implications for the coal mining industry in central Queensland. Ecol Manag Restor 2007;8(1):63–5.
[44] Tam PC. Heavy metal tolerance by ectomycorrhizal fungi and metal amelioration by *Pisolithus tinctorius*. Mycorrhiza 1995;5(3):181–7.
[45] Mexal J, Reid CP. The growth of selected mycorrhizal fungi in response to induced water stress. Can J Bot 1973;51(9):1579–88.
[46] Coleman MD, Bledsoe CS, Lopushinsky W. Pure culture response of ectomycorrhizal fungi to imposed water stress. Can J Bot 1989;67(1):29–39.
[47] Whiting EC, Rizzo DM. Effect of water potential on radial colony growth of *Armillaria mellea* and *A. gallica* isolates in culture. Mycologia 1999;91(4):627–35.
[48] Chen DM, Khalili K, Cairney JW. Influence of water stress on biomass production by isolates of an ericoid mycorrhizal endophyte of *Woollsia pungens* and *Epacris microphylla* (Ericaceae). Mycorrhiza 2003;13(3):173–6.
[49] Duñabeitia MK, Hormilla S, Garcia-Plazaola JI, Txarterina K, Arteche U, Becerril JM. Differential responses of three fungal species to environmental factors and their role in the mycorrhization of *Pinus radiata* D. Don. Mycorrhiza 2004;14(1):11–8.
[50] Ali NA, Jackson RM. Stimulation of germination of spores of some ectomycorrhizal fungi by other micro-organisms. Mycol Res 1989;93(2):182–6.
[51] Melo IS, Faull JL. Scanning electron microscopy of conidia of *Trichoderma stromaticum*, a biocontrol agent of witches' broom disease of cocoa. Braz J Microbiol 2004;35(4):330–2.
[52] Cairney JW. *Pisolithus*–death of the pan-global super fungus. New Phytol 2002;153(2):199–201.
[53] Arai H, Tamai Y, Yajima T, Obase K, Miyamoto T. Ectomycorrhizal fungal communities associated with *Quercus dentata* in a coastal broadleaf forest. Mycosphere 2017;8(4):561–7.

[54] Brundrett M, Bougher N, Dell B, Grove T, Malajczuk N. Working with mycorrhizas in forestry and agriculture. Canberra, Australia: Australian Centre for International Agricultural Research; 1996. p. 344.

[55] Bevege DI. A rapid technique for clearing tannins and staining intact roots for detection of mycorrhizas caused by *Endogone* spp., and some records of infection in Australasian plants. Trans Br Mycol Soc 1968;51(5):808. IN16.

[56] Brundrett MC, Piche Y, Peterson RL. A new method for observing the morphology of vesicular–arbuscular mycorrhizae. Can J Bot 1984;62(10):2128–34.

[57] Comas LH, Eissenstat DM. Patterns in root trait variation among 25 co-existing North American forest species. New Phytol 2009;182(4):919–28.

[58] Olsson PA. Signature fatty acids provide tools for determination of the distribution and interactions of mycorrhizal fungi in soil. FEMS Microbiol Ecol 1999;29(4):303–10.

[59] Ekblad AL, Wallander H, Näsholm T. Chitin and ergosterol combined to measure total and living fungal biomass in ectomycorrhizas. New Phytol 1998;138(1):143–9.

[60] Wallander H, Wickman T. Biotite and microcline as potassium sources in ectomycorrhizal and non-mycorrhizal Pinus sylvestris seedlings. Mycorrhiza 1999;9(1):25–32.

[61] Bidartondo MI, Bruns TD. Extreme specificity in epiparasitic Monotropoideae (Ericaceae): widespread phylogenetic and geographical structure. Mol Ecol 2001;10(9):2285–95.

[62] Mello A, Balestrini R. Recent insights on biological and ecological aspects of ectomycorrhizal fungi and their interactions. Front Microbiol 2018;9:216.

[63] Landeweert R, Hoffland E, Finlay RD, Kuyper TW, van Breemen N. Linking plants to rocks: ectomycorrhizal fungi mobilize nutrients from minerals. Trends Ecol Evol 2001;16(5):248–54.

[64] Clasen BE, Silveira AD, Baldoni DB, Montagner DF, Jacques RJ, Antoniolli ZI. Characterization of Ectomycorrhizal species through molecular biology tools and morphotyping. Sci Agric 2018;75(3):246–54.

[65] Hortal S, Pera J, Parladé J. Tracking mycorrhizas and extraradical mycelium of the edible fungus *Lactarius deliciosus* under field competition with *Rhizopogon* spp. Mycorrhiza 2008;18(2):69–77.

[66] De la Varga H, Águeda B, Martínez-Peña F, Parladé J, Pera J. Quantification of extraradical soil mycelium and ectomycorrhizas of Boletus edulis in a Scots pine forest with variable sporocarp productivity. Mycorrhiza 2012;22(1):59–68.

[67] Daguerre Y, Levati E, Ruytinx J, Tisserant E, Morin E, Kohler A, Montanini B, Ottonello S, Brun A, Veneault-Fourrey C, Martin F. Regulatory networks underlying mycorrhizal development delineated by genome-wide expression profiling and functional analysis of the transcription factor repertoire of the plant symbiotic fungus *Laccaria bicolor*. BMC Genomics 2017;18(1):737.

[68] Heller G, Adomas A, Li G, Osborne J, van Zyl L, Sederoff R, Finlay RD, Stenlid J, Asiegbu FO. Transcriptional analysis of *Pinus sylvestris* roots challenged with the ectomycorrhizal fungus *Laccaria bicolor*. BMC Plant Biol 2008;8(1):19.

[69] Du M, Ding G, Cai Q. The Transcriptomic responses of *Pinus massoniana* to drought stress. Forests 2018;9(6):326.

[70] Peter M, Kohler A, Ohm RA, Kuo A, Krützmann J, Morin E, Arend M, Barry KW, Binder M, Choi C, Clum A. Ectomycorrhizal ecology is imprinted in the genome of the dominant symbiotic fungus *Cenococcum geophilum*. Nat Commun 2016;7:12662.

[71] Zeng X, Li Y, Ling H, Chen J, Guo S. Revealing proteins associated with symbiotic germination of *Gastrodia elata* by proteomic analysis. Botanical Studies 2018;59(1):8.

[72] Finlay RD, Lindahl BD, Taylor AFS. Responses of mycorrhizal fungi to stress. In: Avery S, Stratford M, van West P, editors. Stress in yeasts and filamentous fungi. Amsterdam: Elsevier; 2008. p. 201–20.

[73] Fillinger S, Chaveroche MK, Van Dijck P, de Vries R, Ruijter G, Thevelein J, d'Enfert C. Trehalose is required for the acquisition of tolerance to a variety of stresses in the filamentous fungus *Aspergillus nidulans*. Microbiology 2001;147(7):1851–62.

[74] Ramesh G, Podila GK, Gay G, Marmeisse R, Reddy MS. Different patterns of regulation for the copper and cadmium metallothioneins of the ectomycorrhizal fungus *Hebeloma cylindrosporum*. Appl Environ Microbiol 2009;75(8):2266–74.

[75] Sebastiana M, Neto TL, Pessoa F, Kopka J, Pais MS. New advances on ectomycorrhizal establishment: learning from transcriptome and metabolome analysis. In: Meeting on plant-microbe interactions; 2011. p. 13–4.

[76] Velišek J, Cejpek K. Pigments of higher fungi – a review. Czech J Food Sci 2011;29:87–102.

[77] Zhou Z-Y, Liu J-K. Pigments of fungi (macromycetes). Nat Prod Rep 2010;27(11):1531–70.

[78] Rice MC. Mushrooms for dyes, paper pigments and Myco-StixTM, D.M. Beebee (illustrations). Mushrooms for colour Press; 2007. ISBN-10: 1604618027, ISBN-13: 978-1604618020.

[79] Gill M, Kiefel MJ. Pisoquinone, a new naphthalenoid pulvinic acid from the fungus *Pisolithus arhizus*. Aust J Chem 2011;47(10):1967–77.

[80] Nehls U, Ecke M, Manpp R. Sugar- and nitrogen-dependent regulation of *Amanita muscaria* phenylalanine ammonium lyase gene. J Bacteriol 1999;181(6):1931–3.

[81] Marx DH, Marrs LF, Cordell CE. Practical use of the mycorrhizal fungal technology in forestry, reclamation, arboriculture, agriculture, and horticulture. Dendrobiology 2002;47:27–40.

[82] Marx DH, Cordell CE. The use of specific ectomycorrhizas to improve artificial forestation practices. In: Whipps JM, Lumsden RD, editors. Biotechnology of fungi improving plant growth; 1989. p. 1–25. New York, USA.

[83] Marx DH. Ectomycorrhizal fungus inoculation: a Tool for mproving forest Practices. In: Mikola P, editor. Ectomycorrhiza research. Oxford, UK: Clarendon Press; 1980. p. 13–71.

[84] Selosse MA, Bouchard D, Martin F, Le Tacon F. Effect of *Laccaria bicolor* strains inoculated on Douglas-fir (*Pseudotsuga menziesii*) several years after nursery inoculation. Can J For Res 2000;30:360–71.

[85] Schwartz MW, Hoeksema JD, Gehring CA, Johson NC, Klironomo JN, Abott LK, Pringle A. The promise and the potential consequences of the global transport of mycorrhizal fungal inoculants. Ecol Lett 2006;9:501–15.

[86] Mikola P. Application of Mycorrhizal symbiosis in forestry practices. In: Marks GC, Kozlowski TT, editors. Ectomycorrhizae, their ecology and physiology. New York, USA: Acadamic Press; 1973. p. 383–411.

[87] Castellano MA, Molina R. Mycorrhizae. In: Landis TD, Tinus RW, McDonald SE, Barnett JP, editors. The biological components: nursery pests and mycorrhizae – the container tree nursery manual. Washington, DC, USA: US Department of Agriculture, Forest Service; 1989. p. 101–67.

[88] Theodorou C. Introduction of mycorrhizal fungi into soil by spore inoculation of seed. Aust For 1971;35:23–6.

[89] Marx DH, Ruehle JL, Cordell CE. Methods for studying nursery and field response of trees to specific ectomycorrhiza. In: R Norris J, Read DJ, Varma AK, editors. Methods in microbiology. London, UK: Academic Press; 1991. p. 383–411.

[90] Alves JR, Souza O, Podlech PAS, Giachini AJ, Oliveira VL. Effect of ectomycorrhizal inoculants produced by solid state fermentation on growth of *Eucalyptus dunnii* Maiden. Pesqui Agropecu Bras 2001;36:349–55.

[91] Castellano MA. Outplanting performance of mycorrhizal inoculated seedlings: a review. In: Proceedings of the 8th North American conference on mycorrhizae. Jackson, Wyoming, USA, vol. 49; 1990.

[92] Kuek C, Tommerup IC, Malajczuk N. Hydrogel bead inocula for the production of ectomycorrhizal eucalypts for plantations. Mycol Res 1992;96:273–7.

[93] Rossi MJ, Souza JAR, Oliveira VL. Inoculant production of the ectomycorrhizal fungus *Pisolithus microcarpus* in an airlift bioreactor. Appl Microbiol Biotechnol 2002;59:175–81.

[94] Rossi MJ. Technology of the production of ectomycorrhizal fungi inoculants employing submerged cultivation in an airlift bioreactor. PhD Thesis. Florianópolis, Brazil: Universidade Federal de Santa Catarina; 2006.

[95] Rossi MJ, Furigo A, Oliveira VL. Inoculant production of ectomycorrhizal fungi by solid and submerged fermentations. Food Technol Biotechnol 2007;45(3):277–86.

[96] Le Tacon Jung G, Mugnier J, Michelot P, Mauperin C. Efficiency in a forest nursery of an ectomycorrhizal fungus inoculant produced in a liquid medium and entrapped in a calcium alginate gel. Can J Bot 1985;63:1664–8.

[97] Cordell CE, Mans LF, Farley ME, Marx DH. Mycorrhizal fungi and trees – a successful reforestation alternative for mine land reclamation. In: Forest nursery proceedings, Kona, HI-2000. Fort Collins, US: US Department of Agriculture Forest Service, Rocky Mountain Research Station; 2000. p. 177–87.

[98] Reddy MS, Natarajan K. Coinoculation efficacy of ectomycorrhizal fungi on *Pinus patula* seedlings in a nursery. Mycorrhiza 1997;7:133–8.

Further reading

[1] Thompson GW, Medve RJ. Effects of aluminum and manganese on the growth of ectomycorrhizal fungi. Appl Environ Microbiol 1984;48(3):556–60.

[2] White TJ, Bruns T, Lee SJ, Taylor JL. Amplification and direct sequencing of fungal ribosomal RNA genes for phylogenetics. PCR protocols 1990;18(1):315–22.

[3] Hiremath S, Lehtoma K, Podila GK. Identification of a small heat-shock protein associated with a ras-mediated signaling pathway in ectomycorrhizal symbiosis. In: Res Papers NRS-7. Newton square, PA: U.S. Department of Agriculture, Forest Service, Northern Research Station; 2009. https://doi.org/10.2737/NRS-RP-7. 8pp.

Chapter 20

Photocatalytic and microbial degradation of Amaranth dye

Pranay P. Morajkar*, Amarja P. Naik, Sandesh T. Bugde, Bhanudas R. Naik
Department of Chemistry, Goa University, Taleigao Plateau, Goa, India
**Corresponding author: pranay@unigoa.ac.in.*

20.1 Introduction

Dyes are one of the most fascinating chemical compounds because of their unique chemical structure, presence of specific functional groups called *chromophores*, and a network of extended conjugation system that, together, allows harvesting photon energy from the visible spectrum of solar radiation and re-emit light as complementary colors. These colored dyes have now become an inherent and integral part of our daily food. *Amaranth* is one such cationic *azo* dye having dark red to purple color and is widely used as an additive in the food and cosmetic industries. The name Amaranth is derived from the plant Amaranthus; belongs to the kingdom: *Plantae*; clade: *Angiosperm*; family: Amaranthaceae; and genus: *Amaranthus*. Some of the edible species of Amaranthus are cultivated as leafy vegetables and pseudo cereals whereas others are used as ornamental plants [1]. The chemically synthesized Amaranth dye is commonly referred to as Acid Red 27 and Food Red 9, while the accepted nomenclature as per IUPAC is Trisodium (4E)-3-oxo-4-[(4-sulfonato-1-naphthyl)hydrazono]naphthalene-2,7-disulfonate.

Amaranth dye was traditionally extracted from the Amaranthus plant. In more recent years, it is artificially synthesized on an industrial scale, as a trisodium salt (Fig. 20.1), with improved water solubility and dyeing efficiency for fabrics, leather, paper, etc., and it reached a million-dollar mark. In the 20th century it was one of the most widely used food colorants in the United States [2]. However, the untreated and uncontrolled emission of Amaranth dye waste in river streams, owing to its complex aromatic structure, photo stability, and low biodegradability, resulted in its accumulation in water bodies, leading to adverse effects on living systems. In 1971, the first report of the possible carcinogenic effects of Amaranth appeared in the Soviet Union [3]. Since then, several research groups have evaluated the toxicity impact of azo dyes, and Amaranth in particular. For instance, Tsuda et al. [4] conducted genotoxicity

Advances in Biological Science Research. https://doi.org/10.1016/B978-0-12-817497-5.00020-3

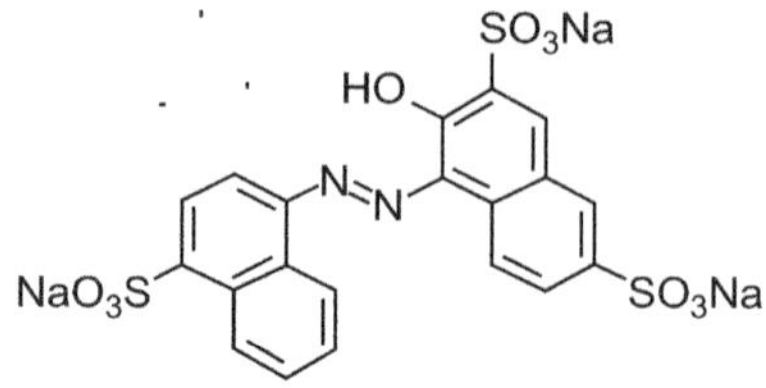

FIGURE 20.1 Chemical structure of amaranth dye.

studies on a group of four pregnant mice with a daily dose of 2000 mg/kg/day and collected samples from various organs such as brain, kidney, urinary bladder, embryo, liver colon, etc. and measured the DNA damages using comet assay. The assay was found to be positive after administration of Amaranth. DNA damages in the colon were observed in the male mice when injected with Amaranth dye starting at 10 mg/kg/day. Another study conducted by Clodeet al. [5], wherein amaranth dye was first incorporated into the food diet and subsequently in utero for long-term evaluation on a group of 90 (control) and 50 (treated) rats with doses from 0 to 250 mg/kg/day. This investigation showed no carcinogenic effects. However, renal calcification and pelvic epithelial hyperplasia with degenerative changes were observed in female rats after 18 months from exposure to amaranth dye.

The metabolites of Amaranth dye, such as sodium naphthionate and R-amino salts, were injected into the stomach tube of Osborne–Mendel pregnant females at doses of 15–200 mg/kg/day by Collins et al. [6]. An increase in fetuses with sternebra abnormalities were observed at 100 mg/kg/day dose level of sodium naphthionate, without inducing any soft tissue abnormalities. Similar observations have been reported in the case of R-amino salts with higher dose of 200 mg/kg/day. DNA interaction and cytogenic in vitro evaluation studies using Amaranth dye (0.02–8 mM) were conducted by Mpountoukas et al. [7] in human peripheral blood cells and they found a high toxic potential of Amaranth dye to human lymphocytes. Mutagenic and genotoxic investigation of amaranth was also conducted by Das et al. [8] using Ames mutagenicity assay and bone marrow of the mouse (in vivo). They concluded that Amaranth doses from 50 to 200 mg/kg body weight do not show any mutagenic or genotoxic effects. Other indirect effects such as allergic and asthmatic issues of Amaranth when given in conjunction with aspirin drug in certain patients have also been reported [9,10].

The Joint European Committee of Food Additives (JECFA), together with FAO/WHO, evaluated toxicological effects of Amaranth in 1972, 1975, 1978 and 1984. The EU Scientific Committee for Food performed independent studies in 1976, 1979 and 1983, and together with JECFA agreed upon an acceptable daily intake (ADI) of 0.5–0.8 mg/kg bw/day, considering a long-term carcinogenicity study in rats [7,11]. The European Food Safety Authority in 2010 revised the ADI to 0.15 mg/kg bw/day considering an uncertainty factor of 100. Although there are disagreements in several studies about carcinogenic and genotoxicity effects of Amaranth dye, all these panels of experts agree on the fact that reduction of azo group of Amaranth in gastrointestinal tract produces free sulfonated aromatic amines, which are known to be toxic, carcinogenic, and mutagenic and can enter the systemic circulation system and cause adverse effects [7,12]. In recent times, the use of Amaranth as food additive has been completely banned in the United States as a suspected carcinogen [3]. Therefore, efforts were channeled towards use of conventional and biological treatment methods such as coagulation, separation using membrane filters or precipitation, advanced photocatalytic degradation, and enzyme/microbial degradation methods for treating waters contaminated with Amaranth dye. An overview of some of these advanced treatment methods is presented in the next section.

20.2 Advanced photocatalytic amaranth degradation using titanium dioxide

Several heterogeneous metal oxides as effective photocatalysts have been synthesized and utilized for degradation of organic azo dyes, such as Amaranth, over the past decade. Pilot-scale wastewater treatment plants are operational in many European countries such as the SOLARDETOX plant in Arganda del Rey, Spain, and the PhoRTex plant in Augsburg, Germany, wherein the water channel tubes/sheets/tanks are painted with photocatalysts and wastewaters contaminated with dyes flow over it under UV/visible light to achieve effective dye degradation. The photocatalytic degradation efficiencies of some of the catalysts reported in the literature [13–20] are presented in Table 20.1. Titanium dioxide (TiO_2) is so far the best-known photo catalyst among the transition metal oxide series, due to its inertness, economic viability, insolubility in water, stability against photo corrosion, and high photocatalytic activity [18,21]. TiO_2 is a wide-band gap semiconductor oxide with band gap around 3.2 eV, which corresponds to the absorption wavelength of 387.45 nm. This effectively means that TiO_2 cannot absorb any significant portion of visible light and is mostly UV active. If the energy of the incident UV light is greater than the band gap energy of TiO_2, excitation of electrons from valence band (V.B.) to conduction band (C.B.) occurs, producing positively charged holes in V.B, together that act as charge carries and drive the photocatalytic action in TiO_2 (Fig. 20.2). However, the major limitation in this process is the inherent recombination of charge carries that severely reduces the formation of reactive oxygen species such as $O_2^{\cdot -}$ and OH radicals on TiO_2 surface. Moreover, the lower surface area of TiO_2 (<50 m^2/g for Degussa P25) also plays a detrimental role in limiting its large-scale application in water treatment technology. Many efforts have been channeled towards increasing the specific surface area of TiO_2, and laboratory-scale studies have achieved this using nanosized/porous TiO_2, with surface areas of 200–300 m^2/g [22–25]. However, large-scale synthesis of such a high

TABLE 20.1 Summary of amaranth photodegradation efficiency of semiconductor catalysts.

Catalyst	Initial dye concentration	UV source	Reaction time (min)	% Degradation	Comment	Reference
Fenton reagent	3×10^{-5} M	200 W tungsten lamp	22	59.0	Rate of degradation for photo Fenton was obtained to be higher than that of fenton reagents.	[13]
N,S-doped titania	3.90×10^{-5} M	200 W tungsten lamp	90	59.3	Higher photocatalytic degradation of amaranth dye due to extended absorption of the catalyst in the visible range.	[14]
TiO_2	Amaranth plant extract	1000 W	15	85.0	Nanoscale TiO_2 nanoparticles were observed to be better than microscale TiO_2 for photocatalytic degradation of amaranth dye.	[15]
$Ti/Ru_{0.3}Ti_{0.7}O_2$ electrode	100 ppm	125 W	90	34.0	Photoelectrochemical degradation of amaranth dye was found to be better as compared to photochemical degradation.	[16]
$N-WO_3$	25 mg/L	160 W Visible and UVA light	120	100.0	Higher photocatalytic degradation of amaranth dye was observed as compared to pristine WO_3 due to photo stability and visible light absorption.	[17]
$N-ZrO_2$	10 ppm	Fluorescent lamp (visible) 24 W	120	26.4	N_2 doping enhanced the photocatalytic degradation of amaranth dye over pristine ZrO_2.	[18]
ZnO	4.4×10^{-5} M	200 W tungsten lamp	300	92.5	Investigation of photo bleaching of amaranth dye using ZnO catalyst.	[19]
$Cu/Ag/TiO_2$	50 ppm	15 W UV tubes	210	93.0	Transition metal doping enhanced the photocatalytic degradation rate of amaranth dye.	[20]

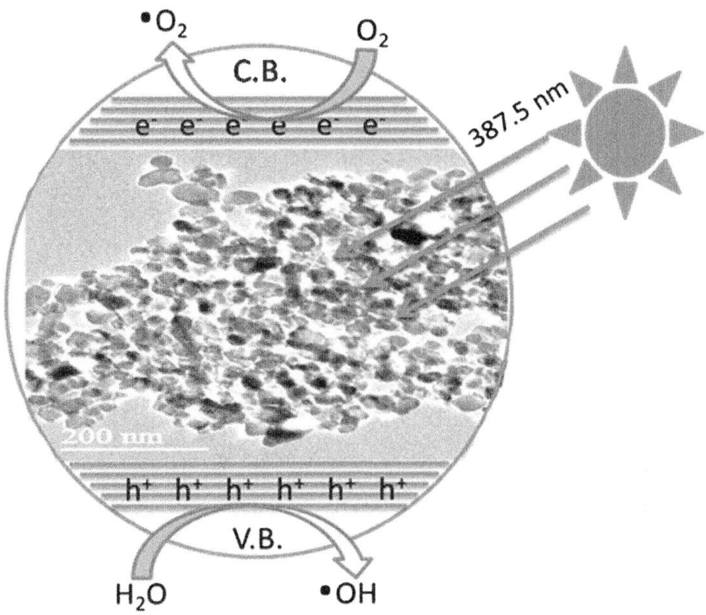

FIGURE 20.2 Formation of photoexcited electrons and positive holes as charge carriers in TiO_2 when exposed to sunlight.

surface area TiO_2 is still limited due to the inherent property of the nanoparticles to agglomerate. One of the efficient ways to overcome this limitation is to disperse the active TiO_2 over inert, large surface area support materials, such as porous glass, zeolites, and mesoporous Al_2O_3 [26–29], using wet impregnation synthesis. The presence of large surface area support not only allows reduction in agglomeration of TiO_2 nanoparticles but also allows greater adsorption of the adsorbate, thus contributing to the enhancement of photo catalyzed degradation of the dye. A brief overview to characterize TiO_2 supported on mesoporous Al_2O_3 catalyst is presented in the next section.

20.2.1 Characterization of TiO_2 supported mesoporous Al_2O_3 catalyst

A typical high-resolution transmission electron microscopy image (Fig. 20.3A) depicts the morphology of the TiO_2 impregnated on mesoporous Al_2O_3 support. The image shows distribution of irregular-sized TiO_2 particles over the porous Al_2O_3 support, thus enhancing the dispersion and reduction in agglomeration of TiO_2 nanoparticles. The selected area electron diffraction image (Fig. 20.3B) of TiO_2 supported on porous Al_2O_3, produces a concentric pattern of bright and dark rings of weak intensity, indicating polycrystalline or dominantly amorphous nature of the synthesized material. Energy dispersive X-ray analysis is routinely used to determine the surface purity of synthesized metal oxides or mixed

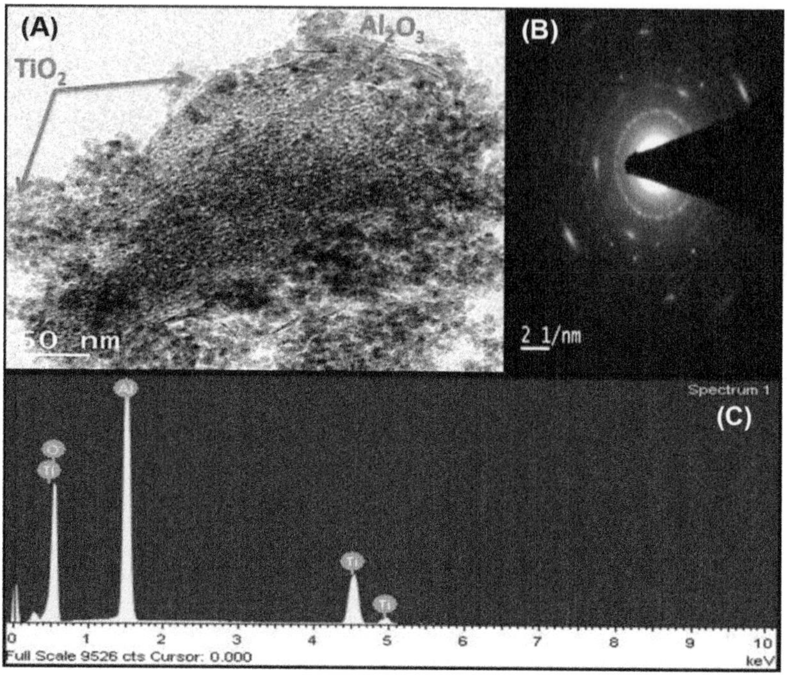

FIGURE 20.3 (A) HRTEM Image, (B) SAED pattern and (C) EDX spectrum of TiO_2 supported on Al_2O_3 catalyst [30].

metal oxides. The presence of Ti, Al, and O peaks with absence of any other impurity peak in Fig. 20.3C, which confirms the chemical purity of synthesized catalyst. The crystal phase of the synthesized catalyst is typically evaluated by recording its powder X-ray diffraction pattern and is compared with corresponding validated pattern of the Joint Committee on Powder Diffraction Standards (JCPDS), which was recently renamed as the International Center for Diffraction Data. Fig. 20.4A presents the X-ray diffraction pattern of TiO_2, Al_2O_3, and TiO_2 supported on Al_2O_3 catalyst, recorded on PXRD, Philips PW-1840 with Cu Kα (k = 0.154 nm) and nickel filter over the 2θ range of 10−80 degrees. The diffraction peaks at 2θ values of 25.3, 37.81, 48.0, 53.9, 55.06, 62.7, 68.7, and 70.2 degrees represents the crystal planes 101, 004, 200, 105, 211, 205, 116, and 220, respectively, in accordance with JCPDS file no. 84-1286, confirming the formation of anatase phase of TiO_2.

The mesoporous Al_2O_3 is largely amorphous and shows limited diffraction peaks at 2θ values of 37.4, 45.8 and 66.8 degrees originating from crystal planes 311, 440 and 400, respectively, thus due to gamma phase of Al_2O_3 nanoparticles (JCPDS file no. 47-1308). The TiO_2 impregnated on Al_2O_3 support shows peaks corresponding to both the metal oxides, indicating no significant phase change during TiO_2 impregnation on Al_2O_3. The porous nature of the synthesized catalyst is typically evaluated by recording N_2

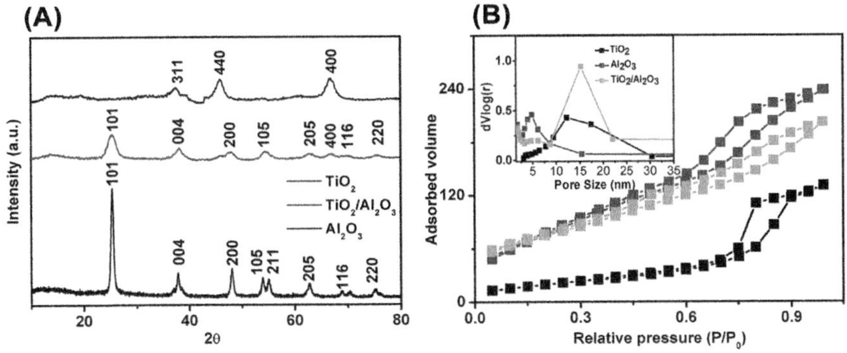

FIGURE 20.4 (A) Powder X-ray diffraction pattern and (B) N_2 adsorption—desorption isotherms of synthesized TiO_2, Al_2O_3 and TiO_2/Al_2O_3 catalyst [30].

TABLE 20.2 BET surface area and pore size distribution of TiO_2, Al_2O_3 and TiO_2 supported on Al_2O_3 catalyst [30].

Catalyst	Surface area (m^2/g)	Pore size (nm)
TiO_2	77	12.2
Al_2O_3	297	4.7
TiO_2/Al_2O_3	264	15.2

adsorption-desorption isotherms and the pore diameter and pore volume are determined using BJH analysis method. Fig. 20.4B depicts the comparative analysis of adsorption-desorption isotherms, i.e., a typical type IV isotherm with H2a hysteresis indicates the presence of complex meso-microporous structure with a network of interconnected pores [31,32]. A shift in the average pore size distribution of Al_2O_3 from 5 to 15 nm is observed as the larger mesopores of TiO_2 aggregates get distributed over Al_2O_3 surface. This is also marked with a slight decrease in the effective surface area of the composite catalyst, as shown in Table 20.2 [30]. Nonetheless, this small decrease in pore area is compensated by reduction in agglomeration of TiO_2, thus increasing turnover frequency and the photocatalytic activity of TiO_2 supported on mesoporous Al_2O_3 catalyst.

20.2.2 Amaranth adsorption versus photocatalytic-degradation kinetics

A typical laboratory-scale dye degradation experiment consists of three parts: (1) estimation of dye adsorption capacity of the catalyst, (2) measuring kinetics of dye degradation on catalyst surface, and (3) determination of degradation products. The dye adsorption process on catalyst surface is the first and most important step in degradation experiments because the rate of catalyzed reaction is directly proportional to the dye concentration available on the catalyst surface for reaction. Typical laboratory adsorption experiments are performed by varying the initial dye concentration (30–200 ppm) as a function of adsorption time in dark [33]. The amount of dye adsorbed on the catalyst surface at equilibrium is calculated by monitoring the decrease in initial absorbance of the dye solution using UV-visible spectrometer, at the maximum absorption wavelength, i.e., 520 nm for Amaranth [34]. The catalyst surface is tailored in such a way that the functional groups such as SO_3H^- present in the Amaranth dye find oppositely charged preferential adsorption site on the TiO_2 catalyst surface. When TiO_2 is dispersed on high surface area support such as mesoporous Al_2O_3, the presence of greater effective surface area enhances the rate of adsorption of the dye as shown in Table 20.3. When this dye adsorbed catalyst surface is exposed to sunlight or UV light source, the electrons from the valence band in TiO_2 are excited to the conduction band, resulting in formation of holes in the valence band (V.B.) and electrons in the conduction band (C.B.). This potential gradient, created due to separation of charge carriers, is responsible for driving the degradation reaction. The holes from V.B. react with adsorbed water to produce reactive OH radicals, while the surface adsorbed oxygen accepts electrons from C.B. to form superoxide radical anion species. Both the reactive oxygen species together initiate the dye degradation process [35,36].

$$TiO_2 + hv \rightarrow e^- + h^+$$

$$(O_2)_{ads} + e^- \rightarrow O_2^{\cdot -}$$

$$H_2O + h^+ \rightarrow OH + H^+$$

TABLE 20.3 Comparison of adsorption and photocatalytic degradation efficiency of uncatalyzed, TiO_2 and TiO_2–Al_2O_3 catalyzed amaranth degradation, respectively [30].

Catalyst	% Adsorption	% Degradation	Adsorption rate constant (min^{-1})	Degradation rate constant (min^{-1})
Without catalyst	–	3.7	–	0.05×10^{-2}
TiO_2	20	75.6	2.0×10^{-3}	2.3×10^{-2}
TiO_2/Al_2O_3	45	98.6	4.9×10^{-3}	6.7×10^{-2}

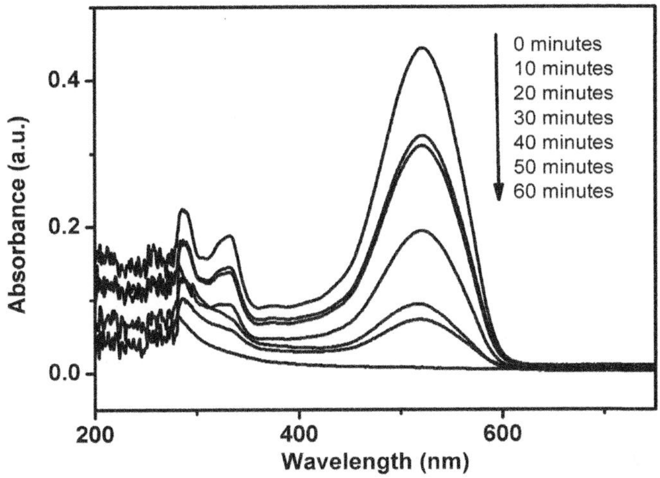

FIGURE 20.5 UV-visible spectra showing a decrease in absorbance of Amaranth dye when exposed to sunlight in presence of $TiO_2-Al_2O_3$ catalyst [30].

The attack of the reactive radical species preferentially to the azo group (N=N) present in the dye results in N=N cleavage. This destruction of the conjugated system of alternating double bonds results in a blue shift, and hence the absorption intensity at 520 nm for Amaranth dye decreases [37]. This decrease in absorbance is a direct measure of the rate of Amaranth degradation induced by photocatalytic action of TiO_2, as shown in Fig. 20.5. The changes in absorbance are then related to the corresponding change in Amaranth concentration as a function of photocatalytic reaction time, using the Beer–Lambert law. The kinetic plots representing enhancement in photocatalytic degradation rate using TiO_2-supported Al_2O_3 can be seen in Fig. 20.6A. The linear regression analysis of these kinetic plots is performed to

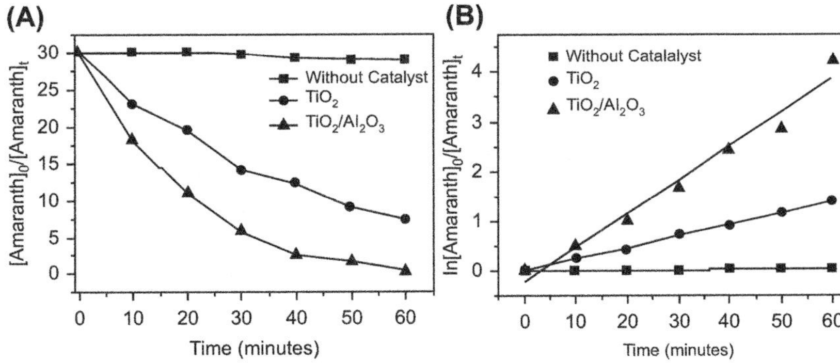

FIGURE 20.6 (A) Kinetic profiles of photocatalytic degradation of Amaranth over TiO_2 and TiO_2/Al_2O_3 catalyst. (B) Linear fits to the logarithmic ratio of (Amaranth) at time 0 and at time t. $(Amaranth)_0 = 30$ ppm, Amount of catalyst = 50 mg, solution pH = 5.7 at 30°C [30].

obtain the rate constant of degradation as shown in Fig. 20.6B, and calculated values are listed in Table 20.3. The enhancement in photocatalytic degradation of Amaranth dye on mesoporous $TiO_2-Al_2O_3$ catalyst is attributed to the synergistic effect of improved dispersion, reduction in agglomeration, and recombination rate of charge carriers in TiO_2 and greater adsorption efficiency due to higher surface area of mesoporous $TiO_2-Al_2O_3$ composite.

20.2.3 Identification of photodegradation products using LC-ESI-HRMS technique

Photocatalytic degradation of azo dyes has been extensively studied over the past few decades. However, less effort has been directed toward identification and quantification of degradation intermediates and products. This is probably because advanced high-resolution analytical techniques are required to achieve this goal. One of the useful and efficient techniques to achieve this is liquid chromatography-high resolution mass spectrometry (LC-HRMS) [37]. The liquid chromatography compartment is used to achieve physical separation between different constituents of the sample by distributing them over two nonmiscible phases (liquid and solid) as a function of differential partition, ion exchange, size exclusion, and adsorption property of the constituents of the sample or analyte. The separated constituents of the sample are then injected into the ionization zone to produce gas phase ionized particles or fragments using electron beam, photon beam, and even chemical modes of ionization. The mass analyzer separates these ions according to their masses, producing a mass spectrum that is used to determine elemental, isotopic, and structural elucidation of the components of the sample. A typical spectrum recorded for a sample aliquot during the photocatalytic degradation of Amaranth dye over TiO_2 catalyst in the literature [38], using the LC-HRMS (Electron Spray Ionization, ESI-Agilent 6520), is shown in Fig. 20.7. The high performance liquid chromatogram (HPLC) analyses are performed using Phenomenex

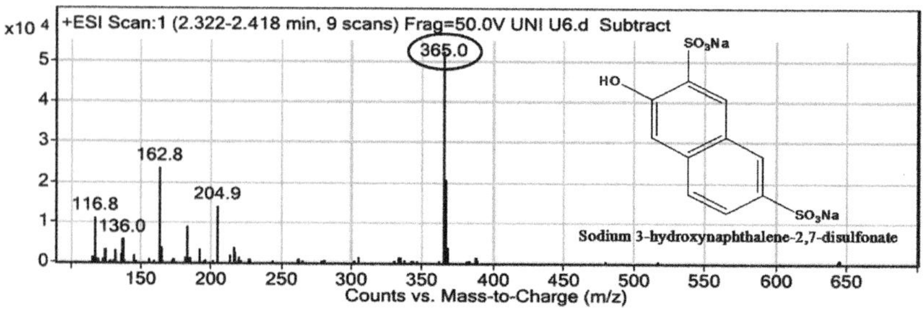

FIGURE 20.7 Typical mass spectrum recorded during photocatalytic amaranth degradation with peak m/z value of 365 corresponding to sodium-3-hydroxynaphthalene-2,7-disulfonate. *Reproduced from Naik AP, Salkar AV, Majik MS, Morajkar PP. Enhanced photocatalytic degradation of amaranth dye on mesoporous anatase TiO_2: evidence of C—N, N=N bond cleavage and identification of new intermediates. Photochem Photobiol Sci 2017;16:1126—1138. With permission from the Journal of Photochemical and Photobiological Sciences and the Royal Society of Chemistry.*

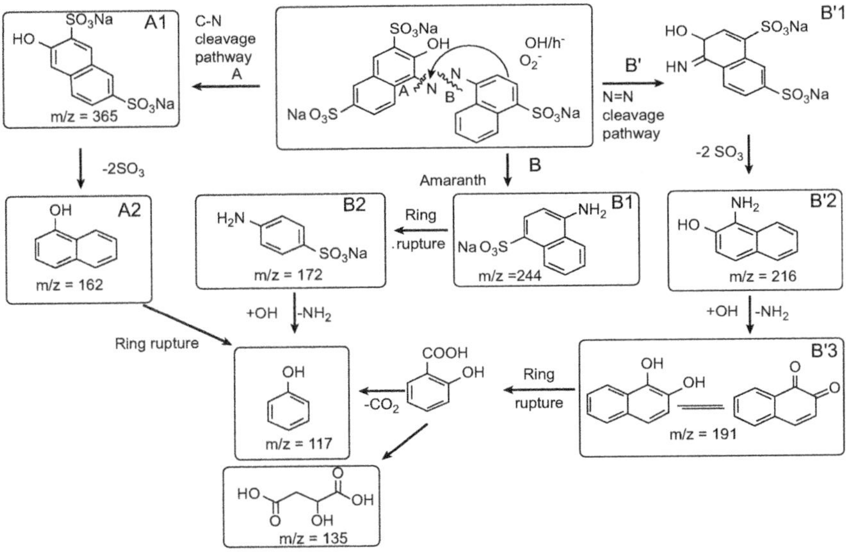

FIGURE 20.8 Plausible mechanism of amaranth degradation on TiO_2 surface. *Reproduced from Naik AP, Salkar AV, Majik MS, Morajkar PP. Enhanced photocatalytic degradation of amaranth dye on mesoporous anatase TiO_2: evidence of C—N, N═N bond cleavage and identification of new intermediates. Photochem Photobiol Sci 2017;16:1126-1138. With permission from the Journal of Photochemical and Photobiological Sciences and the Royal Society of Chemistry.*

Kinetex XB C18 column (length 30 mm, internal diameter 3.0 mm). The mass parameters are optimized using a dual spray (ESI) with positive/negative polarity, operated under the full scan mode with a gas temperature of 350°C and a flow rate of 10 L/min [38].

Several new intermediates of Amaranth degradation, such as 1-aminonaphthalene-2-ol, phenol, 1,2-dihydroxynaphthalene-1,2-diol, sodium-4-aminobenzenesulphonate, sodium-3-hydroxynaphthalene-2,7-disulfonate, 3-hydroxynaphthalene, and malic acid, etc., have been identified in our recent work, and a new plausible mechanism for Amaranth degradation is presented in Fig. 20.8 [38]. This study provided the first evidence of simultaneous C—N and N═N bond cleavage pathways during Amaranth photodegradation over titania-based photo catalyst. Some of these intermediates can be utilized as value-added chemicals and reused with purification. If the photocatalytic degradation process is continued, eventually these intermediates and products either mineralize or transform into carbon dioxide and water, which are nontoxic to the environment.

20.2.4 Toxicity of photodegradation products

In order to implement the photocatalytic technology for safe and efficient treatment of wastewater contaminated with Amaranth dye, it is important to first estimate the toxicity of degradation products relative to the parent dye. An overview of the intermediates and products reported in the earlier section, such as sodium-3-hydroxynaphthalene- 2,7-disulphonate, 3-hydroxynaphthalene, sodium-4-aminonaphthalenesulphonate, sodium-4-aminobenzenesulphonate, phenol, and malic acid, indicates comparatively lower toxicity and significantly less-damaging effects than the parent Amaranth dye. An assessment provided by the Safety Assessment Laboratory, Panapharm Laboratories Co. Ltd., reports that sodium-3-hydroxynaphthalene-2,7-disulphonate is nonmutagenic and nontoxic. 1-Aminonaphthalene-2-ol is a known carcinogen when tested on rats [39,40]. However, since it is formed only as an intermediate and subsequently transformed into phenol or malic acid, which are known to be nongenotoxic [41,42], the photocatalytic treatment method can therefore be considered as an economically viable and largely safer solution for treating wastewaters contaminated with Amaranth dye.

20.3 Bioremediation of amaranth dye

Use of microbial bioremediation methods for the treatment of industrial wastewater containing harmful organic contaminants have become an integral part of wastewater treatment systems due to their economic cost-effectiveness, reduced sludge production, and use of green environmental approaches. The recent advancements in the field of biotechnology and availability of sophisticated analytical instruments for the analysis of degradation products of dyes have resulted in better understanding and growing interest in this area, which is evident by the publication of top-quality research papers by the scientists working in this area [43–45]. Many specialized journals such as the *Journal of Hazardous Materials* and *Chemosphere* publish high-quality papers in this field [44]. Microbial degradation of azo dyes in aerobic/anaerobic conditions by use of free cells, biofilms, and enzymes could be either at intracellular or extracellular levels [46–48]. Immobilized microbes and biofilms have been found to be more efficient degradation systems compared to free cells [47,49,50]. Optimal efficiencies have been obtained from the biological systems for the water treatment by employing different approaches such as variation of incubation time, pH, temperature, concentration of dye, use of specific genus of microbes, use of aerobic/anaerobic conditions, etc. [51–53]. Ghodake et al. [46] studied decolorization of Amaranth dye under static anoxic condition and agitated culture conditions and found the involvement of intracellular and extracellular lignin peroxidase, intracellular laccase, dichlorophenol indophenols reductase and riboflavin reductase in the biodegradation of amaranth dye. Anjaneya et al. [47] have found that biofilm systems are a more promising wastewater treatment method compared to free cells. Biofilms are advantageous due to their ability to work at different dilution rates, reusability of microbial cells, high decolorization efficiency, excellent mechanical stability, and presence of self-sustainable polymeric matrix, which adhere to an inert or living surface, which help in the attachment of cells and dye molecules [47,49,54]. Use of fungal microbial systems have been reported by several scientists and found that degradation of dyes is often initiated under anaerobic conditions and in certain cases under aerobic conditions by enzymatic biotransformations [48,49,51,53]. Table 20.4 compares decolorization of Amaranth dye using different bacterial and fungal species. In most cases decolorization of dyes is initiated by the reductive cleavage of azo linkage by the microbial systems similar to the mechanism presented in Fig. 20.8.

20.4 Coupling of photocatalysis with bioremediation methods

The most recent development in wastewater treatment research embraces the coupling of photocatalysis with biodegradation. This innovative merger of two fields, photocatalysis induced by either UV or visible light and the biodegradation by microbes, act synergistically in the process of efficient treatment of water waste containing biorecalcitrant and inhibitory organic compounds. The amalgamate of photocatalysis and biodegradation for dye degradation was first reported by Marsolek et al. in 2008 and was termed as "intimate coupling of photocatalysis and biodegradation" [55]. This process involves the use of a light source to carry primary degradation, and if bacteria/microbes are present, it could further biodegrade the primary degraded products as soon as they are formed in the reactor. The key towards achieving the synergistic enhancement needs protection of the biofilms (bacteria) from direct contact with UV light source and toxic reactant generated from photolysis. This is achieved by using macroporous cellulosic cubes that act as macroporous carriers to protect the biofilm. A simple schematic representation of a novel photocatalytic circulating-bed biofilm reactor (PCBBR) is shown in Fig. 20.9. M. Huo, S. Dong and coworkers [56] recently investigated the efficiency of PCBBR for degradation of phenol using either visible or UV light sources. The use of visible light showed better activity of 99.8% for phenol degradation compared to 67.2% using UV light. The decrease in degradation efficiency in case of UV light is attributed to its disinfecting property on microbes. Thus, intimate coupling of visible light photocatalysis and biodegradation (Fig. 20.10) has added a new dimension in finding the potential solution towards the treatment of wastewater contaminated with azo dyes [57–59].

TABLE 20.4 Biodegradation of amaranth dye using bacterial and fungal species.

Species/Genus	Initial dye concentration (mg L^{-1})	Incubation time (h)	% Degradation	Comment	Reference
Acinetobacter caloaceticus	100	48	91	Experiment was performed under static anoxic condition.	[46]
A.caloaceticus	100	48	68	Experiment was performed under shaking or agitated condition (120 rpm) and 48 h of incubation.	[46]
Bacillus sp. strain AK1	200	24	99	Batch decolorization of amaranth dye was studied using free cells.	[47]
Bacillus sp. strain AK1(Biofilm)	200	24	100	Batch decolorization of amaranth dye was studied on biofilm.	[47]
Lysinibacillus sp. strain AK2	200	12	99	Batch decolorization of amaranth dye was studied using free cells.	[47]
Lysinibacillus sp. strain AK2 (biofilm)	600	24	100	Batch decolorization of amaranth dye was studied on biofilm.	[47]
Kerstesia sp. Strain VKY1	200	15	99	Batch decolorization of amaranth dye was studied using free cells.	[47]
Kerstesia sp. Strain VKY1(Biofilm)	200	12	99	Batch decolorization of amaranth dye was studied on biofilm.	[47]
Bjerkandera adusta	94	240	100	Cultures were grown at 29°C with shaking at 120 rpm, degradation was achieved due to presence of fungal enzymes.	[51]
Kerstersia sp.	200	24	100	Experiments were carried out aerobically at 40°C and pH 7.0.	[52]
Trametes versicolor	10	1	64	Influence of three different enzymes on the degradation of dye was studied.	[48]

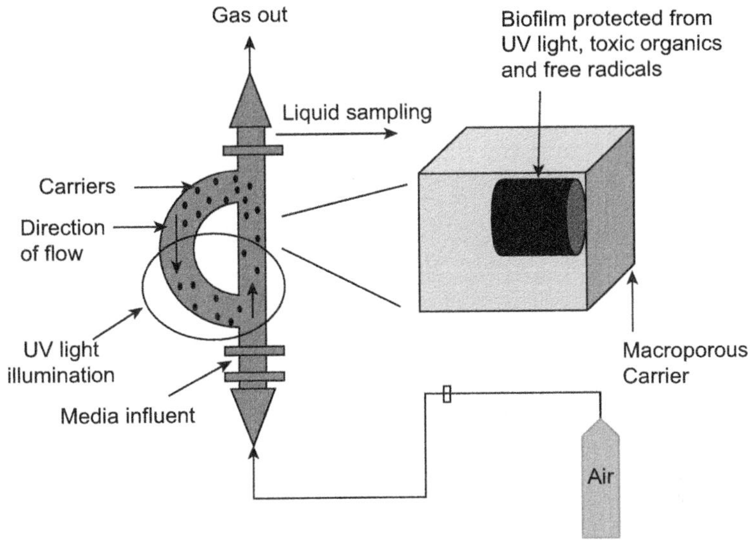

FIGURE 20.9 Schematic representation of photocatalytic circulating-bed biofilm reactor (PCBBR). *Adapted with permission from Marsolek MD, Torres CI, Hausner M, Rittmann BE. Intimate coupling of photocatalysis and biodegradation in a photocatalytic circulating-bed biofilm reactor. Biotechnol Bioeng 2008;101:83–92.*

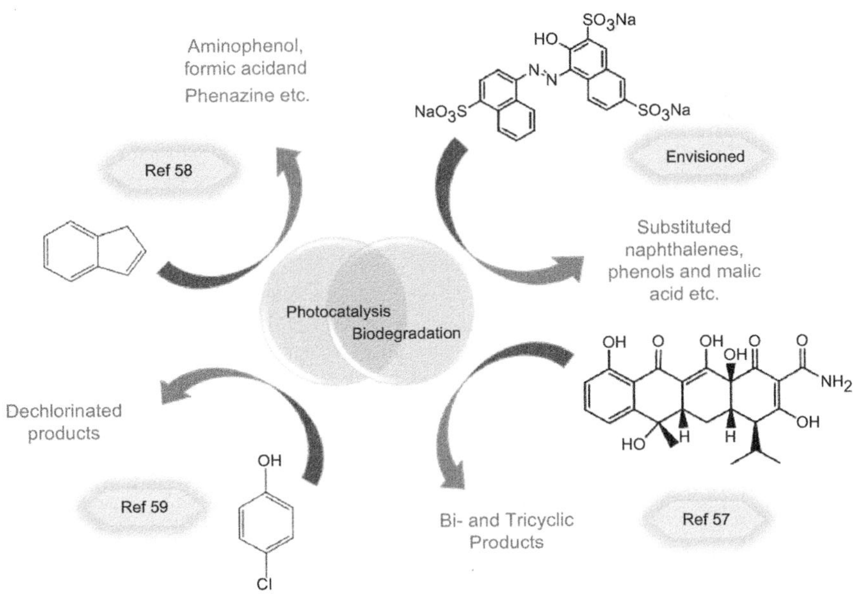

FIGURE 20.10 Schematic representation of application of intimate coupling of photocatalysis and biodegradation (ICPB) for treatment of organic pollutants.

References

[1] Bensch CN, Horak MJ, Paterson D. Interference of redroot pigweed (*Amaranthus retroflexus*), Palmer amaranth (*A. palmeri*), and common waterhemp (*A.rudis*) in soybean. Weed Sci 2003;51(1):37–73.

[2] Death of a Dye. Time Magazine 1976. Retrieved 2009-07-07.

[3] Omaye ST. Food and nutritional toxicology. CRC Press LLC; 2004, ISBN 0-203-48530-0. p. 257.

[4] Tsuda S, Murakami M, Matsusaka N, Kano K, Taniguchi K, Sasaki YF. DNA damage induced by red food dyes orally administered to pregnant and male mice. Toxicol Sci 2001;61(1):92–9.

[5] Clode SA, Hooson J, Grant D, Butler WH. Long-term toxicity study of amaranth in rats using animals exposed in utero. Food Chem Toxicol 1987;25(12):937–46.

[6] Collins TFX, McLaughlin J, Gray GC. Teratology studies on food colourings. Part II. Embryotoxicity of R salt and metabolites of amaranth (FD &C red no. 2) in rats. Food Chem Toxicol 1973;11(3):355–65.

[7] Mpountoukas P, Pantazaki A, Kostareli E, Christodoulou P, Kareli D, Poliliou S, Mourelatos C, Lambropoulou V, Lialiaris T. Cytogenetic evaluation and DNA interaction studies of the food colorants amaranth, erythrosine and tartrazine. Food Chem Toxicol 2010;48(10):2934–44.

[8] Das A, Mukherjee A. Genotoxicity testing of the food colours amaranth and tartrazine. Int J Hum Genet 2004;4(4):277–80.

[9] Zhang G, Ma Y. Mechanistic and conformational studies on the interaction of food dye amaranth with human serum albumin by multispectroscopic methods. Food Chem 2013;136:442–9.

[10] Nevado JJB, Cabanillas CG, Salcedo AMC. Simultaneous spectrophotometric determination of three food dyes by using the first derivative of ratio spectra. Talanta 1995;42:2043–51.

[11] EFSA panel on food additives and nutrient sources added to food (ANS); scientific opinion on the re-evaluation of amaranth (E 123) as a food additive on request from the european commission. EFSA Journal 2010;8(7):1649.

[12] King-Thom C. Mutagenicity and carcinogenicity of aromatic amines metabolically produced from azo dyes. J Environ Sci Health Part C Environ Carcinog Ecotoxicol Rev 2000;18:51–74.

[13] Ali Y, Ameta A. Degradation and decolouration of amaranth Dye by photo-fenton and fenton reagents: a comparative study. Int J Chem Sci 2013;11(3):1277–85.

[14] Jain AK, Sharma S, Ameta R. Enhanced photocatalytic activity of N, S-doped titania for degradation of amaranth. Merit Res J Environ Sci Toxicol 2015;3:25–30.

[15] Elgendy EM, Al-Zahrani NA. Comparative study of natural and synthetic food additive dye amaranth through photochemical reactions. Indian J Sci Res 2015;4:827–32.

[16] Silva CP, Marmitt S, Haetinger C, Stulp S. Amaranth food dye photochemical and photo-electrochemical degradation: experiments and mathematical modelling. WSEAS Trans Syst 2008;7(9):793–802.

[17] Sudrajat H, Babel S. Rapid photocatalytic degradation of the recalcitrant dye amaranth by highly active N-WO_3. Environ Chem Lett 2016;14(2):243–9.

[18] Sudrajat H, Babel S, Sakai H, Takizawa S. Rapid enhanced photocatalytic degradation of dyes using novel n-doped ZrO_2. J Environ Manag 2015;165:224–34.

[19] Ameta R, Jain N, Kothari S. Photocatalytic bleaching of amaranth dye over ZnO nanopowder. Indian J Chem Technol 2004;11:423–6.

[20] Divya N, Bansal A, Jana AK. Photocatalytic activity of transition metal ion doped titania for amaranth dye degradation. Mater Sci Forum 2012;712:85–104.

[21] Shivaraju HP. Removal of organic pollutants in the municipal sewage water by TiO_2 based heterogeneous photocatalysis. Indian J Environ Sci 2011;1:911–23.

[22] Sivaranjani K, Gopinath CS. Porosity driven photocatalytic activity of wormhole mesoporous $TiO_{2-x}N_x$ in direct sunlight. J Mater Chem 2011;21:2639–47.

[23] Devaraji P, Sathu NK, Gopinath CS. Ambient oxidation of benzene to phenol by photocatalysis on $Au/Ti_{0.98}V_{0.02}O_2$: role of holes. ACS Catal 2014;4:2844–53.

[24] Liu R, Wu HS, Yeh R, Lee CY, Hung Y. Synthesis and bactericidal ability of TiO_2 and Ag-TiO_2 prepared by coprecipitation method. Int J Photoenergy 2012;1:1–7.

[25] Thangavelu K, Annamalai R, Arulnandhi D. Preparation and characterization of nanosized TiO_2 powder by sol-gel precipitation route. Int J Emerging Technol Adv Eng 2013;3:636–9.

[26] Mau AWH, Huang CB, Kakuta N, Bard AJ, Campion A, Fox MA, White JM, Webber SE. Hydrogen photoproduction by nafion/cadmium sulfide/platinum films in water/sulfide ion solutions. J Am Chem Soc 1984;106(22):6537–42.

[27] Kuczynski J, Thomas JK. Photophysical properties of cadmium sulfide deposited in porous Vycor glass. J Phys Chem 1985;89(13):2720–2.

[28] Anpo M, Zhang SG, Higashimoto S, Matsuoka M, Yamashita H, Ichihashi Y, Matsumura Y, Souma Y. Characterization of the local structure of the vanadium silicalite (VS-2) catalyst and its photocatalytic reactivity for the decomposition of NO into N_2 and O_2. J Phys Chem B 1999;103:9295–301.

[29] Zhang JL, Minagawa M, Matsuoka M, Yamashita H, Anpo M. Photocatalytic decomposition of NO on Ti-HMS mesoporous zeolite catalysts. Catal Lett 2000;66:241–3.

[30] Naik A, Fernandes W, Salkar A, Morajkar P. Photocatalysed amaranth degradation kinetics over supported TiO_2 nanocatalyst. Goa India: Goa University Taleigao Plateau 403206; 2018. Unpublished Manuscript.

[31] Thommes M, Kaneko K, Neimark AV, Olivier JP, Reinoso FR, Rouquerol J, Sing KSW. Physisorption of gases, with special reference to the evaluation of surface area and pore size distribution (IUPAC Technical Report). Pure Appl Chem 2015:1–19.

[32] Vinodgopal K, Wynkoop DE, Kamat PV. Environmental photochemistry on semiconductor surfaces: photosensitized degradation of a textile azo dye, acid orange 7, on TiO_2 particles using visible light. Environ Sci Technol 1996;30(5):1660–6.

[33] Karkmaz M, Puzenat E, Guillard C, Herrmann JM. Photocatalytic degradation of the alimentary azo dye amaranth mineralization of the azo group to nitrogen. Appl Catal, B 2004;51:183–94.

[34] Steter JR, Barros WRP, Lanza MRV, Motheo AJ. Degradation route for amaranth dye by sonoelectrochemical process using BDD anode. Chemosphere 2014;117:200–7.

[35] Sathish M, Viswanathan B, Viswanath RP, Gopinath CS. Synthesis characterization electronic structure and photocatalytic activity of nitrogen-doped TiO_2 nanocatalyst. Chem Mater 2005;17:6349–53.

[36] Ghande AR, Fernandes JB. A simple method to synthesize N-doped rutile titania with enhanced photocatalytic activity in sunlight. J Solid State Chem 2005;178:2953–7.

[37] Voyksner RD, Strauband R, Keever JT. Determination of aromatic amines originating from Azo Dyes by chemical reduction combined with liquid chromatography/mass spectrometry. Environ Sci Technol 1993;27:665–72.

[38] Naik AP, Salkar AV, Majik MS, Morajkar PP. Enhanced photocatalytic degradation of amaranth dye on mesoporous anatase TiO_2: evidence of C–N, N=N bond cleavage and identification of new intermediates. Photochem Photobiol Sci 2017;16:1126–38.

[39] U.S. EPA. Health and environmental effects profile for 1-amino-2-naphthol and 1-amino-2-naphthol hydrochloride. Washington, D.C: U.S. Environmental Protection Agency; 1986. EPA/600/X-87/029 (NTIS PB89120315).

[40] Field FE, Roberts G, Hallowes RC, Palmer AK, Williams KE, Lloyd JB. Trypan blue: identification and teratogenic and oncogenic activities of its coloured constituents. Chem Biol Interact 1977;16:69–88.

[41] Xu H, Heinze TM, Chen S, Cerniglia CE, Chen H. Anaerobic metabolism of 1-amino-2-naphthol-based azo dyes (Sudan dyes) by human intestinal microflora. Appl Environ Microbiol 2007;73:7759–62.

[42] Weisburger JH. Mutagenic carcinogenic and chemopreventive effects of phenols and catechols. ACS Symp Ser 1992;507:35–47.

[43] Chengalroyen MD, Dabbs ER. The microbial degradation of azo dyes: mini review World. J Microbiol Biotechnol 2013;29:389–99.

[44] Singh P, Borthakur A. A review on biodegradation and photocatalytic degradation of organic pollutants: a bibliometric and comparative analysis. J Clean Prod 2018;196:1669–80.

[45] Pandey A, Singh P, Iyengar L. Bacterial decolorization and degradation of azo dyes. Int Biodeterior Biodegrad 2007;59:73–84.

[46] Ghodake G, Jadhav U, Tamboli D, Kagalkar A, Govindwar S. Decolorization of textile dyes and degradation of mono-azo dye amaranth by *Acinetobacter calcoaceticus* NCIM 2890. Indian J Microbiol 2011;51:501–8.

[47] Anjaneya O, Shrishailnath SS, Guruprasad K, Nayak AS, Mashetty SB, Karegoudar TB. Decolourization of amaranth dye by bacterial biofilm in batch and continuous packed bed bioreactor. Int Biodeterior Biodegrad 2013;79:64–72.

[48] Gavril M, Hodson PV, McLellan J. Decolourization of amaranth dye by bacterial biofilm in batch and continuous packed bed bioreactor. Can J Microbiol 2007;53:313–26.

[49] Susana RC. Dye removal by immobilized fungi. Biotechnol Adv 2009;27:227–35.

[50] Talha MA, Goswami, Giri BS, Sharma A, Rai BN, Singh RS. Bioremediation of congo red dye in immobilized batch and continuous packed bed bioreactor by *Brevibacillus parabrevis* using coconut shell bio-char. Bioresour Technol 2018;252:37–43.

[51] Gomi N, Yoshida S, Matsumoto K, Okudomi M, Konno H, Hisabori T, Sugano Y. Degradation of the synthetic dye amaranth by the fungus *Bjerkandera adusta* Dec 1: inference of the degradation pathway from an analysis of decolorized products. Biodegradation 2011;22:1239–45.

[52] Halaburg V, Karegoudar T. Oxidative degradation of amaranth dye by a new genus *Kerstersia* sp. Biocatal Biotransform 2016;34(6):265–71.

[53] Saroj S, Dubey S, Agarwal P, Prasad R, Singh P. Evaluation of the efficacy of a fungal consortium for degradation of azo dyes and simulated textile dye effluents. Sustain Water Resour Management 2015;1:233–43.

[54] Yiguo H, Jun G, Zhicheng X, Cuiyun M, Meiying X, Guoping S. Reduction and partial degradation mechanisms of naphthylaminesulfonic azo dye amaranth by *Shewanella* decolouration is S12. Appl Microbiol Biotechnol 2007;7:5647–54.

[55] Marsolek MD, Torres CI, Hausner M, Rittmann BE. Intimate coupling of photocatalysis and biodegradation in a photocatalytic circulating-bed biofilm reactor. Biotechnol Bioeng 2008;101:83–92.

[56] Zhou D, Xu Z, Dong S, Huo M, Dong S, Tian X, Cui B, Xiong H, Li T, Ma D. Intimate coupling of photocatalysis and biodegradation for degrading phenol using different light types: visible light vs UV light. Environ Sci Technol 2015;49:7776–83.

[57] Xiong H, Zou D, Zhou D, Dong S, Wang J, Rittmann B. Enhancing degradation and mineralization of tetracycline using intimately coupled photocatalysis and biodegradation (ICPB). Chem Eng J 2017;316:7–14.

[58] Shan S, Zhang Y, Zhang Y, Hui L, Shi W, Zhang Y, Rittmann BE. Comparison of sequential with intimate coupling of photolysis and biodegradation for benzotriazole. Front Environ Sci Eng 2017;11(6):1–9.

[59] Zhang C, Fu L, Xu Z, Xiong H, Zhou D, Huo M. Contrasting roles of phenol and pyrocatechol on the degradation of 4-chlorophenol in a photocatalytic–biological reactor. Environ Sci Pollut Res 2017;24:24725–31.

Chapter 21

Role of nanoparticles in advanced biomedical research

R.K. Kunkalekar[1], Umesh B. Gawas[2]

[1]*Department of Chemistry, Goa University, Taleigao Plateau, Goa, India;* [2]*Department of Chemistry, Dnyanprassarak Mandal's College and Research Centre, Assagao, Goa, India*

21.1 Introduction

Nanotechnology refers to the branch of science and engineering dedicated to materials with dimensions in the range 1−100 nm. Nanoparticles (NPs) are of great scientific interest as they are effectively a bridge between bulk materials and atomic or molecular structures. Morphological and topographical features of NPs play important roles in their potential applications in various fields like biomedical, optical, storage system, magnetic separation, targeted drug delivery, electronics, etc. Consequently, with a wide range of applications available, NPs have the potential to make a significant impact on society. The characteristic feature of NPs is that the physical and chemical properties are significantly different from bulk counterparts [1]. In the case of magnetic nanomaterials, the quantum size effects and large surface area dramatically changes their magnetic behavior and they exhibit superparamagnetic phenomena with quantum tunneling of magnetization because at nanodimensions below a critical size each particle behaves as a single magnetic domain [2]. The research associated with nanomaterials in the biological field is mostly directed toward their use in medical diagnosis and treatment of cancer-related ailments. The potential for drug delivery systems involving NPs offers several advantages such as (1) the ability to target specific locations in the body; (2) reduction of the drug quantity needed to attain a particular concentration in the vicinity of the target cells; and (3) reduction of the concentration of the drug in the normal cells to minimize the severity of size effects [3]. This chapter gives an overview of NPs in advanced biomedical research. The biomedical roles of metallic, nonmetallic (metal oxide NPs), and carbon-based nanomaterials with special reference to cancer diagnosis and treatment have been discussed.

Advances in Biological Science Research. https://doi.org/10.1016/B978-0-12-817497-5.00021-5

21.2 Cancer therapy

Cancer refers to a group of diseases characterized by abnormal cell growth with the potential to invade or spread to other parts of the body with the exception of benign tumors which do not spread to other parts of the body. With severe health consequences, cancers are a major cause of death worldwide [4,5]. According to the World Health Organization, worldwide deaths due to cancer are estimated to be 8.2 million, which is about 13% of the total deaths, and this is expected to rise to 22 million by 2030 [6,7]. In the United States alone, 1,735,350 new cancer cases and 609,640 deaths due to different types of cancers are projected to occur in 2018. In India, the estimated number of people living with cancer is around 2.5 million, and every year over seven lakhs are diagnosed for cancer out of which 5,56,400 deaths occur due to cancer-related diseases [8]. The most common types of cancer include breast cancer affecting females, which is the leading cause of cancer mortality next to lung cancer [9]. There is no single cause for cancer; it is caused due to the interaction of many factors together, which may be genetic, environmental, or constitutional characteristics of individuals. The treatments for cancer depend on its type, advancement in the body, all of which have some limitations and side effects [10]. The commonly used cancer treatments include: surgery, radiation therapy, chemotherapy, immunotherapy, hormone therapy, stem cell transplant, and precision medicine. The major disadvantage of the existing cancer therapies is their inability to deliver specific drugs to the target, causing drugs to act on both cancerous and healthy cells, leading to systemic toxicity, which also prevents sufficient drug concentration to be delivered to tumor sites, thus making the cancer treatment deemed to be ineffective. Most of anticancer agents used in conventional methods for cancer treatment are hydrophobic with low solubility and high metabolism, consequently, the bioavailability of drug decreases [11]. Furthermore, the available conventional chemotherapeutic treatments are limited in their solubility, selectivity toward tumor cells and are increasingly multidrug resistant (MDR), and hence, the resistance of tumors against anticancer drug increases [12]. This has generated the necessity for alternative methods for cancer treatment, and one approach that has evolved in the recent past is targeted drug delivery. It works selectively on cancer cells with minimum side effects on normal cells, tissues, and the body as a whole. For an effective cancer treatment, it is desirable to increase the efficacy of anticancer drugs, which can be achieved by specific targeted drug delivery, thereby minimizing the side effects. The recent development of nanomedicines offers tremendous opportunities in specific targeted drug delivery for anticancer therapy. The binding ability and specificity of NPs to bind malignant tumor cells can be enhanced by conjugation with suitable biomolecular ligands and high surface area that may be utilized as carriers for therapeutic and diagnostic agents [13]. Among metallic NPs, colloidal silver, in addition to its antitumor activity exhibits excellent in vivo

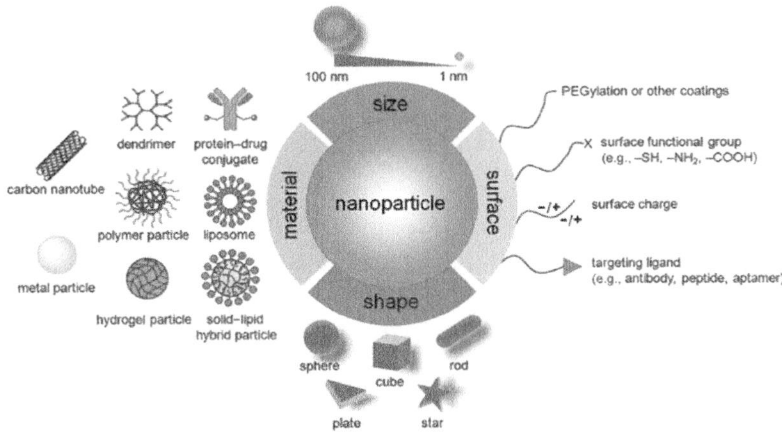

FIGURE 21.1 Physicochemical properties of nanomaterials.

distribution and very low toxicity [14,15]. The in vitro antiproliferative activity of silver NPs (AgNPs) is reported on breast cancer cell lines (MCF-7, MDA-MB-231) [16]. Gold NPs (AuNPs) display unique physicochemical properties that can be utilized in building multifunctional platforms for transport of low solubility and poor pharmacokinetic profile therapeutics to the tumor target as well as to sensitize cells and tissues to the treatment [17]. Superparamagnetic iron oxide nanoparticles (SPIONs), due to their low in vivo toxicity and high tolerability, offer great advantage in cancer diagnosis as well as anticancer therapy by hyperthermia [18]. Carbon-based nanomaterials such as graphene oxide (GO), reduced graphene oxide (rGO), and carbon nanotubes (CNTs) have been explored as promising drug carriers for targeted drug delivery systems [19]. GO has attracted more attention due to its effective endocytosis, biocompatibility, and large surface area for drug loading. Furthermore, GO and rGO display good dispersing capability in water and physiological environments because of the surface functionalities that may help in forming hydrogen bonds with the associated drug molecules [20,21]. This hydrogen bonding, in addition to the π-π stacking and hydrophobic interaction, can assist drug loading on the nanocarriers [22]. Physicochemical properties of various nanomaterials are illustrated in Fig. 21.1.

21.3 Metal nanoparticles as drug delivery and anticancer agents

Metal nanoparticles (MNPs) have enthralled researchers for over a century and are now profoundly utilized in material science, catalysis, fuel cells, biomedical sciences, electrochemical sensors, and biosensors. MNPs' surface can be suitably modified with different functional groups that help them to conjugate with antibodies, ligands, and drugs of interest and thus open a wide range of potential biomedical applications. The size- and shape-controlled synthesis of MNPs with high surface area is important in present-day cutting-edge materials. NPs of noble metals such as Pt, Au, Ag, and Cu have potential applications in catalysis and other fields [23–27]. Biomedical functions and applications of noble metals like Au and Ag are presented in the following section.

21.3.1 Gold nanoparticles

Gold nanoparticles are being looked upon as an ideal candidate for various biomedical applications because of their characteristic properties such as small size, unique photophysical features, easy surface functionalization, and biocompatibility (Fig. 21.2). These properties render AuNPs as a versatile nanoplatform for emerging biomedical applications like cell imaging, ultrasensitive detection, transfection, drug transport and delivery system, antiviral agent, efficient material for photothermal ablation, etc. [28,29]. X. Zheng et al. [30] have obtained the renal clearable ~2 nm glutathione-coated AuNPs. The zwitterionic coating was found to minimize nonspecific MPS uptake in balb/c mice. Further, the pharmacokinetic measurements in animal model have indicated rapid distribution and circulation with a blood-elimination half-life

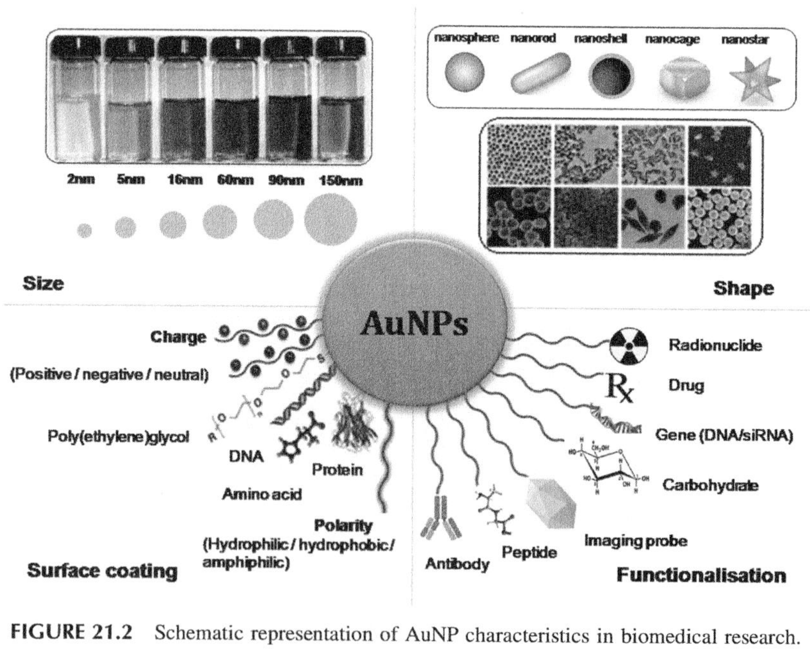

FIGURE 21.2 Schematic representation of AuNP characteristics in biomedical research.

of 12.7 hr. The long blood retention time signifies the ability of these coated AuNPs to passively target tumor-bearing nude mice [31]. K. Huang et al. [32] observed superior cancer cell penetration and in vivo tumor accumulation for N-(2-mercatopropionyl)glycine-coated ultrasmall AuNPs (2 nm) as compared with 6 and 15 nm particles. P. Xu et al. [33] have studied remote control drug release employing supramolecular assembly system and synergistic chemo-photothermal therapy for cancer treatment. They have used cucurbit (CB) [7] uril-stabilized gold nanostar (GNS) to encapsulate anticancer drug camptothecin (CPT) via host-guest chemistry. Importantly, the drug release was triggered using near infrared (NIR) light and CB [7] performs a dual role, that is, acting as surfactant to improve stability of GNS in aqueous solution and as cage for intermolecular assembly of CPT molecules. S. Gulla et al. [34] have evaluated the bioactivity of novel tumor vasculature targeting noncytotoxic Au-CGKRK nanoconjugates. They observed >70% enhancement in overall survivability in melanoma-bearing mice by intraperitoneal administration of the Au-CGKRK NPs complexed with both PD-L1siRNA and STAT3siRNA; while, the biodistribution study using NIR dye-loaded Au-CGKRK nanoconjugates has revealed the accumulation of dye in tumor site in the mice. S. Ke et al. [35] have observed higher potency in promoting apoptosis for AuNPs combined with tumor necrosis factor—related apoptosis-inducing ligand (TRAIL) in nonsmall-cell lung cancer cells. The AuNPs-TRAIL combination causes excessive mitochondrial fragmentation in cancer cells, which is accompanied by dramatic increase in mitochondrial recruitment of dynamin-related protein 1 (Drp1), mitochondrial dysfunction, and enhancement of autophagy induction resulting in increased apoptosis in exposed cells.

21.3.2 Silver nanoparticles

Silver NPs are one of the most vital and fascinating nanomaterials extensively explored in biological and medical research. AgNPs are being employed in various fields such as medical, food, health care, consumer, as well as industrial purposes due to their unique optical, electrical, thermal, and biological properties [36]. Commercially, AgNPs find use as antibacterial agents in industrial, household, health care—related products, in consumer products, medical device coatings, optical sensors, and cosmetics. As well, AgNPs also find biomedical applications such as drug delivery and anticancer agents in targeted drug therapy [37,38]. P. Roychoudhary et al. [39] investigated the antiproliferative activity of AgNPs against leukemic cell lines (K562, MOLT-3, REH) through MTT assay. AgNPs synthesized by *Lyngbya majuscula* displayed dose- and time-dependent activity in REH cells and 4′,6-diamidino-2-phenylindole (DAPI) staining clearly revealed the fragmentation of cancer cells due to treatment of AgNPs. J. Blanco et al. [40] have investigated effects of AgNPs on a human lung carcinoma cell line (A549). They have observed decrease in p53, p21, MDM2, and caspase-3 expression after low dosage daily and high single dosage exposure to AgNPs; the opposite effect was noticed with changed frequency of doses/administration, which clearly indicate time- and dose-dependent antiproliferative activity. M. Khan et al. [41] observed smallest IC_{50} values for A549 with higher cytotoxic activity than the reference tamoxifen for AgNPs decorated with highly reduced graphene oxide. Further, the cytotoxic effect was found to be proportional to the concentration of AgNPs and cell death mechanism due to cell cycle arrest at G0/G1 phase and apoptosis induction. P. Yuan et al. [42] reported the synthesis and cytotoxic studies of GO-AgNPs nanocomposite against human neuroblastoma cancer cells (SH-SY5Y). The GO-AgNPs nanocomposite displayed significant cytotoxicity at lower concentrations. Further, they have established the molecular mechanism of cytotoxicity. P. Netchareonsirisuk et al. [43] have reported the antiproliferative activity of AgNPs capped with sodium alginate and poly(4-styrensulfonic acid-comaleic acid) sodium salt (PSSMA) against human normal skin fibroblast (CCD-986SK) and malignant melanoma (A375). The sodium alginate–capped AgNPs displayed high selectivity against A375 cell line through apoptosis and necrosis, while PSSMA-capped AgNPs exhibited nonselective toxicity. R. Bhanumathi et al. [44] investigated drug discharge capacity and anticancer effect of folic acid (FA) and berberine (BBR), an isoquinoline alkaloid–loaded AgNPs. They have used BBR encapsulated on citrate-capped AgNPs in conjugation with polyethylene glycol (PEG)-functionalized FA. The in vivo antitumor efficiency of NP-encapsulated drug showed significant restraint of tumor progression. The toxicities behavior of FA-PEG@BBR-AgNPs against different organs was established by histopathological observations.

21.4 Metal oxide nanoparticles as drug delivery and anticancer agent

Metal oxide nanoparticles (MONPs) are important compounds in the materials chemistry field, attracting considerable interest due to the potential technological applications of these compounds. The importance of these materials in different areas such as medicine, information technology, catalysis, energy storage, piezoelectric devices, corrosion protection coatings, and sensing, etc. has driven much research in developing synthetic pathways to such nanostructures. MONPs with different morphological features such as nanorods, nanotubes, nanospheres, nano-hollow spheres, and nanofibrous materials can be conveniently synthesized using different techniques such as hydrothermal, precursor, sol-gel, etc. MONPs have a unique structure, high surface area, unusual redox properties, good mechanical stability, and biocompatibility. For these reasons, MONPs have attracted considerable interest in the fields of biomedical therapeutics, bioimaging, and biosensing and are important components in medical implants, cancer diagnosis and therapy, and in neurochemical monitoring [45–50].

21.4.1 Iron oxide nanoparticles

Iron oxide NPs (IONPs) have been investigated for magnetic properties, which find several important applications such as magnetically mediated hyperthermia for cancer treatment [51], contrast agent MRI [52], treatment of anemia [53], etc. However, the recent studies have indicated that the ability of NPs to generate reactive oxygen species (ROS) can be used in cancer therapy [54,55]. Among iron-based NPs, SPIONs are being explored extensively, especially for their therapeutic and diagnostic applications. SPIONs, in particular magnetite (Fe_3O_4) and maghemite (γ-Fe_2O_3), are considered as means of transport for targeted drug delivery as they can be guided to the tumor site through an external magnetic field and this prevents drug diffusion to the rest of the body. Due to their superparamagnetic nature, SPIONs lose their magnetism and enter the blood circulation when the external magnetic field ceases [56,57]. The effectiveness of NPs in diagnostic or therapeutic applications can be increased by surface coating, which enhances the physicochemical and biological properties of nanomaterials by providing protection against corrosion and environmental degradation. Also, the biocompatibility and colloidal stability of NPs increases, thereby enhancing the drug release at therapeutic site [58]. N. Mallick et al. [59] have investigated the chondroitin-4-sulphate (CS)-capped SPIONs for loading of the anticancer agent doxorubicin hydrochloride (DOX). The in vitro drug release profile indicated 96.77% of DOX release within 24 hr and MTT assay in MCF7 cells has revealed significantly higher toxicity for CS-SPIONs-DOX with IC_{50} value 6.294 ± 0.4169. S. Mondal et al. [60] have investigated hydroxyapatite (HAp)-coated IONPs synthesized using solvothermal and chemical precipitation for magnetic hyperthermia-mediated cancer therapy. The nontoxic nature of IONPs-HAp was established by trypan blue and MTT assay. The hyperthermia study performed on osteosarcoma cells (MG-63) displayed excellent hyperthermia effect with specific absorbance rate value 85 W/g. They have achieved hyperthermia temperature of about 45°C within 3 min, which could kill nearly all the studied cancer cells within 30 min. C. Saikia et al. [61] have studied FA-tagged aminated starch/ZnO-coated SPIONs for targeted delivery of anticancer drug, curcumin. The cytotoxicity study of the drug-loaded NPs was analyzed by MTT assay in human lymphocytes, liver cancer cells (HepG2), and MCF7, wherein NPs were found to be compatible with human lymphocyte cells and reduce the cell viability up to 61% with 0.5% ZnO concentration in HepG2 and MCF7 cells. The cell uptake efficiency and ROS generation was studied using HepG2 cell lines. The ROS generation was found to enhance with increasing ZnO concentration in the system. N. Moghadam et al. [62] reported improved antiproliferative effect of the nevirapine (Nev) on cancer cell line (Hela) by loading onto chitosan-coated magnetic iron oxide nanoparticles (MIONPs). The in vitro ct-DNA-binding study has revealed DNA aggregation on Nev-loaded MIONPs through groove-binding mode.

21.4.2 Miscellaneous

Photocatalyzed titanium dioxide NPs (TiO_2NPs) have been shown to eradicate cancer cells. However, the required in situ introduction of UV light limits the use of such therapy in humans. Hydrophilized TiO_2 NPs (HTiO_2NPs) produce ROS in vivo when activated by ultrasound to eradicate tumor. HTiO_2 NPs-based sonodynamic therapy generates a high level of ROS both in vitro and in vivo that can cause destruction of the tumor [63]. Cerium oxide NPs (CONPs) are a novel and very interesting material for radiation therapy, possessing the "smart" capacity to selectively induce the death of irradiated cancer cells by increasing oxidative stress and apoptosis, while protecting the surrounding tissue from radiation-induced damage and oxidative stress. Therefore, CONPs have the unique feature of acting as radio-protecting, as well as radio-sensitizing agents simultaneously [64]. Application of zinc oxide NPs (ZnONPs) has shown that they are most efficacious on cancer cells (T98G), moderately effective on tumor cell line (KB), and least toxic on normal human (HEK) cells. These results demonstrated that treatment with ZnONPs sensitizes T98G cells by increasing both mitotic (linked to cytogenetic damage) and interphase (apoptosis) death [65]. Copper oxide NPs can be used to kill human liver cancer cells. The small spherical NPs generate ROS (such as superoxide anions, hydroxyl radical, and hydrogen peroxide) that damages the membranes and the DNA of the cancer cells, eventually inducing their death. ROS also turned on several death-triggering genes, driving the cancer cells to commit mass suicide. Also, these NPs have shown cytotoxic effects on A549 cells [66]. A. Wani et al. [67] reported that PEGylation of mesoporous silica nanorods (MSNRs) prevented dose-dependent hemolysis in concentration range 0–10 mg/L, improved colloidal stability of MSNR, and increased mitoxantrone (MTX) release. Also, decrease in the IC_{50} of MTX and MTX-loaded MSNR was observed under hypoxic conditions.

21.5 Carbon-based nanoparticles as drug delivery and anticancer agents

Graphene, carbon nanotubes, and fullerenes are important classes of carbon-based nanoparticles. They have a unique pore structure, adsorptive capacity, electronic properties, and acidity. This has generated tremendous interest in these nanomaterials for applications in fields such as physics, chemistry, biology, and medicine. The basic building blocks of all the carbon nanostructures are a single layer of graphite that consists of hexagonally aligned sp^2 hybridized carbon atoms forming a hexagonal honeycomb-like lattice. Graphene is the thinnest two-dimensional single layer of hexagonal packed carbon atoms that has attracted researchers all over the world [68–71]. Graphene oxide is a product of chemical exfoliation of graphite. It is typically synthesized by reacting graphite powders with strong oxidizing agents such as

$KMnO_4$ in concentrated sulfuric acid. GO is described as a graphene sheet modified with different types of organic functionalities such as carboxyl (-COOH), carbonyl (-CO), hydroxyl (-OH), and epoxy (C—O—C). The presence of these functionalities renders GO hydrophilicity and the potential to improve the solubility of some water-insoluble drugs. Hence, GO is preferred over pristine graphene for drug delivery application [72]. Reduced GO is another most commonly explored graphene material for biomedical applications. It is generally obtained by chemical, thermal, and electrochemical reduction of GO [73]. Both GO and rGO hold great potential in biomedicine such as polymer composites, biological sensors, bioimaging, targeted drug delivery, and photothermal therapy (PTT) [74]. CNTs are allotropes of carbon, which consist of hexagonally aligned sp^2 hybridized carbon atoms interlinked with each other to form a tubular shape with an outer diameter ranging from 4 to 30 nm. Structurally, they are similar to graphite sheets that are rolling upon themselves. The rolled sheets can be single, double, or many walls, and therefore they are named as single-walled, double-walled, or multiwalled carbon nanotubes (MWNTs), respectively. Due to their unique physical, chemical, and mechanical characteristics, these materials have wide applications in various areas including polymer science, biomedical research, energy storage, electrodes, gas sensors, catalyst support, etc. [75,76].

21.5.1 Graphene oxide/reduced graphene oxide for drug delivery

Functionalized GO has been extensively investigated for anticancer therapy because of its high water solubility and biocompatibility. Z. Rao et al. [77] investigated complex of amino-modified GO with carboxymethycellulose as a carrier of DOX. A cumulative drug release of 65.2% was observed at pH 5. The cytotoxicity studies on Hela cell and mouse fibroblasts (NIH-3T3) cells by MTT assay have indicated good biocompatibility with no cytotoxicity. N. Duran et al. [78] have developed hybrids by coupling small interfering RNA (siRNA)-GO-PEG (6ARM-poly(ethylene glycol)amine-PEI (polyethylenimine) as a carrier to administer DOX in nonmuscle invasive bladder cancer (NMIBC) treatment. In vivo studies revealed 60% normal bladder diagnosis for the association of GO-COOH-DOX and GO-PEG-PEI/siRNA. A. Deb et al. [79] demonstrated use of the anticancer drug camptothecin (CPT) loaded onto GO nanomaterial with PEG and FA against MCF-7 by MTT assay. They observed higher cytotoxicity to the cancer cell with CPT loaded onto GO-PEG-FA in comparison with free drug. Also, the nontoxic nature of the drug composite was confirmed by cell viability assay wherein 95% of cell count was observed after 24 hr incubation. N. Hussien et al. [80] investigated aptamer-conjugated magnetic graphene oxide (MGO) nanocarrier for targeted drug delivery cancer treatment. They used nanosize magnetite particles on GO layer with aptamer as a targeting moiety and paclitaxel (PAC) as an anticancer agent. In vitro results have indicated 95.75% entrapment efficiency and pH-sensitive drug release. Cytotoxicity studies have shown biocompatibility of MGO nanocarrier with over 80% cell viability for fibroblast cell line (L-929) with high cytotoxic effect of PAC-loaded MGO on MCF-7. S. Kiew et al. [81] studied dextrin-conjugated GO nanocarrier as drug delivery system to respond to a tumor-associated stimulus, α-amylase. They observed 1.5-fold higher release and 2-fold increase in the cytotoxic effect of DOX in dextrin-conjugated GO. Also, higher permeabilities through fenestrated endothelial barrier were observed for GO-based nanocarrier. R. Sousa et al. [82] developed hyaluronic acid (HA) functionalized-rGO for cancer PTT. The rGO was obtained by greener route employing L-ascorbic acid as reducing agent and functionalized using hyaluronic acid–based amphiphilic polymer. The functionalization with amphiphile improved its thermal stability, cytocompatibility, and internalization by CD44 overexpressing cells, which indicates its potential for targeted cancer therapy.

21.6 Conclusions

Nanotechnology has provided novel and powerful materials that may be used in the treatment and diagnosis of cancer. However, there are still limitations due to the heterogeneity of the cells used for each tumor model in vitro and/or in vivo, which make it difficult to do a comparison between the different studies. Another limitation is the formation of protein corona when NPs reach the blood and interact with the plasma proteins, affecting in vivo distribution and clearance. Nonetheless, the majority of products, reagents, and drugs being used for the development of these nanoscale theranostic agents have still to be approved by the main supervising agencies, such as the FDA and EMA. Research continues in this area, and more information about the distribution, biocompatibility, and low toxicity for normal tissues is necessary prior to clinical trials. Thus far, there are some questions whose answers still provide no clear understanding about the design and application of NPs, such as pharmacokinetics, biodistribution, and side effects of the nanotherapies, and safety profile of NPs before and after conjugation and toxicity. Even though there is no general mechanism for making NPs universally “nontoxic” to all living cells and all organisms, there are important findings that can be applied for increasing nanoparticle biocompatibility and reducing cytotoxic interactions in vivo and in vitro. Although both metallic and nonmetallic NPs have shown potential to be powerful tools against cancer, they still need further optimization and characterization for complete understanding of therapeutic mechanisms. It is now time to start translating these promising nanoplatforms to the clinical settings toward widespread effective therapy strategies in the fight against cancer.

Acknowledgments

Author UBG would like to acknowledge the Research Center of Dnyanprassarak Mandal’s College for providing financial support via grant no. DNY/CC/2014-15/03/1198.

References

[1] Babes L, Denizot B, Tanguy G, Jeune L, Jallet P. Synthesis of iron oxide nanoparticles used as MRI contrast agents: a parametric study. J Colloid Interface Sci 1999;212:474–82.

[2] Goya GF, Berquo TS, Fonseca FC, Morales MP. Static and dynamic magnetic properties of spherical magnetite nanoparticles. J Appl Phys 2003;94(5):3520–8.

[3] Arruebo M, Pacheco RF, Ibarra MR, Santamaría J. Magnetic nanoparticles for drug delivery. Nano Today 2007;2:22–32.

[4] Madani SY, Naderi N, Dissanayake O, Tan A, Seifalian AM. A new era of cancer treatment: carbon nanotubes as drug delivery tools. Int J Nanomed 2011;6:2963–79.

[5] Wicki A, Witzigmann D, Balasubramanian V, Huwyler J. Nanomedicine in cancer therapy: challenges, opportunities, and clinical applications. J Contr Release 2015;200:138–57.

[6] Siegel RL, Miller KD, Jemal A. Cancer statistics. CA Cancer J Clin 2018;68(1):7–30.

[7] GBD 2015 mortality and causes of death collaborators. Global, regional, and national life expectancy, all-cause mortality, and cause-specific mortality for 249 causes of death, 1980–2015: a systematic analysis for the global burden of disease study 2015. Lancet 2016;388:1459–544.

[8] Reddy KS. Global burden of disease study 2015 provides GPS for global health 2030. Lancet 2016;388:1448–9.

[9] DeSantis C, Siegel R, Bandi P, Jemal A. Breast cancer statistics, 2011. CA Cancer J Clin 2011;61(6):409–18.

[10] Nandakumar A. National cancer registry programme. Indian council for medical research, consolidated report of the population based cancer registries 1990–96. New Delhi: Indian Council of Medical Research; 2009.

[11] Probst CE, Zrazhevskiy P, Bagalkot V, Gao X. Quantum dots as a platform for nanoparticle drug delivery vehicle design. Adv Drug Deliv Rev 2013;65(5):703–18.

[12] Chidambaram M, Manavalan R, Kathiresan K. Nanotherapeutics to overcome conventional cancer chemotherapy limitations. J Pharm Pharmaceut Sci 2011;14:67–77.

[13] Nguyen VL, Cao MT, Masayuki N. The recent patents and highlights of functionally engineered nanoparticles for potential applications in biology, medicine, and nanomedicine. Curr Phys Chem 2014;4:173–94.

[14] Stensberg MC, Wei Q, McLamore ES, Porterfield DM, Wei A, Sepúlveda MS. Toxicological studies on silver nanoparticles: challenges and opportunities in assessment, monitoring and imaging. Nanomedicine 2011;6:879–98.

[15] Munger MA, Radwanski P, Hadlock GC, Stoddard G, Shaaban A, Falconer J. In vivo human time-exposure study of orally dosed commercial silver nanoparticles. Nanomedicine 2014;10:1–9.

[16] Rao PV, Nallappan D, Madhavi K, Rahman S, Wei LJ, Gan SH. Phytochemicals and biogenic metallic nanoparticles as anticancer agents. Oxid Med Cell Longev 2016. https://doi.org/10.1155/2016/3685671.

[17] Dreaden EC, Austin LA, Mackey MA, El-Sayed MA. Size matters: gold nanoparticles in targeted cancer drug delivery. Ther Deliv 2012;3:457–78.

[18] Gobbo OL, Sjaastad K, Radomski MW, Volkov Y, Prina-Mello A. Magnetic nanoparticles in cancer. Theranostics 2015;5:1249–63.

[19] John AA, Subramanian AP, Vellayappan MV, Balaji A, Mohandas H, Jaganathan SK. Carbon nanotubes and graphene as emerging candidates in neuroregeneration and neurodrug delivery. Int J Nanomed 2015;10:4267–77.

[20] Zhou T, Zhou X, Xing D. Controlled release of doxorubicin from graphene oxide based charge-reversal nanocarrier. Biomaterials 2014;35:4185–94.

[21] Yang X, Wang Y, Huang X, Ma Y, Huang Y, Yang R, Duan H, Chen Y. Multifunctionalized graphene oxide based anticancer drug-carrier with dual-targeting function and pH-sensitivity. J Mater Chem 2011;21:3448–54.

[22] Goenka S, Sant V, Sant S. Graphene-based nanomaterials for drug delivery and tissue engineering. J Contr Release 2014;173:75–88.

[23] Wei D, Qian W. Facile synthesis of Ag and Au nanoparticles utilizing chitosan as a mediator agent. Colloids Surf B Biointerfaces 2008;62:136–42.

[24] Nadagouda MN, Varma RS. Green synthesis of Ag and Pd nanospheres, nanowires, and nanorods using vitamin b_2: catalytic polymerisation of aniline and pyrrole. J Nanomater 2008;1–8.

[25] Fkiri A, Mezni A, Robert C, Caps V, Smiri LS. Synthesis of monodisperse gold octahedral in polyol: selective oxidation of stilbene. Colloids Surf, A 2017;530:85–92.

[26] Hatakeyama Y, Morita T, Takahashi S, Onishi K, Nishikawa K. Synthesis of gold nanoparticles in liquid polyethylene glycol by sputter deposition and temperature effects. J Phys Chem C 2011;115(8):3279–85.

[27] Panda BR, Chattopadhyay A. Synthesis of Au nanoparticles at "all" pH by H_2O_2 reduction of $HAuCl_4$. J Nanosci Nanotechnol 2007;7(6):1911–5.

[28] Giljohann DA, Seferos DS, Daniel WL, Massich MD, Patel PC, Mirkin CA. Gold nanoparticles for biology and medicine. Angew Chem Int Ed 2010;49:3280–94.

[29] Dreaden EC, Alkilany AM, Huang X, Murphy CJ, El-Sayed MA. The golden age: gold nanoparticles for biomedicine. Chem Soc Rev 2012;41:2740–79.

[30] Zhang XD, Chen J, Luo Z, Wu D, Shen X, Song SS. Enhanced tumor accumulation of sub-2 nm gold nanoclusters for cancer radiation therapy. Adv Healthc Mater 2014;3:133–41.

[31] Zhang XD, Luo Z, Chen J, Song S, Yuan X, Shen X. Ultrasmall glutathione-protected gold nanoclusters as next generation radiotherapy sensitizers with high tumor uptake and high renal clearance. Sci Rep 2015;5:8669.

[32] Huang K, Ma H, Liu J, Huo S, Kumar A, Wei T. Size-dependent localization and penetration of ultrasmall gold nanoparticles in cancer cells, multicellular spheroids, and tumors in vivo. ACS Nano 2012;6:4483–93.

[33] Xu P, Feng Q, Yang X, Liu S, Xu C, Huang L, Chen M, Liang F, Cheng Y. Near infrared light triggered cucurbit [7] uril-stabilized gold nanostars as a supramolecular nanoplatform for combination treatment of cancer. Bioconjug Chem 2018;29:2855–66.

[34] Gulla SK, Kotcherlakota R, Nimushakavi S, Nimmu NV, Khalid S, Patra CR, Chaudhuri A. Au-CGKRK nanoconjugates for combating cancer through t- cell-driven therapeutic RNA interference. ACS Omega 2018;3:8663–76.

[35] Ke S, Zhou T, Yang P, Wang Y, Zhang P, Chen K, Ren L, Ye S. Gold nanoparticles enhance TRAIL sensitivity through Drp1-mediated apoptotic and autophagic mitochondrial fission in NSCLC cells. Int J Nanomed 2017;12:2531–51.

[36] Li WR, Xie XB, Shi QS, Zeng HY, Ou-Yang YS, Chen YB. Antibacterial activity and mechanism of silver nanoparticles on *Escherichia coli*. Appl Microbiol Biotechnol 2010;8:1115–22.

[37] Chernousova S, Epple M. Silver as antibacterial agent: ion, nanoparticle, and metal. Angew Chem Int Ed 2013;52:1636–53.

[38] Gurunathan S, Park JH, Han JW, Kim JH. Comparative assessment of the apoptotic potential of silver nanoparticles synthesized by Bacillus tequilensis and Calocybe indica in

MDA-MB-231 human breast cancer cells: targeting p53 for anticancer therapy. Int J Nanomed 2015;10:4203–22.

[39] Roychoudhury P, Gopal PK, Paul S, Pal R. Cyanobacteria assisted biosynthesis of silver nanoparticles-a potential antileukemic agent. J Appl Phycol 2016;28:3387–94.

[40] Blanco J, Lafuente D, Gómez M, Garcıá T, Domingo JL, Sánchez DJ. Polyvinyl pyrrolidone-coated silver nanoparticles in a human lung cancer cells: time- and dose-dependent influence over p53 and caspase-3 protein expression and epigenetic effects. Arch Toxicol 2016;91(2):651–66.

[41] Khan M, Khan M, Al-Marri AH, Al-Warthan A, Alkhathlan HZ, Siddiqui MR. Apoptosis inducing ability of silver decorated highly reduced graphene oxide nanocomposites in A549 lung cancer. Int J Nanomed 2016;7:873–83.

[42] Yuan YG, Wang YH, Xing HH, Gurunathan S. Quercetin-mediated synthesis of graphene oxide-silver nanoparticle nanocomposites: a suitable alternative nanotherapy for neuroblastoma. Int J Nanomed 2017;16:5819–39.

[43] Netchareonsirisuk P, Puthong S, Dubas S, Palaga T, Komolpis K. Effect of capping agents on the cytotoxicity of silver nanoparticles in human normal and cancer skin cell lines. J Nanoparticle Res 2016;18:322.

[44] Bhanumathi R, Manivannan M, Thangaraj R, Kannan S. Drug-carrying capacity and anticancer effect of the folic acid- and berberine-loaded silver nanomaterial to regulate the akt-erk pathway in breast cancer. ACS Omega 2018;3:8317–28.

[45] Kunkalekar RK, Prabhu MS, Naik MM, Salker AV. Silver-doped manganese dioxide and trioxide nanoparticles inhibit both gram positive and gram negative pathogenic bacteria. Colloids Surf B Biointerfaces 2014;113:429–34.

[46] Kunkalekar RK, Naik MM, Dubey SK, Salker AV. Antibacterial activity of silver-doped manganese dioxide nanoparticles on multidrug-resistant bacteria. J Chem Technol Biotechnol 2013;88:873–7.

[47] Shen W, Li Z, Wang H, Liu Y, Guo Q, Zhang Y. Photocatalytic degradation for methylene blue using zinc oxide prepared by codeposition and sol–gel methods. J Hazard Mater 2008;152:172–5.

[48] Wang LC, He L, Liu YM, Cao Y, He HY, Fan KN, Zhuang JH. Effect of pretreatment atmosphere on CO oxidation over α-Mn_2O_3 supported gold catalysts. J Catal 2009;264:145–53.

[49] Gawas SG, Gawas UB, Verenkar VMS, Kothawale MM, Pednekar R. Structural and magnetic studies of cu-substituted nanocrystalline ni–zn ferrites obtained via hexamine–nitrate combustion route. J Supercond Nov Magnetism 2017;30:1447–52.

[50] Gawas UB, Verenkar VMS, Meena SS, Bhatt P. Influence of Mn substitution on mössbauer and magnetic properties of ni-zn ferrites nanoparticles. J Supercond Nov Magnetism 2017;30:3241–6.

[51] Abenojar EC, Wickramasinghe S, Bas-Concepciona J, Samia. Structural effects on the magnetic hyperthermia properties of iron oxide nanoparticles. Prog Nat Sci 2016;26:440–8.

[52] Savla R, Minko T. Nanoparticle design considerations for molecular imaging of apoptosis: diagnostic, prognostic, and therapeutic value. Adv Drug Deliv Rev 2017;113:122–40.

[53] McCormack PL. Ferumoxytol in iron deficiency anaemia in adults with chronic kidney disease. Drugs 2012;72:2013–22.

[54] Hauser AK, Mitov MI, Daley EF, McGarry RC, Anderson KW, Hilt JZ. Targeted iron oxide nanoparticles for the enhancement of radiation therapy. Biomaterials 2016;105:127–35.

[55] Zou Z, Li H, Wang S. Induction of reactive oxygen species: an emerging approach for cancer therapy. Apoptosis 2017;22(11):1321–35.

[56] Li W, Zaloga J, Ding Y, Liu Y, Janko C, Pischetsrieder M, Alexiou C, Boccaccini AR. Facile preparation of multifunctional superparamagnetic PHBV microspheres containing SPIONs for biomedical applications. Sci Rep 2016;6:23140.

[57] Lee KJ, An JH, Shin JS, Kim DH, Chung KH. Tumor-targeting hederagenin-loaded magnetic nanoparticles for anticancer drug delivery. Adv Mater Lett 2016;7:366–70.

[58] Malekzadeh AM, Ramazani A, Tabatabaei Rezaei SJ, Niknejad H. Design and construction of multifunctional hyperbranched polymers coated magnetite nanoparticles for both targeting magnetic resonance imaging and cancer therapy. J Colloid Interface Sci 2017;490:64–73.

[59] Mallick N, Anwar M, Asfer M, Mehdi SM, Rizvi MMA, Panda AK. Chondroitin sulfate-capped super-paramagnetic iron oxide nanoparticles as potential carriers of doxorubicin hydrochloride. Carbohydr Polym 2016;151:546–56.

[60] Mondal S, Manivasagan P, Bharathiraja S, Moorthy MS, Nguyen VT, Kim HH, Nam SY, Lee KD, Oh J. Hydroxyapatite coated iron oxide nanoparticles: a promising nanomaterial for magnetic hyperthermia cancer treatment. Nanomaterials 2017;7:426.

[61] Saikia C, Das MK, Ramteke A, Maji TK. Evaluation of folic acid tagged aminated starch/ZnO coated iron oxide nanoparticles as targeted curcumin delivery system. Carbohydr Polym 2017;157:391–9.

[62] Moghadam NH, Salehzadeh S, Rakhtshah J, Tanzadehpanah H, Moghadam AH, Hajibabaei F, Sharifinia S, Asl SS, Saidijam M. Improving antiproliferative effect of the nevirapine on Hela cells by loading onto chitosan coated magnetic nanoparticles as a fully biocompatible nano drug carrier. Int J Biol Macromol 2018;15:1220–8.

[63] Zhang AP, Sun YP. Photocatalytic killing effect of TiO_2 nanoparticles on Ls- 174-t human colon carcinoma cells. World J Gastroenterol 2004;10(21):3191–3.

[64] Colon J, Hsieh N, Ferguson A, Kupelian P, Seal S, Jenkins DW, Baker CH. Cerium oxide nanoparticles protect gastrointestinal epithelium from radiation-induced damage by reduction of reactive oxygen species and upregulation of superoxide dismutase 2. Nanomedicine 2010;6(5):698–705.

[65] Wahab R, Kaushik NK, Kaushik N, Choi EH, Umar A, Dwivedi S, Musarrat J, Al-Khedhairy AA. ZnO nanoparticles induces cell death in malignant human T98G gliomas, KB and non-malignant HEK cells. J Biomed Nanotechnol 2013;9(7):1181–9.

[66] Sankar R, Maheswari R, Karthik S, Shivashangari KS, Ravikumar V. Anticancer activity of *Ficus religiosa* engineered copper oxide nanoparticles. Mater Sci Eng C 2014;44:234–9.

[67] Wani A, Savithra GHL, Abyad A, Kanvinde S, Li J, Brock S, Oupický D. Surface PEGylation of mesoporous silica nanorods (MSNR): effect on loading, release, and delivery of mitoxantrone in hypoxic cancer cells. Sci Rep 2017;7(1):2274.

[68] Bhuyan MSA, Uddin MN, Islam MM, Bipasha FA, Hossain SS. Synthesis of graphene. Int Nano Lett 2016;6:65–83.

[69] Bagotia N, Choudhary V, Sharma DK. A review on the mechanical, electrical and EMI shielding properties of carbon nanotubes and graphene reinforced polycarbonate nanocomposites. Polym Adv Technol 2018;29:1547–67.

[70] Power AC, Gorey B, Chandra S, Chapman J. Carbon nanomaterials and their application to electrochemical sensors; a review. Nanotechnol Rev 2017;7(1):1–48.

[71] Khan I, Saeed K, Khan I. Nanoparticles: properties, applications and toxicities. Arabian J Chem 2018. https://doi.org/10.1016/j.arabjc.2017.05.011.

[72] McCallion C, Burthem J, Rees-Unwin K, Golovanov A, Pluen A. Graphene in therapeutics delivery: problems, solutions and future opportunities. Eur J Pharm Biopharm 2016;104:235–50.

[73] Stobinski L, Lesiak B, Malolepszy A, Mazurkiewicz M, Mierzwa B, Zemek J, Jiricek P, Bieloshapka I. Graphene oxide and reduced graphene oxide studied by the XRD, TEM and electron spectroscopy methods. J Electron Spectrosc Relat Phenom 2014;195:145–54.
[74] Davami K, Jiang Y, Cortes J. Tuning the mechanical properties of vertical graphene sheets through atomic layer deposition. Nanotechnology 2016;27(15):155701.
[75] Popov VN. Carbon nanotubes: properties and application. Mater Sci Eng 2004;43:61–102.
[76] Mishra NS, Kuila A, Nawaz A, Pichiah S, Leong KH, Jang M. Engineered carbon nanotubes: review on the role of surface chemistry, mechanistic features, and toxicology in the adsorptive removal of aquatic pollutants. Chem Select 2018;3:1040–55.
[77] Rao Z, Ge H, Liu L. Carboxymethyl cellulose modified graphene oxide as pH-sensitive drug delivery system. Int J Biol Macromol 2018;107:1184–92.
[78] Durán N, Fávaro W. Nanopharmaceuticals and their applications in bladder cancer therapy: a mini review. J Braz Chem Soc 2018;29(5):973–85.
[79] Deb A, Vimala R. Camptothecin loaded graphene oxide nanoparticle functionalized with polyethylene glycol and folic acid for anticancer drug delivery. J Drug Deliv Sci Technol 2018;43:333–42.
[80] Hussien NA, Işıklan N, Türk M. Aptamer-functionalized magnetic graphene oxide nanocarrier for targeted drug delivery of paclitaxel. Mater Chem Phys 2018. https://doi.org/10.1016/j.matchemphys.2018.03.015.
[81] Kiew SF, Ho YT, Kiew LV, Kah JCY, Lee HB, Imae T, Chung LY. Preparation and characterization of an amylase-triggered dextrin-linked graphene oxide anticancer drug nanocarrier and its vascular permeability. Int J Pharm 2017;534:297–307.
[82] Sousa RL, Melo-Diogo D, Alves CG, Costa EC, Ferreira P, Louro RO, Correia IJ. Hyaluronic acid functionalized green reduced graphene oxide for targeted cancer photothermal therapy. Carbohydr Polym 2018;200:93–9.

Chapter 22

Iron-oxygen intermediates and their applications in biomimetic studies

Sunder N. Dhuri, Sarvesh S. Harmalkar
Department of Chemistry, Goa University, Taleigao Plateau, Goa, India

22.1 Introduction

The controlled oxidation of organic compounds by activating dioxygen carried out by metals is required for the key metabolic functions, such as the repair of RNA and DNA, in plants the desaturation of fatty acids, hydroxylation of methane in methanotrophs, biosynthesis of β-lactam antibiotics, and the formation of blood vessels for sensing of hypoxia [1−5]. Developing oxidation reactions that use renewable and environment friendly oxidants has become very important as oxidation reactions are critical for numerous chemical transformations [6,7]. A variety of biological reactions carried out by metalloenzymes utilize dioxygen (O_2) molecule. These enzymes perform oxidative reactions like epoxidation, halogenation, *cis*-dihydroxylation, and N-dealkylation by activating O_2 using a variety of active sites containing nonheme and heme copper, iron, and other metals [8].

A large number of enzymes employ iron centers to carry out metabolically vital oxidative transformations by activation of O_2 molecule [1,3]. Iron center coordinated to 2-His-1-carboxylate facial triad motifs [9−12] are utilized by most of these enzymes to catalyze substrate oxidations, including epoxidation, desaturation, *cis*-dihydroxylation, and halogenation. Commonly used reductants like NADH (nicotinamide adenine dinucleotide), tetrahydrobiopterin (THB) [13,14], α-ketoglutarate (α-KG) [3], and ascorbate [15,16] are used as electron donors for O_2 activation. For these enzymes, a $Fe^{IV}(O)$ species is postulated as an active oxidant; for Rieske dioxygenase, which catalyzes *cis*-dihydroxylation of arene, *cis*-HO-$Fe^{V}O$ is proposed as the oxidant [1,17,18].

The iron containing cytochromes P450 (CYP 450) of the heme-based monooxygenase are an important and versatile group with wide applications for human health and known to perform biosynthesis of hormones, the

Advances in Biological Science Research. https://doi.org/10.1016/B978-0-12-817497-5.00022-7

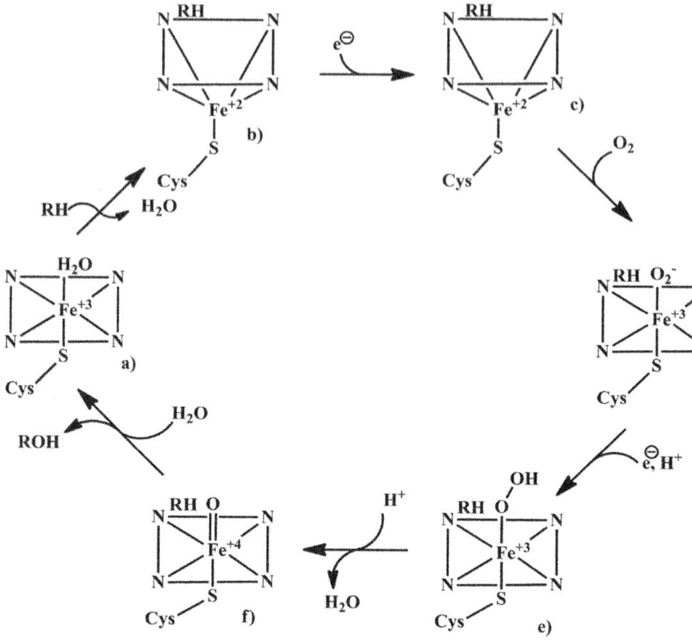

FIGURE 22.1 Catalytic cycle of cytochrome P450 enzymes.

metabolism and biodegradation of toxic compounds [10,19–24]. This group of enzymes reacts via oxygen atom transfer to substrate utilizing molecular oxygen at its heme center, leading to C=C epoxidation, aliphatic C–H hydroxylation, heteroatom oxidation and aromatic hydroxylation [25]. The heme active site of CYP 450 is linked via a thiolate bridge from a cysteine residue to the protein part [26,27]. The proposed catalytic cycle of CYP 450 is shown in Fig. 22.1. This cycle starts with the water molecule occupying the sixth binding site of the iron center. In the second step, the substrate binds to the active site resulting in the release of a water molecule and a five-coordinate high-spin ferric species with a vacant site. The ferric species is then reduced by putidaredoxin to afford a five-coordinate high-spin ferrous species to which dioxygen binds to give a ferric(III)-superoxo form, which then picks a proton and an electron forms a ferric-hydroperoxo species which is further deprotonated to give an iron(IV)-oxoheme cation radical oxidant called as Compound I, which is the active species of the enzyme. This species was very difficult to trap due to its high reactivity and short half-life and it was only characterized by spectroscopy. The CYP P450s have highly versatile structures owing to the diverse functions they perform. Despite this limitation, they contain conserved structural features that have been highlighted from the reported crystal structure coordinates of thousands of isozymes [28]. All the iron-oxygen

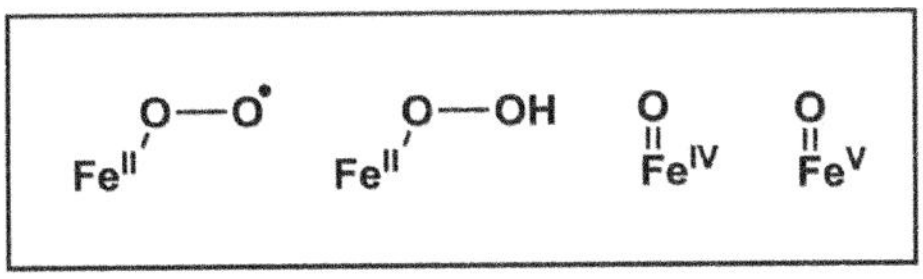

FIGURE 22.2 The different high valent iron-oxygen species.

intermediates (Fig. 22.2) are essential in nature and a lot of work has been carried out on them.

Spectroscopic characterization along with reaction rate measurements for deuterium and hydrogen substituted substrates of Compound I by Rittle and Green has established Compound I as the active oxidant for CYP P450s [29]. This experimental evidence was further supported by computational modeling and identified Compound I as a versatile oxidant [30].

Some of the metalloenzymes utilize dioxygen to catalyze a wide range of reactions required to oxidize a cosubstrate in conjugation with dioxygen activation to produce the high-valent intermediate to react with the target substrate [31–37]. One such example is nonheme iron(II)-dioxygenase, which uses α-KG as a substrate and is oxidatively decarboxylated to succinate and carbon dioxide via O_2 activation to generate an $Fe^{IV}(O)$ as the active site of the enzyme. This species can then attack the primary substrate, usually a C–H bond having a high bond dissociation energy (BDE) [31]. The αKG-dependent oxygenases contain a double-stranded β-helix consisting of eight β-strands arranged in two β-sheets in opposite direction to generate a barrel shape around the active site cavity that supports a 2-His-1-carboxylate metal binding motif [38–40]. A nonheme iron(II) ion is bound to the facial triad through the imidazole rings of two histidine amino acid residues and one carboxylate of either a glutamate or aspartate residue occupying one face of the octahedral coordination of the nonheme iron(II) sphere. Thus the adjacent face is available for binding of α-KG and an open site for O_2 coordination.

Since the catalytic cycle of intermediates in enzymes are short lived, it is very difficult to isolate and characterize them and hence many research groups are investigating the active site of the enzymes by preparing the biomimetic compounds with similar structural features without the protein part. By studying the structural properties and reactivities of biomimetic compounds of the heme and nonheme iron enzymes that catalyze dioxygen activation reactions have aided our understanding of the iron-oxygen intermediates to the large extend. In recent years iron-oxygen compounds such as mononuclear $Fe^{II}(O_2)^-$, $Fe^{III}(O_2)^{2-}$, $Fe^{III}(OOH)$, $Fe^{IV}(O)$, and $Fe^{V}(O)$ stabilized by the non-heme macrocyclic and acyclic ligands have been well documented (Fig. 22.3) [41–45]. Wieghardt and coworkers reported the first spectroscopic evidence for mononuclear iron(IV)-oxo species in the ozonolysis of $[Fe^{III}(cyclamacetato)(CF_3SO_3)]^+$ (cyclamacetato = 1-(carboxymethyl)-1,4,8,11-tetraazacyclotetradecane) [46].

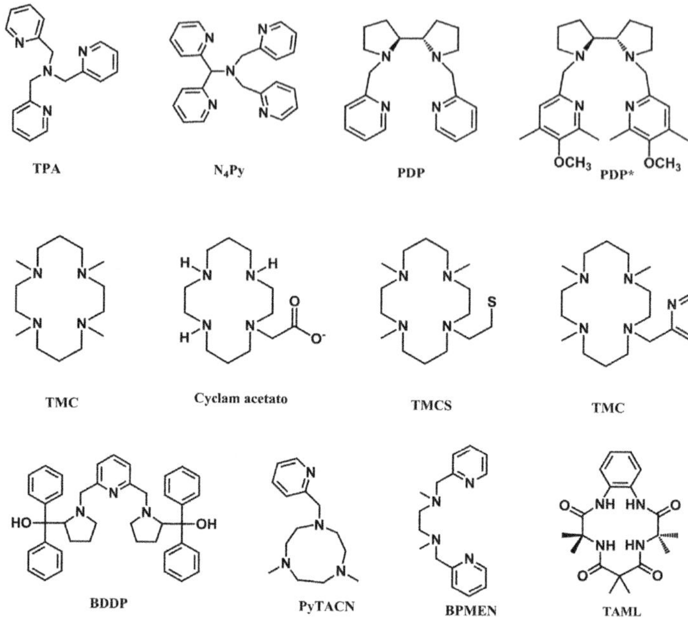

FIGURE 22.3 Chemical structures of few nonheme ligands that stabilize iron-oxygen intermediates.

The first structurally characterized iron-oxygen intermediate was the mononuclear nonheme iron(IV)-oxo species, viz., $[14\text{-TMC})Fe^{IV}(O)]^{2+}$ supported by 1,4,8,11-tetramethyl-1,4,8,11-tetraazacyclotetradecane (14-TMC) [47]. The structural characterization of this metastable intermediate has inspired the isolation of several new intermediates using various ancillary nonheme ligands.

As shown in (Fig. 22.4 (i)), the iron enzymes containing high-spin iron(II) active center with heme as well as nonheme ligands tend to form a bond with dioxygen molecule forming $Fe^{II}(O_2)^-$ (A) species. Further reduction by

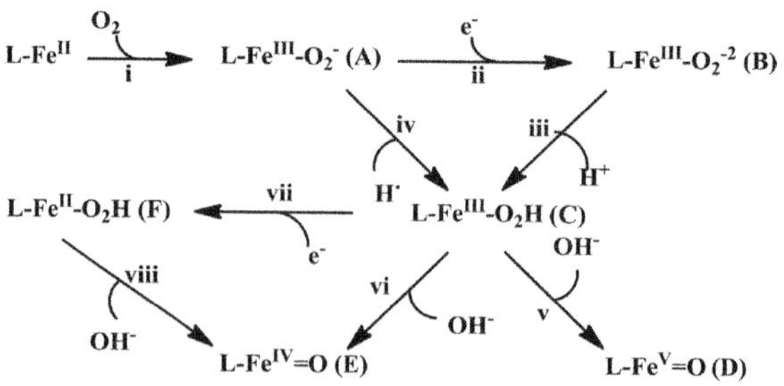

FIGURE 22.4 (i-viii). Pathways showing dioxygen activation by heme and nonheme iron containing enzymes.

another electron results in a new $Fe^{III}(O_2)^{2-}$ species (B) (Fig. 22.4 (ii)). The protonation of peroxo species $Fe^{III}(O_2)^{2-}$ then leads to the formation of Fe^{III}(OOH) species (C) (Fig. 22.4 (iii)). It has been also reported that this Fe^{III}(OOH) species can be generated by a hydrogen atom abstraction from the substrates, such as alkanes (Fig. 22.4 (iv)) [48]. This reaction, however, is found to occur only in nonheme iron enzymes. High-valent iron(V)-oxo (D) species is generated by the heterolytic cleavage of the O—O bond in iron(III)-hydroperoxo species (Fig. 22.4 (v)), whereas the homolytic cleavage of O—O bond in iron(III)-hydroperoxo generates a high-valent iron(IV)-oxo species (E) (Fig. 22.4 (vi)). The Fe^{IV}(O) species (Fig. 22.4 (viii)) has been alternatively prepared by one-electron reduction of Fe^{III}(OOH) species via proposed species Fe^{II}(OOH) (F) (Fig. 22.4 (vii)).

22.2 Mononuclear nonheme iron(III)-superoxo complexes

The lower-valent ferric-superoxo species are not well characterized as compared to the high-valent ferryl-oxo species. However, recently iron-superoxo species are thought to be an alternatie electrophilic oxidant, as in the cases of hydroxyethylphosphate dioxygenase, isopenicillin N-synthase, tryptophan 2,3-dioxygenase (TOD), catechol dioxygenases, nitric oxide dioxygenase, and indoleamine 2,3-dioxygenase (IDO) [41,48—53].

Since iron(III)-superoxo species are invoked as the active oxidant in the oxygen atom transfer (OAT) and C—H bond activation reactions, such studies have attracted the attention by the chemist community. There are only a few reports about synthetic nonheme iron superoxo compounds even though iron(III)-superoxo species in heme system with crystal structure were well characterized as early as the 1970's. The dioxygen-bound iron intermediate was first structurally reported from synthetic heme compounds of oxyhemoglobin and myoglobin [54]. In heme containing proteins as well as artificially synthesized iron compounds, the superoxo intermediate binds in an end-on fashion to iron(III). Alternately, in nonheme iron-superoxo compounds O_2 binds to the iron center by side-on mode. A first such example was given by Karlsson et al. in 2003 [55]. The structure of napthalene dioxygenase (NDO) containing side-on bound dioxygen molecule to iron was determined. NDO is a Rieske nonheme iron dioxygenase that oxidizes naphthalene to *cis*-(1R,2S)-dihydroxy-1,2-dihydronaphthalene. Later in 2007 another iron superoxo intermediate was structurally characterized for homoprotocatechuate 2,3-dioxygenase [56]. In both of these cases, the intermediates were generated by soaking the crystals of the dioxygenase in a solution of substrate in oxygen atmosphere with oxygen binding to the iron active site in a side-on fashion in both cases. The dinuclear Fe(II) precursors have been also used for the generation of Fe^{II}-superoxo compounds [57,58].

In the year 2014, Nam and coworkers reported the first structurally and spectroscopically characterized mononuclear iron(III)superoxido species

$[Fe^{III}(O_2)(TAML)]^{2-}$ having side-on η^2 binding mode of $(O_2)^-$ with iron(III) (TAML = tetraamido macrocyclic ligand) [59]. It was prepared in the presence of 2,2,2-cryptand by the reaction of potassium superoxide with sodium salt of Fe(III)(TAML)$^-$ at 5 °C in CH_3CN. Since the species was very stable, it was investigated by spectroscopic methods such as UV−Vis, IR-Raman, electron paramagnetic resonance (EPR), and Mössbauer spectroscopy. Electrospray ionization mass and single crystal X-ray data further confirmed its real structure. The structure showed bonding between Fe(III) and dioxygen where it bound in side-on fashion (η^2). The O−O bond length of this species and that of reported $Fe^{II}(O_2)^-$ found in homoprotocatechuate 2,3-dioxygenase were ~1.32 Å and 1.34 Å, respectively [56]. The species showed reactivity in both electrophilic as well as nucleophilic reactions. The positive Hammett ρ value of 1.4 suggested the nucleophilic reactivity for the iron(III)−superoxo complex. The reactivity of this species was also studied in intermolecular O_2 transfer. In this reaction Fe^{III}-O_2^{2-} on reaction with Mn^{III}-(TAML)$^-$ forms Mn^{IV}-O_2^{2-}.

Another iron-superoxo species, (BDPP)$Fe^{III}(O_2)$ (BDPP-2,6-bis (((S)-2-(diphenylhydroxymethyl)-1-pyrrolidinyl)methyl)pyridine), was also generated but at a lower temperature (−80°C) due to its low stability [60]. Although there was no structure reported, this species was characterized by the spectroscopic techniques like resonance Raman EPR and Mössbauer spectroscopies. This species showed the reactivity in hydrogen atom abstraction reactions.

22.3 Mononuclear nonheme iron(III)-peroxo complex

Iron-peroxo complexes are frequently invoked as the key intermediates in heme and nonheme iron enzyme catalyzed oxidation reactions [3,61,62]. Iron-peroxo species act as the active oxidant in many cytochrome P450 catalyzed reactions like the cleavage of the C-17 side chain of progesterone to form androstenedione by progesterone 17α-hydroxylase, 17,20-lyase and the aromatization of androgen to estrogen by cytochrome P450 aromatase [63,64]. Mechanistic studies of the reactions of enzymes and synthetic iron-peroxo porphyrin compounds have shown the nucleophilic character for these compounds as they attack an aldehyde carbon. Iron-peroxo species was also proposed to be an active oxidant in aromatic *cis*-dihydroxylation catalyzed by Rieske dioxygenases [3,17,65]. The crystal structure of iron-peroxo intermediate was obtained in naphthalene dioxygenase (NDO) with molecular oxygen species bound to the mononuclear iron in a side-on fashion [55]. The first example of crystal structure of $Fe^{III}(O_2)^{2-}$ was reported by Nam et al., which was stabilized by 14-TMC. In this structure the iron(III) ion bonded to dioxygen molecule in a side-on fashion (η^2) and the O−O bond length of 1.463 Å [66] was similar to that reported for the NDO (O−O bond length

(1.45 Å) [55]. These species were found to be reactive in nucleophilic oxidation reactions [44,67].

Recently there has been great interest to understand the role of redox-inactive metal ions in controlling the reactivities of metal-oxygen intermediates in biomimetic and enzymatic reactions. Redox-inactive metal ions bound to mononuclear nonheme iron-peroxo compounds, [(14-TMC)-$Fe^{III}(O_2)]^{+}$-M^{n+}, where $M^{n+} = Sr^{2+}$, Ca^{2+}, Zn^{2+}, Lu^{3+}, Y^{3+} and Sc^{3+} were synthesized by reacting various redox-inactive metal ions with nonheme iron(III)-peroxo complex [(14-TMC)-$Fe^{III}(O_2)]^{+}$ [68,69]. The Lewis acidity of the redox-inactive metal ions bound to the iron center governed the redox potentials and reactivities of intermediates with electron donors and acceptors. With increasing Lewis acidity of the redox inactive metal, the redox potentials and one-electron oxidation increased. Complex binding to Ca^{+2} and Sr^{+2} ions showed similar reactivities in one-electron oxidation and reduction reactions along with identical redox potentials. It was shown that [(14-TMC)$Fe^{III}(O_2)]^{+}$-M^{n+} on reaction with electron donors leads to the formation of [(14-TMC) $Fe^{IV}(O)]^{2+}$, and the rates for this conversion were dependent on the oxidation potentials of electron donors and the redox-inactive metal ions acidity [68a]. The effects on the reactivities due to the Lewis acidity of the redox inactive metal [(14-TMC)-$Fe^{III}(O_2)]^{+}$-M^{n+} were investigated by one-electron reduction and oxidation reactions in the presence of ceric ammonium nitrate (CAN) and reductant 1-benzyl-1,4-dihydronicotinamide dimer. In the oxidation transfer reaction of [(14-TMC)$Fe^{III}(O_2)]^{+}$-M^{n+} using CAN, O_2 was released when $M^{n+} = Ca^{+2}$ and Sr^{+2} whereas when $M^{n+} = Zn^{+2}$, Lu^{+2}, Y^{+3}, and Sc^{+3} there was no observation of such reaction. Similarly, [(14-TMC)$Fe^{III}(O_2)]^{+}$-M^{n+} in one-electron reduction gave [(14-TMC)$Fe^{IV}(O)]^{2+}$ when $M^{n+} = Zn^{+2}$, Lu^{+2}, Y^{+3}, and Sc^{+3}, whereas when $M^{n+} = Ca^{+2}$ and Sr^{+2} no reaction occurred and [(14-TMC)$Fe^{III}(O_2)]^{+}$-M^{n+} remained intact.

22.4 Mononuclear nonheme iron(III)-hydroperoxo complex

Iron(III)-hydroperoxo (Fe^{III}-OOH) species have been proposed to be the "second electrophilic oxidant" in oxygenation reactions [1,10,29,33,49b]. It is investigated that the Fe^{III}-OOH species can act as an active oxidant in nonheme iron systems such as Rieske dioxygenases and other synthetic compounds of iron [18,71].

The protonation of iron(III)-peroxo or the reaction of iron (II/III) compounds with H_2O_2 leads to the formation of iron(III)-hydroperoxo species [71b,72,73]. The formation of iron(III)-hydroperoxo intermediate by H-atom abstraction of substrate C—H bonds by iron-superoxo also attracted attention of researchers in nonheme iron enzymes and their models [48,67]. The formation of iron(IV)-oxo, [(14-TMC)$Fe^{IV}(O)]^{2+}$ species was observed by activation of dioxygen molecule when an alkene was present by the corresponding iron(II) complex [74]; whereas no formation of such a species was observed in the absence of an alkene. According to the mechanism proposed for this reaction, the iron(III)-peroxo species was formed by an H-atom abstraction of the allylic C–H bonds by $[(14\text{-TMC})Fe^{III}(O_2)]^{2+}$. The O–O bond cleavage of this hydroperoxo species leads to the formation of the iron(IV)-oxo species. Although there was no direct evidence for the peroxo and hydroperoxo species in the reaction, there was indirect evidence for the presence of peroxo species that was obtained by trapping the peroxo species when the reaction was carried out in basic medium.

Iron(III)-hydroperoxo species reactivity in oxidation reactions like alkane hydroxylation and sulfoxidation has been a long-standing controversy [75,76]. Although there is a general agreement that in heme reactions iron-hydroperoxides are sluggish oxidant [76,77], the reactivities of hydroperoxo compounds are not very clear and have to be investigated thoroughly. For example, the synthetic low-spin Fe^{III}-OOH species are dull in oxidation of olefins, aldehydes, and sulfides [78]; however, the high-spin Fe^{III}-OOH showed good reactivity in nucleophilic and electrophilic reactions [66]. The high-spin Fe^{III}-OOH in $[LFe^{III}(OOH)]^{2+}$ (L = N,N-bis(2-pyridylmethyl)-N bis(2-pyridyl)methylamine) showed high reactivity in the oxidation of alkyl aromatic compounds, in H-atom abstraction [79].

The mechanism of O–O bond cleavage of iron(III)-hydroperoxo in heme and nonheme iron enzymatic and biomimetic reactions is another debate (heterolysis vs. homolysis). There have been extensive studies of the O–O bond cleavage mechanisms in heme systems over many years; such studies of the mechanism of O–O cleavage in nonheme iron(III)-hydroperoxo species is only recently being investigated. It was thought that the O–O bond cleavage could have occurred by two ways in $[(14\text{-TMC})Fe^{III}(O_2H)]^{2+}$ to give $[(14\text{-TMC})Fe^{IV}(O)]^{2+}$. In one, the O–O bond homolysis gives a hydroxyl radical and $[(14\text{-TMC})Fe^{IV}(O)]^{2+}$ [66]. While in the other one, heterolytic O–O bond cleavage gives $[(14\text{-TMC})Fe^{V}(O)]^{3+}$, which upon one-electron reduction gives $[(14\text{-TMC})Fe^{IV}(O)]^{2+}$ [80]. More recently, it was shown that $[(14\text{-TMC})Fe^{IV}(O)]^{2+}$ can also be prepared by one-electron reduction of $[(14\text{-TMC})Fe^{III}(O_2H)]^{2+}$ by derivatives of ferrocene [81]. It was observed that the conversion rates were directly related to the oxidation potentials as well as the concentration of the electron donors; the rate-determining step was proposed to be the one-electron reduction of iron(III)-hydroperoxo species to give iron(II)-hydroperoxo.

22.5 Mononuclear high-valent iron(IV)-oxo complex

Iron(IV)-oxo intermediates are involved in essential processes in nature and therefore act as the key intermediates in biology [41,82]. The iron(IV)-oxo intermediates are active species in the catalytic cycle of the metalloenzymes. One of the most important factors that controls the chemical reactivity and biochemical properties is the iron coordination environment. For instance, in heme monoxygenase like the cytochrome P450 the iron(IV)-oxo intermediate acts as an active oxidant, whereas in the structurally analogous peroxidases it acts as a catalyst in detoxification of hydrogen peroxide to water [19,21,23,83]. On the other hand, oxygen transfer reactions by nonheme iron dioxygenase containing $Fe^{IV}(O)$ species are utilized in the catalytic cycle to transfer oxygen atoms to substrates and have functions ranging from biosynthesis of antibiotics to DNA base repair [1,3,84]. Many studies have been carried out to prepare and characterize these short-lived species using natural enzymes [24,29,85] and model compounds [45,47,67,86–89]. Many of such mononuclear high-valent iron(IV)-oxo species supported by corrole [86], porphyrin [87], phthalocyanine [88], and nonheme [89] ligands have been synthesized and characterized at low temperatures.

The X-ray single crystal structure of $[(14\text{-TMC})\text{-}Fe^{IV}(O)]^{2+}$, was first reported by Nam and coworkers in 2003 [47]. It was obtained by reacting $[Fe^{II}(TMC)(CF_3SO_3)_2]$ with PhIO (iodosylbenzene) in CH_3CN -40 °C. In 2005, a thiolate-ligated iron(IV)-oxo intermediate was reported using $[Fe^{II}(TMCS)](PF_6)$ (where TMCS is a monoanion of 1-mercaptoethyl-4,8,11-trimethyl-1,4,8,11-tetraaza cyclotetradeane) [90]. Reactivity of this species in H-atom abstraction and oxygen transfer reactions was studied and it was suggested that it favors H-atom abstraction chemistry over the oxygen transfer reaction. In 2008 another iron(IV)-oxo complex $[Fe^{IV}(O)(TMC\text{-}py)](OTf)_2$ (one of the TMC methyl groups is replaced by a 2-pyridylmethyl arm to get TMC-py) was structurally reported. The reaction of $[Fe^{II}(TMC\text{-}py)](OTf)_2$ with PhIO in CH_3CN solution at room temperature gave pale brown-green complex $[Fe^{IV}(O)(TMC\text{-}py)](OTf)_2$, which can also be synthesized with 3 equiv H_2O_2 [89d].

It is to be noted that more than 60 nonporphyritic ligand-supported iron(IV)-oxo complexes have been so far characterized along with their reactivities after the first report on the crystal structure of Fe(IV)-oxo by Nam et. al. [34,42,43,45,91]. The various effects such as the geometry and topology of ligands, auxiliary, and axial ligands, the spin state of iron(IV)-oxo species, and the type of redox-inactive metal ions on the reactivity of the synthetic nonheme iron(IV)-oxo have been well studied [34,42]. The density functional theory (DFT) calculations have been widely used to obtain the information on the reaction mechanisms, the reactivity pathway, and the transition state structure [91,92]. Thus, more than heme analogues, our understanding on the mononuclear nonheme iron(IV)-oxo over the past few years has been widely advanced.

22.6 Mononuclear nonheme iron(V)-oxo complex

Iron(V)-oxo plays a key role in hydrocarbon oxidation reactions. It is used in cytochrome P450, which is an O_2-activating heme enzyme to functionalize strong C—H bonds in substrates [83b,49b,93a]. In Rieske dioxygenase, which catalyzes a remarkable number of transformations such as C—H bond hydroxylation, *cis*-dihydroxylations of C=C bonds, C—C bond formation and N-demethylation, a nonheme iron(V)-oxo intermediate is proposed to be involved [93b,93c,93d]. An unprecedented level of regio-, chemo-, and stereoselectivity exhibited by such enzymes have inspired chemists to investigate this field thoroughly, and hence a wide range of nonheme catalysts have been developed and remarkable activity has been reported. All the C—H and C-C bond oxidation processes catalyzed by iron(II) catalyst are believed to follow a common mechanism. Fe^{III}-OOH or Fe^{III}-OOC(O)CH_3 intermediates are formed initially, which undergo O—O bond cleavage to generate iron(V)-oxo species, which are responsible for the substrate functionalization [94].

There have been extensive efforts to trap this iron(V)-oxo species. Investigation by variable temperature cryospray mass spectrometer detected a peak corresponding to $\{[Fe(O)\ (OH)\ (^{Me,H}PyTACN)](OTf)\}^+$ (^{Me,H}PyTACN = 1-(2′-pyridylmethyl)-4,7-dimethyl1,4,7-triazacyclo-nonane, OTf = OSO_2CF_3), indicating an iron(V)-oxo species in hydroxylation of *cis*-1,2-dimethylcyclohexane [89b]. Later, in a reaction of $[(CH_3CN)_2Fe^{II}(TPA)]$ (TPA = Tris(2-pyridylmethyl)amine) and $[(CH_3CN)_2 Fe^{II}(BPMEN)]$ (BPMEN = N,N′-dimethyl-N,N′-bis(2-pyridylmethyl)-1,2-diaminoethane), a species with S = 1/2 with a large *g*-anisotropy ($g_{max} = 2.58-2.74$ and $\Delta g = 0.85-1.2$) was detected by EPR and assigned to Fe^V(O)(OC(O)R) (R = H or t Bu) [95]. Similar species were also found for the nonheme iron catalysts supported by ligands such as PDP (PDP = N,N′- bis(2-pyridylmethyl)-2,2′-bipyrrolidine) and PDP* (PDP* = N,N′-bis(3,5-dimethyl-4-methoxy-2-pyridylmethyl)-2,2′-bipyrrolidine). In a reaction of $[Fe^{II}(PyNMe_3)(CF_3SO_3)_2]$($PyNMe_3$ = 3,6,9,15- tetraazabicyclo[9.3.1]pentadeca-1(15),11,13-triene-3,6,9-trimethyl) and $CH_3C(O)OOH$ resulted in the formation of $[Fe^V(O)(OC(O)CH_3)(PyNMe_3)]^{2+}$, which was spectroscopically characterized [96]. In recent kinetic study it is seen that Fe^{III}-OOC(O)CH_3 and Fe^V(O)(OC(O)CH_3) exist in dynamic equilibrium [97]. Newcomb et al. reported photolysis of iron-corrole compound resulting in transient species Fe(V) (O) [86b].

Another way to produce iron(V)-oxo species is by one-electron oxidation of iron(IV)-oxo species. The first complex using this method was reported for $[Fe^V(O)(TAML)]^-$ (TAML = tetraamido macrocyclic ligand), reported by Collins and coworkers, featuring a low-spin ferryl center [89a]. In 2007, Collins et al. gave evidence of an Fe(V)(O) species using chemical and spectroscopic methods [89a]. One-electron oxidation was also performed on $[Fe^{IV}(O)(TMC)(NCCH_3)]^{2+}$ (TMC = 1,4,8,11-tetramethyl-1,4,8,11-tetraazacyclotetradecane) employing *tert*-butyl hydroperoxide with a strong base to generate $[Fe^V(O)(TMC)\text{-}(NC(O)CH_3)]^+$ and its conjugate acid $[Fe^V(O)(TMC)(NC(OH)CH_3)]^{2+}$ [98].

22.7 Application of iron-oxygen intermediates in biomimetics

Mononuclear iron-oxygen species detected in heme and nonheme iron enzymes are key intermediates involved in the catalysis of many oxidation reactions. The iron-oxygen species performs versatile reactivities such as oxygen atom transfer to substrates containing S and P, the C—H bond activation of alkyl hydrocarbons, *cis*-dihydroxylation of olefins etc (Fig. 22.5). Another biomimetic application of iron-oxygen species that they have been used is in oxidation of cyclohexane to adipic acid. Nonheme iron enzymes carry out a variety of reactions that are metabolically relevant to dioxygen-dependent transformations, like hydroxylation of alkanes and arenes, aliphatic and aromatic C—C bond cleavage, oxidation of organosulfur compounds, halogenation, epoxidation and dihydroxylation of olefins, etc.

22.8 Summary

In this chapter we have presented a recent account which focus on the overview of reactivity and mechanisms of iron-oxo intermediates in biomimetic reactions. Learning from the evolution of high-valent metal intermediates from biology, there have been various attempts to mimic the metal biological molecules and later use them with biomimetic approach for performing various transformations. Of all the probable iron-oxygen intermediates known, the iron-superoxo as the first member in the catalytic cycle has been well characterized and used in several important oxidation reactions. Iron-peroxo species has also been characterized and widely used in nucleophilic reactions. The iron-hydroperoxy species are know for their sluggish behaviour

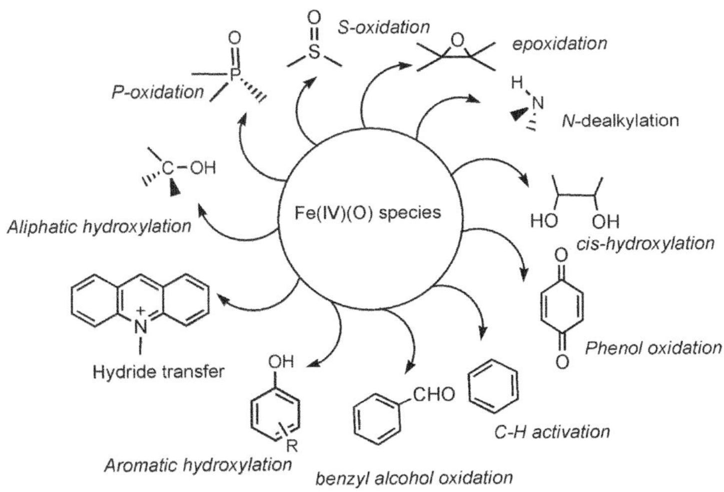

FIG. 22.5 Biomimetic applications of high-valent iron(IV)-oxo compounds.

and therefore not very well characterized and explored. Meanwhile, a large number of compounds containing Fe(IV)-oxo and Fe(V)-oxo centers have been well characterized and shown interesting properties and reactivities. As seen from the available literature, the lot of efforts are being by the researchers to understand the chemistry of, iron-oxygene compounds, however, there are still unanswered questions in this field, and therefore this field is open for extensive research.

Acknowledgments

SND thank Wonwoo Nam, a Distinguished Professor, Department of Chemistry, Ewha Womans University, Seoul, Korea for introducing him to the chemistry of high valent metal-oxygen intermediates. Authors also thank Goa University for creating conducive atmosphere for performing research.

References

[1] Solomon EI, Brunold TC, Davis MI, Kemsley JN, Lee S-K, Lehnert N, et al. Geometric and electronic structure/function correlations in non-heme iron enzymes. Chem Rev 2000;100(1):235–49.
[2] Hausinger RP. Fe(II)/α-Ketoglutarate-Dependent hydroxylases and related enzymes. Crit Rev Biochem Mol Biol 2004;39:21–68.
[3] Costas M, Mehn MP, Jensen MP, Que Jr L. Dioxygen activation at mononuclear nonheme iron active sites: enzymes, models, and intermediates. Chem Rev 2004;104:939–86.
[4] Kovaleva EG, Lipscomb JD. Versatility of biological non-heme Fe(II) centers in oxygen activation reactions. Nat Chem Biol 2004;4(3):186–99.
[5] Loenarz C, Schofield CJ. Expanding chemical biology of 2-oxoglutarate oxygenases. Nat Chem Biol 2008;4(3):152–6.
[6] Sheldon RA, Kochi JK. Metal-catalyzed oxidations of organic compounds. In: Mechanistic principles and synthetic methodology including biochemical processes. New York: Academic Press; 1981.
[7] Hill CL. Homogeneous catalysis: controlled green oxidation. Nature 1999;401(6752):436–7.
[8] Franz G, Sheldon RA. Oxidation. Ullmann's encyclopedia of industrial chemistry. Weinheim: Wiley-VCH; 2004.
[9] Punniyamurthy T, Velusamy S, Iqbal J. Recent advances in transition metal catalyzed oxidation of organic substrates with molecular oxygen. Chem Rev 2005;105(6):2329–64.
[10] Nam W. Dioxygen activation by metalloenzymes and models. Acc Chem Res 2007;40(7):465.
[11] Hegg EL, Que Jr L. The 2-His-1-carboxylate facial triad–an emerging structural motif in mononuclear non-heme iron(II) enzymes. Eur J Biochem 1997;250(3):625–9.
[12] Que Jr L. One motif–many different reactions. Nat Struct Biol 2000;7(3):182–4.
[13] Kappock TJ, Caradonna JP. Pterin-dependent amino acid hydroxylases. Chem Rev 1996;96(7):2659–756.
[14] Fitzpatrick PF. Tetrahydropterin-dependent amino acid hydroxylases. Annu Rev Biochem 1999;68(1):355–81.
[15] Rocklin AM, Tierney DL, Kofman V, Brunhuber NMW, Hoffman BM, Christofferson RE, et al. Role of the nonheme Fe(II) center in the biosynthesis of the plant hormone ethylene. Proc Natl Acad Sci USA 1999;96:7905–9.

[16] Mirica LM, Klinman JP. The nature of O_2 activation by the ethylene-forming enzyme 1-aminocyclopropane-1-carboxylic acid oxidase. Proc Natl Acad Sci USA 2008;105(6):1814–9.

[17] Wolfe MD, Lipscomb TD. Hydrogen peroxide-coupled cis-diol formation catalyzed by Napthalene 1,2 dioxygenase. J Bol Chem 2003;278(2):829–35.

[18] Neibergall MB, Stubna A, Mekmouche Y, Münck E, Lipscomb JD. Hydrogen peroxide dependent cis-dihydroxylation of benzoate by fully oxidized benzoate 1,2-dioxygenase. Biochemistry 2007;46(27):8004–16.

[19] Sono M, Roach MP, Coulter ED, Dawson JH. Heme-containing oxygenases. Chem Rev 1996;96(7):2841–88.

[20] Guengerich FP. Common and uncommon cytochrome P450 reactions related to metabolism and chemical toxicity. Chem Res Toxicol 2001;14(6):611–50.

[21] Groves JT. The bioinorganic chemistry of iron in oxygenases and supramolecular assemblies. Proc Natl Acad Sci USA 2003;100(7):3569–74.

[22] Munro AW, Girvan HM, McLean KJ. Variations on a (t)heme–novel mechanisms, redox partners and catalytic functions in the cytochrome P450 superfamily. Nat Prod Rep 2007;24(3):585–609.

[23] Ortiz de PR, editor. Cytochrome P450: structure, mechanism and biochemistry montellano. 3rd ed. New York: Kluwer Academic/Plenum Publishers; 2004.

[24] de Visser SP. In: Kumar, editor. Iron-containing enzymes: versatile catalysts of hydroxylation reaction in nature. Cambridge UK: RSC Publishing; 2011.

[25] Groves JT. In: Ortiz de PR, editor. Cytochrome P450: structure, mechanism and biochemistry montellano. 3rd ed. New York: Kluwer Academic/Plenum Publishers; 2004.

[26] Schlichting I, Berendzen J, Chu K, Stock AM, et al. The catalytic pathway of cytochrome P450cam at atomic revolution. Science 2000;287:1615–22.

[27] Poulos TL, Finzel BC, Gunsalus IC, Wagner GC, Kraut J. The 2.6-A crystal structure of Pseudomonas putida cytochrome P-450. J Biol Chem 1885;260(30):16122–8.

[28] (a) Nelson DR. The cytochrome p450 homepage. Hum Genom 2009;4(1):59–65;
(b) Mestres J. Structure conservation in cytochromes P450. Proteins Struct. Funct. Bioinf 2004;58(3):596–609.

[29] Rittle J, Green MT. Cytochrome P450 compound I: capture, characterization, and C-H bond activation kinetics. Science 2010;330(6006):933–7.

[30] (a) Shaik S, Kumar D, de Visser SP, Altun A, Thiel W. Theoretical perspective on the structure and mechanism of cytochrome P450 enzymes. Chem Rev 2005;105(6):2279–328;
(b) Blomberg MRA, Borowski T, Himo F, Liao R-Z, Siegbahn PEM. Quantum chemical studies of mechanisms for metalloenzymes. Chem Rev 2014;114(7):3601–58.

[31] Hausinger RP. In: Schofield CJ, Hausinger RP, editors. Royal Society of Chemistry Cambridge U.K; 2015.

[32] Krebs C, Galonic-Fujimori DP, Walsh CT, Bollinger Jr JM. Non-heme Fe(IV)–Oxo intermediates. Acc Chem Res 2007;40(7):484–92.

[33] Lundberg M, Siegbahn PE, Morokuma K. The mechanism for isopenicillin N synthase from density-functional modeling highlights the similarities with other enzymes in the 2-His-1-carboxylate family. Biochemistry 2008;47(3):1031–42.

[34] Baldwin JE, Bradley M. Isopenicillin N synthase: mechanistic studies. Chem Rev 1990;90(7):1079–88.

[35] Lipscomb JD. Mechanism of extradiol aromatic ring-cleaving dioxygenases. Curr Opin Struct Biol 2008;18(6):644–9.

[36] Straganz GD, Diebold AR, Egger S, Nidetzky B, Solomon EI. Kinetic and CD/MCD spectroscopic studies of the atypical, three-his-ligated, non-heme Fe^{2+} center in diketone dioxygenase: the role of hydrophilic outer shell residues in catalysis. Biochemistry 2010;49(5):996–1004.

[37] Joseph CA, Maroney MJ. Cysteine dioxygenase: structure and mechanism. Chem Commun 2007;0:3338–49.

[38] Hangasky JA, Taabazuing CY, Valliere MA, Knapp MJ. Imposing function down a (cupin)-barrel: secondary structure and metal stereochemistry in the αKG-dependent oxygenases. Metallomics 2013;5(4):287–301.

[39] Koehntop KD, Emerson JP, Que Jr L. The 2-His-1-carboxylate facial triad: a versatile platform for dioxygen activation by mononuclear non-heme iron(II) enzymes. J Biol Inorg Chem 2005;10(2):87–93.

[40] Neidig ML, Brown CD, Light KM, Fujimori DG, Nolan EM, Price JC, et al. CD and MCD of CytC3 and taurine Dioxygenase: role of the facial triad in α-KG-dependent oxygenases. J Am Chem Soc 2007;129(46):14224–31.

[41] Nam W. High-valent iron(IV)–Oxo complexes of heme and non-heme ligands in oxygenation reactions. Acc Chem Res 2007;40(7):522–31.

[42] Que Jr L. The road to non-heme oxoferryls and beyond. Acc Chem Res 2007;40(7):493–500.

[43] Hohenberger J, Ray K, Meyer K. The biology and chemistry of high-valent iron–oxo and iron–nitrido complexes. Nat Commun 2012;3:3720–33.

[44] Cho J, Sarangi R, Nam W. Mononuclear metal–O_2 complexes bearing macrocyclic N-tetramethylated cyclam ligands. Acc Chem Res 2012;45(8):1321–30.

[45] Nam W, Lee Y-M, Fukuzumi S. Tuning reactivity and mechanism in oxidation reactions by mononuclear nonheme iron(IV)-Oxo complexes. Acc Chem Res 2014;47(4):1146–53.

[46] Grapperhaus CA, Mienert B, Bill E, Weyhermüller T, Wieghardt K. Mononuclear (Nitrido) iron(V) and (Oxo)iron(IV) complexes via photolysis of $[(\text{cyclam-acetato})Fe^{III}(N_3)]^+$ and ozonolysis of $[(\text{cyclam-acetato})Fe^{III}(O_3SCF_3)]^+$ in water/acetone mixtures. Inorg Chem 2000;39(23):5306–17.

[47] Rohde J-U, In J-H, Lim MH, Brennessel WW, Bukowski MR, Stubna A, et al. Crystallographic and spectroscopic characterization of a nonheme Fe(IV)=O complex. Science 2003;299(5609):1037–9.

[48] van der Donk WA, Krebs C, Bollinger JM. Substrate activation by iron superoxo intermediate. Curr Opin Struct Biol 2010;20(6):673–83.

[49] (a) Siegbahn PEM, Borowski T. Modeling enzymatic reactions involving transition metals. Acc Chem Res 2006;39(10):729–38;

(b) Shaik S, Cohen S, Wang Y, Chen H, Kumar D, Thiel W. Hydrocarbon hydroxylation by cytochrome P450 enzymes. Chem Rev 2010;110(2):932–48.

[50] (a) Bollinger Jr JM, Diao Y, Matthews ML, Xing G, Krebs C. myo-Inositol oxygenase: a radical new pathway for O_2 and C–H activation at a nonheme diiron cluster. Dalton Trans 2009;0:905–14;

(b) Cicchillo RM, Zhang H, Blodgett JAV, hitteck JT, Li G, Nair SK. An unusual carbon-carbon bond cleavage reaction during phosphinothricin biosynthesis. Nature 2009;459(7248):871–4.

[51] (a) Forouhar F, Anderson JLR, Mowat CG, Vorobiev SM, Hussain A, Abashidze M, et al. Molecular insights into substrate recognition and catalysis by tryptophan 2,3-dioxygenase. Proc Natl Acad Sci USA 2007;104(2):473–8;

(b) Sugimoto H, Oda S-I, Otsuki T, Hino T, Yoshida T, Shiro Y. Crystal structure of human indoleamine 2,3-dioxygenase: catalytic mechanism of O2 incorporation by a heme-containing dioxygenase. Proc Natl Acad Sci USA 2006;103(8):2611–6.

[52] (a) Lai W, Shaik S. Can ferric-superoxide act as a potential oxidant in P450cam? QM/MM investigation of hydroxylation, epoxidation, and sulfoxidation. J Am Chem Soc 2011;133(14):5444–52;

(b) Hirao H, Morokuma K. Ferric superoxide and ferric hydroxide are used in the catalytic mechanism of hydroxyethylphosphonate dioxygenase: a density functional theory investigation. J Am Chem Soc 2010;132(50):17901–9.

[53] Gardner PR, Gardner AM, Martin LA, Salzman AL. Nitric oxide dioxygenase: an enzymic function for flavohemoglobin. Proc Natl Acad Sci USA 1998;95(18):10378–83.

[54] Collman JP, Gagne RR, Reed CA, Robinson WT, Rodley GA. Structure of an iron(II) dioxygen complex; a model for oxygen carrying hemeproteins. Proc Natl Acad Sci USA 1974;71(4):1326–9.

[55] Karlsson A, Parales JV, Parales RE, Gibson DT, et al. Crystal structure of Napthalene Dioxygenase: Side-on binding of Dioxygen to Iron. Science 2003;299:1039–42.

[56] Kovaleva EG, Lipscomb JD. Crystal structures of Fe^{2+} dioxygenase superoxo, alkylperoxo, and bound product intermediates. Science 2007;316(5823):453–7.

[57] Zhao M, Helms B, Slonkina E, Friedle S, Lee D, DuBois J, et al. Iron complexes of dendrimer-appended carboxylates for activating dioxygen and oxidizing hydrocarbons. J Am Chem Soc 2008;130(13):4352–63.

[58] Shan X, Que Jr L. Intermediates in the oxygenation of a nonheme diiron(II) complex, including the first evidence for a bound superoxo species. Proc Natl Acad Sci USA 2005;102(15):5340–5.

[59] Hong S, Sutherlin KD, Park J, Kwon E, Siegler MA, Solomon EI, et al. Crystallographic and spectroscopic characterization and reactivities of a mononuclear non-haem iron(III)-superoxo complex. Nat Commun 2014;5:5440–4.

[60] Chiang C-W, Kleespies ST, Stout HD, Meier KK, Li P-Y, Bominaar EL, et al. Characterization of a paramagnetic mononuclear nonheme iron-superoxo complex. J Am Chem Soc 2014;136(31):10846–9.

[61] (a) Wertz DL, Valentine JS. Nucleophilicity of iron-peroxo porphyrin complexes. Struct Bond 2000;97:37–60;

(b) Momenteau M, Reed CA. Synthetic heme-dioxygen complexes. Chem Rev 1994;94(3):659–98.

[62] Girerd J-J, Banse F, Simaan AJ. Characterization and properties of non-heme iron peroxo complexes. Struct Bond 2000;97:145–77.

[63] (a) Akhtar M, Calder MR, Corina DL, Wright JN. The status of oxygen atoms in the removal of C-19 in oestrogen biosynthesis. J Chem Soc Chem Commun 1981;129–30;

(b) Akhtar M, Calder MR, Corina DL, Wright JN. Mechanistic studies on C-19 demethylation in oestrogen biosynthesis. Biochem J 1982;201(3):569–80;

(c) Vaz ADN, Roberts ES, Coon MJ. Olefin formation in the oxidative deformylation of aldehydes by cytochrome P-450. Mechanistic implications for catalysis by oxygen-derived peroxide. J Am Chem Soc 1991;113(15):5886–7.

[64] (a) Akhtar M, Corina DL, Miller S, Shyadehi AZ, Wright JN. Mechanism of the acyl-carbon cleavage and related reactions catalyzed by multifunctional P-450s: studies on cytochrome P-45017.alpha. Biochemistry 1994;33(14):4410–8;

(b) Swinney DC, Mak AY. Androgen formation by cytochrome P450 CYP17. Solvent isotope effect and pL studies suggest a role for protons in the regulation of oxene versus peroxide chemistry. Biochemistry 1994;33(8):2185–90.

[65] (a) Gibson DT, Parales RE. Aromatic hydrocarbon dioxygenases in environmental biotechnology. Curr Opin Biotechnol 2000;11(3):236–43;

(b) Bassan A, Blomberg MR, Siegbahn PM. A theoretical study of the cis-dihydroxylation mechanism in naphthalene 1,2-dioxygenase. J Biol Inorg Chem 2004;9(4):439–52.

[66] Cho J, Jeon S, Wilson SA, Liu LV, Kang EA, Braymer JJ, et al. Structure and reactivity of a mononuclear non-haem iron(III)-peroxo complex. Nature 2011;478(3730):502–5.

[67] Ray K, Pfaff FF, Wang B, Nam W. Status of reactive non-heme metal–oxygen intermediates in chemical and enzymatic reactions. J Am Chem Soc 2014;136(40):13942–58.

[68] (a) Lee Y-M, Bang S, Kim YM, Cho J, Hong S, Nomura T, et al. A mononuclear nonheme iron(III)-Peroxo complex binding redox-inactive metal ions. Chem Sci 2013;4:3917–23;
(b) Bang S, Lee Y-M, Hong S, Cho KB, Nishida Y, Seo MS, et al. Redox-inactive metal ions modulate the reactivity and oxygen release of mononuclear non-haem iron(III)–peroxo complexes. Nat Chem 2014;6(10):934–40.

[69] Li F, Van Heuvelen KM, Meier KK, Mìnck E, Que Jr L. Sc^{3+}-Triggered oxoiron(IV) formation from O_2 and its non-heme iron(II) precursor via a Sc^{3+}–peroxo–Fe^{3+} intermediate. J Am Chem Soc 2013;135(28):10198–201.

[70] Van Eldik R. Introduction: inorganic and bioinorganic mechanisms. Chem Rev 2005;105(6):1917–22.

[71] (a) Liu LV, Bell CB, Wong SD, Wilson SA, et al. Defination of the intermediates and mechanism of the anticancer drug belmycin using nuclear resonance vibrational spectroscopy and related methods. Proc Natl Acad Sci USA 2010;107(52):22419–24;
(b) Wada A, Ogo S, Nagatomo, Kitagawa T, et al. Reactivity of hydroperoxide bound to a mononuclear non-heme Iron site. Inorg Chem 2002;41(6):616–8.

[72] Jensen KB, McKenzie CJ, Nielsen LP, Pedersen JZ, Svendsen HM. Deprotonation of low-spin mononuclear iron(III)–hydroperoxide complexes give transient blue species assigned to high-spin iron(III)–peroxide complexes. Chem Commun 1999;(0):1313–4.

[73] Bukowski MR, Comba P, Limberg C, Merz M, Que Jr L, Wistuba T. Bispidine ligand effects on iron/hydrogenperoxide chemistry. Angew Chem Int Ed 2004;43:1283–7.

[74] Lee Y-M, Hong S, Morimoto Y, Shin W, Fukuzumi S, Nam W. Dioxygen activation by a non-heme iron(II) complex: formation of an iron(IV)–Oxo complex via C–H activation by a putative iron(III)–Superoxo species. J Am Chem Soc 2010;132(44):10668–70.

[75] Nam W, Ryu YO, Song WJ. Oxidant types in copper-dioxygen chemistry: the ligand coordination defines the Cu(n)-O2 structure and subsequent reactivity. J Biol Inorg Chem 2004;9(6):654–60.

[76] Shaik S, de Visser SP, Kumar D. One oxidant, many pathways: a theoretical perspective of monooxygenation mechanisms by cytochrome P450 enzymes. J Biol Inorg Chem 2004;9(6):661–8.

[77] Wang B, Li C, Cho K-B, Nam W, Shaik S. The $Fe^{III}(H_2O_2)$ complex as a highly efficient oxidant in sulfoxidation reactions: revival of an underrated oxidant in cytochrome P450. J Chem Theor Comput 2013;9(6):2519–25.

[78] Park MJ, Lee J, Suh Y, Kim J, Nam W. Reactivities of mononuclear non-heme iron intermediates including evidence that iron(III)–Hydroperoxo species is a sluggish oxidant. J Am Chem Soc 2006;128(8):2630–4.

[79] Liu LV, Hong S, Cho J, Nam W, Solomon EI. Comparison of high-spin and low-spin nonheme Fe^{III}–OOH complexes in O–O bond homolysis and H-atom abstraction reactivities. J Am Chem Soc 2013;135(28):3286–99.

[80] Li F, Meier KK, Cranswick MA, Chakrabarti M, et al. Characterisation of high-spin nonheme FeIII-OOH intermediate and its Quantitative conversion to an FeIV=O complex. J Am Chem Soc 2011;133:7256–9.

[81] Bang S, Park S, Lee Y-M, Hong S, Cho K-B, Nam W. Demonstration of the heterolytic O-O bond cleavage of putative nonheme iron(II)-OOH(R) complexes for Fenton and enzymatic reactions. Angew Chem Int Ed 2014;53(30):7843–7.

[82] Bollinger Jr JM, Krebs C. Stalking intermediates in oxygen activation by iron enzymes: motivation and method. J Inorg Biochem 2006;100(4):586–605.

[83] (a) Meunier B, de Visser SP, Shaik S. Mechanism of oxidation reactions catalyzed by cytochrome P450 enzymes. Chem Rev 2004;104(9):3947–80;
(b) Denisov IG, Makris TM, Sligar SG, Schlichting I. Structure and chemistry of cytochrome P450. Chem Rev 2005;105(6):2253–64.

[84] (a) Bugg TDH. Oxygenases: mechanisms and structural motifs for O2 activation. Curr Opin Chem Biol 2001;5(5):550–5;
(b) Bugg TDH, Ramaswamy S. Non-heme iron-dependent dioxygenases: unravelling catalytic mechanisms for complex enzymatic oxidations. Curr Opin Chem Biol 2008;12(2):134–40;
(c) Kryatov SV, Rybak-Akimova EV, Schindler S. Kinetics and mechanisms of formation and reactivity of non-heme iron oxygen intermediates. Chem Rev 2005;105(6):2175–226.

[85] Jung C, de Visser S, Schünemann V. Spectroscopic characterization of cytochrome P450 Compound I. Arch Biochem Biophys 2011;507(1):44–55.

[86] (a) McGown AJ, Kerber WD, Fujii H, Goldberg DP. Catalytic reactivity of a meso-N-substituted corrole and evidence for a high-valent Iron–Oxo species. J Am Chem Soc 2009;131(23):8040–8;
(b) Harischandra DN, Zhang R, Newcomb M. Photochemical generation of a highly reactive Iron–Oxo intermediate. A true iron(V)–Oxo species? J Am Chem Soc 2005;127(40):13776–7.

[87] (a) Groves JT, Haushalter RC, Nakamura M, Nemo TE, Evans BJ. High-valent iron-porphyrin complexes related to peroxidase and cytochrome P-450. J Am Chem Soc 1981;103(10):2884–6;
(b) Wolter T, Meyer-Klaucke W, Müther M, Mandon D, Winkler H, Trautwein AX, Weiss R. Generation of oxoiron(IV) tetramesitylporphyrin π-cation radical complexes by m-CPBA oxidation of ferric tetramesitylporphyrin derivatives in butyronitrile at −78°C. Evidence for the formation of six-coordinate oxoiron(IV) tetramesitylporphyrin π-cation radical complexes FeIV=O(tmp)X (X=Cl−, Br−), by Mössbauer and X-ray absorption spectroscopy. J Inorg Biochem 2000;78(2):117–22;
(c) Gold A, Jayaraj K, Doppelt P, Weiss R, Chottard G, Bill E, et al. Oxoferryl complexes of the halogenated (porphinato)iron catalyst [tetrakis(2,6-dichlorophenyl)porphinato]iron. J Am Chem Soc 1988;110(17):5756–61;
(d) Wang X, Peter S, Kinne M, Hofrichter M, Groves JT. Detection and kinetic characterization of a highly reactive heme–thiolate peroxygenase compound I. J Am Chem Soc 2012;134(31):12897–900.

[88] Afanasiev P, Kudrik EV, Albrieux F, Briois V, Koifman OI, Sorokin AB. Generation and characterization of high-valent iron oxo phthalocyanines. Chem Commun 2012;48:6088–90.

[89] (a) Tiago de Oliveira F, Chanda A, Banerjee D, Shan X, Mondal S, Que Jr L, et al. Chemical and spectroscopic evidence for an FeV-oxo complex. Science 2007;315(5813):835–8;
(b) Prat I, Mathieson JS, Güell M, Ribas X, Luis JM, Cronin L, et al. Observation of Fe(V)=O using variable-temperature mass spectrometry and its enzyme-like C-H and C=C oxidation reactions. Nat Chem 2011;3(10):788–93;

(c) Lyakin OY, Bryliakov KP, Talsi EP. EPR, ^1H and ^2H NMR, and reactivity studies of the iron–oxygen intermediates in bioinspired catalyst systems. Inorg Chem 2011;50(12):5526–38;

(d) Thibon A, England J, Martinho M, Young VG, et al. Proton- and Reductant-Assisted Dioxygen Activation by a Nonheme Iron(II) Complex to Form an Oxoiron(IV) Intermediate. Angew Chem Int Ed 2008;47(37):7064–7.

[90] Bukowski MR, Munnck E, Collins TJ. Chemical and spectroscopic evidance for an FeV-oxo complex. Science 2007;315:835–8.

[91] Shaik S, Chen H, Janardanan D. Exchange-enhanced reactivity in bond activation by metal-oxo enzymes and synthetic reagents. Nat Chem 2011;3(1):19–27.

[92] Usharani D, Janardanan D, Li C, Shaik S. A theory for bioinorganic chemical reactivity of oxometal complexes and analogous oxidants: the exchange and orbital-selection rules. Acc Chem Res 2013;46(2):471–82.

[93] (a) Schöneboom JC, Cohen S, Lin H, Shaik S, Thiel W. J Am Chem Soc 2004;126(12):4017–34;

(b) Ferraro DJ, Gakhar L, Ramaswamy S. Rieske business: structure-function of Rieske non-heme oxygenases. Biochem Biophys Res Commun 2005;338(1):175–90;

(c) Wolfe MD, Parales VJ, Gibson DT, Lipscomb JD. Single turnover chemistry and regulation of O2 activation by the oxygenase component of naphthalene 1,2-dioxygenase. J Biol Chem 2001;276(3):1945–53;

(d) Barry SM, Challis GL. Mechanism and catalytic diversity of Rieske non-heme iron-dependent oxygenases. ACS Catal 2013;3(10):2362–70.

[94] Oloo WN, Que Jr L. Bioinspired nonheme iron catalysts for C–H and C–H;C bond oxidation: insights into the nature of the metal-based oxidants. Acc Chem Res 2015;48(9):2612–21.

[95] Lyakin OY, Bryliakov KP, Britovsek GJP, Talsi EP. EPR spectroscopic trapping of the active species of nonheme iron-catalyzed oxidation. J Am Chem Soc 2009;131(31):10798–9.

[96] (a) Lyakin OY, Ottenbacher RV, Bryliakov KP, Talsi EP. Asymmetric epoxidations with H_2O_2 on Fe and Mn aminopyridine catalysts: probing the nature of active species by combined electron paramagnetic resonance and enantioselectivity study. ACS Catal 2012;2(6):1196–202;

(b) Cusso O, Garcia-Bosch I, Ribas X, Lloret-Fillol J, Costas M. Asymmetric epoxidation with H_2O_2 by manipulating the electronic properties of non-heme iron catalysts. J Am Chem Soc 2013;135(39):14871–8;

(c) Lyakin OY, Prat I, Bryliakov KP, Costas M, Talsi EP. EPR detection of Fe(V)=O active species in nonheme iron-catalyzed oxidations. Catal Commun 2012;291:05–8.

[97] Fan R, Serrano-Plana J, Oloo WN, Draksharapu A, Delgado-Pinar E, Company A, et al. Spectroscopic and DFT characterization of a highly reactive nonheme FeV–oxo intermediate. J Am Chem Soc 2018;140(11):3916–28.

[98] Van Heuvelen KM, Fiedler AT, Shan X, De Hont RF, Meier KK, Bominaar EL, et al. One-electron oxidation of an oxoiron(IV) complex to form an $[O{=}Fe^V{=}NR]^+$ center. Proc Natl Acad Sci USA 2012;109(30):11933–8.

Chapter 23

Frontiers in developmental neurogenesis

Shanti N. Dessai
Department of Zoology, Goa University, Taleigao Plateau, Goa, India

23.1 Introduction to neurogenesis

Neurogenesis is the formation of new neurons in the brain. It is a crucial phenomenon during embryonic development and also tends to continue in certain brain regions after birth and throughout our lifespan. Neurogenesis is broadly categorized into two types, developmental neurogenesis and adult neurogenesis.

23.1.1 Developmental neurogenesis

Developmental neurogenesis is defined as formation of neurogenic cells, their proliferation, migration, and differentiation into neurons that covers prenatal (germinal, embryonic, and fetal) as well as postnatal (neonatal, infancy, childhood, and adolescence) development (Fig. 23.1). Remaining neurogenic events that form part of postnatal development during adult phase form adult neurogenesis.

Discovery of stem cells in parts of adult brain in the 1990s has changed the misconception about incapability of the central nervous system, including brain, to undergo neurogenesis and regeneration. The changeover of proliferative and multipotent neuronal stem cells (NSCs) to fully differentiated neurons and glia denotes neurogenesis and gliogenesis, respectively. Developmental neurogenesis involves generation of neurons from early embryonic development until early postnatal stages, with only a few neurogenic zones remaining active in the adult, representing adult neurogenesis [1]. Developmental gliogenesis starts during late embryogenesis and continues in postnatal stages, with low but widespread production of both astrocytes and oligodendrocytes also occurring throughout the adult brain, representing adult gliogenesis [2]. Regulated neurogenesis during embryonic development marks the incredible diversity of neurons in the brain. During developmental neurogenesis, neural progenitor cells (NPCs) initially proliferate symmetrically to

Advances in Biological Science Research. https://doi.org/10.1016/B978-0-12-817497-5.00023-9

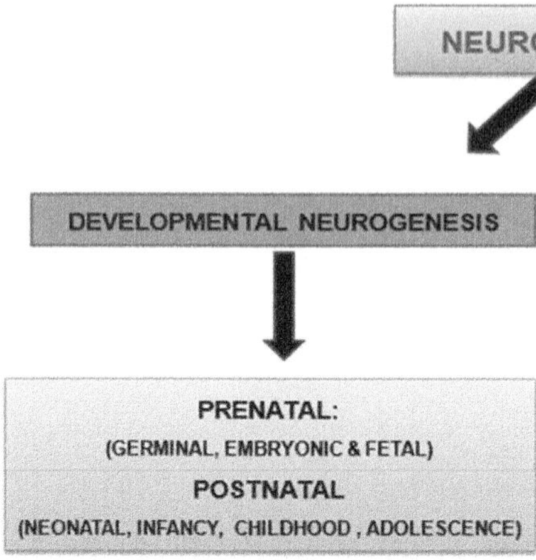

FIGURE 23.1 Classification of neurogenesis: developmental and adult neurogenesis.

increase their pool and then show asymmetric neurogenic divisions to form a number of specialized cell types at specific times and regions in the brain [1]. Thus, cautious regulation of spatial and temporal balance between NPC proliferation and differentiation results in normal developmental neurogenesis.

Despite the considerable progress in understanding the molecular mechanisms behind neurodevelopment, much still remains undiscovered. Hundreds of interacting signaling pathways likely lie behind even the most simple neurodevelopmental events, and only further analysis of animal models, basic biochemistry, and human diseases will bring us closer to elucidating these mechanisms.

This chapter reviews significant molecular parameters to be undertaken for better understanding of developmental neurogenesis and also provides some of the emerging tools that may be applied for such studies, hopefully enabling better planning and approaches to understand molecular-level depths in developmental neurogenesis studies using in vitro models.

23.2 Signaling pathway cross talk of developmental neurogenesis

Cells sense and respond to their environment through signaling pathways that involve a chain of regulatory and physical interactions among receptors, intermediate signaling proteins, transcription factors, and respective target genes. Sometimes stimulation of one pathway's receptors also elicits regulatory effects on the target genes of a second, biochemically distinct pathway, and such an involvement or phenomenon is referred to as "pathway cross talk." Developmental neurogenesis also involves such signaling pathway cross talk to shape up the nervous system in the growing embryo. More such signaling

TABLE 23.1 Summary of some of the significant signaling pathways involved in developmental neurogenesis discussed in this chapter.

Notch	NSC differentiation and self-renewal, activation promotes gliogenesis of NSCs, deactivation promotes neurogenesis of NSCs.
Wnt	Patterning: Early developmental neural tube formation and its differentiation, dorsoventral patterning. Proliferation: Throughout development, NPC proliferation. Neural morphology: Axonal and dendritic sizing, branching, and synapse formation and functionality.
Shh	Proliferative and morphogenic action on NSCs and NPCs, dorsoventral patterning, promotes symmetric proliferation of RGCs, neuronal differentiation, axonal path finding, and synapse formation and its functioning (in synchronization with many other signaling pathways).
FGFs	Anterior-posterior patterning, in association with other signaling pathways, brings about progenitor division regulation, cell survival, cell identity, neuronal differentiation, and axonal growth regulation.
NPC environment	Manipulation of NPC behavior by means of symmetric proliferation, asymmetric neurogenic division, and differentiation with the help of extracellular matrix including cell adhesion molecules, receptors, meningeal retinoic acid, blood vessels, etc.

pathway cross talk needs to be investigated in both healthy and diseased states across multiple tissues and in several species. Signaling pathways such as Notch, Wnt, Shh, and fibroblast growth factor (FGF) pathways trigger at plasma membrane and are also involved in proliferation with substantial cross talk among them during developmental neurogenesis [1] (Table 23.1). Analysis of signaling pathways and their cross talk is the need of the hour for better understanding of systems biology at the molecular level. Development of XTALKDB (Software database of signaling pathway cross talk) [3] based on published research data is one such software instrument that will aid in development of better insight in signaling pathway cross talk.

23.2.1 Notch

Notch signaling governs essential events of developmental neurogenesis. It supports NSC self-renewal and also regulates gliogenesis and neurogenesis via collaboration with other growth factors such as FGF, Wnt, Shh, and TGF-b. It is found that activation of Notch signaling pathway directly promotes gliogenesis of NSC or cross talk with CNTF, LIF, and bone morphogenic protein

(BMP) signaling pathway to facilitate gliogenesis. However, decreased Notch signaling activity promotes neurogenesis of NSC [1].

Recent findings also suggest that Notch signaling pathway could be associated with pathogenesis of neurodegenerative disorders [4]. Therefore, further detailed investigations on roles of Notch signaling pathways in NSC biology are not only bound to disclose the underlying molecular mechanisms of Notch signals in the process of NSC differentiation and self-renewal but also will provide new therapeutic interventions to treat neurodegenerative disorders.

23.2.2 Wingless/Integrated

Wingless/integrated (Wnt) signaling pathway, being the most conserved pathway, is involved in regulation of cellular fates, their migration, polarity inductions, patterning, and many more phenomena in the developing brain. Developmental neurogenesis Wnt pathways include both canonical or Wnt/β-catenin-dependent pathway and the noncanonical or β-catenin-independent pathway (planar cell polarity and the Wnt/Ca^{2+} pathways). Some of the critical processes of developmental neurogenesis regulated by Wnt pathway are as follows:

1. **Patterning:** Wnt signaling promotes development of the neural tube during early developmental neurogenesis and further patterns the neural tube by establishing signaling centers such as the midbrain–hindbrain boundary, and restricting rhombomere boundaries in the developing hindbrain. Wnt signaling is also essential for dorsal/ventral patterning of the neural tube, including telencephalon [5].
2. **Proliferation:** Research data also suggests the role of Wnt signaling in maintaining the proliferation of neural precursor cells throughout developmental neurogenesis [5].
3. **Neural morphology:** In addition to regulating patterning and cell proliferation, Wnts promote neurite outgrowth and influence axon size and branching, as well as growth cone size, complexity, and remodeling. Wnts and β-catenin also promote dendrite growth and branching and promote synapse formation and plasticity [5]. Symmetric radial glial cell (RGC) proliferation and delayed genesis of intermediate progenitor (IP) are also known to be regulated by Wnt pathway [6]. However, it promotes IP formation and neuronal differentiation during later phases of neurogenesis [7]. Wnt signaling is known to play a crucial role throughout brain development. Therefore, altered Wnt signaling during neural development can give rise to neurological dysfunction and sometimes may also lead to birth defects [8]. Hence it is important to have a thorough knowledge about the physiological outcomes of Wnt signaling in the developing brain so that new therapeutic strategies can be worked upon for neurological diseases.

23.2.3 Hedgehog/Sonic hedgehogs

Hedgehog/sonic hedgehog (Shh), a secreted morphogenic determinant, regulates dorsoventral patterning of the vertebrate central nervous system. Shh signaling is initiated due to binding of Shh ligand to the patched receptor, which then leads to ciliary accumulation of smoothened receptors (Class Frizzled: G protein-coupled receptor) thus leading to activation of transcription factors that intervene downstream gene transcription [9]. In the absence of Shh, the Gli proteins are repressed. Shh signaling also regulates kinetics of RGC cell cycle and IP production [10]. RGC symmetric proliferative divisions are also influenced by Shh signaling through transcription of the Notch transcription factors [11]. Many such studies now indicate that Shh signaling occurs in synchronization with the many other signaling pathways cross talk, thus contributing toward proliferative and morphogenic actions on neural stem cells and progenitors. It also governs neuronal differentiation, axonal path finding and synapse formation and its functionality [1]. This diverse expression is dependent on maturational state of the responsive cell, cross talk with other signaling pathways regulating neural cell function, and the environmental cues surrounding target cells. Shh signaling dynamism in the nervous system ranges from canonical transcription dependent, to noncanonical and localized, to axonal growth cones [9]. The wide range of Shh signaling functions account for its malfunctions in neurodevelopmental diseases, thus making it a target of interest for therapeutic applications.

23.2.4 Fibroblast growth factor

Anterior–posterior patterning of the brain and proliferation of RGCs by symmetric division through downstream activation of Hes1-mediated transcription marks the importance of FGF signaling [12]. A wide range of cellular processes such as cell survival, proliferation, and specification of cell identity, neuronal differentiation, and axon growth are regulated by isthmic FGFs [13]. The various isoforms of the four FGFRs through integral components of the FGF ligand-receptor complex have the potential to diversify its signaling activities. The activation of FGF receptor complexes are known to trigger several signal transduction cascades and cross talk with other pathways, such as the synergistic and antagonistic interactions with Wnts, EGF, retinoic acid, and Notch, they regulate progenitor divisions, and further expand the range of cellular responses to FGFs. In addition to this multiplicity of signaling mechanisms, the response of neural tissues to the same FGF signal seems to vary with respect to space and time [14]. Such spatial and temporal competence factors are likely to play an important role in the diversification of FGF functions and elucidate how they modulate the cellular response to FGF signaling, thus giving it an exciting direction for future research.

23.2.5 Neuronal progenitor cell environment

Besides the thus far discussed extracellular signals, many other factors in the NPC environment manipulate NPC behavior. A variety of ECM molecules such as laminin and syndecan-1 are present at the ventricular surface, which regulate via integrin receptors the apical adhesion and proliferation of RGCs, and these are also governed by ephrin B1 [1]. The interaction of the NPC basal process with basal lamina ECM is found to be significant for the self-renewing potential of RGCs and bRGs [15]. Meningeal retinoic acid signal from the basal side shifts RGCs from symmetric proliferative to asymmetric neurogenic divisions at the onset of neurogenesis. Many other environmental cues, such as the presence of blood vessels near Ips, appear to regulate their proliferation [16]. In addition, already differentiated nonneuronal cell types like microglia have been shown to regulate maintenance of the RGC population during neurogenesis [17]. Cross talk also involves postmitotic neurons that provide feedback to RGCs. The Cajal–Retzius cells, early differentiated first type of neurons, secrete the glycoprotein reelin that mediates neuronal migration and regionalization through expression of the cell adhesion molecules called nectins [18]. Also, their signaling cross talk influences RGC proliferation through reelin, which amplifies Notch signaling in early RGCs to promote symmetric proliferations that postpone neurogenesis [19]. Expression of signaling molecules such as neurotrophin 3 and Fgf9 by cortical neurons in later phases regulate cell fates and proliferation of RGCs to undergo astrogenesis [20]. Transient glutamatergic neurons of ventral telencephalon also provide cross talk signals to bring about tangential migration into the dorsal telencephalon [21].

23.3 Tools to study developmental neurogenesis

Establishment of complex neuronal circuitry is facilitated by neuronal cells themselves due to their ability to sense, integrate, and respond to various stimuli in a timely and highly organized manner during neurogenesis. These phenomena are also guided by other neural cell types such as astrocytes and oligodendrocytes along with nonneural cells such as microglia, endothelial, fibroblast, and blood cells. They also play a crucial role in supporting neuronal cells to achieve their appropriate function in a homeostatic balanced microenvironment [22]. Hence, central nervous system diseases and disorders are associated with neuronal cell loss and dysfunction [23]. Considering the complex architecture and circuit of the nervous system, investigating all such parameters in vivo is a daunting task. Therefore, to facilitate such investigations, in vitro assays and methodologies have been developed to simulate/recapitulate these in vivo processes. This will also serve as a reductionist approach by decreasing some of the associated complexities if they were to be carried out in vivo [22]. Thus this section reviews some of the latest available in vitro cell-based neurogenesis models and methodologies.

23.3.1 In vitro models

Pluripotent stem cells such as embryonic stem cells (ESCs) and induced pluripotent stem cells (iPSCs) [24,25] are used to create in vitro neurogenesis models. The cell culture systems used to create neural cells from the pluripotent stem cells include embryoid body (EB) formation, coculture with cells, such as bone marrow stromal cells or their conditioned medium that potentiate neuralization processes. As well, monolayer culture systems are used to produce neural cells [22].

1. Embryoid body
   In this culture system (Figs. 23.2A and 23.3), pluripotent stem cells are first spontaneously differentiated into different cell lineages followed by improvements in purity of neural cells through extracellular factors such as retinoic acid. Because of the similarities of EB and pregastrulation embryo, involvement of spontaneous differentiation (EB method) is widely used in examining the effect of the cell signaling pathways and the effect of chemicals during early embryogenesis [26].

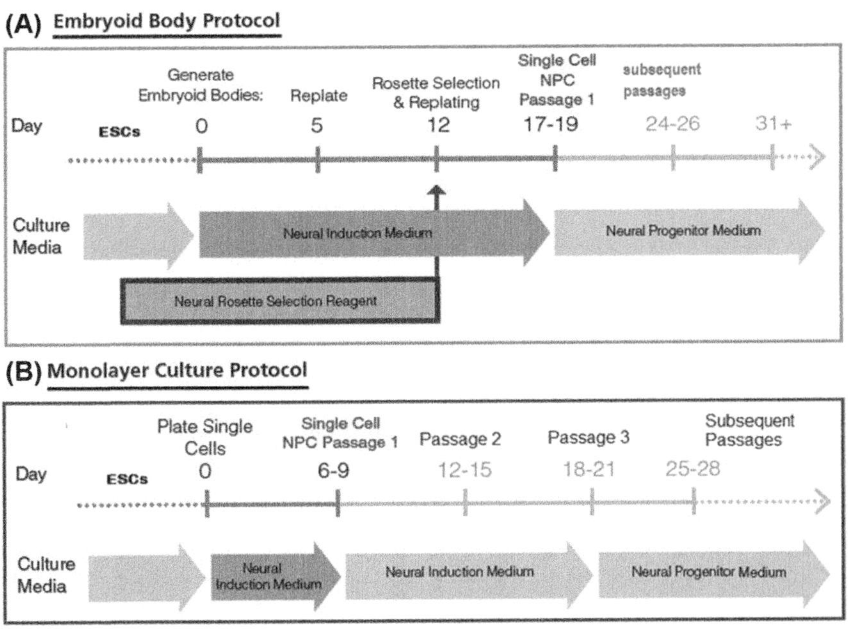

FIGURE 23.2 Time line and schematic representation of embryoid body and monolayer culture system: (A) Embryoid body protocol and (B) Monolayer culture protocol for initiation and cultures from embryonic stem cells.

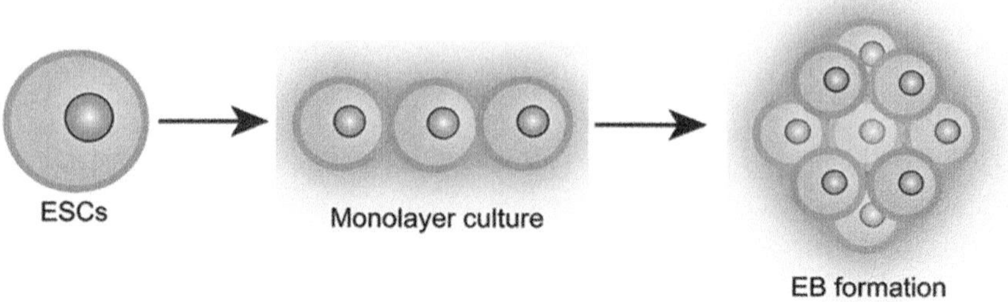

FIGURE 23.3 Flowchart of embryoid body formation from embryonic stem cells.

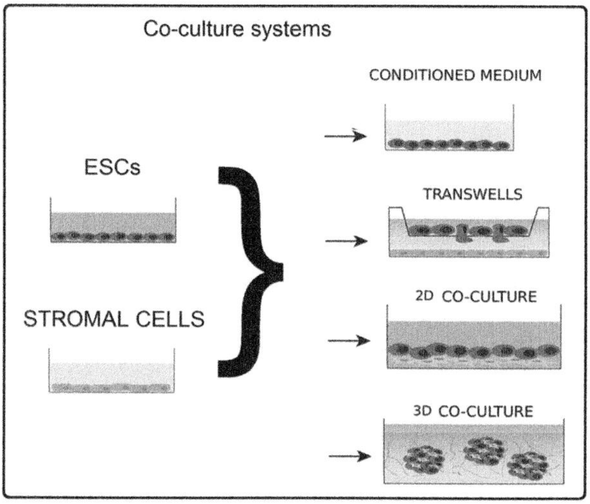

FIGURE 23.4 Schematic representation of different kinds of coculture system that can be adopted to study developmental neurogenesis.

2. Coculture with stromal cells
   This culture system (Fig. 23.4) involves induction of neural identity in ESCs using stromal cells or their conditioned medium that reduces the length of the neuralization process and hence can be also designated as a serum-free culture system [27]. Although, generation of neuroepithelial cells from murine and human ESCs through this methodology is simple, fast, and efficient, the approach is dependent on feeder cells and their unidentified stromal-derived inducing activity, thus suggesting its minimal use for research work to study greater depths of signaling pathways and more usage toward metabolite regulated pathways of neurogenesis.
3. Adherent monolayer culture system
   In such culture system of pluripotent stem cells (Fig. 23.2B), neural differentiation is dependent on early neural induction involving both

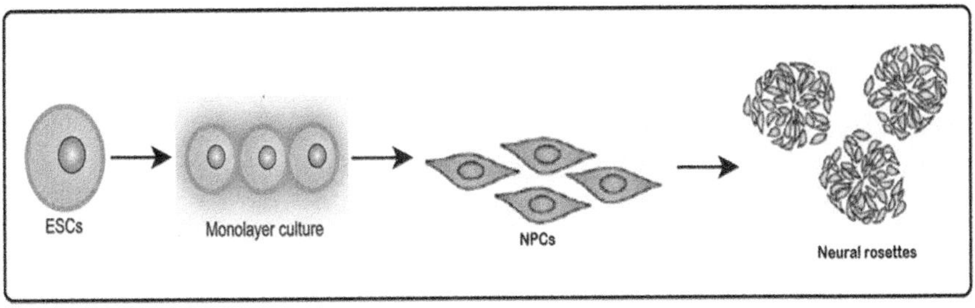

FIGURE 23.5 Flowchart of monolayer culture of embryonic stem cells depicting formation of neuronal progenitor cells and neural rosettes.

fibroblast growth factor signaling and the inhibition of BMP signaling [28]. Monolayer adherent-type culture systems of pluripotent stem cells recapitulate the process of neural tube formation in vitro, which is observed as neural tube-like structures called neural rosettes [22] (Fig. 23.5). This in vitro culture system provides a unique substrate to dissect out signaling pathways and complex molecular mechanisms underlying pluripotent stem-cell-derived neurogenesis as it involves usage of chemically defined serum-free medium [29]. Also, this monolayer culture system allows possible visual insights of neural conversions from ES cells, thus providing researchers with an opportunity to study various factors and signaling pathways underlying neurogenesis.

4. Organoid/3D models
Development of in vitro cell, tissue systems and advent of pluripotent stem cells have led to new frontiers in usage of three-dimensional models of developing neural tissue that model the developing brain with remarkable reliability and recapitulation features that are observed in developing brain in vivo [30]. Unprecedented levels of research studies on developmental processes and disorders can now be carried out by combining the organoid/3D culture method with recently developed gene editing techniques such as CRISPR (clustered regularly interspaced short palindromic repeats), and further methodological improvements could influence the field of developmental neurogenesis and also the development of new therapies [31].

23.3.2 Time-lapse analysis

Dynamism of developmental neurogenesis phenomena to develop neural circuitry demands using time-lapse microscopy and other image analysis techniques that will best explore the tracking of moving cells during gastrulation and neurulation, processes that transform the central nervous system to normal architecture. Adherent-type clonal cell cultures derived from different embryonic ages are found to recapitulate generation of progenitors as well as distinct neuronal subtypes and glia, with timing and order that are observed in vivo [32]. This can be proved with the help of specific markers and correlative staining to tag the daughter cells generated from such isolated progenitor embryonic tissues [33]. These studies may be further refined with the help of fluorescence-activated cell sorting that will provide complete registry of individual cell divisions and behavior and transparent analysis of intrinsic mechanisms involved in cell fate. The only drawback of such clonal analysis of isolated progenitors lies in fact that they do not encounter the permissive microenvironments, maintenance of polarity and orientation they would have in the intact brain [34]. Use of tissue slices for such assays may overcome these limitations of isolated progenitor cultures.

23.3.3 Transcriptome, metabolomics, and single-cell "omics"

Each event during developmental neurogenesis is regulated by a variety of intrinsic mechanisms and cell-extrinsic cues well regulated through defined transcripts and key morphogenic signaling pathways. Transcriptome studies of distinct progenitor cells indicated the importance of their metabolic states in the governance of developmental neurogenesis [35]. Observation of highest glycolysis in proliferating stem cells that accounts for its stemness and highest oxidative phosphorylation encountered in differentiated cells speak about the importance of metabolism of progenitors during development. As well, the lipid metabolism is also found to help maintain proliferation and neurogenesis. Cellular metabolites of progenitors will definitely offer an important unexplored direction in understanding the pluripotency of stem cells [36] besides slowly emerging studies related to population-based systems biology methods and single-cell "omics" [37]. Progenitor cells of developmental neurogenesis include different states of cells from simple quiescent stage to variety of differentiation stages, which definitely demands single-cell metabolic investigations. For single-cell metabolomics, mass spectrometry serves as the best tool with its finest detection sensitivity at mole (10^{-18} mol) concentration range. These studies may be approached through nanoelectrospray ionization, microfluidic chips, and sample arrays for monodispersed sample droplets as sample sources for MS data acquisition [38]. Detection of single-cell metabolites is also a focus of emerging imaging mass spectrometry (secondary ion mass spectrometry), with spatial resolution at the micron level [39].

23.3.4 Real-time analysis of progenitors in both embryonic and postnatal studies by tissue explants/slice assays

The real-time tracking involving targeted expression of fluorescent protein markers of individual progenitors along with their progeny can be captured by time-lapse confocal imaging of tissue slice cultures. It also provides an objective-based choice of an experimental approach by targeting single or multiple progenitor cells in desired proliferative regions of both embryonic as well as postnatal tissue samples that covers developmental neurogenesis.

Available fluorescent gene expressions mostly rely on electroporation techniques either in utero or ex utero, viral transduction, and genetic manipulation of germ line, each having its own benefits. Progenitor cohorts are best handled with electroporation and genetic manipulations, whereas tracking of isolated cells can be worked out with viral transduction. Electroporation next followed by live-cell imaging serves as the effective method for tracking dynamism of developmental neurogenesis. Especially, available fluorescent markers make it possible even to track interkinetic nuclear movement of neuronal progenitors in both normal and disease models [40]. Further, combination of fluorescent labeling, genetic manipulation followed by coelectroporation of mutated genes or short hairpin RNA constructs can be utilized to analyze the molecular mechanisms regulating progenitor morphology, proliferation, and neurogenesis. Postnatal studies can also be worked out with the help of electroporation [41]. Such approaches can significantly facilitate our ability to study developmental neurogenesis and the response of neural progenitors to different types of abnormal conditions.

23.4 Conclusion

Signaling molecules and pathway cross talk are recently studied phenomena underlying signaling biology that open many avenues for further research to unfold many more such pathways and their cross talk in healthy and disease conditions across various tissues and in different species during development.

References

[1] Paridaen JT, Huttner WB. Neurogenesis during development of the vertebrate central nervous system. EMBO Rep 2014;15:351–64.

[2] Guérout N, Li X, Barnabé-Heider F. Cell fate control in the developing central nervous system. Exp Cell Res 2014;321:77–83.

[3] Sam SA, Teel J, Tegge AN, Bharadwaj A, Murali TM. XTalkDB: a database of signaling pathway crosstalk. Nucleic Acids Res 2017;45(Database issue):D432–9.

[4] Osumi O, Shinohara H, Numayama-Tsuruta K, Maekawa M. Concise review: pax6 transcription factor contributes to both embryonic and adult neurogenesis as a multifunctional regulator. Stem Cell 2008;26:1663–72.

[5] Valvezan AJ, Klein PS. GSK-3 and wnt signaling in neurogenesis and bipolar disorder. Front Mol Neurosci 2012;5:1–13.

[6] Wrobel CN, Mutch CA, Swaminathan S, Taketo MM, Chenn N. Persistent expression of stabilized beta-catenin delays maturation of radial glial cells into intermediate progenitors. Dev Biol 2007;309:285–97.

[7] Fang WQ, Chen WW, Fu AK, Ip NY. Axin directs the amplification and differentiation of intermediate progenitors in the developing cerebral cortex. Neuron 2013;79:665–79.

[8] Noelanders R, Vleminckx K. How wnt signaling builds the brain: bridging development and disease. Neuroscientist 2016;23:314–29.

[9] Belgacem YH, Hamilton AM, Shim S, Spencer KA, Borodinsky LN. The many hats of sonic hedgehog signaling in nervous system development and disease. J Dev Biol 2016;4:1–18.

[10] Wang H, Ge G, Uchida Y, Luu B, Ahn S. Gli3 is required for maintenance and fate specification of cortical progenitors. J Neurosci 2011;31:6440–8.

[11] Dave RK, Ellis T, Toumpas MC, Robson JP, Julian E, Adolphe C, Bartlett PF, Cooper HM, Reynolds BA, Wainwright BJ. Sonic hedgehog and notch signaling can cooperate to regulate neurogenic divisions of neocortical progenitors. PLoS One 2011;6:e14680.

[12] Rash BG, Lim HD, Breunig JJ, Vaccarino FM. FGF signaling expands embryonic cortical surface area by regulating notch-dependent neurogenesis. J Neurosci 2011;31:15604–17.

[13] Partanen J. FGF signaling pathways in development of the midbrain and anterior hindbrain. J Neurochem 2007;101:1185–93.

[14] Guillemot F, Zimmer C. From cradle to grave: the multiple roles of fibroblast growth factors in neural development. Neuron 2011;71:574–88.

[15] Fietz SA, Kelava I, Vogt J, Wilsch-Brauninger M, Stenzel D, Fish JL, Corbeil D, Riehn A, Distler W, Nitsch R, Huttner WB. OSVZ progenitors of human and ferret neocortex are epithelial-like and expand by integrin signalling. Nat Neurosci 2010;13:690–9.

[16] Javaherian A, Kriegstein A. A stem cell niche for intermediate progenitor cells of the embryonic cortex. Cerebr Cortex 2009;19:i70–7.

[17] Cunningham CL, Martinez-Cerdeno V, Noctor SC. Microgliaregulate the number of neural precursor cells in the developing cerebral cortex. J Neurosci 2013;33:4216–33.

[18] Griveau A, Borello U, Causeret F, Tissir F, Boggetto N, Karaz S, Pierani A. A novel role for Dbx1-derived Cajal-Retzius cells in early regionalization of the cerebral cortical neuro-epithelium. PLoS Biol 2010;8:e1000440.

[19] Lakoma J, Garcia-Alonso L, Luque JM. Reelin sets the pace of neocortical neurogenesis. Development 2011;138:5223–34.

[20] Seuntjens E, Nityanandam A, Miquelajauregui A, Debruyn J, Stryjewska A, Goebbels S, Nave KA, Huylebroeck D, Tarabykin V. Sip1 regulates sequential fate decisions by feed-back signaling from postmitotic neurons to progenitors. Nat Neurosci 2009;12:1373–80.

[21] Teissier A, Waclaw RR, Griveau A, Campbell K, Pierani A. Tangentially migrating transient glutamatergic neurons control neurogenesis and maintenance of cerebral cortical progenitor pools. Cerebr Cortex 2012;22:403–16.

[22] Azari H, Reynolds BA. In vitro models for neurogenesis. Cold Spring Harbor Perspect Biol 2016;8:a021279.

[23] Amor S, Puentes F, Baker D, van der Valk P. Inflammation in neurodegenerative diseases. Immunology 2010;129:154–69.

[24] Fathi A, Rasouli H, Yeganeh M, Salekdeh GH, Baharvand H. Efficient differentiation of human embryonic stem cells toward dopaminergic neurons using recombinant LMX1A factor. Mol Biotechnol 2015;57:184–94.

[25] Velasco I, Salazar P, Giorgetti A, Ramos-Mejia V, Castano J, Romero-Moya D, Menendez P. Concise review: generation of neurons from somatic cells of healthy individuals and neurological patients through induced pluripotency or direct conversion. Stem Cell 2014;32:2811–7.

[26] Liyang G, Abdullah S, Rosli R, Nordin N. Neural commitment of embryonic stem cells through the formation of embryoid bodies (EBs). Malays J Med Sci 2014;21:8–16.

[27] Vazin T, Chen J, Lee CT, Amable R, Freed WJ. Assessment of stromal-derived inducing activity in the generation of dopaminergic neurons from human embryonic stem cells. Stem Cell 2008;26:1517–25.

[28] Linker C, Stern CD. Neural induction requires BMP inhibition only as a late step, and involves signals other than FGF and Wnt antagonists. Development 2004;131:5671–81.
[29] Ying QL, Stavridis M, Griffiths D, Li M, Smith A. Conversion of embryonic stem cells into neuroectodermal precursors in adherent monoculture. Nat Biotechnol 2003;21:183–6.
[30] Lancaster MA, Knoblich JA. Organogenesis in a dish: modelling development and disease using organoid technologies. Science 2014;345:1247125.
[31] Kelava I, Lancaster MA. Dishing out mini-brains: current progress and future prospects in brain organoid research. Dev Biol 2016;420:199–209.
[32] Shen Q, Wang Y, Dimos JT, Fasano CA, Phoenix TN, Lemischka IR, Ivanova NB, Stifani S, Morrisey EE, Temple S. The timing of cortical neurogenesis is encoded within lineages of individual progenitor cells. Nat Neurosci 2006;9:743–51.
[33] Al-Kofahi O, Radke RJ, Goderie SK, Shen Q, Temple S, Roysam B. Automated cell lineage construction: a rapid method to analyze clonal development established with murine neural progenitor cells. Cell Cycle 2006;5:327–35.
[34] Lu B, Roegiers F, Jan LY, Jan YN. Adherens junctions inhibit asymmetric division in the Drosophila epithelium. Nature 2001;409:522–5.
[35] Knobloch M, Jessberger S. Metabolism and neurogenesis. Curr Opin Neurobiol 2017;42:45–52.
[36] Yanes O, Clark J, Wong DM, Patti GJ, Sánchez-Ruiz A, Benton HP, Trauger SA, Desponts C, Ding S, Siuzdak G. Metabolic oxidation regulates embryonic stem cell differentiation. Nat Chem Biol 2010;6:411–7.
[37] Zenobi R. Single-cell metabolomics: analytical and biological perspectives. Science 2013;342:1243–59.
[38] Heinemann M, Zenobi R. Single cell metabolomics. Curr Opin Biotechnol 2011;22:26–31.
[39] Klerk LA, Dankers PY, Popa ER, Bosnian AW, Sanders ME, Reedquist KA, Heeren RM. TOF-secondary ion mass spectrometry imaging of polymeric scaffolds with surrounding tissue after *in vivo* implantation. Anal Chem 2010;82:4337–43.
[40] Yokota Y, Kim WY, Chen Y, Wang X, Stanco A, Komuro Y, Snider W, Anton ES. The adenomatous polyposis coli protein is an essential regulator of radial glial polarity and construction of the cerebral cortex. Neuron 2009;61:42–56.
[41] Higginbotham H, Yokota Y, Anton ES. Strategies for analysing neuronal progenitor development and neuronal migration in the developing cerebral cortex. Cerebr Cortex 2010;21(7):1465–74.

Chapter 24

Analytical methods for natural products isolation: principles and applications

Mahesh S. Majik[1], Umesh B. Gawas[2], Vinod K. Mandrekar[3]
[1]*Department of Chemistry, Goa University, Taleigao Plateau, Goa, India;* [2]*Department of Chemistry, Dnyanprassarak Mandal's College and Research Centre, Assagao, Goa, India;* [3]*Department of Chemistry, St. Xavier's College, Mapusa, Goa, India*

24.1 Introduction

Natural products extracted from natural sources were found to display impressive therapeutic potential, some of which have been utilized in traditional as well as modern medicines [1]. These chemicals are known as secondary metabolites, which are synthesized biochemically by plants, microbes, marine animals, insects, and amphibians for their survival [2]. In recent years, marine organisms and associated microbes have been established as a good source of novel structural skeletons/new medicines for treatment of different diseases [3,4]. Marine organisms like sponges and corals synthesize various chemicals to: (1) defend themselves against predators; (2) prevent other animals from moving into their space; (3) find a mate; and, (4) protect themselves from infections [5]. The biodiversity of the oceans is reflected in their chemo diversity, which shows the potential of marine organisms for isolation of novel drugs. However, much of these sources remain unexplored to date and hence this opens the door for young researchers to undertake projects in this interesting research area [6]. With less than 1% of the marine microbial world being explored to date, the modernization in cultivation and extractions from marine habitat has offered access to an unexplored basin of genetic and metabolic diversity [7]. Moreover, the advances in chromatographic and spectroscopic techniques, especially in high performance liquid chromatography (HPLC), tandem mass spectrometry and their hyphenation, have made isolation, purification, and characterization of biologically active molecules much easier and quicker.

Advances in Biological Science Research. https://doi.org/10.1016/B978-0-12-817497-5.00024-0

24.2 Extraction techniques

The natural products present in marine sources/biomasses need to be extracted preserving their original identity using suitable extraction technique prior to isolation and purification. The choice of extraction technique depends on the source employed for extraction, nature of the product, and stability of compounds to be extracted from the sources. Prior to selection of a particular method, it is necessary to establish the target of the extraction. The diverse groups of the natural compounds present in sources will possess specific physicochemical properties (e.g., solubility, volatility). The basic principles involved in the separation by using different methods are depicted in (Table 24.1) [8]. Therefore, it is necessary to define objectives clearly before selection of particular method to get efficient extraction from the natural source under investigation. Amongst various methods available, the solvent-extraction method is considered to be efficient and the best protocol for initial and bulk laboratory-scale extraction of natural products from plant and microbial sources [9].

The extraction method uses two immiscible phases to transfer a solute from one phase into another. The distribution of a solute between two phases is an equilibrium described by partition theory. A good example of extraction protocol in daily life is boiling tea powder/leaves in water giving color, wherein compounds such as tannins, theobromine, and caffeine are extracted out of the leaves into the water. In solvent extraction technique, the principle of "like dissolves like" is employed, i.e., extraction of organic compounds

TABLE 24.1 Methods for separation of mixtures.

Different methods of separation of mixtures	
Methods	Principle involved in the separation
Mechanical phase separation	
1. precipitation/Filtration	Solubility differences for different analytes
2. distillation	Volatility differences for different analytes
3. ***Extraction***	***Solubility differences for different analytes in two immiscible solvents***
4. Ion exchange	Interaction differences of different analytes with ion exchange resins
Chromatography	***Migration rate differences of analyte on stationary phase and mobile phase***
Electrophoresis	Migration rate differences of charged analyte on aqueous medium under electric field

(natural products) out of an aqueous phase and into an organic phase. The two different approaches often employed for extraction of biochemicals from natural sources are: (1) extraction by soaking (solid-liquid extraction); and (2) fractionation using different solvents (liquid-liquid extraction).

In solid-liquid extraction, the solvent such as methanol or methanol/chloroform mixture, extracts the different components present in the sample. The organic extract is then analyzed by thin-layer chromatography (TLC) to evaluate complexity of the natural components present in the extract. Depending on complexity of the components in crude extract, the extraction procedure is simplified further by fractionation in three different solvents like nonpolar, medium polar, and highly polar using liquid-liquid extraction (Fig. 24.1) [10]. The pure natural compounds are obtained from these simplified fractions (four fraction includes, petroleum ether, chloroform, n-butanol, and water) using different separation techniques. The analytical techniques that are used to achieve this target are discussed in the next section of this chapter.

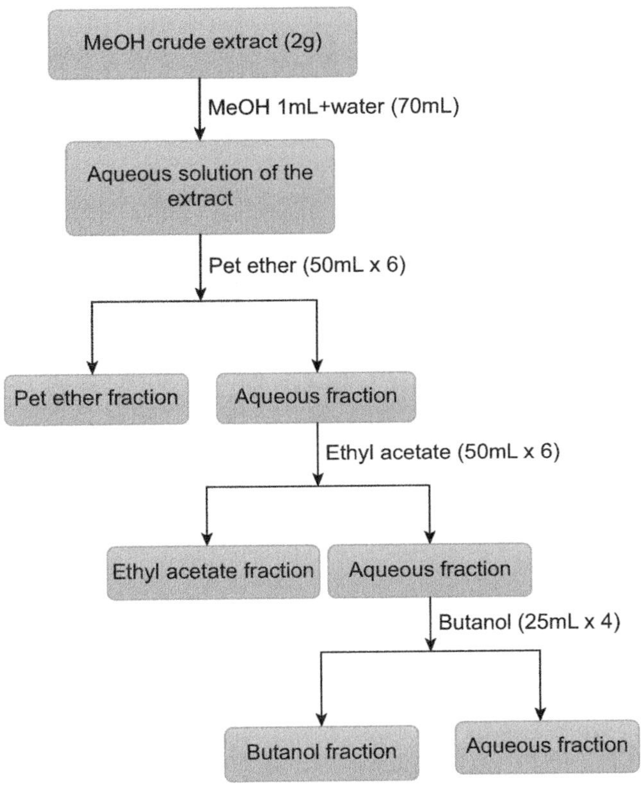

FIGURE 24.1 Flow diagram for extraction protocol for the fractionation of complex crude mixture.

24.3 Isolation and purification techniques

The most important factor that has to be considered before selection of the purification techniques is the nature of the target compounds present in the crude extract or fractions. Prior knowledge of the physiochemical properties of the target compounds in extract may help in designing the extraction protocol, thus making the separation process easier. The principles involved in separation of biomolecules are based on their solubility, acid-base properties, charge, stability, and molecular size. Isolation of known compounds from the same or a different natural source is quite simple since the literature information is easily available for determining the chromatographic method. However, the procedure becomes more difficult and complicated when the components present in natural extract are unknown, wherein it is necessary to carry out qualitative analysis for phytochemicals, e.g., phenolics, steroids, alkaloids, flavonoids, etc. to generate preliminary information about the unknown components. Subsequently, chromatographic techniques can be employed for the purification of individual compounds.

The chromatographic techniques used in the isolation of various types of natural products can be broadly classified into the following two categories [11]:

1. **Planar Chromatography:** The stationary phase is supported by a flat surface (e.g., a glass plate, a plastic sheet, paper, etc.). The mobile phase and analyte pass through the stationary phase by capillary action and/or gravity. The sample is spotted onto the paper/TLC plate as shown in Fig. 24.2, and the bottom edge of the piece of paper/TLC plate is dipped into an appropriate liquid. The liquid mobile phase moves up the thin layer of adsorbent (silica gel or alumina powder) by capillary action, and components of the mixture are separated into different spots depending on their relative attraction to the stationary phase or nature of solvents that make up the mobile phase. The Rf value obtained will be different for

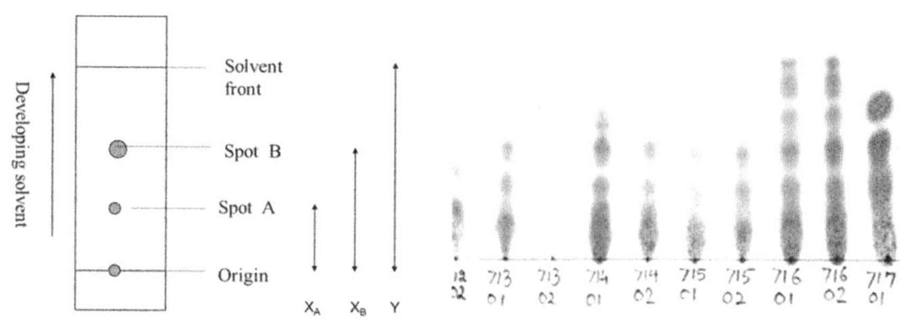

$$R_f = \frac{\text{distance from the origin migrated by a compound}}{\text{distance from origin migrated by solvent}}$$

FIGURE 24.2 TLC plate depicting separation and detection of components from extract.

different compounds. Hence, different solvent systems (mobile phases) are scanned to derive information about a number of components present in the crude sample (the number of spots on TLC/accounts for the number of components present in sample). The TLC plate is visualized either by nondestructive methods such as UV-visible light/iodine vapors, or by using destructive methods such as spraying 5% H_2SO_4 in methanol, ninhydrine, etc.

As the mobile phase rises up the TLC plate, the individual components move up at different rates depending on intermolecular forces between the components with stationary phase and the components with the mobile phase. These adsorption strengths of compounds increase with the increasing polarity of functional group (Fig. 24.3).

The separation of two analytes on the chromatogram can be expressed as resolution. Hence, the composition or chemical nature of mobile phase is changed to get the optimum resolution in which the separation between two components is in the Rf range of 0.2–0.5. This Rf value can be used to aid in the identification of given components in the sample by comparing with the standards. However, Rf value is not a physical constant and the comparison should be made only between spots on the same sheet, run at a same time under identical experimental conditions.

2. **Column Chromatography:** In column chromatography, a column made up of glass is packed with a stationary phase such as silica gel or alumina and equilibrated by adding mobile phase. The analyte mixture to be separated is transferred to the top of the column using either wet-loading or dry-loading method. Mobile phase is continuously fed from the top to flush the analytes through the column under gravity. If some gas pressures are used to force the liquid through a column, then the technique is referred to as flash chromatography, which is often used by synthetic chemists to separate relatively large amounts of reaction products. As the components pass through the column, depending on their affinity they are adsorbed on stationary phase and get separated in the form of bands. Analytes with lower affinity spend less time on the column and hence pass through quickly, while those with higher affinity for the stationary phase spend more time on the column and hence are retained for a longer time. The pictorial representation of conventional column chromatography is depicted in (Fig. 24.4).

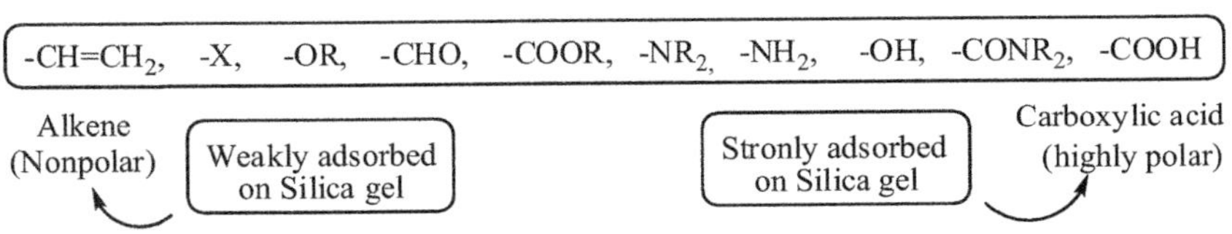

FIGURE 24.3 Adsorption strength of different solute based on functional groups.

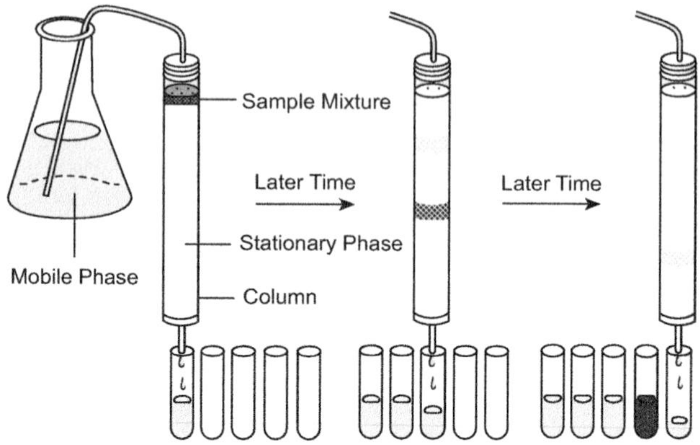

FIGURE 24.4 Column chromatography demonstrating the separation of bands of each component.

In column chromatographic separation, the resolution of the components varies with the particle size of adsorbent and better separations are obtained with smaller size particles. However, smaller particles slow down the elution rate of mobile phase. The elution rate can be increased by using pumps that generate lower, medium to high pressures; this separation technique is called HPLC.

24.4 High-performance liquid chromatography

Different types of liquid chromatographic techniques are available, depending on stationary phase and mobile phase combinations [12]. The different types of packing material are used to give particular types of HPLC, including partition, adsorption, ion exchange, size exclusion, etc. The major advantages of HPLC over other chromatographic methods are: (1) utility of stationary phases with very small particle size with large surface area, and (2) utility of high pressure with improvement in the elution rates. The separation of analytes occurs within the column in form of bands that are then passed through the online detector to give the chromatogram as depicted in (Fig. 24.5).

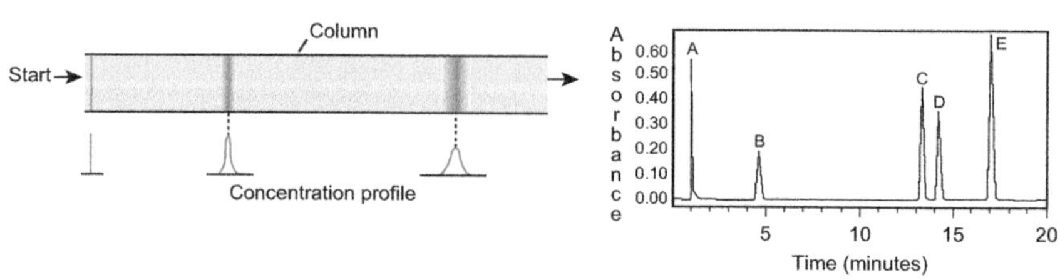

FIGURE 24.5 Conversion of bands on column to corresponding peaks on the chromatogram.

TABLE 24.2 Comparison between RPLC and NPLC.

	Comparison of RPLC and NPLC		
Type	**Stationary phase**	**Weak mobile phase**	**Strong mobile phase**
RPLC	Nonpolar	Polar liquid	More nonpolar
NPLC	Polar	Nonpolar liquid	Polar liquid

In HPLC separation of complex mixtures, the two popular modes of chromatography used are (1) normal-phase liquid chromatography (NPLC) and (2) reverse-phase column chromatography (RPLC) [13]. The comparison of these two modes is represented in Table 24.2.

24.4.1 Analysis of chromatograms obtained from HPLC/GC

1. Retention time of peaks is useful for qualitative identification of specific species in mixtures.
2. Under a given set of conditions, a single analyte will always have the same retention time.
3. A control sample of analyte is run to measure its retention time; the presence of analyte in a complex, unknown mixture is usually determined by the presence of a peak identical in retention time to the control sample.
4. It is always advisable to repeat the analysis with a different set of chromatographic parameters. If the control sample and complex, unknown mixture have a peak that matches under more than one set of chromatographic conditions, it is usually safe to assume that they are the same species.
5. Peak heights and areas under peak are used to derive quantitative data from chromatograms.
6. The concentration of an analyte is proportional to its area (or height) under the peak observed in the chromatogram.
7. To quantify an analyte, a series of control runs with varying amount of standard analyte must be run to prepare a standard curve (a plot of peak area vs. amount of standard run).
8. The unknown sample is then run under conditions identical to the control, and the area under peak (or height) of the analyte peak is used to determine the quantity.

24.5 Spectroscopic methods for characterization

The complete chemical identity of natural product isolated by chromatographic separation can be established only after detailed spectroscopic characterization. In most cases of natural product extraction and isolation, the objective may be either identification of the compound, qualitative analysis, or structure determination, which is generally a complex and time-consuming process. The structural characteristics of pure sample are derived by studying the interactions of analyte molecules with different types of electromagnetic radiations. The following spectroscopic techniques [14] are generally used for the structure determination of natural products.

24.5.1 Ultraviolet-visible spectroscopy

The absorption of ultraviolet/visible radiation by a molecule leads to transition among the electronic energy level of the molecules and hence it is used to measure the extent of multiple bond or aromatic conjugation within molecules. The strength of this spectroscopy lies in its ability to provide information on chromophores present in the molecule. Natural products belonging to classes of flavonoids, isoquinoline alkaloids, and coumarins, to name a few, can be primarily characterized (chemical class) from characteristic absorption peaks.

24.5.2 Infrared spectroscopy

When the infrared (IR) light is passed through organic molecules, some of the frequencies of light are absorbed by the molecule [15]. Different bonds in the molecule have different vibrational frequencies, and hence IR spectroscopic analysis is used to determine the characteristic functional groups present in a molecule. The IR spectrum is divided into five zones and a fingerprint region as shown in Table 24.3.

24.5.3 Mass spectrometry

The mass spectrometer is used to determine the exact molecular formula and to measure the relative molecular weight of the compounds. Moreover, the fragmentation pattern obtained by hard ionization technique provides more structural information about the molecule and the substituents on it. The representative example of mass spectrum along with the predicted fragmentation pattern of hydrocinnamaldehyde is given in (Fig. 24.6).

24.5.4 Nuclear magnetic resonance spectroscopy

This spectroscopy is the heart of structural elucidation of unknown compounds. The main goal of nuclear magnetic resonance (NMR) data is to give number and types of protons and carbons (and other elements like nitrogen, fluorine, etc.) present in the molecule, and the relationships among these atoms. The chemical shift values obtained are assigned to the proton and carbon present in the molecules. The chemical shift value scale given in Fig. 24.7 can be used as a reference guide to assign the signals for the type of protons in the molecule.

TABLE 24.3 Characteristic stretching frequencies: the five zones in IR spectra.

Bond	Stretching Frequency(cm^{-1})	Intensity and shape
ZONE 1: 3700–3200 cm^{-1}		
Alcohol O–H	3650–3200	Usually strong and broad
Alkyne ≡C–H	3340–3250	Usually strong and broad
Amine or Amide N–H	3500–3200	Medium; often broad
ZONE 2: 3200–2700 cm^{-1}		
Aryl or Vinyl sp2 C–H	3100–3000	Variable
Alkyl sp3 C–H	2960–2850	Variable
Aldehyde	~2900 and ~2700	Medium; two peaks
Carboxylic acid O–H	3000–2500	Usually strong; very broad
ZONE 3: 2300–2000 cm^{-1}		
Alkyne C≡C	2260–2000	Variable and sharp
Nitrile C≡N	2260–2220	Variable and sharp
ZONE 4: 1850–1650 cm^{-1}		
Ketone C=O	1750–1705	Strong
Ester C=O	1750–1735	Strong
Aldehyde C=O	1740–1720	Strong
Carboxylic acid C=O	1725–1700	Strong
Amide C=O	1690–1650	Strong
ZONE 5: 1680–1450 cm^{-1}		
Alkene C=C	1680–1620	Variable
Benzene ring C=C	~1600 and ~1500	~1600 often has 2 peaks

The structural information obtained from the above-mentioned techniques (1–4) is combined with X-ray crystallographic data to arrive at the correct structure of the isolated natural product. The chiral molecules are analyzed by using chiral column in HPLC, while the optical rotations are studied using polarimetry.

24.6 Chemical profiling of marine sponges: case studies

The first step in extraction of components from any natural source is maceration of the sample with solvents such as methanol or chloroform, followed by

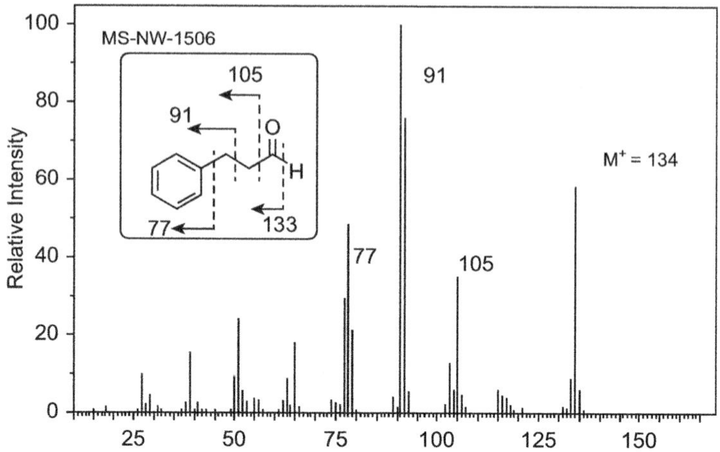

FIGURE 24.6 Mass spectrometry fragmentation pattern of hydrocinnamaldehyde.

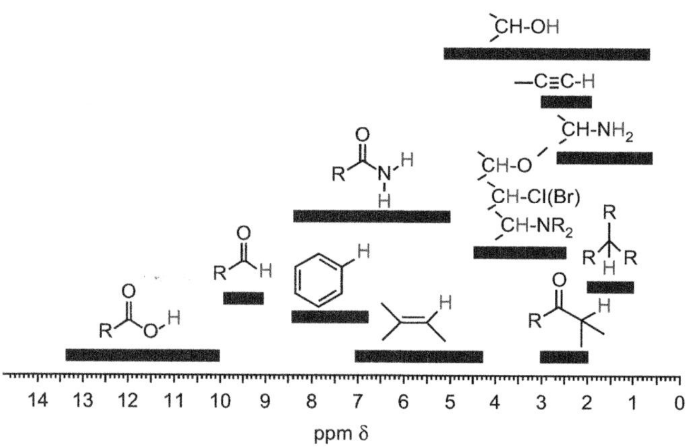

FIGURE 24.7 Chemical shift values of different functional groups.

filtration or centrifugation. However, the use of solvent mixtures with increasing polarity is also common to achieve a certain degree of fractionation. The solid residue recovered after filtration is returned to the extraction container and subjected to multiple extractions by repeating the above protocol. This extraction process is repeated until the extract is free from components present in the source sample, which is confirmed by running TLC. In certain cases, the surface area of sample is increased by cutting it into small pieces, followed by extraction so that all tissues of the sample get exposed to the solvent. In recent years, sonication or stirring have been used to increase the efficiency of extraction (Fig. 24.8).

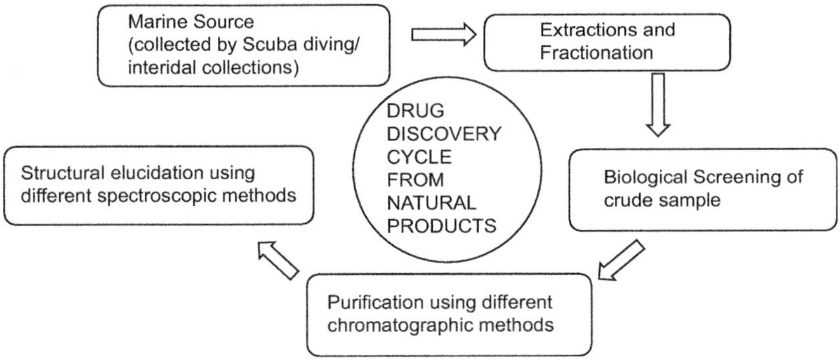

FIGURE 24.8 General protocol for isolation of natural products for drug discovery application.

24.6.1 Marine sponge, *Haliclona cribricutis*

Tilvi, S. et al. [16] demonstrated the utility of advanced mass spectroscopy for the chemical investigation of marine sponge, *Haliclona cribricutis* (phylum: *Porifera*; class: Demospongiae; subclass: Ceractinomorpha; order: Haplosclerida; family: Haliclonidae) collected from the southeast coast of India. The crude extract was obtained using methanol, which was further fractionated using increasing polarities of solvent such as pet ether, chloroform, and n-butanol fractions. After analyzing TLC pattern in different solvent systems, the chloroform fraction was chosen for further chemical investigation. The silica gel column chromatography using 35% ethyl acetate:petroleum ether as mobile phase gave partially pure sample, which was further purified by using size exclusion chromatography. The pure compound recovered as white solid as indicated by single spot in TLC was characterized using FTIR, ^{1}H and ^{13}C NMR, and tandem-MS. The flow sheet diagram for extraction is depicted in (Fig. 24.9).

24.6.2 Marine sponge, *Fasciospongia cavernosa*

Majik, M. et al. [17] have reported isolation, purification and characterization of metabolites from the marine sponges, *Fasciospongia cavernosa*, collected from Rameshwaram, Tamil Nadu. The methanol extract of sponge was subjected to fractionation using different solvent systems, and n-butanol fraction was repeatedly chromatographed on silica gel column (60–120 mesh) using chloroform in methanol as eluent. The partially purified fraction was further purified employing semipreparative reversed-phase HPLC (C18 X-Bridge column, waters, 5 mm, 250 × 21 mm; gradient of acetonitrile-water 70:30 to 100:0 over 60 min as eluent, flow rate 10 mL/min; UV detection at 254 nm) which led to the isolation of three pure compounds. The structures of active compounds secomanoalide (1), 4E, 6E-dehydromanoalide (2), and cavernosine (3) were assigned by comparing the spectroscopic data with reported literature (Fig. 24.10).

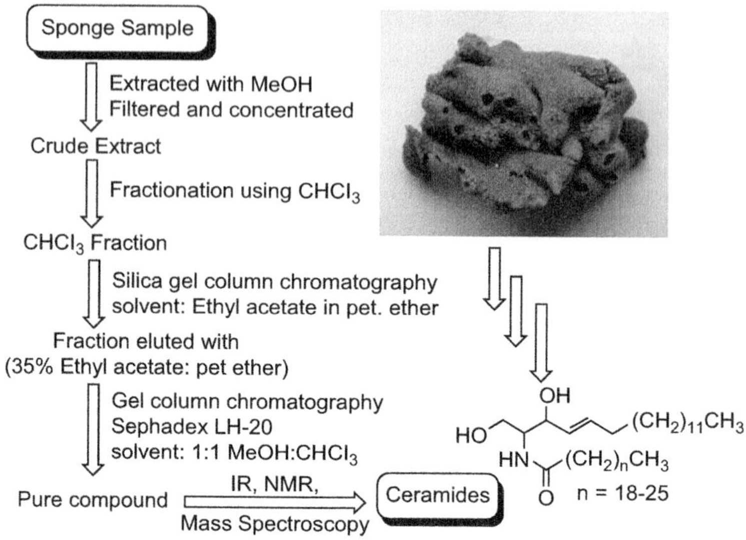

FIGURE 24.9 Marine sponge, *Haliclona cribricutis*, collected from southeast coast of India and the protocol for extraction of secondary metabolites.

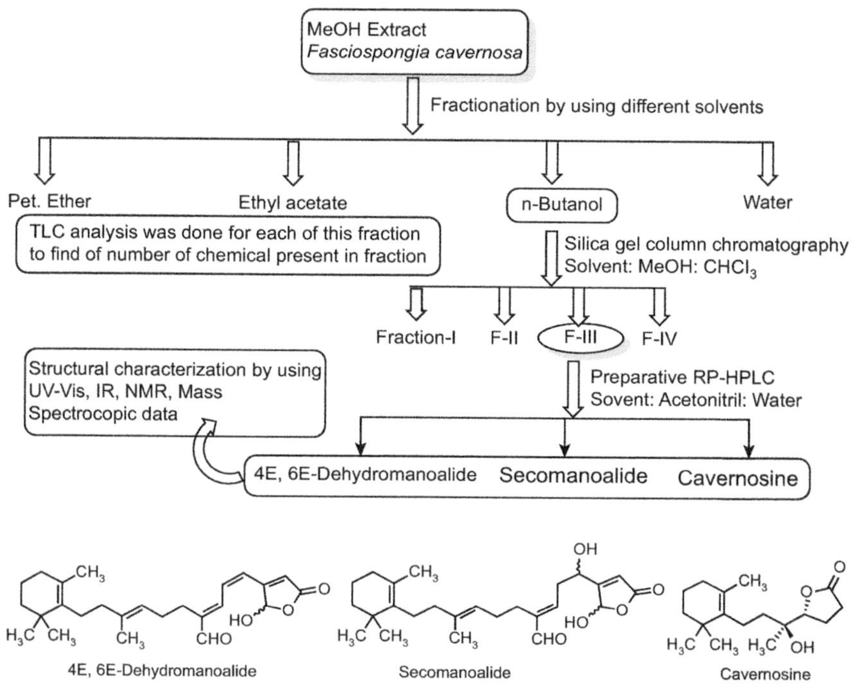

FIGURE 24.10 Structure of metabolites isolated from marine sponge, *Fasciospongia cavernosa.*

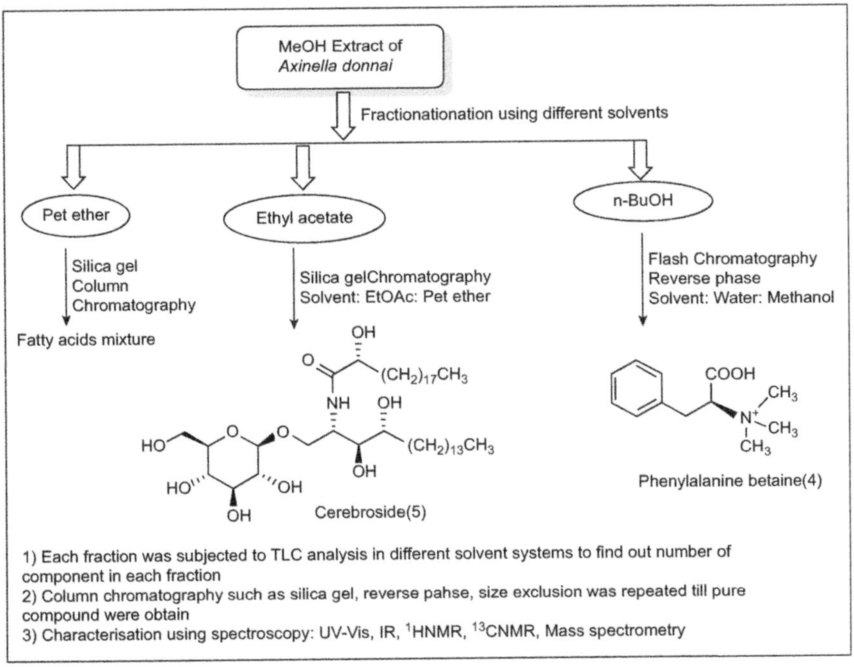

FIGURE 24.11 Structure of metabolites isolated from marine sponge, *Axinella donnai.*

24.6.3 Marine sponge, *Axinella donnani*

The methanolic extract of the sponge *Axinella donnani* collected from Mandapam, Tamil Nadu, was fractionated with petroleum ether, ethyl acetate, and butanol [17]. The fractions were further purified by using repetitive column chromatography. The butanol fraction was subjected to reversed-phase combiflash column (octadecyl silane column) using gradient elution with water and methanol to give subfractions. The subfractions on further purification by flash chromatography gave zwitterionic compound, phenylalanine betaine (4). The purity of compound 4 was confirmed by TLC and the chemical structure was assigned based on the ^{1}H and ^{13}C NMR spectra recorded in methanol-d4. The purification of ethyl acetate fraction on silica gel column (60–120 mesh) with ethyl acetate in petroleum ether yielded cerebroside 5 (Fig. 24.11).

24.7 Conclusion

The separation of natural products from various sources is a challenging area that requires the proper understanding about different analytical techniques. An appropriate protocol for efficient extraction and purification can be designed, if the objectives are well defined. Nevertheless, different approaches involving separation technology could be randomly used for extraction and isolation of natural products. The TLC shall be considered as an important step for choosing the methods of further separation and HPLC for a final cleanup of extract fractions to get to a purified natural product. The various modes of HPLC such as normal-phase, reversed-phase, size exclusion, and ion-exchange could be combined in different ways to purify most classes of natural products. The combination of the various spectroscopic data for assigning the structure to the isolated natural products is a crucial step in isolation chemistry. The proposed structures of natural products are later used as synthetic target to explore its detailed biological activity studies in medicinal chemistry, and thus used in drug discovery.

Acknowledgments

The author MSM thanks DST-SERB (New Delhi, India) (project no. YSS/2014/000776) for Young Scientist Research grant and funding. VKM would like to thank UGC (project No.F.47-1186/14/WRO) for research funding. Author UBG would like to acknowledge Research Center of Dnyanprassarak Mandal's College (project no. DNY/CC/2014-15/03/1198) for financial support.

References

[1] Stonik VA. Marine natural products: a way to new drugs. Acta Naturae 2009;2:15–25.
[2] Faulkner DJ. Highlights of marine natural products chemistry (1972–1999). Nat Prod Rep 2000;17(1):1–6.
[3] König GM, Kehraus S, Seibert SF, Abdel-Lateff A, Müller D. Natural products from marine organisms and their associated microbes. Chembiochem 2006;7(2):229–38.
[4] Bhatnagar I, Kim SK. Immense essence of excellence: marine microbial bioactive compounds. Mar Drugs 2010;8:2673–701.
[5] Fusetani N. Antifouling marine natural products. Nat Prod Rep 2011;28:400–10.
[6] Cragg GM, Newman DJ. Natural product drug discovery in the next millennium. Pharm Biol 2001;39:8–17.
[7] Cragg GM, Newman DJ. Medicinals for the millennia-the historical record. Ann N Y Acad Sci 2001;953:3–25.
[8] Christian GD. Analytical chemistry. 6 edn. John Wiley; 2004.
[9] Santos TR, Duarte AC. Comprehensive analytical chemistry, 1st ed.
[10] Majik MS, Adel H, Shirodkar D, Tilvi S, Furtardo J. Isolation of stigmast-5, 24-dien-3-ol from marine brown algae Sargassum tenerrimum and its antipredatory effect. RSC Adv 2015;5:51008–11.
[11] Skoog DA, West DM, Hollar FJ, Crouch SR. Fundamentals of analytical chemistry. Cengage learning; 9 edn.
[12] Gunzler H, Williams A. Handbook of analytical techniques. 1st ed. WILEY-VCH Verlag GmbH; 2001.
[13] Snyder LR, Kirkland JJ, Dolan JW. Introduction to modern liquid chromatography. 3rd ed.
[14] Kemp W. Organic spectroscopy, 3rd ed. Palgrave.
[15] Silverstein RM, Webster FX. Spectrometric identification of organic compounds, 6th ed. Wiley- India.

[16] Tilvi S, Majik MS, Naik CG. A tandem mass spectrometric approach for determining the structure of molecular species of ceramide in the marine sponge, *Haliclona cribricutis*. Eur J Mass Spectrom 2005;11:345–51.

[17] Majik MS, Shirodkar D, Rodrigues C, D'Souza L, Tilvi S. Evaluation of single and joint effect of metabolites isolated from marine sponges, *Fasciospongia cavernosa* and *Axinella donnani* on antimicrobial properties. Bioorg Med Chem Lett 2014;24:2863–6.

Chapter 25

Advanced bioceramics

Kiran Suresh Naik
Department of Chemistry, P.E.S.'s R.S.N. College of Arts & Science, Farmagudi, Ponda, Goa, India

25.1 Introduction

Biomaterials or bioceramics can be defined as "synthetic or natural materials that are used to replace human body parts which are connected intimately to the living tissues" [1] or "A biomaterial is a substance that has been engineered to take a form which, alone or as part of a complex system, is used to direct, by control of interactions with components of living systems, the course of any therapeutic or diagnostic procedure, in human or veterinary medicine" [2–4]. The definitions are further refined by many authors according to the conceptual ideas and biological performances of ceramic materials [5–7].

Bioceramics have been known and used since 1000 AD, when plaster of Paris (calcium sulfate) was widely used for setting broken bones [7]. Due to exponential growth in population, demand for biomaterials is intensifying day by day and there is always a need of new discoveries and development of new biomaterials. In the field of medical studies, biologically active drugs as well as materials play a very important role [8]. When it comes to bioceramics, thermal stability, mechanical strength, biocompatibility, and their similarities with bone tissues are the factors that influence their implementation in different applications. Because of their unique properties, bioceramics promote quality of human life and their applications have been extended to the treatment of disorders of human organs. The increase in demand for bioceramics is uncountable and incomparable with respect to other medical treatments. Among the potential value of bioceramics is a huge demand for orthopedic implants. Bioceramics are used in various treatments such as dental fillings, periodontal treatments, joint and limb replacements, craniomaxillofacial reconstruction, artificial arteries and skin substitutes, contact lenses and corneal tissue-engineered scaffolds, bone substitutes, and so on [6,7,9]. This chapter will review classification, applications, and properties of different types of bioceramics.

Advances in Biological Science Research. https://doi.org/10.1016/B978-0-12-817497-5.00025-2

25.2 Classification of biomaterials

The major reason for using ceramics in biomedical applications is because of their biocompatibility. Biocompatibility is the material property that refers to the material's ability to interact with bone, surrounding tissues, and organs in human body. Based on biocompatibility, the bioceramics are generally classified into three different categories as shown in Fig. 25.1.

Bioactive: The biocompatible ceramics that are chemically active and favor bond formation between new bone tissues and the surface of ceramics. These materials have structural similarities with the bone tissues and can be used as a replacement for them. This includes various calcium phosphates, bioactive glasses, and bioactive glass ceramics. Bioactive ceramics have been used to repair and regenerate bone tissues [10,11].

Bioinert: These are the ceramics or materials that do not or only partially react with the surrounding tissues or organs in the human body. These materials are used for biomedical application in the form of supporting media or inert materials. These materials do not interfere with the bone tissues or human organs. As the bioceramics are basically ionic oxides and have strong ionic forces of attraction, they mostly tend to be nontoxic, noncarcinogenic, and nonallergenic. Some of the most common bioinert ceramics are alumina, stainless steel, titanium, zirconia, and alumina-zirconia composite. [6,12].

Bioresorbable: These are the materials that are implanted in the human body, and after implant they react and dissolve with the body fluids and are further metabolized to form new tissues. Generally, tissues are capable of regenerating themselves and need only a temporary support of bioceramic for tissue replacement. Examples of such bioceramics are calcium phosphate and biodegradable polymers. These materials are used only when temporary support is required for tissue growth (e.g., artificial skin, cartilage repair,

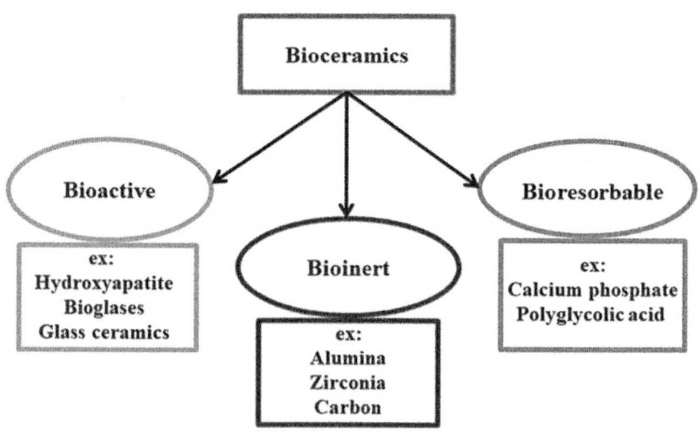

FIGURE 25.1 Classification of bioceramics.

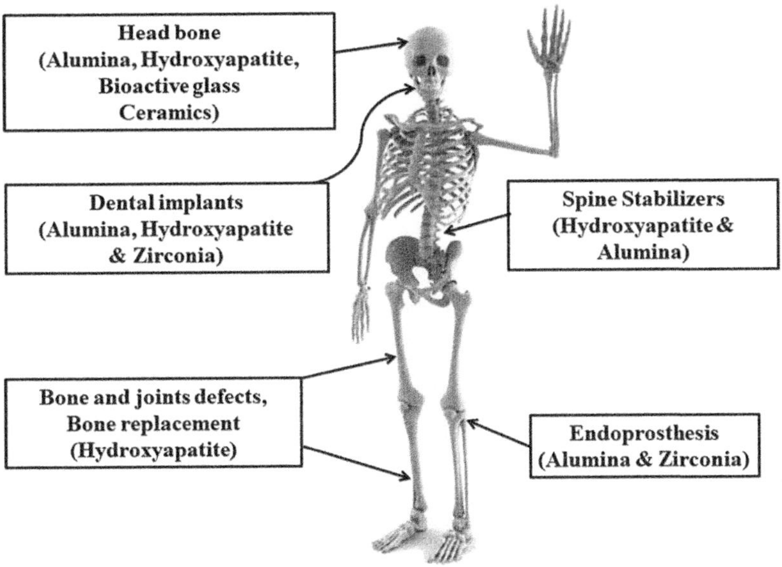

FIGURE 25.2 Applications of bioceramics in human body.

peripheral nerve repair) and not in the case where permanent implant is required (e.g., artificial heart, kidney, and liver) [13,14] (Fig. 25.2).

25.3 Applications and properties of bioceramics

25.3.1 Hydroxyapatite

Hydroxyapatite (HA) is one type of calcium phosphate ceramic ranked in the category of bioactive ceramics. It crystallizes into a hexagonal dipyramidal structure with a molecular formula of $Ca_5(PO_4)_3OH$ [15]. The most important criterion of HA is the atomic ratio of calcium to phosphate (1.67); slight deviation in this Ca/P ratio destabilizes the crystal structure, which in turn affects the physical and chemical properties of HA [11].

Synthesis of HA is reported in many different ways. The most common method of synthesis is by wet precipitation method [16], where phosphate solution (H_3PO_4 or $(NH_4)_2HPO_4$) is added dropwise to the aqueous suspension of calcium ($Ca(OH)_2$) or ($Ca(NO_3)_2$) at pH $\sim$11 maintained by addition NH_4OH or NaOH [11].

From the 20th century onwards, HA has been reported to be used as bone filler for reconstruction of defected bone [17]. HA can also be used in the form of heterogeneous catalyst; the replacement of Ca^{+2} sites with divalent or trivalent cations leads to improved catalytic activity [15]. It has also been observed that Cu^{2+}, Zn^{2+}, and Ag^{+} substituted HA show antimicrobial effect [18]. HA is known to be highly bioactive but its weaker mechanical strength constrains its use as a bioceramic. However, a composite of HA with bioinert ceramics has increased its mechanical strength. It was recently reported that the composite of HA and yttrium-stabilized zirconia can be used as a replacement for dentin substitutes for dental purposes [19].

25.3.2 β-Tricalcium phosphate (β-TCP)

Among the tricalcium phosphate family, β-TCP is widely used as bioresorbable ceramic for biomedical application. It has stoichiometric formula of $Ca_3(PO_4)_2$ with Ca/P molar ratio of 1.5 and crystalizes into rhombohedral crystal structure [11]. β-TCP is a stable phase at lower temperature and transforms into α-TCP at a temperature of ~1125°C [20]. Of the two phases, β-TCP has wider use in tissue engineering applications.

The synthesis of β-TCP follows a similar procedure as that of HA. The powder prepared by such methods as solid-state reaction, precipitation method, or by some other method is used to make granules or scaffolds. Higher temperatures destabilized the structure, while doping with cations (e.g., Mg, Sr) enhanced the stability and sintering characteristic of β-TCP [20]. It is used as absorbable material for developing tissue engineering scaffolds [21–23]. However, composite of β-TCP/HA enhances the biocompatibility and biodegradation properties of HA [24].

25.3.3 Alumina (Al_2O_3)

Alumina falls into the category of bioinert ceramics, with a corundum crystal structure. It has unique mechanical and thermal properties because of the strong ionic and covalent bond between Al^{+3} and O^{-2} ions [10,12]. Various methods for synthesis of alumina have been reported, and mechanical strength of alumina has a direct relation with grain size and porosity [6]. So synthesis plays a very important role in the preparation of nano-alumina bioceramic. The bioinert nature of alumina makes it suitable for dental and orthopedic applications. In addition to this, alumina is also used as porous spacer, coatings for femoral stems ,and for hip prostheses [6].

25.3.4 Zirconia

Zirconia exists in three different forms of cubic, monoclinic, and tetragonal. Zirconia is stabilized to tetragonal or cubic form by doping oxides of calcium, yttrium, magnesium, etc. [25]. It comes into a family of bioinert ceramics, with excellent biocompatibility. In addition, zirconia has a high affinity for bone tissues and lacks any potential oncogenic effect. One more important feature is that it serves as a nucleation site for the development of calcium phosphate ceramics [26]. The use of zirconia is observed in hip joint replacement and also in dental applications. Addition of HA to the zirconia matrix (zirconia/HA composite) enhances the bioactivity and corrosive resistance of zirconia [25].

25.3.5 Bioglass and glass ceramics

Bioglass is composed of silicon oxide, Na_2O, K_2O, CaO, and MgO, and its biological properties are determined by the composition of bioglass. The concentration of silica classifies bioglass into different categories. For bioactive glass ceramics, the silica content is between 40% and 60%, whereas in bioinert glass ceramics, the silica content is >60% [10]. Bioactive glasses can be processed into microspheres, fibers, and porous implants. It can be also used in the form of a composite with hydroxyapatite. Because of attractive mechanical and thermal stability, glass ceramics are suitable for the orthopedic applications. Bioglass favors the formation of new bone by the deposition of calcium and phosphorous, which enables its use in bone grafting, reconstruction, and as a scaffold in tissue engineering [10].

25.4 Conclusion and future perspectives

Bioceramics display outstanding contributions in the field of bone implantation and tissue engineering. Based on their biocompatibility, bioceramics are classified into bioactive, bioinert, and bioresorbable materials and are utilized in different medical applications. Recently, it was found that bioresorbable ceramics have immense potential in the field of biomedical applications. They act as supports and drive the normal tissue growth mechanism and also undergo decomposition after its use. Most of these bioceramics are in an early stage of development and need further experimental investigation.

Future research should focus on fabrication of new types of bioceramics that can have healing effects on the surrounding tissue at the implant site as well as assisting in bone regeneration. A possible approach in creating such bioceramics is by doping them with certain biomaterials capable of boosting their properties as well as biocompatibility. Recent use of bioactive ceramics and bioglasses demonstrated promising impact on bone regeneration, healing, and regeneration of soft and hard tissues. This area needs to be focused on, with further systematic analysis. The strength of bioceramics is the next important criteria. In the application of tissue engineering, mixtures of organic and inorganic biocomposites have ability to increase the mechanical strength in biomaterials, and this offers a compelling challenge for future research work.

Acknowledgments

The author would like to acknowledge Dr. Shambhu S. Parab and Ms. Vijayashri R. Naik for their helpful suggestions while writing this chapter.

References

[1] Grainger DW. The Williams dictionary of biomaterials. Mater Today 1999;2(3):29.

[2] Williams DF. On the nature of biomaterials. Biomaterials 2009;30(30):5897–909.

[3] Williams DF. On the mechanisms of biocompatibility. Biomaterials 2008;29(20):2941–53.

[4] Williams DF. Definitions in biomaterials: proceedings of a consensus conference of the European Society for Biomaterials. 1987.

[5] Peppas NA, Langer R. New challenges in biomaterials. Mater Sci 1994;263:1715–20.

[6] Shanmugam K, Sahadevan R. Bioceramics—An introductory overview. In: Fundamental biomaterials: ceramics. Elsevier Ltd.; 2018. p. 1–46.

[7] Baino F, Vitale-Brovarone C. Bioceramics in ophthalmology. Acta Biomater 2014;10(8):3372–97.

[8] Rödel M, Meininger S, Groll J, Gbureck U. Bioceramics as drug delivery systems. In: Fundamental biomaterials: ceramics. Elsevier Ltd.; 2018. p. 153–94.

[9] Dorozhkin SV. Calcium orthophosphate bioceramics. Eurasian Chem J 2010;12(3–4):247–58.

[10] Mala R, Ruby Celsia AS. Bioceramics in orthopaedics: a review. In: Fundamental biomaterials: ceramics. Elsevier Ltd.; 2018. p. 195–221.

[11] El-Ghannam A, Ducheyne P. 1.9 bioactive ceramics. In: Comprehensive biomaterials II. Elsevier; 2017. p. 204–34.

[12] Pezzotti G. Bioceramics are not bioinert. Mater Today 2017;20(8):395–8.

[13] Pertici G. Introduction to bioresorbable polymers for biomedical applications. In: Bioresorbable polymers for biomedical applications: from fundamentals to translational medicine. Elsevier Ltd; 2016. p. 3–29.

[14] Kanno T, Sukegawa S, Furuki Y, Nariai Y, Sekine J. Overview of innovative advances in bioresorbable plate systems for oral and maxillofacial surgery. Jpn Dent Sci Rev 2018;54(3):127–38.

[15] Fihri A, Len C, Varma RS, Solhy A. Hydroxyapatite: a review of syntheses, structure and applications in heterogeneous catalysis. Coord Chem Rev 2017;347:48–76.

[16] Aal NA, Bououdina M, Hajry A, Chaudhry AA, Darr JA, Al-ghamdi AA, et al. Synthesis, characterization and electrical properties of hydroxyapatite nanoparticles from utilization of biowaste eggshells. Biomater Res 2011;15:52–9.

[17] Oonishi H. Orthopaedic applications of hydroxyapatite. Biomaterials March 1991;12(2):171–8.

[18] Karthika A. Aliovalent ions substituted hydroxyapatite coating on titanium for improved medical applications. Mater Today Proc 2018;5(2):8768–74.

[19] Ayoub G, Veljovic D, Zebic ML, Miletic V, Palcevskis E, Petrovic R, et al. Composite nanostructured hydroxyapatite/yttrium stabilized zirconia dental inserts – the processing and application as dentin substitutes. Ceram Int 2018;44(15). 1–9.

[20] Rahaman MN. Bioactive ceramics and glasses for tissue engineering. In: Tissue engineering using ceramics and polymers. 2nd ed. 2014. p. 67–114.

[21] Shi YZ, Liu J, Yu L, Zhong LZ, Jiang HB. β-TCP scaffold coated with PCL as biodegradable materials for dental applications. Ceram Int 2018;44(13):15086–91.

[22] Maji K, Dasgupta S, Pramanik K, Bissoyi A. Preparation and characterization of gelatin-chitosan-nanoβ-TCP based scaffold for orthopaedic application. Mater Sci Eng C 2018;86:83–94. Elsevier.

[23] Tanaka T, Komaki H, Chazono M, Kitasato S, Kakuta A, Akiyama S, et al. Recherche fondamentale et application clinique du bêta-tricalcium phosphate (β-TCP). Morphologie 2017;101(334):164–72.

[24] Pillai RS, Frasnelli M, Sglavo VM. HA/β-TCP plasma sprayed coatings on Ti substrate for biomedical applications. Ceram Int 2018;44(2):1328–33.

[25] Soon G, Pingguan-Murphy B, Lai KW, Akbar SA. Review of zirconia-based bioceramic: surface modification and cellular response. Ceram Int 2016;42(11):12543–55.

[26] Stefanic M, Kosmač T. β-TCP coatings on zirconia bioceramics: the importance of heating temperature on the bond strength and the substrate/coating interface. J Eur Ceram Soc 2018;38(15):5264–9.

Chapter 26

Production of polyhydroxyalkanoates by extremophilic microorganisms through valorization of waste materials

Bhakti B. Salgaonkar, Judith M. Bragança
Department of Biological Sciences, Birla Institute of Technology and Science (BITS) Pilani, K K Birla, Goa Campus, Zuarinagar, Goa, India

26.1 Introduction

Conventional plastics such as polyethylene, polypropylene, and polyvinyl alcohol are petroleum-based plastics obtained from nonrenewable petrochemical resources. For the past few years, anthropogenic activity such as increased use of both plastic and petroleum has resulted in environmental pollution. Hence there is an increasing need for eco-friendly biobased green plastics derived from renewable resources [1]. Renewable resources are organic materials obtained from natural environments such as agricultural products that are replenishable [2]. Sustainable conversions of readily available agroindustrial byproducts/waste to value-added products through biological transformations have been studied. Microbial valorization of agricultural waste has proved to be an excellent process resulting in the synthesis of a great array of products with varied functionalities and economic value [3].

Polyhydroxyalkanoates (PHAs) are homo/heteropolyesters of hydroxyalkanoic acid synthesized by microorganisms as reserve material under nitrogen-limiting conditions and excess of carbon [4]. PHAs are potentially emerging as the next generation of environmentally friendly materials and have myriads of applications. PHA accumulation by many gram-positive and gram-negative bacteria and a few members of archaea have been investigated [5]. Standardization of every single fermentation parameter is a prerequisite for the successful execution of commercial PHA production at industrial scale. The final

Advances in Biological Science Research. https://doi.org/10.1016/B978-0-12-817497-5.00026-4

price of the PHA ultimately depends on the cost of the substrate (50%), PHA yield, and the effectiveness of the downstream process. PHA yield implies (1) high amount of PHA to the cell dry mass (CDM) and (2) high PHA productivity in terms of gram of PHA per unit volume and time. The cost of PHA using the natural wild-type producer *Alcaligenes eutrophus* (currently known as *Cupriavidus necator*), a gram-negative bacillus, is US$ 16 kg^{-1}, which is ~18 times more expensive than conventional fossil plastics such as polypropylene, which is US$ ~0.88 kg^{-1} [6]. Currently, microbial fermentations using recombinant *Escherichia coli* are being used for PHA production, the market price of which is US$ ~4 kg^{-1} when 4.63 g/L h is the volumetric productivity [7].

The selection of microbial strain, raw materials as carbon sources, efficient PHA production strategy, and recovery process should be the guidelines considered for a successful PHA production process [8]. Carbon sources represent half of the PHA fermentation cost. High production cost and low yields are the major hurdles for commercial production and application of PHA, making microbial synthesis of PHAs 5–10 times more expensive than the petroleum-derived polymers [9]. Therefore, designing efficient biotechnological processes employing microorganisms that utilize renewable inexpensive agroindustrial wastes as substrates are the need of the hour. India produces ~960 million tones (MT) of solid wastes per annum, out of which 350 MT is organic wastes from agricultural sources, which has resulted in major environmental and ecological problems [10]. Naranjo et al. (2013) reported the valorization of crude glycerol generated as a coproduct of biodiesel, by transforming into polyhydroxybutyrate P(3HB) using *Bacillus megaterium*. The glycerol was used as a carbon source, substituting commercial sugars for P(3HB) biosynthesis [3]. Di Donato and colleagues investigated the production of microbial biomass, enzymes, and biopolymers [P(3HB)] through the valorization of vegetable wastes such as tomato, lemon, carrot, and fennel, from vegetable processing industries as sole carbon sources by thermophilic and halophilic microbial strains [11]. Valorization of the wastes by utilizing them as a cheap-carbon source for PHAs production will be of great advantage as it will help in waste management as well as obtaining value-added products such as novel biomolecules to be exploited in various biotechnological processes [12].

PHA production is carried by liquid/submerged fermentation (LF/SmF) and solid-state fermentation (SSF) processes wherein the former employs liquid wastes such as whey, molasses, vinasse, etc. and latter uses solid wastes such as the sugarcane bagasse (SCB), cassava waste (CW), etc. [2]. In SmF there is rapid utilization of liquid substrates and therefore this is favorable for microorganisms requiring high-moisture content/water activity. Also, the recovery of the microbial biomass and their products is easier. Industrial processes based on SSF are advantageous, as being eco-friendly there is less wastewater generation, low energy consumption, product yields are a high, and there is low risk of bacterial contamination. SSF is best suited for microorganisms requiring low water activity, such as extremely halophilic archaea that inhabit salt-saturated brines where the water activity is very low. SSF takes a longer time as there is very slow and steady utilization of agroindustrial waste substrates by microorganisms. Employing extremophilic microbial strains requiring high salt/temperature for optimum growth may reduce the cost involved in sterilization and PHA recovery/downstream process, thereby making the fermentation process more cost-effective [7].

26.2 Synthesis of polyhydroxyalkanoates

PHAs are synthesized by microorganisms as reserve material under conditions of nitrogen limitation and excess of carbon. PHAs help the microorganisms to withstand and survive stress conditions by acting as carbon-dense reservoirs. Moreover, PHAs are known to protect the cells from adverse environmental stress such as heat, ultraviolet (UV) irradiation, osmotic shock, etc. and are therefore advantageous for microorganisms thriving under polyextremophilic stress [13,14].

Microorganisms employ various pathways for the metabolism and/or transformation of carbohydrates and fatty acids into a diverse range of PHAs with main intermediate being either acetyl-CoA and/or acyl-CoA and conclude with monomer polymerization by PHA synthases (Fig. 26.1A) [15,16]. The ability of microorganisms to synthesize a particular form of PHAs is mainly due to the substrate specificity of PHA synthases, and these enzymes are divided into four classes [17]. Class I and II synthases are encoded by *PhaC* gene. Interestingly, class I PHA synthases utilize CoA thioesters of 3-hydroxyalkanoates (3-HAs) comprising 3–5 carbon atoms, whereas, class II polymerases are also specific to CoA thioesters of 3-HAs comprising 6–14 carbon atoms [18]. Class III and IV synthases are composed of two genes (*PhaC* and *PhaE*) and (*PhaC* and *PhaR*), respectively. However, class III synthases possess substrate specificities similar to class I and IV synthases and utilize 3-HA monomers with 3–5 carbon atoms [17].

The *PhaC*, *PhaA*, and *PhaB* genes are localized on the chromosomes of microorganisms and encode three cytoplasmic proteins/enzymes: (1) *PhaA* gene encodes acetyl-CoA C-acetyl transferase, the 393 amino acids protein; (2) *PhaB* gene encodes acetoacetyl-CoA reductase, 246 amino acids protein; and (3) *PhaC* gene encodes poly-β-hydroxybutyrate polymerase/synthase, 589 amino acids protein. PHA synthase is the key enzyme for the PHA biosynthesis, which polymerizes monomers of (R)-3-hydroxyacyl-CoA thioester into polyester [19].

PHA synthesis begins with binding of PHA synthase (PhaC) to the periplasmic membrane of the cell (Fig. 26.1B) [20]. Phasin is the most abundant granule-associated amphiphilic protein that is responsible for regulating PHA accumulation and forms proteinaceous barriers between surfaces of adjacent polymer cores, thereby preventing blending of PHA granules [21]. Along with phospholipids, there are a number of other proteins associated with the surface

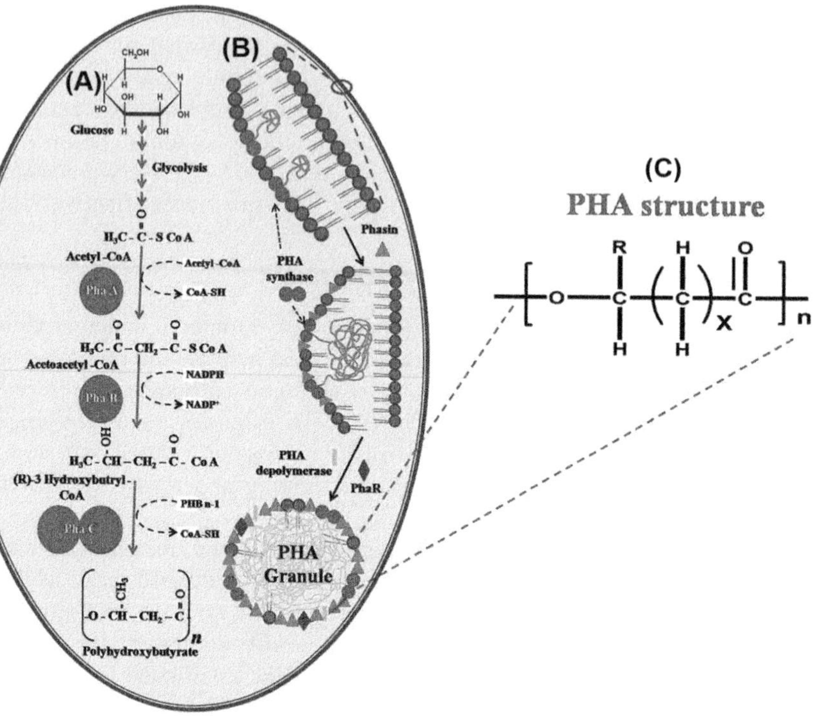

FIGURE 26.1 General scheme of PHA synthesis in a microbial cell: (A) Biosynthetic pathway for synthesis of poly(3-hydroxybutyrate) (PHB) from glucose via Embden–Meyerhof pathway; (B) Mechanism of PHA synthesis in association with the cytoplasmic membrane by budding model; and (C) General structure of polyhydroxyalkanoates (PHA). *Adapted from Ref. Madison LL, Huisman GW. Metabolic engineering of poly (3-hydroxy alkanoates): from DNA to plastic. Microbiol Mol Biol Rev 1999;63(1):21–53.*

of PHA granule and form monolayer of ~4 nm around the granule. It is hypothesized that these proteins protect the hydrophobic core from the cytoplasmic enzymes and also play a major role in regulating the size and number of PHA granules synthesized. The PhaR is a regulatory protein that binds to PHA granules. After attaining a critical granule concentration inside the cell cytoplasm, and when no more binding sites are available for PhaR, it binds to gene promoters of *PhaP1* and *PhaR* thereby downregulating transcription and hence reducing the rate of PHA synthesis. PHA depolymerase is granule associated and plays a role is depolymerizing the PHA chains. However, its proper function is yet to be fully understood [22].

The homopolymer of P(3HB) is mostly obtained if monosaccharide sugars, such as glucose (hexose) and fructose (pentose), are processed via glycolysis/ Embden–Meyerhof pathway. However, copolymers could be produced if fatty acids or sugars are metabolized by other pathways, such as fatty acid β-oxidation, fatty acid de novo biosynthesis, or pentose phosphate pathway [23]. Interestingly, there are only a limited number of bacteria belonging to the genus *Pseudomonas*, such as *Pseudomonas putida* strain CA-3, which are reported to produce PHA heteropolymer containing both aliphatic and aromatic monomers by utilizing aromatic hydrocarbons as a carbon and energy source [24].

26.3 Classification of PHAs

The general structure of PHA is represented in Fig. 26.1C. PHAs are polymers of carbon, oxygen, and hydrogen and are made from a single monomer or a combination of the monomers. Microbial PHAs can be classified according to the various criteria listed in Fig. 26.2 [25].

26.3.1 Biosynthetic origin

Microorganisms accumulating polyhydroxyalkanoates directly from the substrates without utilization of any additional specialized precursors are referred to as natural PHAs, e.g., synthesis of homopolymer P(3HB) by *Bacillus megaterium* strain H16 [26] and heteropolymer/copolymer P(3HB-*co*-3HV) by an archaeal genera *Halogeometricum*, *Haloferax* [27,28]. Semisynthetic PHAs are produced by addition of unusual precursors such as 3-mercaptopropionic acid to a general substrate, e.g., biosynthesis of poly(3-hydroxybutyrate-*co*-3-mercaptopropionate) [P(3HB-*co*-3MP)] [29].

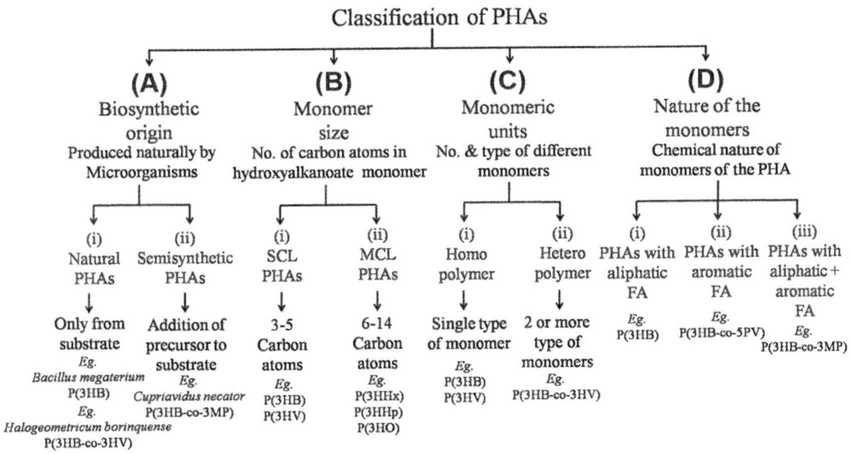

FIGURE 26.2 Pictorial view of the various criteria listed for the classification of microbial PHAs. FA, fatty acid; MCL, medium-chain length; P(3HB), poly(3hydroxybutyrate); P(3HB-*co*-3HV), poly(3hydroxybutyrate-*co*-3hydroxyvalerate); P(3HB-*co*-3MP), poly(3-hydroxybutyrate-*co*-3-mercaptopropionate); P(3HB-*co*-5PV), poly(3-hydroxy-5-phenylvalerate); P(3HHp), poly(3-hydroxyheptanoate); P(3HHx), poly(3hydroxyhexanoate); P(3HO), poly(3hydroxyoctanoate); P(3HV), poly(3hydroxyvalerate); PHAs, polyhydroxyalkanoates; SCL, short-chain length.

26.3.2 Monomer size

PHAs are classified depending on the number of carbon atoms in hydroxyalkanoate monomer: (1) short-chain-length PHAs (PHA_{scl}) in which the monomer contains 3–5 carbon atoms, e.g., P (3HB) and P(3HV); (2) medium-chain-length PHAs (PHA_{mcl}) in which the monomer contains 6–14 carbon atoms, such as PHHx, PHHp, PHO, PHN, etc. [11,25].

26.3.3 Monomers units

PHAs are classified based on the number and type of different monomers: (1) homopolymers that are composed of a single type of monomeric unit and the linkage is between ester bonds of one monomer to the carboxylic group of adjacent monomer to give polymers such as P(3HB), P(3HV), etc.; and (2) heteropolymers are formed by linkage of two or more different type of monomeric units, such as a copolymer of poly(3-hydroxybutyrate-*co*-3-hydroxyvalerate) [P(HB-*co*-HV)] [6].

26.3.4 Nature of the monomers

PHAs are classified by studying the chemical nature of the monomers: (1) PHAs containing aliphatic fatty acids, e.g., P(3HB); (2) PHAs containing aromatic fatty acids such as poly(3-hydroxy-5-phenylvalerate) PHPV; and (3) PHAs heteropolymers containing both aliphatic and aromatic fatty acids, i.e., P(3HB-*co*-3MP) [24].

26.4 Screening, extraction, and characterization of polyhydroxyalkanoates

26.4.1 Screening for PHA

The most routinely used methods of screening for PHA-producing microorganisms are colony staining [30]and cell staining [31] using lipophilic dye, Nile blue A or Nile red (oxazone of Nile blue): (1) In colony staining, the dye is directly added into the growth medium and the culture exhibits bright orange fluorescence upon exposure to UV light, indicating the accumulation of PHAs (Fig. 26.3A,B); (2) In cell staining, the dye is added to smeared cells and bright orange fluorescent granules within the cells are observed using fluorescence microscopy under propidium iodide (PI) filter at an excitation wavelength of 460 nm, indicating PHA accumulation by the cells (Fig. 26.3C); (3) The PHA granules can also be detected inside the cell cytoplasm by transmission electron microscopy (TEM) analysis. These granules are visible as white inclusion bodies and each microbial cell mostly contains ≥ 1 PHA granules of $\sim 0.05-0.30$ μm in diameter [32]. A TEM image of cells of *Haloarcula* strain BS2 with cytoplasmic PHA granules is represented in Fig. 26.3D; and

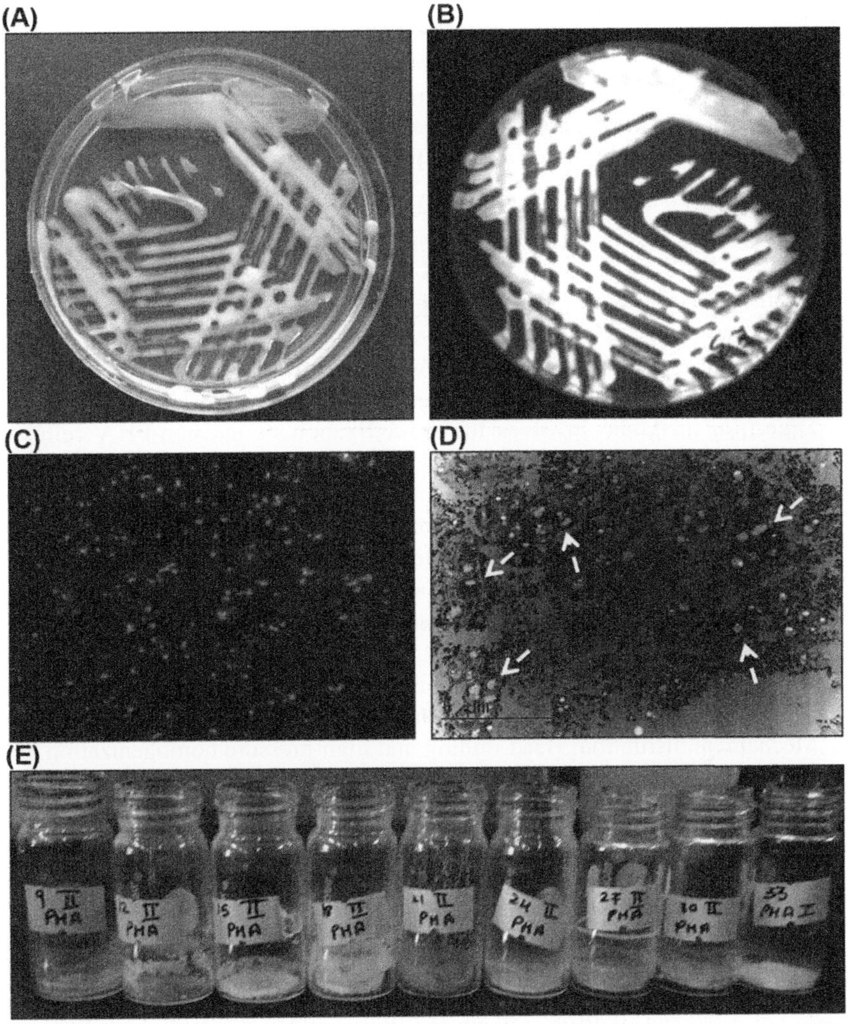

FIGURE 26.3 (A) *Halogeometricum borinquense* strain E3 grown on minimal medium supplemented with 2% (w/v) glucose and Nile red dye; (B) Fluorescence exhibited by *Hgm. borinquense* strain E3 after exposure of culture plate to UV light; (C) Fluorescence exhibited by PHA accumulating cells on staining with Nile red dye and analyzed using propidium iodide (PI) filter of fluorescence microscope; (D) Transmission electron microscopy (TEM) image of cells of *Haloarcula* strain BS2 representing PHA granules (white); (E) PHA extracted using sodium hypochlorite from *Bacillus megaterium* strain H16, after various time intervals.

(4) Polymerase chain reaction (PCR) with specific primers for the locus of the gene responsible for the key enzyme, PHA synthase (*phaC* gene) [33]. Lu et al., (2008) cloned the gene cluster ($phaEC_{Hme}$) encoding a PHA synthase involved in the synthesis of copolymer P(3HB-*co*-3HV) in halophilic archaeon *Haloferax mediterranei* CGMCC 1.2087 using PCR [34].

26.4.2 PHA extraction

Polydroxyalkanoates are accumulated by microbial cells inside the cell cytoplasm and therefore one must rupture the cell for the isolation of PHA. The various methods for the effective recovery of PHA from the microbial cells are described by Kunasundari and Sudesh [35].

1. Solvent extraction: Employs solvents such as chloroform and is the most extensively used method in the laboratory to recover PHA from the biomass. It is a two-step process: PHA release and solubilization is effected by modification of cell membrane permeability followed by precipitation using methanol and ethanol. The extraction of PHAs from CDM can be achieved using soxhlet apparatus with chloroform or other chlorinated hydrocarbons for complete extraction [27].
2. Digestion methods: Involves the solubilization of the non-PHA cellular mass (NPCM) surrounding the PHA granules and can be classified into either chemical digestion or enzymatic digestion. The most commonly used method is chemical digestion, which employs strong oxidizing reagents such as sodium hypochlorite to digest the NPCM and this releases PHA (Fig. 26.3E) [26]. Sodium hypochlorite is employed singularly or in combination with other surfactants such as sodium dodecyl sulfate, Triton X-100, etc. Enzymatic digestion for PHA recovery involves solubilization of non-PHA cell components by heat treatment and enzymatic hydrolysis using proteases, followed by surfactant washing [36].
3. Mechanical disruption: Bead milling and high-pressure homogenization are the two methods mostly employed for extraction of PHAs at commercial scale in industries. Bead mill disruption systems are based on the mechanical shearing of the cells using beads (diameter in μm). High-pressure homogenization systems are based on disruption of cell suspension under high pressure [37].

26.4.3 PHA characterization

There are a myriad of PHA monomeric units ($\geq$150) available and different compositions of these monomers in PHA polymers result in their diverse chemical composition and material properties [5]. Law and Slepecky (1961) developed a convenient spectrophotometric assay for the quantitative conversion of poly-β-hydroxybutyric acid to crotonic acid [38]. In this method, the P(3HB) was converted to crotonic acid by heating with concentrated sulfuric acid, which gave a characteristic absorption maximum at 235 nm. This method is routinely used in the laboratories for the preliminary characterization of the biomaterial as PHAs and has worked for homopolymer P(3HB) as well as a copolymer of P(3HB-*co*-3HV) obtained from halophilic bacteria and archaea [26,27]. Some of the most routinely used techniques for the detailed characterization of the PHA are listed below.

1. **Monomeric composition:** PHAs as an intact polymer are difficult to analyze, and therefore, the polymer should be depolymerized first with chemical derivatization and then analyzed for its monomeric units using techniques such as (a) gas chromatography-mass spectrometry, in which mild acid/alkaline hydrolysis of the polymer to 3HB methyl ester is analyzed by gas chromatography; (b) nuclear magnetic resonance (NMR) spectroscopy (both ^{1}H-NMR and ^{13}C-NMR) is used to analyze the chemical composition of the monomeric units of the polymer. The polymer is dissolved in deuterated chloroform ($CDCl_3$), followed by a recording of NMR spectrum [24,39].
2. **Molecular Mass:** The polymer's average (a) molecular mass (Mw), (b) molecular mass distribution (Mn), and (c) polydispersity index (PDI; Mw/Mn) can be determined using gel permeation chromatography system [5].
3. **Thermal properties**: Determines the temperature conditions at which the polymer can be processed and utilized. (a) Differential scanning calorimetry (DSC) and (b) differential thermal analysis determine properties such as glass transition temperature (Tg), melting temperature (Tm), and thermo-degradation temperature (Td) [5,27]. (c) Thermogravimetric analysis is used to obtain the Td value of PHA, where a sample is heated in a controlled atmosphere at a defined rate while sample mass loss is measured [26].
4. **Crystallinity:** PHA polymers can be noncrystalline to highly crystalline with crystallinity values between 0% and 70% [40]. The crystallinity of PHA polymer could be measured by instruments such as Fourier transform infrared (FTIR), X-ray diffraction (XRD), and DSC analysis. (a) FTIR analysis of PHAs shows prominent characteristic absorption bands at certain wave numbers, which can be correlated to crystallinity [41]. (b) DSC analysis of the polymer, melting enthalpy (ΔHm), provides an estimated value for the heat of fusion (ΔHf), which could be related to PHA's crystallinity [42]. (c) XRD analysis could be used to measure absolute crystallinity of the polymer and provides information on the polymer's rate of crystallinity, atomic structures (chemical bonds), and their disorder. Crystallinity percentage can be calculated according to semicrystalline and amorphous polymer areas in the diffractogram using Lorentzian and Gauss functions [43].
5. **Mechanical properties:** The mechanical properties that are commonly evaluated for PHA polymers are Young's modulus, elongation at break, and tensile strength, and these can be performed with tensile tester instrument by standardized test methods such as the ones recommended by American Society for Testing and Materials standards [44]. (a) The Young's modulus/tensile/elastic strength provides a measure of PHA's stiffness and ranges from the very ductile mcl-PHA (0.008 MPa) to the stiffer scl-PHA (3.5×103 MPa) [45]. (b) The elongation at break/fracture strain measures the extent that a material will stretch before it breaks and is expressed as a percentage of the material's original length. PHA polymers

can either be a soft elastomeric material or a hard-rigid material, thereby displaying a wide elongation at break values between 2% and 1000% [46]. (c) The tensile strength measures the amount of force required to pull a material until it breaks, and for PHA polymers it falls in the range of 8.8–104 MPa [45].

26.5 Advances in the applications of PHAs

The application of PHAs, both in quantity and type, has expanded particularly over the past 3 to 4 decades [47]. Food, medicine, and agriculture are the three main areas employing biodegradable materials. The large-scale application of PHAs is limited due to the high production cost and brittle nature of the homopolymers. Therefore, continuous research is being conducted for improving the microbial strains' ability to produce PHA by changing the monomeric composition to obtain better thermal and mechanical properties through approaches such as metabolic engineering through recombinant DNA technology and process development [48].

26.5.1 Food industry

The use of commercial polymers for food packaging has caused environmental hazards due to their nonrenewable and nonbiodegradable nature ultimately resulting in depletion of petrochemical sources for their production, as well as high amounts of plastic waste persistence after their use [1]. These facts have led to an increasing concern on the use of more sustainable polymers that are biobased and biodegradable. PHAs can be processed to excellent packaging films via thermoforming; this can be done using PHAs as the sole material or in combination with other compatible synthetic or biobased polymers, thus creating composite materials or blends. Bucci and colleagues (2007) reported the potential use of P(3HB) as packaging for food products [49]. The main use of PHAs are in packaging such as milk cartons, for cosmetic materials, shampoo containers, bottles, cardboards/papers cover/coatings, foils, films, etc. [50]. Also, combination polylactic acid–P(3HB) blends for usage as short-term food packaging have been reported [51].

Studies on P(3HB) as a food supplement has demonstrated growth-promoting effects on young fish and crustaceans. Recent studies by De Schryver et al. (2010) and Situmorang et al. (2016) reported P(3HB)-supplemented diet of juveniles of European sea bass (*Dicentrarchus labrax*) [52] and Nile tilapia (*Oreochromis niloticus*) [53] resulted in a significant increase in weight and resistance against pathogenic infection.

26.5.2 Medical industry

The biodegradable and biocompatible properties of PHAs make them attractive biomaterial for applications in conventional medical devices, drug delivery, and tissue engineering [54]. PHAs are thermoprocessable biomaterial and can be fabricated into porous matrices and implants for medical use [15]. PHA pellets, microspheres, and capsules have been explored for drug delivery. PHA meshes and fibers for sutures show promise for surgical use. PHA films have been used in postsurgery recovery [54]. PHA prepared as nanoparticles (NPs) is used for efficient uptake by the target eukaryotic cell to be used as sustained drug carrier or delivery vehicle. Due to the small size of PHA nanoparticles, they penetrate deeply into tissues, and hence PHA NPs loaded with a drug either as nano- or microspheres or capsules have proven their potential in cancer therapy [55]. P(3HB), P(3HB-*co*-3HHX), and their blends are used in designing scaffolds for regeneration of skin/nerve, soft tissue, artificial esophagus, drug delivery [56].

26.5.3 Agricultural industry

PHAs have been demonstrated as promising new source carrier material for long-term slow release of insecticides/pesticides/herbicides [57]. Applications of PHA-containing material in agriculture do not require expensive extraction and purification process from bacterial biomass [58]. Environmentally friendly, PHA films/microparticles loaded with herbicide/pesticide are employed in sustained-release systems for agrochemical purposes. In order to improve the herbicidal action and reduce environmental toxicity, Grillo and colleagues (2011) developed a modified and sustained release system by encapsulating the herbicide ametryn in microparticles of biopolymer of P(3HB) or P(3HB-*co*-3HV) [59]. Another study by Prudnikova et al. (2013) demonstrated the use of P(3HB-*co*-3HV) to construct eco-friendly PHA-herbicide systems in the form of films and microgranules, which helped in sustained release of the herbicide Zellek Super and could be placed into the soil together with seeds [57]. In the marine agriculture sector, PHAs are used as support for marine cultures and also used in making ropes and fishing nets [15].

Industries such as the Metabolix (Cambridge, MA, USA) and Procter and Gamble (USA) are known to manufacture PHAs such as BIOPOL and Nodax [15]. BIOPOL is a copolymer of poly(3-hydroxybutyrate-*co*-3-hydroxyvalerate) whereas Nodax is a copolymer consisting of 3-hydroxybutyrate and a comparatively small quantity of medium-chain-length PHA (mcl-PHA) monomers such as 3-hydroxyhexanoate, 3-hydroxyoctanoate, and 3-hydroxydecanoate. This polymer can be applied to coat paper/paperboards, blow molding, film production, electric and electronic packaging, biomedical applications, fishing nets, and ropes [15]. Depending on the applications, the mechanical properties or biocompatibility of PHA polymer can be modified either by blending or doing the surface modification and/or making composites with other polymers, enzymes, or even inorganic materials [56].

26.6 Extremophilic microorganisms

Microorganisms inhabit certain econiches with high salinity, high pressure, high/low temperature, and pH that may be deleterious for otherwise mesophilic microbes usually thriving in normal conditions. These organisms are termed as extremophiles, and they thrive in environments like hypersaline lakes, solar salterns, hydrothermal deep-sea vents, permafrost/high-altitude ice sheets, hot springs, and methane clarets [60,61]. Extremophiles comprise organisms ranging from bacteria, archaea to eukaryotes like fungi and algae. These can be classified into major categories of halophiles (organisms requiring at least 2M NaCl and thriving in up to 5.1M NaCl), methanogens (organisms thriving in anoxic conditions producing methane), acidophiles (organisms thriving in acidic pH conditions of 3 or below), alkaliphiles (organisms thriving in alkaline pH conditions of 9 or above), thermophiles (organisms that thrive between 45 and 122°C; organisms that grow between in 80–122°C are known as hyperthermophiles, and psychrophiles (organisms that grow and reproduce between−20 and 10°C). Some organisms may possess more than one trait of extremophiles, like growing in high salinity or tolerating high temperatures, and are termed as polyextremophiles [62].

26.7 Extremophilic microorganisms producing PHAs

Extremophiles thrive in unnatural habitats, possess unusual metabolic pathways, and produce unique biotechnological products. These biological adaptations not only allow the organisms to survive in extreme conditions but also to thrive and multiply [63]. For instance, these microorganisms contain unique lipid composition for protecting against high temperature, high salinity, and variable pH conditions. Similarly, the proteins of extremophiles have certain modifications like highly charged surface groups, increased hydrophobic interactions, and the presence of osmolytes, for the functioning of the organisms in extreme conditions. This opens an interesting prospect of employing secondary metabolites produced by these extremophiles in harsh conditions, otherwise the mesophilic metabolites would fail. Some of the enzymes produced by extremophiles, called extremozymes, such as amylase, protease, lipase, and xylanase, have been employed in paper, detergent, and food processing industries [62]. Other biotechnological applications include biomining, bioremediation, as biosensors, and in the pharmaceutical industry.

One of the promising applications of extremophiles is the accumulation of biopolymers or PHAs [12]. Among extremophiles, halophiles and thermophiles have been reported to accumulate a substantial amount of PHAs [61]. Employing halophiles for PHA production has a distinct advantage because of the high salinity, the process can be operated in nonsterile conditions, and also their lysis in distilled water may reduce the downstream cost in production [64]. Similarly, thermophiles are cultivated at high temperatures (~50–55°C), the process does not require additional sterilization. Reports are available on optimization of PHA production by batch and fed-batch mode of cultivation using the moderate halophilic bacteria belonging to *Halomonas* genus [64].

Yue et al., employed halophilic *Halomonas campaniensis* strain LS21 to developed long-lasting open and continuous PHA production process based on artificial seawater and used mixed substrates such as starch, lipids, cellulose, and proteins [65]. PHA production by *H. campaniensis* LS21 wild-type and recombinant strains were compared by growing for 65 days undisturbed at 37°C in presence of 2.7% NaCl, alkaline pH ~10 and was found to accumulate ~26% and 70% P(3HB). Among thermophiles, *Thermus thermophiles, Chelatococcus* sp., *Aneurinibacillus* sp., and *Cladimonas* sp. have been reported to accumulate PHAs in the presence of commercial carbon substrates like glucose [8,66–68]. Apart from being a reserve carbon, PHAs have other roles to play in microbes. For an instance, psychrophilic bacteria *Pseudomonas* sp. accumulates PHA to combat the oxidative stress conditions at cold temperatures [69]. In another psychrophilic bacteria, *Pseudomonas extremaustralis*, the PHA is responsible for the motility and survival of planktonic cells in the biofilm formed by the bacteria [70].

Koller and colleagues provided case examples on how microbiological strain selection and low-cost raw materials and their processing to be used as microbial feedstocks can be optimized for sustainable production of biobased and biodegradable PHAs [71,72]. Screening of halophiles has revealed that halophilic organisms belonging to the archaeal family *Halobacteriaceae*, genera *Halogranum*, *Halococcus*, *Halorubrum*, *Natronobacterium*, *Natronococcus*, *Halopiger*, *Haloarcula*, *Halobiforma*, *Haloferax*, and *Halobacterium* [32,63,73,74] and eubacterial family *Halomonadaceae*, genus *Halomonas* [16] are capable of accumulating PHA. Some halophilic archaea such as *Haloterrigena hispanica*, *Haloquadratum walsbyi*, *Halorhabdus tiamatea*, *Halorhabdus utahensis*, and *Natrinema altinense* have also been reported to synthesize PHA. However, the polymer has neither been quantified nor characterized [27]. Extremely halophilic archaea of family Halobacteriaceae contains 48 genera and 177 species (as of November 2014), which is the largest archaeal group [75]. Much of the work on PHAs has been carried out in halophilic archaea with *Hfx. mediterranei* being the best PHA producer among extremophiles, accumulating P(3HB-*co*-3HV) with 10 mol% 3HV monomer [63].

Recent studies by Han and colleagues (2017, 2015) reported the synthesis of the random poly(3-hydroxybutyrate-*co*-3-hydroxyvalerate) (R–PHBHV) and higher-order PHBHV (O–PHBHV) polyesters with diversified mechanical properties using engineered haloarchaeon *Hfx. mediterranei* strain ES1 by cofeeding valerate with glucose [28,76]. The O-PHBV film had a rough surface that exhibited increased platelet adhesion thereby accelerating blood clotting [28]. The films of the polyesters, R–PHBHV and O–PHBHV, exhibited excellent biocompatibility when screened for the attachment and proliferation using rat fibroblast L929 and osteoblast MC3T3-E1 cell lines. The degradation of polyester films did not release any toxic constituents when implanted in rabbits [76]. Another report by Danis et al. (2015) employed *Natrinema pallidum* for P (HB-*co*-HV) produced, which was used to prepare biocompatible films for drug delivery (rifampicin) by blending with polyethylene glycol [77]. Zhao et al. (2015) compared the properties of copolymer P (3HB-*co*-3HV) from gram-negative bacterium *Ralstonia eutropha* (PHBV−B) and halophilic archaea *Halogranum amylolyticum* (PHBV−H). The polymer films PHBV-H and PHBV-B, obtained from both the cultures, though having similar 3HV content, they varied in surface properties. PHBV-H was reported to exhibit better hemocompatibility, revealing its potential to be used as a biomaterial for blood contact [73].

Moderately halophilic bacteria, belonging to the genus *Halomonas*, such as *H. boliviensis* LC1, *H. nitroreducens*, and *H. salina*, have been reported to accumulate 56.0%, 33.0%, and 55.0% (wt/wt)of CDW of homopolymer of hydroxybutyrate (HB), P(3HB) by utilizing versatile substrates such as starch hydrolysate, glucose, and glycerol, respectively [78−80]. Similarly, Van-Thuoc et al. (2012) reported the ability of halophilic bacteria *Bacillus* sp. ND153 to accumulate P(3HB)/P(3HB-*co*-3HV) of 65.0/71.0% (w/w) of CDW when glucose and/or glucose along with propionate was provided as the carbon source [81]. However, Kulkarni et al. (2010) reported halophilic bacteria *H. campisalis* MCM B-1027 to synthesize copolymer P(HB-*co*-HV), without the addition of fatty acid precursors like propionic/valeric acid in the culture medium. Additions of the precursor fatty acids for obtaining copolymers add on to the production cost and are also toxic to the microbial strains [82]. Shrivastav et al. (2010) reported the utilization of *Jatropha* biodiesel by-product as substrate by halophilic *Halomonas hydrothermalis* strain SM-P-3M for P(3HB) production of 75.0% (w/w) of CDW [83].

26.8 PHAs from renewable resources and agroindustrial wastes

To date, various carbon-rich renewable materials, such as lignocellulosic waste, starch waste, molasses, whey from the dairy industry, glycerol phase from biodiesel production, slaughterhouse waste, and waste oils have been tested effective for PHA production [2,12]. There are few reports on PHA production using various carbon-rich renewable materials as substrates by extremophilic microorganisms mostly employing extremely halophilic archaea and countable number of thermophilic microorganisms (Table 26.1). Van-Thuoc et al. reported the production of P(3HB) by moderately halophilic *Halomonas boliviensis* using agroindustrial byproducts like wheat bran and potato wastes [16]. Among extremely haloarchaeal, *Hfx. mediterranei* is the most widely studied and is reported to produce P(HB-*co*-HV) using various renewable agroindustrial wastes like the extruded corn starch, rice bran, wheat bran, hydrolyzed whey, waste stillage from rice-based ethanol industry, and vinasse [84−87].

TABLE 26.1 Accumulation of PHA by extremely halophilic archaea and bacteria using inexpensive agroindustrial wastes.

Extremophilic microorganisms	Substrate	PHA type	Yield	Reference
Halophiles				
Halogeometricum borinquense strain E3	25% SCB	PHBV	1.6[b]	[88]
	50% SCB	PHBV	1.9[b]	
	10% CW	PHBV	1.52[b]	[89]
Natrinema pallidum 1KYS1	2% CS	PHBV	0.075[b]	[77]
	2% Whey waste		0.091[b]	
	2% Melon waste		0.039[b]	
	2% Apple waste		0.077[b]	
	2% Tomato waste		0.464[b]	
	Starch		0.81[b]	
Haloferax mediterranei DSM 1411	25% PTV	PHBV	19.7[b]	[87]
	50% PTV		17.4[b]	
	Stillage FS	PHBV	16.42[b]	[86]
	Stillage RS		16.25[b]	
Hfx. mediterranei ATCC 33500	ERB/ECS	PHBV	77.8[b]	[84]
	EWB/ECS		52.7[b]	
	NWB/ECS		28.0[b]	
Hfx. mediterranei	Hydrolyzed whey	PHBV	5.5[b]	[91]
Hfx. mediterranei	Hydrolyzed whey	PHBV	12.2[b]	[85]
	Hydrolyzed whey + GBL	PHBVB	14.7[b]	
Haloarcula sp. IRU1	Petrochemical wastewater	PHB	46.0[a]	[92]
Haloarcula marismortui MTCC 1596	10% RV	PHB	2.8[b]	[94]
	100% PTV		4.5[b]	
Haloarcula japonica strain T5	Molasses	PHB	1.0[a]	[95]
Haloterrigena hispanica	Carrot waste	PHB	0.13[a]	[11]

Continued

TABLE 26.1 Accumulation of PHA by extremely halophilic archaea and bacteria using inexpensive agroindustrial wastes.—cont'd

Extremophilic microorganisms	Substrate	PHA type	Yield	Reference
Bacillus sonorensis SM-P-1S	*Jatropha* biodiesel by-product	PHB	71.8[a]	[83]
Halomonas hydrothermalis SM-P-3M			75.0[a]	
Halomonas boliviensis	WBH	PHB	34.0[a]	[16]
Halomonas sp. SK5	OPTS	PHB	1.6[b]	[96]
Halomonas campaniensis strain LS21(wild type)	Kitchen waste	PHB	26.0[a]	[65]
H. campaniensis strain LS21(Recombinant)			70.0[a]	
Thermophiles				
Caldimonas taiwanensis	Starch	PHB	71.0[a]	[8]
Thermus thermophilus HB8	Whey	Heteropolymer (HV-HHp-HN-HU)	35.0[a]	[66]
Chelatococcus sp. MW10	Glu	PHB	2.9[b]	[67]
Aneurinibacillus sp.	Glu/Pep/YE	PHBV	0.1[b]	[68]
Synechococcus sp. strain MA19	Acetate	PHB	6.5[b]	[97]
Chlorogloeopsis fritschii PCC 6912	Acetate	PHB	6.2[b]	[97]

CDW, cell dry weight; CS, corn starch; CW, cassava waste; ECS, extruded cornstarch; ERB, extruded rice bran; EWB, extruded wheat bran; GBL, γ-butyrolactone; Glu, glucose; HHp, hydroxyheptanoate; HN, hydroxynanoate; HU, hydroxyundecanoate; HV, hydroxyvalerate; NWB, native wheat bran; OPTS, oil palm trunk sap; pep, peptone; PHB, poly (3-hydroxybutyrate); PHBV, poly (3-hydroxybutyrate-*co*-3-hydroxyvalerate); PHBVB, poly(3-hydroxybutyrate-*co*-3-hydroxyvalerate-*co*-4-hydroxybutyrate); PTV, pretreated vinasse, FS, fresh salts; RS, recovered salts; SCB, sugarcane bagasse; WBH, wheat bran hydrolysate; YE, yeast extract.
[a] *% w/w of cell dry weight.*
[b] *g/L.*

Koller (2015) reported PHA production by the halophilic archaeon *Hfx. mediterranei* DSM1411 from whey as a carbon source and medium containing 15% (w/w) NaCl [12]. The study also assessed the recyclability of cells debris and spent fermentation broth, and found that the *Hfx. mediterranei* DSM1411 could accumulate maximum P(3HB-*co*-3HV) of 7.2 and 2.28 g/L from original fermentation and recycling experiments, respectively.

Bhattacharyya et al., used the waste stillage from rice-based ethanol agroindustry for the production of P(3HB-*co*-3HV) by *Hfx. mediterranei* DSM 1411 and also demonstrated the recovery of salts from the spent medium and their reuse. The culture was reported to grow and produce a P(3HB-*co*-3HV) yield of 71% and 69% (w/w) of CDW, in minimal medium containing 20% (w/v) NaCl and recovered salts, respectively. Growing extremely halophilic archaeal strains involves the use and disposal of high-salt-containing spent medium in the environment, which could be a major problem as it will overall increase the total dissolved solids. This study has proved that the salts from the spent medium could be recovered and reused without drastically affecting the polymer yield [86].

Salgaonkar and coauthors reported the biosynthesis of P(3HB-*co*-3HV) comprising 13.29% and 19.65% 3HV by *Halogeometricum borinquense* strain E3 using readily available carbon-rich waste bagasse from sugarcane (*Saccharum officinarum*) and cassava (*Manihot esculenta*), respectively. The strain E3 was reported to accumulate maximum PHA of 4.15 and 1.52 g/L of CDM using crude SCB and CW hydrolysate without any additional carbon source [88,89].

Using inexpensive crude glycerol phase (CGP) from biodiesel production as a carbon source, Hermann-Krauss and colleagues revealed that halophilic archaeon *Hfx. mediterranei* DSM 1411could synthesize copolymer PHBV 16.2 g/L in medium containing 15% (w/v) NaCl. The PHBV produced using CGP was much higher than 13.4 g/L produced using pure glycerol [90]. Interestingly, the study also revealed the ability of the strain to synthesize 11.1 g/L terpolyesters poly[(*R*)-3-hydroxybutyrate-*co*-(*R*)-3-hydroxyvalerate-*co*-4-hydroxybutyrate)] and P(3HB-*co*-3HV-*co*-4HB) when the medium was supplemented with adequate 4HB precursors in the form of γ-butyrolactone (GBL) along with CGP.

The homopolymers of PHAs are brittle in nature, which limits their range of application. Hence, research is conducted on controlling the repeating unit composition of PHA homopolymer so as to obtain polymers with diverse (more than one) monomeric units with improved material properties such as decreased crystallinity and increased flexibility. Studies have demonstrated that switching from one-stage to two- or multistage continuous production processes provides the possibility to fine-tune the PHA properties during PHA biosynthesis by microorganisms [93]. A multistage process, such as cascades of continuously stirred tank reactors (CSTRs), is employed in which PHA accumulation occurs under continuous supply of carbon source and nutrient-limiting condition. CSTRs are used to produce block polymers (b-PHA) where modulated soft or hard segments can be incorporated into matrices of homopolymers (3HB) [72]. Tripathi and colleagues studied the biosynthesis of a block copolymer consisting of soft poly(4-hydroxybutyrate) (P4HB) block with a strong poly(3-hydroxypropionate) (P3HP) block using a recombinant *E. coli* strain [98].

Halophilic archaea are known to naturally synthesize copolymer with various HV units by feeding on various renewable agroindustrial wastes and without the addition of any precursor fatty acids. P(3HB-*co*-3HV) with HV content of 10%, 15.4%, 12.36%, and 14.09% were naturally obtained from extremely halophilic archaeon *Hfx. mediterranei* by utilizing renewable wastes such as CGP, whey, stillage from the ethanol industry, and pretreated vinasse 25% and 50%, respectively [12,86,87,90]. The P(3HB-*co*-3HV) production from various substrates could be due to class III type PHA synthases of *Hfx. mediterranei* composed of $PhaC_{Hme}$ and $PhaE_{Hme}$, which is encoded by the gene cluster ($phaEC_{Hme}$) [18].

There are limited studies on thermophilic microorganisms producing PHA. Pantazaki and colleagues reported the ability of thermophilic bacterium *Thermus thermophilus* HB8 to utilize whey when grown at 70°C for production of novel heteropolymer comprising of the short-chain length with 38 mol% 3-hydroxyvalerate (3HV) and the medium-chain length, 9.89 mol% 3-hydroxyheptanoate (3HHp), 16.59 mol% 3-hydroxynanoate (3HN) and 35.42 mol% 3-hydroxyundecanoate (3HU), all having unique properties [66].

Sheu et al. studied the ability of thermophilic bacteria *Caldimonas taiwanensis* requiring a temperature of 55°C for growth and accumulated P(3HB) 71% (w/w) of CDW from starch, which is one of the most easily available and abundant renewable carbon sources [8]. The strain was reported to synthesize copolymer P(3HB-*co*-3HV) when feeding with mixed substrates such as starch and fatty acid valerate. The amount of P(3HB-*co*-3HV) varied with the kind of starch provided, which was 67% > 65% > 55% > 52% > 42% (w/w) of CDM for starch from cassava, corn, potato, sweet potato, and wheat, respectively. *Caldimonas taiwanensis* strain was also reported to grow under nitrogen-limited condition in presence of glycerol, fructose, gluconate, and maltose and accumulate P(3HB) 52%, 62%, 70%, and 60% (w/w) of CDW, respectively [8].

Hai et al., studied the PHA-synthesizing ability of photoautotrophic cyanobacteria both mesophilic and thermophilic by growing the strains in acetate containing nitrogen-free BG11 medium at 37°C and 50°C temperature [97]. The study found that mesophilic cyanobacteria such as *Cyanothece* sp., *Gloeocapsa* sp., *Stanieria* sp., *Synechococcus* sp. PCC 6715 growing at 37°C accumulated P(3HB) ~ 0.7%–2.7% (w/w) of CDM. However, the thermophilic cyanobacteria *Synechococcus* sp. strain MA19 and *Chlorogloeopsis fritschii* PCC 6912 was reported to accumulate ~6.5% and 6.2% (w/w) of CDW of P(3HB). Their study concluded that thermophilic cyanobacterial strains are better producers of P(3HB) than the mesophilic ones.

26.9 Conclusions

Extremophilic archaea, bacteria, and cyanobacteria are a promising group of microorganisms that accumulate PHAs. They are promising candidates for PHA production at an industrial level as they can utilize versatile renewable agroindustrial wastes as substrates. An added advantage is their unique qualities of surviving under extreme conditions such as high NaCl/temperature, which eliminates the risk of contamination by their nonhalophilic/mesophilic counterparts, thereby eliminating the need for sterilization, hence lowering the fermentation costs. Halophilic microorganisms are known to produce extracellular hydrolytic enzymes that are active in extreme conditions and could be used for hydrolysis of complex substrates, thereby converting them into PHA. Haloarchaea belonging to genera *Haloferax*, *Haloarcula*, and *Halogeometricum* are reported to naturally synthesize copolymers such as P(3HB-*co*-3HV) without providing precursors like propionic/valeric acid, which their mesophilic counterparts mostly fail to do. Since worldwide importance is given to the development of agriculture and agrobased industries, large amounts of agroindustrial wastes are being generated. Employing extremophilic microorganisms to utilize the agroindustrial wastes as substrates for the production of an economically value-added product such as polyhydroxyalkanoates will help in both managing the agroindustrial wastes and cutting down the costs of commercial substrates. Considering this, more attention should be given to isolation of extremophilic microorganisms from diverse econiches and screening them for novel metabolites. In-depth studies should focus on understanding the molecular basis of the PHA production to explore the possibility of engineering their metabolic pathways and applying these potential microorganisms for fermentation processes at the industrial level. In case of valorization of waste, the availability and consistency of the raw waste materials and storage, the uniformity of its composition need to be cautiously analyzed.

Acknowledgments

The authors acknowledge the financial help provided by the various funding agencies: (1) Council of Scientific and Industrial Research (CSIR) India for Research Associateship (RA; Ref No: 09/919(0030)/2016-EMR-I); (2) CSIR, India for Senior Research Fellowship (SRF; Ref No: 09/919(0016)/2012-EMR-I; (3) Birla Institute of Technology and Science (BITS), Pilani for the BITS Seed Grant 2013; and (4) University Grants Commission (UGC), India for Major Research Project No: 34–500/2008(SR).

References

[1] Chen GQ. Introduction of bacterial plastics PHA, PLA, PBS, PE, PTT, and PPP. Plastics from bacteria. Berlin, Heidelberg: Springer; 2010.

[2] Castilho LR, Mitchell DA, Freire DM. Production of polyhydroxyalkanoates (PHAs) from waste materials and by-products by submerged and solid-state fermentation. Bioresour Technol 2009;100(23):5996–6009.

[3] Naranjo JM, Posada JA, Higuita JC, Cardona CA. Valorization of glycerol through the production of biopolymers: the PHB case using *Bacillus megaterium*. Bioresour Technol 2013;133:38–44.

[4] Chee JY, Yoga SS, Lau NS, Ling SC, Abed RM, Sudesh K. Bacterially produced polyhydroxyalkanoate (PHA): converting renewable resources into bioplastics. Curr Res Technol Edu Topics Appl Microbiol Microbial Biotechnol 2010;2:1395–404.

[5] Tan GY, Chen CL, Li L, Ge L, Wang L, Razaad IM, Li Y, Zhao L, Mo Y, Wang JY. Start a research on biopolymer polyhydroxyalkanoate (PHA): a review. Polymers 2014;6(3):706–54.

[6] Reddy CS, Ghai R, Kalia V. Polyhydroxyalkanoates: an overview. Bioresour Technol 2003;87(2):137–46.

[7] Quillaguamán J, Guzmán H, Van-Thuoc D, Hatti-Kaul R. Synthesis and production of polyhydroxyalkanoates by halophiles: current potential and future prospects. Appl Microbiol Biotechnol 2010;85(6):1687–96.

[8] Sheu DS, Chen WM, Yang JY, Chang RC. Thermophilic bacterium *Caldimonas taiwanensis* produces poly (3-hydroxybutyrate-*co*-3-hydroxyvalerate) from starch and valerate as carbon sources. Enzym Microb Technol 2009;44(5):289–94.

[9] Sudesh K, Bhubalan K, Chuah JA, Kek YK, Kamilah H, Sridewi N, Lee YF. Synthesis of polyhydroxyalkanoate from palm oil and some new applications. Appl Microbiol Biotechnol 2011;89(5):1373–86.

[10] Pappu A, Saxena M, Asolekar SR. Solid wastes generation in India and their recycling potential in building materials. Build Environ 2007;42(6):2311–20.

[11] Di Donato P, Fiorentino G, Anzelmo G, Tommonaro G, Nicolaus B, Poli A. Re-use of vegetable wastes as cheap substrates for extremophile biomass production. Waste Biomass Valorization 2011;2(2):103–11.

[12] Koller M. Recycling of waste streams of the biotechnological poly (hydroxyalkanoate) production by *Haloferax mediterranei* on whey. Int J Polym Sci 2015;2015.

[13] Salgaonkar BB, Mani K, Bragança JM. Accumulation of polyhydroxyalkanoates by halophilic archaea isolated from traditional solar salterns of India. Extremophiles 2013;17(5):787–95.

[14] Sedlacek P, Slaninova E, Koller M, Nebesarova J, Marova I, Krzyzanek V, Obruca S. PHA granules help bacterial cells to preserve cell integrity when exposed to sudden osmotic imbalances. N Biotech 2019;49:129–36.

[15] Philip S, Keshavarz T, Roy I. Polyhydroxyalkanoates: biodegradable polymers with a range of applications. J Chem Technol Biotechnol 2007;82(3):233–47.

[16] Van-Thuoc D, Quillaguaman J, Mamo G, Mattiasson B. Utilization of agricultural residues for poly (3-hydroxybutyrate) production by *Halomonas boliviensis* LC1. J Appl Microbiol 2008;104(2):420–8.

[17] Bernd HA. Polyester synthases: natural catalysts for plastics. Biochem J 2003;376(1):15–33.

[18] Han J, Hou J, Liu H, Cai S, Feng B, Zhou J, Xiang H. Wide distribution among halophilic archaea of a novel polyhydroxyalkanoate synthase subtype with homology to bacterial type III synthases. Appl Environ Microbiol 2010;76(23):7811–9.

[19] Madison LL, Huisman GW. Metabolic engineering of poly (3-hydroxy alkanoates): from DNA to plastic. Microbiol Mol Biol Rev 1999;63(1):21–53.

[20] Stubbe J, Tian J. Polyhydroxyalkanoate (PHA) homeostasis: the role of the PHA synthase. Nat Prod Rep 2003;20(5):445–57.

[21] Banki MR, Gerngross TU, Wood DW. Novel and economical purification of recombinant proteins: intein - mediated protein purification using *in vivo* polyhydroxybutyrate (PHB) matrix association. Protein Sci 2005;14(6):1387–95.

[22] Hiraishi T, Komiya N, Matsumoto N, Abe H, Fujita M, Maeda M. Degradation and adsorption characteristics of PHB depolymerase as revealed by the kinetics of mutant enzymes with amino acid substitution in the substrate-binding domain. Biomacromolecules 2009;11(1):113–9.

[23] Steinbüchel A, Lütke-Eversloh T. Metabolic engineering and pathway construction for biotechnological production of relevant polyhydroxyalkanoates in microorganisms. Biochem Eng J 2003;16(2):81–96.

[24] Ward PG, O'Connor KE. Bacterial synthesis of polyhydroxyalkanoates containing aromatic and aliphatic monomers by *Pseudomonas putida* CA-3. Int J Biol Macromol 2005;35(3–4):127–33.

[25] Loo CY, Sudesh K. Polyhydroxyalkanoates: bio-based microbial plastics and their properties. Malaysian Polymer J 2007;2(2):31–57.

[26] Salgaonkar BB, Mani K, Braganca JM. Characterization of polyhydroxyalkanoates accumulated by a moderately halophilic salt pan isolate *Bacillus megaterium* strain H 16. J Appl Microbiol 2013;114(5):1347–56.

[27] Salgaonkar BB, Bragança JM. Biosynthesis of poly (3-hydroxybutyrate-*co*-3-hydroxyvalerate) by *Halogeometricum borinquense* strain E3. Int J Biol Macromol 2015;78:339–46.

[28] Han J, Wu LP, Hou J, Zhao D, Xiang H. Biosynthesis, characterization, and hemostasis potential of tailor-made poly (3-hydroxybutyrate-*co*-3-hydroxyvalerate) produced by *Haloferax mediterranei*. Biomacromolecules 2015;16(2):578–88.

[29] Yu F, Dong T, Zhu B, Tajima K, Yazawa K, Inoue Y. Mechanical properties of comonomer-compositionally fractionated Poly [(3-hydroxybutyrate)-co-(3-mercaptopropionate)] with low 3-mercaptopropionate unit content. Macromol Biosci 2007;7(6):810–9.

[30] Spiekermann P, Rehm BH, Kalscheuer R, Baumeister D, Steinbüchel A. A sensitive, viable-colony staining method using Nile red for direct screening of bacteria that accumulate polyhydroxyalkanoic acids and other lipid storage compounds. Arch Microbiol 1999;171(2):73–80.

[31] Ostle AG, Holt JG. Nile blue A as a fluorescent stain for poly-beta-hydroxybutyrate. Appl Environ Microbiol 1982;44(1):238–41.

[32] Legat A, Gruber C, Zangger K, Wanner G, Stan-Lotter H. Identification of polyhydroxyalkanoates in *Halococcus* and other haloarchaeal species. Appl Microbiol Biotechnol 2010;87(3):1119–27.

[33] Montenegro EM, Delabary GS, Silva MA, Andreote FD, Lima AO. Molecular diagnostic for prospecting polyhydroxyalkanoate-producing bacteria. Bioengineering 2017;4(2):52.

[34] Lu Q, Han J, Zhou L, Zhou J, Xiang H. Genetic and biochemical characterization of the poly (3-hydroxybutyrate-*co*-3-hydroxyvalerate) synthase in *Haloferax mediterranei*. J Bacteriol 2008;190(12):4173–80.

[35] Kunasundari B, Sudesh K. Isolation and recovery of microbial polyhydroxyalkanoates. Express Polym Lett 2011;5(7).

[36] Kapritchkoff FM, Viotti AP, Alli RC, Zuccolo M, Pradella JG, Maiorano AE, Miranda EA, Bonomi A. Enzymatic recovery and purification of polyhydroxybutyrate produced by *Ralstonia eutropha*. J Biotechnol 2006;122(4):453–62.

[37] Tamer IM, Moo-Young M, Chisti Y. Disruption of *Alcaligenes latus* for recovery of poly (β-hydroxybutyric acid): comparison of high-pressure homogenization, bead milling, and chemically induced lysis. Ind Eng Chem Res 1998;37(5):1807–14.
[38] Law JH, Slepecky RA. Assay of poly-β-hydroxybutyric acid. J Bacteriol 1961;82(1):33–6.
[39] Strazzullo G, Gambacorta A, Vella FM, Immirzi B, Romano I, Calandrelli V, Nicolaus B, Lama L. Chemical-physical characterization of polyhydroxyalkanoates recovered by means of a simplified method from cultures of *Halomonas campaniensis*. World J Microbiol Biotechnol 2008;24(8):1513–9.
[40] Chanprateep S. Current trends in biodegradable polyhydroxyalkanoates. J Biosci Bioeng 2010;110(6):621–32.
[41] Chen S, Liu Q, Wang H, Zhu B, Yu F, Chen GQ, Inoue Y. Polymorphic crystallization of fractionated microbial medium-chain-length polyhydroxyalkanoates. Polymer 2009;50(18):4378–88.
[42] Barham PJ, Keller A, Otun EL, Holmes PA. Crystallization and morphology of a bacterial thermoplastic: poly-3-hydroxybutyrate. J Mater Sci 1984;19(9):2781–94.
[43] Sánchez RJ, Schripsema J, da Silva LF, Taciro MK, Pradella JG, Gomez JG. Medium-chain-length polyhydroxyalkanoic acids (PHAmcl) produced by *Pseudomonas putida* IPT 046 from renewable sources. Eur Polym J 2003;39(7):1385–94.
[44] Wu CS, Liao HT. The mechanical properties, biocompatibility and biodegradability of chestnut shell fibre and polyhydroxyalkanoate composites. Polym Degrad Stabil 2014;99:274–82.
[45] Rai R, Keshavarz T, Roether JA, Boccaccini AR, Roy I. Medium chain length polyhydroxyalkanoates, promising new biomedical materials for the future. Mater Sci Eng R Rep 2011;72(3):29–47.
[46] Chen Z, Wilmanns M, Zeng AP. Structural synthetic biotechnology: from molecular structure to predictable design for industrial strain development. Trends Biotechnol 2010;28(10):534–42.
[47] Keshavarz T, Roy I. Polyhydroxyalkanoates: bioplastics with a green agenda. Curr Opin Microbiol 2010;13(3):321–6.
[48] Nikel PI, de Almeida A, Melillo EC, Galvagno MA, Pettinari MJ. New recombinant *Escherichia coli* strain tailored for the production of poly (3-hydroxybutyrate) from agroindustrial by-products. Appl Environ Microbiol 2006;72(6):3949–54.
[49] Bucci DZ, Tavares LB, Sell I. Biodegradation and physical evaluation of PHB packaging. Polym Test 2007;26(7):908–15.
[50] Khosravi-Darani K, Bucci DZ. Application of poly (hydroxyalkanoate) in food packaging: improvements by nanotechnology. Chem Biochem Eng Q 2015;29(2):275–85.
[51] Arrieta MP, Samper MD, Aldas M, López J. On the use of PLA-PHB blends for sustainable food packaging applications. Materials 2017;10(9):1008.
[52] De Schryver P, Sinha AK, Kunwar PS, Baruah K, Verstraete W, Boon N, De Boeck G, Bossier P. Poly-β-hydroxybutyrate (PHB) increases growth performance and intestinal bacterial range-weighted richness in juvenile European sea bass, *Dicentrarchus labrax*. Appl Microbiol Biotechnol 2010;86(5):1535–41.
[53] Situmorang ML, De Schryver P, Dierckens K, Bossier P. Effect of poly-β-hydroxybutyrate on growth and disease resistance of Nile tilapia *Oreochromis niloticus* juveniles. Vet Microbiol 2016;182:44–9.
[54] Brigham CJ, Sinskey AJ. Applications of polyhydroxyalkanoates in the medical industry. Int J Biotechnol Wellness Ind 2012;1(1):52–60.
[55] Shrivastav A, Kim HY, Kim YR. Advances in the applications of polyhydroxyalkanoate nanoparticles for novel drug delivery system. BioMed Res Int 2013;2013.

[56] Valappil SP, Misra SK, Boccaccini AR, Roy I. Biomedical applications of polyhydroxyalkanoates, an overview of animal testing and *in vivo* responses. Expert Rev Med Devices 2006;3(6):853–68.

[57] Prudnikova SV, Boyandin AN, Kalacheva GS, Sinskey AJ. Degradable polyhydroxyalkanoates as herbicide carriers. J Polym Environ 2013;21(3):675–82.

[58] Ivanov V, Stabnikov V, Ahmed Z, Dobrenko S, Saliuk A. Production and applications of crude polyhydroxyalkanoate-containing bioplastic from the organic fraction of municipal solid waste. Int J Environ Sci Technol 2015;12(2):725–38.

[59] Grillo R, Santo Pereira AD, de Melo NF, Porto RM, Feitosa LO, Tonello PS, Dias Filho NL, Rosa AH, Lima R, Fraceto LF. Controlled release system for ametryn using polymer microspheres: preparation, characterization and release kinetics in water. J Hazard Mater 2011;186(2–3):1645–51.

[60] Yin J, Chen JC, Wu Q, Chen GQ. Halophiles, coming stars for industrial biotechnology. Biotechnol Adv 2015;33(7):1433–42.

[61] F Bosma E, van der Oost J, M de Vos W, van Kranenburg R. Sustainable production of bio-based chemicals by extremophiles. Curr Biotechnol 2013;2(4):360–79.

[62] Ma Y, Galinski EA, Grant WD, Oren A, Ventosa A. Halophiles 2010: life in saline environments. Appl Environ Microbiol 2010;76(21):6971–81.

[63] Poli A, Di Donato P, Abbamondi GR, Nicolaus B. Synthesis, production, and biotechnological applications of exopolysaccharides and polyhydroxyalkanoates by archaea. Archaea 2011;2011.

[64] Stanley A, Kumar HP, Mutturi S, Vijayendra SN. Fed-batch strategies for production of PHA using a native isolate of *Halomonasvenusta* KT832796 strain. Appl Biochem Biotechnol 2018;184(3):935–52.

[65] Yue H, Ling C, Yang T, Chen X, Chen Y, Deng H, Wu Q, Chen J, Chen GQ. A seawater-based open and continuous process for polyhydroxyalkanoates production by recombinant *Halomonas campaniensis* LS21 grown in mixed substrates. Biotechnol Biofuels 2014;7(1):108.

[66] Pantazaki AA, Papaneophytou CP, Pritsa AG, Liakopoulou-Kyriakides M, Kyriakidis DA. Production of polyhydroxyalkanoates from whey by *Thermus thermophilus* HB8. Process Biochem 2009;44(8):847–53.

[67] Ibrahim MH, Steinbüchel A. High-cell-density cyclic fed-batch fermentation of a poly (3-hydroxybutyrate)-accumulating thermophile, *Chelatococcus* sp. strain MW10. Appl Environ Microbiol 2010;76(23):7890–5.

[68] Xiao Z, Zhang Y, Xi L, Huo F, Zhao JY, Li J. Thermophilic production of polyhydroxyalkanoates by a novel *Aneurinibacillus* strain isolated from Gudao oilfield, China. J Basic Microbiol 2015;55(9):1125–33.

[69] Ayub ND, Tribelli PM, López NI. Polyhydroxyalkanoates are essential for maintenance of redox state in the Antarctic bacterium *Pseudomonas* sp. 14-3 during low temperature adaptation. Extremophiles 2009;13(1):59–66.

[70] Tribelli PM, López NI. Poly (3-hydroxybutyrate) influences biofilm formation and motility in the novel Antarctic species *Pseudomonas extremaustralis* under cold conditions. Extremophiles 2011;15(5):541.

[71] Koller M, Maršálek L, de Sousa Dias MM, Braunegg G. Producing microbial polyhydroxyalkanoate (PHA) biopolyesters in a sustainable manner. New biotechnol 2017;37:24–38.

[72] Koller M, Muhr A. Continuous production mode as a viable process-engineering tool for efficient poly (hydroxyalkanoate) (PHA) bio-production. Chem Biochem Eng Q 2014;28(1):65–77.

[73] Zhao Y, Rao Z, Xue Y, Gong P, Ji Y, Ma Y. Biosynthesis, property comparison, and hemocompatibility of bacterial and haloarchaeal poly (3-hydroxybutyrate-*co*-3-hydroxyvalerate). Sci Bull 2015;60(22):1901–10.

[74] Lynch EA, Langille MG, Darling A, Wilbanks EG, Haltiner C, Shao KS, Starr MO, Teiling C, Harkins TT, Edwards RA, Eisen JA. Sequencing of seven haloarchaeal genomes reveals patterns of genomic flux. PLoS One 2012;7(7). e41389.

[75] Oren A. Taxonomy of halophilic Archaea: current status and future challenges. Extremophiles 2014;18:825–34.

[76] Han J, Wu LP, Liu XB, Hou J, Zhao LL, Chen JY, Zhao DH, Xiang H. Biodegradation and biocompatibility of haloarchaea-produced poly (3-hydroxybutyrate-*co*-3-hydroxyvalerate) copolymers. Biomaterials 2017;139:172–86.

[77] Danis O, Ogan A, Tatlican P, Attar A, Cakmakci E, Mertoglu B, Birbir M. Preparation of poly (3-hydroxybutyrate-*co*-hydroxyvalerate) films from halophilic archaea and their potential use in drug delivery. Extremophiles 2015;19(2):515–24.

[78] Quillaguaman J, Hashim S, Bento F, Mattiasson B, Hatti-Kaul R. Poly (β-hydroxybutyrate) production by a moderate halophile, *Halomonas boliviensis* LC1 using starch hydrolysate as substrate. J Appl Microbiol 2005;99(1):151–7.

[79] Cervantes-Uc JM, Catzin J, Vargas I, Herrera-Kao W, Moguel F, Ramirez E, Rincón-Arriaga S, Lizama-Uc G. Biosynthesis and characterization of polyhydroxyalkanoates produced by an extreme halophilic bacterium, *Halomonas nitroreducens*, isolated from hypersaline ponds. J Appl Microbiol 2014;117(4):1056–65.

[80] Mothes G, Schubert T, Harms H, Maskow T. Biotechnological coproduction of compatible solutes and polyhydroxyalkanoates using the genus *Halomonas*. Eng Life Sci 2008;8(6):658–62.

[81] Van-Thuoc D, Huu-Phong T, Thi-Binh N, Thi-Tho N, Minh-Lam D, Quillaguaman J. Polyester production by halophilic and halotolerant bacterial strains obtained from mangrove soil samples located in Northern Vietnam. Microbiol open 2012;1(4):395–406.

[82] Kulkarni SO, Kanekar PP, Nilegaonkar SS, Sarnaik SS, Jog JP. Production and characterization of a biodegradable poly (hydroxybutyrate-*co*-hydroxyvalerate)(PHB-*co*-PHV) copolymer by moderately haloalkalitolerant *Halomonas campisalis* MCM B-1027 isolated from Lonar Lake, India. Bioresour Technol 2010;101(24):9765–71.

[83] Shrivastav A, Mishra SK, Shethia B, Pancha I, Jain D, Mishra S. Isolation of promising bacterial strains from soil and marine environment for polyhydroxyalkanoates (PHAs) production utilizing Jatropha biodiesel byproduct. Int J Biol Macromol 2010;47(2):283–7.

[84] Huang TY, Duan KJ, Huang SY, Chen CW. Production of polyhydroxyalkanoates from inexpensive extruded rice bran and starch by *Haloferax mediterranei*. J Ind Microbiol Biotechnol 2006;33(8):701–6.

[85] Koller M, Atlić A, Gonzalez-Garcia Y, Kutschera C, Braunegg G. Polyhydroxyalkanoate (PHA) biosynthesis from whey lactose. In: Macromolecular symposia, vol. 272. Weinheim: Wiley-VCH Verlag; 2008. p. 87–92.

[86] Bhattacharyya A, Saha J, Haldar S, Bhowmic A, Mukhopadhyay UK, Mukherjee J. Production of poly-3-(hydroxybutyrate-*co*-hydroxyvalerate) by *Haloferax mediterranei* using rice-based ethanol stillage with simultaneous recovery and re-use of medium salts. Extremophiles 2014;18(2):463–70.

[87] Bhattacharyya A, Pramanik A, Maji SK, Haldar S, Mukhopadhyay UK, Mukherjee J. Utilization of vinasse for production of poly-3-(hydroxybutyrate-*co*-hydroxyvalerate) by *Haloferax mediterranei*. Amb Express 2012;2(1):34.

[88] Salgaonkar BB, Bragança JM. Utilization of sugarcane bagasse by *Halogeometricum borinquense* strain E3 for biosynthesis of poly (3-hydroxybutyrate-*co*-3-hydroxyvalerate). Bioengineering 2017;4(2):50.

[89] Salgaonkar BB, Mani K, Bragança JM. Sustainable bioconversion of cassava waste to poly (3-hydroxybutyrate-co-3-hydroxyvalerate) by *Halogeometricum borinquense* strain E3. J Polym Environ 2018:1–10.

[90] Hermann-Krauss C, Koller M, Muhr A, Fasl H, Stelzer F, Braunegg G. Archaeal production of polyhydroxyalkanoate (PHA) co-and terpolyesters from biodiesel industry-derived by-products. Archaea 2013;2013.

[91] Koller M, Hesse P, Bona R, Kutschera C, Atlić A, Braunegg G. Potential of various archae- and eubacterial strains as industrial polyhydroxyalkanoate producers from whey. Macromol Biosci 2007;7(2):218–26.

[92] Taran M. Utilization of petrochemical wastewater for the production of poly (3-hydroxybutyrate) by *Haloarcula* sp. IRU1. J Hazard Mater 2011;188(1–3):26–8.

[93] Koller M, Braunegg G. Biomediated production of structurally diverse poly (hydroxyalkanoates) from surplus streams of the animal processing industry. Polimery 2015;60.

[94] Pramanik A, Mitra A, Arumugam M, Bhattacharyya A, Sadhukhan S, Ray A, Haldar S, Mukhopadhyay UK, Mukherjee J. Utilization of vinasse for the production of polyhydroxybutyrate by *Haloarcula marismortui*. Folia Microbiol 2012;57(1):71–9.

[95] Nicolaus B, Lama L, Esposito E, Manca MC, Improta R, Bellitti MR, Duckworth AW, Grant WD, Gambacorta A. *Haloarcula* spp able to biosynthesize exo-and endopolymers. J Ind Microbiol Biotechnol 1999;23(6):489–96.

[96] Rathi DN, Amir HG, Abed RM, Kosugi A, Arai T, Sulaiman O, Hashim R, Sudesh K. Polyhydroxyalkanoate biosynthesis and simplified polymer recovery by a novel moderately halophilic bacterium isolated from hypersaline microbial mats. J Appl Microbiol 2013;114(2):384–95.

[97] Hai T, Hein S, Steinbüchel A. Multiple evidence for widespread and general occurrence of type-III PHA synthases in cyanobacteria and molecular characterization of the PHA synthases from two thermophilic cyanobacteria: *Chlorogloeopsis fritschii* PCC 6912 and *Synechococcus* sp. strain MA19. Microbiology 2001;147(11):3047–60.

[98] Tripathi L, Wu LP, Meng D, Chen J, Chen GQ. Biosynthesis and characterization of diblock copolymer of P (3-hydroxypropionate)-block-P (4-hydroxybutyrate) from recombinant *Escherichia coli*. Biomacromolecules 2013;14(3):862–70.

Chapter 27

Techniques for the mass production of Arbuscular Mycorrhizal fungal species

James Dsouza
St. Xavier College, Mapusa, Goa, India

27.1 Introduction

Arbuscular Mycorrhizal is a "universal symbiosis" and the beneficial effects of these associations are well studied. Arbuscular Mycorrhizal (AM) fungi are obligate symbionts belonging to phylum Glomeromycota, which is currently comprised of around 260 species distributed throughout the world. This symbiosis includes three components, viz., host plant, fungal component, and rhizosphere. They are identified by the presence of unique structures, viz., arbuscules and vesicles. Arbuscules are branched haustoria-like structures within the cortical cells [1], which aid in the transfer of nutrients to the host plant [2]. Vesicles are storage organs differing in shapes ranging from globular to bean shape [3]. Besides arbuscules, some genera such as *Gigaspora* and *Scutellospora* produce auxiliary cells. Other morphological characters used for identification of AM fungi are sporocarp, spore, subtending hyphae, auxiliary cells, spore wall layers, and spore germination [4].

Arbuscular Mycorrhizal fungi are an important component of nutrient management programs that help in the establishment of the plant species. Recent research studies suggest that AM fungi could contribute to plant health and productivity independently by enhancing nutrient uptake. AM fungi absorb various nutrients, such as phosphorus (P), nitrogen (N), potassium (K), calcium (Ca), sulfur (S), manganese (Mn), copper (Cu), and zinc (Zn), from the soil and then translocate these nutrients to the plants. The most consistent and important role is to improve uptake of immobile nutrients such as P, Cu, and Zn [5,6]. Another advantage to associated plants is improved maintenance of a balanced supply of nutrients. The ability of AM fungi to reduce plants' external P requirement has an important environmental benefit [7].

Advances in Biological Science Research. https://doi.org/10.1016/B978-0-12-817497-5.00027-6

Several culture techniques for the mass production of AM fungi are available for the commercial production of the inoculum. The potential of AM fungi as biofertilizers and bioprotectors to enhance plant productivity has been widely recognized but not fully exploited because of the obligate biotrophic nature of AM fungi and other reasons such as the unstable performance of mycorrhizal fungi in plant production systems. Since AM fungi are increasingly considered in agriculture, horticulture, and forestry programs as well as for environmental reclamations, as biofertilizers and bioprotectors, it is necessary to screen and select efficient AM fungal spices. Besides this, selection of suitable host plant species and standardization of various parameters are indispensable for the optimization of the AM fungal inoculums. There is a need to develop low-cost technologies for intensifying the production of these eco-friendly and economically important microorganisms. In this chapter, various strategies are discussed that need to be looked on for maximizing the AM fungal inoculum production.

27.2 Pot/substrate-based mass production system

This method is also known as the traditional method for production of AM fungal species. Here the symbiotic partners are cultured by using various substrates. Pure or mixed substrates are used for mass production of selected, sterilized AM fungal spores (Fig 27.1). Different substrates used are pure sand peat [8], glass beads [9,10], vermiculite [11], perlite, and compost [12]; besides optimizing the mass production of AM fungal species, substrates are amended with different materials and inert substrates as a carrier medium. Pretreatment of the substrates and autoclaving is necessary to minimize the contamination of undesirable soil microorganisms. This full assembly of setup is maintained in carefully controlled conditions (light, temperature,

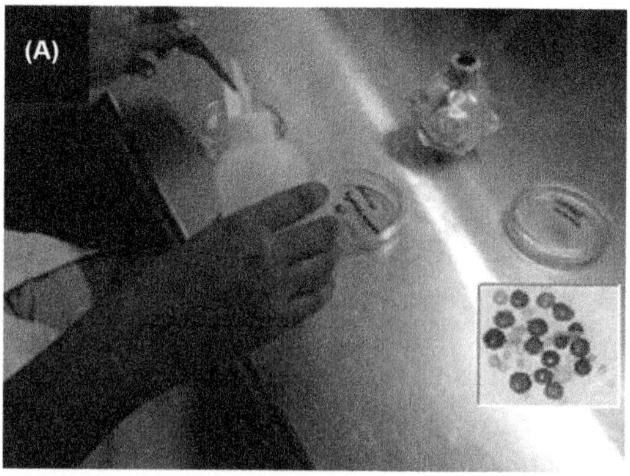

FIGURE 27.1 Arbuscular Mycorrhizal fungal spore sterilization.

humidity, etc). Depending on the host plant species, controlled conditions are varied to suit the other symbiont (host plant). Care should be taken to select appropriate particle size of the substrate, which is necessary for drainage, humidity, and aeration that influence AM fungal sporulation [13,14].

27.3 The AM host plants

The selection of the host plant used to propagate AM fungal species in pot cultures influences AM fungal sporulation and higher inoculum levels [15]. Basic considerations for choosing the host plants requires the ability to tolerate the growing conditions that must suit the selected AM fungal species. Selected host plant species should have a short life cycle, adequate root system development, colonization ability by the large number of AM fungi, and tolerance to relatively low levels of phosphorus. Other relevant characteristics are low susceptibility to pathogens and a wide range of temperature tolerance [15], which are important factors for optimization of inoculum production (Fig 27.2).

The type of AM fungal inoculum (spore or mixed inoculum) that producers aim to promote partly determines the host plant/fungus association chosen. The choice of the host plant may influence the colonization levels of some AM fungal species [16] and possibly also impact intensity of AM fungal sporulation due to its varied nutrient requirements. Dsouza et al. 2017 (unpublished) in their observed *Eleusine coracana* were best suited for mass production of selected AM fungal species. Also, the higher number of entry points for AM fungi was observed, which aids in enhancing root colonization in the well-controlled culture conditions. Previous studies by Guar & Adholeya, 2002; Dodd et al 1999 [17,18] carried out mass multiplication of AM fungi species using the different host plant species, however, lower spore density was reported compared to the present study, suggesting *E. coracana* was the best host for mass multiplication of *Acaulospora* species.

FIGURE 27.2 Plant as inoculum.

FIGURE 27.3 Monospecific culture.

27.4 Root trap cultures

Requirements—Root samples, sterilized sand: soil (1:1), pots, alcohol, trypan blue, absorbent cotton, Hoagland's solution, zip-lock polythene bags.

Procedure—Roots of selected plant species are collected from field conditions, cut into pieces (1 cm), and placed at a depth of 5–8 cm in sterilized sand:soil (1:1) mix in 10–20 cm diameter. Pots are thoroughly prewiped with cotton dipped in absolute alcohol. Surface sterilized selected host plant species cuttings (7 cm tall) are planted in the pots. Depending on host plant species, 28–45 days of plant growth, the roots are assessed for AM fungal colonization. The plants are maintained for a period of 58–190 days to establish AM fungal colonization and sporulation. Then watering is stopped, allowing the plants to dry, after which the shoot is cut off at the soil surface. Later on, the roots are cut into fine pieces and mixed with the soil. The mixture is then placed in a zip-lock polythene bag, labeled, and stored at 4°C. This was used as inoculum for the preparation of monospecific AM fungal cultures (Fig. 27.3).

27.5 Plant trap cultures

Requirements—Herbaceous plants or seedlings of shrubs or trees, sterilized sand:soil (1:1), pots, alcohol, trypan blue, absorbent cotton, Hoagland's solution, zip-lock polythene bags.

Procedure—Small herbaceous plants or seedlings of shrubs or trees are carefully uprooted from their natural habitat. Fine roots are thoroughly washed of adhering soil before planting in the sterilized sand:soil (1:1) mix using 20 cm diam. After 35–70 days of growth, AM fungal root colonization is checked by staining the sample with 0.05% trypan blue. The plants are maintained for a period of 50–150 days (depending on the type of selected AM fungal species) to established colonization and optimized AM fungal sporulation.

27.6 Soil as inoculum

Rhizosphere soil from the root zone of a plant hosting AM fungi was used as soil inoculum. Soil may not be a reliable inoculum unless one knows about the abundance, diversity, and activity of the indigenous AM fungal species. The effectiveness of the AM fungal inoculum, relative to that of a crude inoculum, of surface soils, depends on various factors such as sporulation timing, stage of the plant, and functionality of the inoculum. To maximize the AM fungal mass multiplication, one should make use of the inoculum that suits both the symbiotic partners rather than generalizing and using only soil inoculum. It may include soil, dried root fragments, AM fungi spores, sporocarps, and hyphae.

27.7 Microenvironment

To enhance AM mass production special microenvironment was designed. Here simple polythene bag was used to cover the potted plant. On one side of the polythene bag at the upper half, a cut approximately 5–6 cm in length is made. The open end of polythene bag is sealed. Specific conditions are provided (temperature 25°C) with reduced watering (only once in a week). Since potted cultures are covered with polythene bags, water droplets accumulate that are sufficient to perform the daily process of photosynthesis, thereby reducing the water intake of the host plant. This is maintained in polyhouse conditions. In the present study, *E. coracana* was used as a host. Maximum spore density and AM colonization were recorded in plants subjected to microenvironment as compared to the plants without microenvironment. Maximum AM fugal colonization was observed in plants subjected to microenvironment than the plants grown without microenvironment. Spore density was also maximum in plants grown in microenvironment (Fig. 27.4).

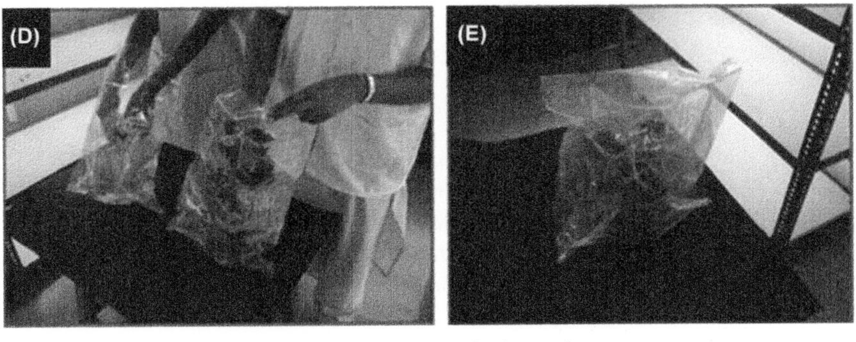

FIGURE 27.4 Preparation of microenvironment.

27.8 Conclusion

Arbuscular mycorrhizal inocula are produced under polyhouse conditions using open pot culture method and the sand as a substrate using single AM fungal species. More studies are needed to test the inoculum potential of the multiple AM fungal species in the same experimental setup. Apart from this, most of the studies are using only spores as inoculum/propagule, which might be not sufficient. To ensure mass production of the quality AM fungal inoculum, priority should be given to screening of the AM fungal species from varying habitats before being used for mass production techniques. Although advanced techniques like autotrophic culture system, other in vitro systems are available, the above-mentioned low-cost techniques for mass production are needed. This is the reason why the potential of AM fungal species as bioprotectors, biofertilizers to increase plant productivity is not exploited. There is a need to identify potential AM fungal species with wider tolerance range to adapt the specific habitat. Secondly, most AM fungal mass production techniques use only spores as inocula. Also, there is a need to use other sources of AM inocula such as vesicles, hyphae, and root fragments to maximize AM fungal production. Selection of a specific inoculum of selected AM fungal species may also lead to using various other forms of inocula and a higher number of AM fungal species as biofertilizers.

References

[1] Smith SE, Read DJ. Mycorrhizal symbiosis. 3rd ed. London: Academic; 2008.

[2] Akhtar MS, Panwar J. Arbuscular mycorrhizal fungi and opportunistic fungi: efficient root symbionts for the management of plant parasitic nemathods. Adv Sci Eng Med 2011;3:165–75.

[3] Smith SE, Read DJ. Mycorrhizal symbiosis. 2nd ed. San Diego: Academic Press; 1997.

[4] Muthukumar T, Radhika KP, Vaingankar J, D'Souza J, Dessai S, Rodrigues BF. Taxonomy of AM fungi – an update. In: Rodrigues BF, Muthukumar T, editors. Arbuscular Mycorrhizae of Goa: a manual of identification protocols, vol 79. India: Goa University Publication; 2009. p. 115.

[5] Manjunath A. Habte M the development of vesicular-arbuscular mycorrhizal infection and the uptake of immobile nutrients in *Leucaena leucocephala*. Plant Soil 1988;106:97–103.

[6] Pacovsky RS. Micronutrient uptake and distribution in mycorrhizal or phosphorus-fertilized soybeans. Plant Soil 1986;95:379–88.

[7] Akhtar MS, Siddiqui ZA. Arbuscular mycorrhizal fungi as potential bioprotectants against plant pathogen. In: Siddiqui ZA, Aktar MS, Futai K, editors. Mycorrhizae: sustainable agriculture and forestry. Dordrecht, The Netherlands: Springer; 2008. p. 61–98.

[8] Ma N, Yokoyama K, Marumoto T. Effect of peat on mycorrhizal colonization and effectiveness of the arbuscular mycorrhizal fungus *Gigaspora margarita*. Soil Sci Plant Nutr 2007;53:744–52. https://doi.org/10.1111/j.1747-0765.2007.00204.x.

[9] Lee YJ, George E. Development of a nutrient film technique culture system for arbuscular mycorrhizal plants. Horticulture Science 2005;40:378–80.

[10] Neumann E, George E. Extraction of extraradical arbuscular mycorrhizal mycelium from compartments filled with soil and glass beads. Mycorrhiza 2005;15:533–7. https://doi.org/10.1007/s00572-0050361-6.

[11] Douds Jr DD, Nagahashi G, Pfeffer E, Reider C, Kayser WM. On-farm production of AM fungus inoculum in mixtures of compost and vermiculite. Bioresour Technol 2006;97:809–18. https://doi.org/10.1016/j.biortech. 2005.04.015.

[12] Douds Jr DD, Nagahashi G, Pfeffer PE, Kayser WM, Reider C. On-farm production and utilization of arbuscular mycorrhizal fungus inoculum. Can J Plant Sci 2005;85:15–21. https://doi.org/10.4141/P03-168.

[13] Gaur A, Adholeya A. Effects of the particle size of soil-less substrates upon AM fungus inoculum production. Mycorrhiza 2000;10:43–8. https://doi.org/10.1007/s005720050286.

[14] Millner PD, Kitt DG. The Beltsville method for soilless production of vesicular– arbuscular mycorrhizal fungi. Mycorrhiza 1992;2:9–15. https://doi.org/10.1007/BF00206278.

[15] Struble S, Skipper HD. Vesicular–arbuscular mycorrhizal fungal spore production as influenced by plant species. Plant Soil 1988;109:277–80.

[16] Hawkins HJ, George E. Hydroponic culture of the mycorrhizal fungal Glomus mosses with *Linum usitatissimum* L., *Sorghum bicolor* L. and *Triticum aestivum* L. Plant Soil 1997;196:143–9. https://doi.org/10.1023/A:1004271417469.

[17] Gaur A, Adholeya A. Arbuscular-mycorrhizal inoculation of five tropical fodder crops and inoculum production in marginal soil amended with organic matter. Biol Fertil Soils 2002;35:214–8. https://doi.org/10.1007/s00374-002-0457-5.

[18] Dodd JC, Arias I, Koomen I, Hayman DS. The management of populations of vesicular–arbuscular mycorrhizal fungi in acid infertile soils of savanna ecosystem. I. The effect of pre-cropping and inoculation with VAM-fungi on plant growth and nutrition in the field. Plant Soil 1990;122:229–40.

Chapter 28

Metagenomics: a gateway to drug discovery

Flory Pereira
PES's Ravi Sitaram Naik College of Arts and Science, Department of Microbiology, Ponda, Goa, India

28.1 Introduction

One of the most exciting milestones in the history of medicine has been the discovery of antibiotics. Since this breakthrough in 1928 by Sir Alexander Fleming, antibiotics have become an indispensable tool in the treatment of bacterial infections [1]. The majority of modern drugs have been isolated from natural sources, since nature presents the best and most abundant reservoir, harboring a plethora of microorganisms. However, in spite of this, microorganisms have remained largely untapped because of the constraints of traditional isolation and cultivation, resulting in the "great plate count anomaly" [2], where less than 0.1% of the microorganisms observed in different habitats have been cultivated, and about 99.9% of microbial species still remain uncultivated [3]. Advances in isolation and cultivation techniques have helped marginally in improving the bioprospecting potential of novel bioactive compounds. Over the years even these antibiotics have been rendered useless, because their overuse and misuse has resulted in the emergence of pathogens showing global antibiotic resistance to almost all classes of antibiotics. In the wake of growing antibiotic resistance, coupled with the slow pace of discovery of new antibiotic producers, there is a growing demand for novel approaches for the mining of the uncultivated biodiversity existing in challenging environments and competitive communities, as a potential source of biologically active compounds. In the last 2 decades, with the inception of culture-independent techniques, the extraction of nucleic acids directly from environmental samples has undergone a revolutionary change, opening new avenues for exploring the vast untapped reservoir of genetic and metabolic diversity. A two-pronged approach of either improving the sensitivity of detection methods or the use of metagenomics for evaluating unexplored strains could be the solution. Herein, we review the various methods available for the study of environmental DNA, with examples of industrially relevant antibiotics discovered using metagenomes.

Advances in Biological Science Research. https://doi.org/10.1016/B978-0-12-817497-5.00028-8

28.2 Approaches to accelerate antibiotic discovery

28.2.1 Mining unusual habitats as a source of novel secondary metabolites

Over the years the interest in the search for new metabolites from traditional sources has declined and there is a renewed interest in mining microorganisms, especially those associated with algae and invertebrates such as mollusks, sponges, tunicates, coelenterates, fish, and crustaceans. Bioprospecting bacteria from extreme environments, or which have a symbiotic or endophytic relationship with higher organisms, poses a challenge because of the low density of organisms, slow cell growth, complexity in sampling due to restricted access, and problems in situ analyses. Soil, deserts, marine habitats, hot springs, deep ocean tube worms, sponges, mangroves and coastal microbial mat ecosystems are recognized as high biodiversity hotspots [4,5]. Cragg and Newman [6] reported that the majority of antibiotics and cytotoxic compounds used in the treatment of cancers were obtained from soil-dwelling bacteria. The human body is another promising, underexplored habitat. Metagenomic methods provide a means for the direct extraction and sequencing of DNA from environmental samples, but screening for novel antibiotics from such uncultivable bacteria still remains a challenge because of the genomic complexity of most metagenomes. This can however be overcome simply by surveying for the cluster of antibiotic biosynthesis genes for polyketide synthase (PKS) and nonribosomal peptide synthetase (NRPS) enzymes, which are the hallmark of antibiotic producers [6].

Some successful examples where metagenomics has been used for the isolation of novel drug producers are *Salinispora* isolated from marine sediments of Guam, Palau, and from the Red Sea [7]; an unusual nasal-colonizing organism, *Staphylococcus lugdunensis*, a producer of a novel peptide antibiotic [8]; and nonactinomycetal bacteria *Clostridium beijerinckii*, which produces the pentacyclic polyphenol clostrubin, effective against diverse pathogenic bacteria [9,10].

28.2.2 Revolutionary cultivation techniques

One major obstacle in mining for novel species is that many bacteria cannot be cultured under conventional laboratory conditions. However, new and revolutionary cultivation techniques have emerged that now allow the growth of a broad range of bacteria that so far have been uncultivable. Some of these techniques are discussed next.

28.2.2.1 High-throughput cultivation of microorganisms using microcapsules technique

Here single cells are encapsulated in microcapsules and cultivated under low nutrient flux conditions to form microcolonies, which are detected using flow cytometry. Individual microcolonies are then grown in microtiter plates containing organic-rich medium. This technique shows a high efficiency and can cultivate in excess of 10,000 bacterial and fungal isolates per environmental sample [11].

28.2.2.2 Microfluidic bioreactor cultivation

Combines the small volumes of microtiter plates with the elaborate control mechanisms of a normal bench-type bioreactor with integrated sensor. Zang et al. [12] reported successful growth of up to 600,000 pure soil-derived *Actinobacteria* in microfluidic droplets per hour. This rapid high-throughput, cost-effective technique particularly caters to the slow-growing strains with a potential to produce promising natural compounds under normal conditions, which otherwise would be outcompeted by fast growers, e.g., protein synthesis inhibitor orthoformimycin [13].

28.2.2.3 Diffusion chamber in situ cultivation

This technique uses a membrane-enclosed chamber placed in its simulated native environment, but in a laboratory setting. It permits only the passage of nutrients and growth factors across the semipermeable membrane to the bacteria. Kaeberlein et al. [14] grew up to 40% of species from marine biofilms as pure culture microcolonies.

28.2.2.4 The "isolation chip" or "ichip"

The ichip consists of a multichannel device composed of several hundred miniature diffusion chambers, each inoculated with a single environmental cell and placed back in its natural environment. This leads to a 5- to 300-fold increased microbial recovery of uncultivable organisms effectively cultivated under laboratory conditions [15,16]. Using this method, Ling et al. [17] could obtain teixobactin from a novel species of β-proteobacteria, *Eleftheria terrae*.

28.2.2.5 Hollow-fiber membrane chamber

This chamber, developed by Aoi et al. [18], is similar to the ichip, except that it utilizes fibers, which are inoculated with microorganisms and placed in a natural or engineered environment, allowing the exchange of metabolites with the environment.

28.2.2.6 I-TIP

The I-TIP is based on micropipette tips used in the labs, which were utilized to cultivate unculturable bacteria associated with sponges [19].

28.2.2.7 Co-culture technique

Genomic studies have shown that a major portion of the genome of some bacteria is related to biosynthesis pathways that cannot be expressed under currently known conditions, but if two cultures are mixed together to promote natural competition, it stimulates antibiotic production [20]. Hence, the ichip and co-culture technique can be combined into a single step [21]. Thus innovative culturing techniques for discovery of new antibiotics have facilitated a new era in antimicrobial discovery from underexplored environments.

28.2.3 Next-generation sequencing techniques in mining for bioactive compounds

Next-generation sequencing uses shorter reads of 50–400 base pairs, which allows for quicker sequencing of multiple specimens [22] from the environment using single-cell sequencing, target sequencing, whole-genome shotgun sequencing, or metagenomic sequencing technologies.

28.2.3.1 Single-cell genome sequencing

Single-cell genome sequencing determines the sequence of target bacteria at single cell levels (Fig. 28.1). Single cells are first isolated from environmental samples using flow cytometry, serial dilution, micromanipulation, or microfluidics. DNA is extracted, a gene library constructed, amplification of 16S rRNA gene is done via polymerase chain reaction (PCR) and multiple displacement amplification (MDA), followed by sequencing and data analysis. The advantage of this technique over metagenomic sequencing is that it can generate a high-quality genome for species with low abundance in relation to their function. The disadvantage is that the isolation of single cells can be expensive, time consuming, and prone to contamination, besides producing increased chimeric reads. Using single-cell genome sequencing, Wilson et al. [27] isolated DNA from *Entotheonella*, a coinhabitant of marine sponge *Theonella swinhoei*. They showed that the organism was a vast source of polyketides and peptides.

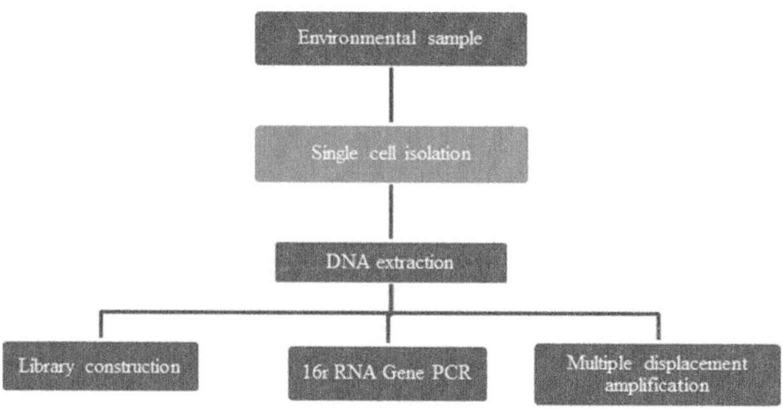

FIGURE 28.1 Single-cell genome sequencing.

28.2.3.2 Target sequencing or amplicon sequencing

Target sequencing or amplicon sequencing is useful in DNA studies because it permits the 16S rRNA gene, which is a specific marker gene of microbes, to be amplified via PCR and then sequenced by next-gen sequencing platforms (Fig. 28.2). Analysis and comparison is done using 16S rRNA database such as green genes [23], SILVA [24], or RDP [25]. However, the resolution is not enough for identification at species or strain level and the functions of these microbes also cannot be directly determined [26].

28.2.3.3 Whole-genome shotgun sequencing

Whole-genome shotgun sequencing bypasses the time-consuming mapping and cloning steps that make clone-by-clone sequencing slow. The whole genome is broken into fragments varying in size from 2 to 300 kb, which are sequenced to determine the order of bases, then assembled together using computer programs by finding regions of overlap (Fig. 28.2). The advantage of

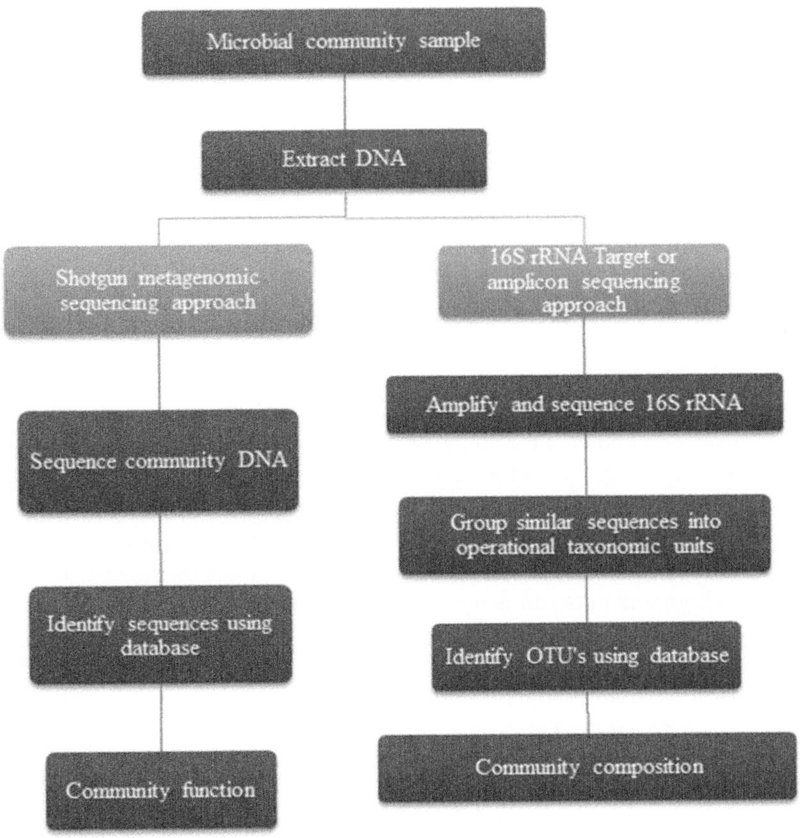

FIGURE 28.2 Target or amplicon sequencing and shotgun metagenomics approaches.

shot gun sequencing is that it requires only a fraction of DNA required in clone-by-clone sequencing and does not need the mapping stage. The disadvantage of this method is that in the absence of a reference genetic map, it becomes difficult to assemble and match the genome. Shotgun metagenomic sequencing finds use in the identification of antibiotic resistance genes (AMRs) and could even play a role in surveillance of metagenomes of several ecological niches for AMR genes such as the human gastrointestinal tract, water, animals, and others [28].

28.3 Metagenomic or environmental or community genomic sequencing

Metagenomic sequencing offers a powerful, culture-independent tool for discovery of new antibiotics because it relies on whole-genome sequencing of all microbes in the environment (Fig. 28.3). The first sequences of environmental samples generated with next-generation sequencing technique were published by Robert Edwards and colleagues with the Roche 454 pyrosequencing instrument [29]. Metagenomics permits both the discovery of genes as well as elucidates the biosynthetic pathways of both cultivable and uncultivable organisms. Either sequence based or function based can be used [30–33].

28.3.1 Sequence-based metagenomics

This involves screening of DNA from environmental samples for genes of interest, by hybridization with labeled DNA probes or by PCR [34]. The PCR primers are designed based on conserved DNA sequences of known genes. Clusters of PCR products called polymerase colonies or "polonies" are derived from a single molecule of nucleic acid. Enzymatic reactions done in parallel on all the polonies help to sequence the DNA using the sequencing machines Illumina GAII and Roche 454 [35]. Assembling of overlapping short sequences is done using computer algorithms, into longer contiguous sequences called contigs. Contigs can be combined into large scaffolds to build complete genomes. Sequence-based metagenomics studies can be used for genome assembly that assists in the understanding of the role of genes in microbial adaptation to a particular environment; to identify genes of interest; to study variation within and between genomes; to find complete biosynthetic pathways from eDNA cosmid libraries; and to compare the numbers and degree of diversity of organisms in different communities.

28.3.2 Function-based metagenomics

This is used to identify biosynthetically active clones by looking for gene products in a heterologous host. The procedure involves a simple extraction of DNA from any sample, amplification of DNA by PCR, sequencing of the DNA

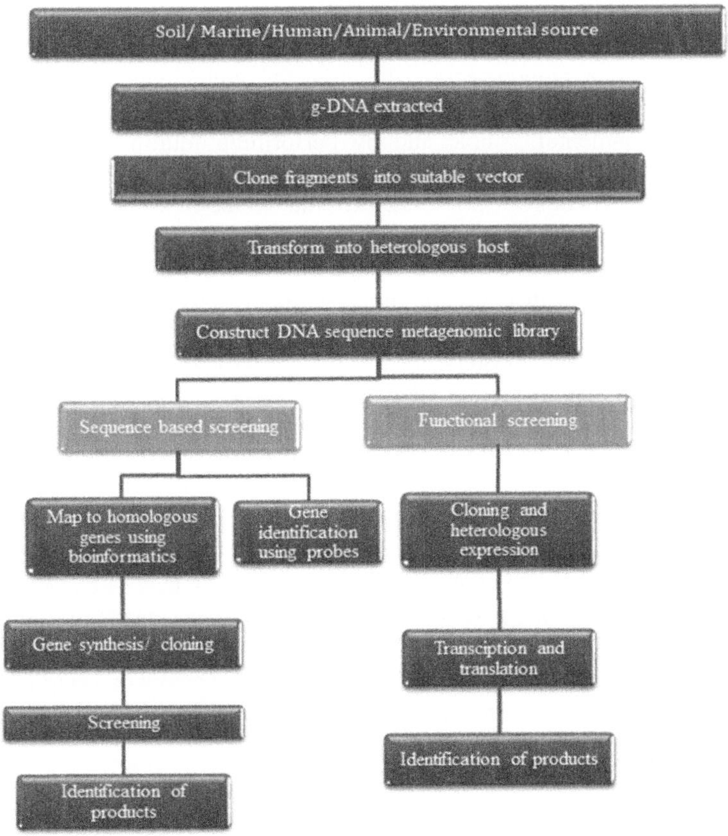

FIGURE 28.3 Sequence-based and functional metagenomics.

to determine the location of the genes, and annotating the reads to a known functional gene database such as NR, KEGG [36], or eggNOG [37] to determine the functional potential of the genes [38,39]. The metagenomics library is cloned into vector DNA and transformed into a suitable host. The obtained clones are screened for the sequence of interest [40]. The drawback of this approach is that since functional annotation relies on previously identified genes and their functions, completely novel sequences may not be annotated correctly or may be totally ignored due to lack of known homologs [38,39,41].

Function-based metagenomics has made it possible to identify novel antibiotics, as well as proteins involved in antibiotic resistance. Since genes are capable of horizontal transfer, the genes for antibiotic synthesis, regulation, and resistance are clustered together on bacterial chromosomes. The genetic clusters often need manipulation to activate production of small molecule in a heterologous host. This fact makes functional metagenomic sequencing a particularly appealing tool in the search for novel gene classes that encode previously known or unknown functions [42–44]. Ekkers et al. [45] showed that the chances of bioprospecting the genes of interest can be increased by reducing the complexity of community by selecting a defined source rather than an environmental sample and by performing multiple extractions to obtain sufficient quantities of DNA [34]. Also, since gene clusters can vary in size from between ca. 10–15 to >120 kb, the choice of vectors is crucial for successful identification of activity [46]. It is necessary to choose a cloning vector that can accommodate large fragments of DNA, so as to capture the majority of pathways in one contiguous segment of DNA. Plasmids have a small insert size (<15 kb) and so are suitable for single genes and not whole operons. For large gene clusters, fosmid (25–35 kb)/cosmid (25–40 kb), and bacterial artificial chromosomes (BACs, 100–200 kb) are more suitable [45]. Also the level of gene expression in the laboratory may be negligible. Choice of expression host will influence expression pattern of the gene product. Aakvik et al. [34] reported *Escherichia coli* as the host of choice, whereas Angelov et al. [47] were successful in transferring a gene library from *E. coli* to *Thermus thermophilus* using a two-host fosmid system, and showed the magnitude of variations in the expression patterns of these two organisms. Hain et al. [48] developed BAC shuttle vectors that were capable of replicating in both Gram-negative as well as Gram-positive bacteria. In libraries made from environments with mixed communities, shuttle vectors carrying multiple origins of replication are particularly suitable for library expression in multiple hosts [49].

28.4 How metagenomics facilitates drug discovery

Since functional metagenomics does not require prior knowledge about genes or their hosts, it is a potent tool for discovery of useful gene products (Table 28.1) from the 99% [69] of the uncultured microbial world. In addition, it is less time consuming, more cost-effective, and facilitates the detection of silent gene clusters that are actually biosynthetically functional but not expressed under standard laboratory conditions. For example, stambomycin production could only be seen after overexpressing a pathway-specific regulator [70]. Among all environmental samples, soil is estimated to have the most extensive genetic diversity, holding up to 10^5 unique species per gram [71], making it the most promising source for antibiotic discoveries. Hemala et al. [72] suggested that prospecting sites that are contaminated and under extremes of environmental conditions hold promise as sources of antibiotics due to selective pressure. Proteobacteria, Bacteroidetes, Firmicutes, and Cyanobacteria are some of the dominant antibiotic-producing classes reported from such sites. The occurrence of resistance genes has been detected by functional metagenomics even in pristine regions such as Alaskan soils [73]. Knowledge of molecular mechanisms of antibiotic resistance in microorganisms helps to design strategies for

TABLE 28.1 Bioactive compounds discovered using metagenomics.

Antibiotic	DNA source	Heterologous host	Screening method	Reference
Fluoroquinolone	Polluted aquatic environment	Plasmid in *Escherichia coli*	Sequence-based	[50]
Malacidin	Soil	*Streptomyces albus*	Next-gen sequencing	[51]
Salinosporamide A	Marine sediment	*Salinospora*	Conventional culturing	[7]
Beta-lactamases	Soil	Fosmid		[52]
Fasamycin A and B	Soil	Cosmid		[53]
Terragine	Soil	Fosmid		[54]
Indirubin	Soil	Cosmid		[55]
Turbomycins A and B	Soil	BAC		[56]
Violacein	Soil	Cosmid		[57]
Onnamide A	Bacterial symbiont *Theonella swinhoei*		Sequence-based	[58]
Bryostatin	Bacterial symbiont *Bugula neritina*		Sequence-based	[59]
Minimide	*Didemnum molle*		Sequence-based	[60]
ApratoxinA	*Lyngbya bouillonii*		Sequence-based	[61]
Patellamides	*Prochloron didemni*		Functional	[62]
Palmitoylputrescine	Bromeliad tank water		Functional	[63]
Antibiotic resistance genes	*Acinetobacter baumannii, Staphylococcus aureus, Enterococcus faecium, Saccharomyces cerevisiae, Candida glabrata* and *Mycobacterium*		Sequence-based	[64]
Borregomycin A, B–D	Desert sand soil	*Streptomyces albus*	Sequence-based	[65]
Fluostatins (F, G, and H)	Desert sand soil	Cosmid clones using TAR in yeast	Sequence-based	[66]

Continued

TABLE 28.1 Bioactive compounds discovered using metagenomics.—cont'd

Antibiotic	DNA source	Heterologous host	Screening method	Reference
Erdacin	Desert sand soil	Cosmid in *Streptomyces albus*		[67]
Indigoidine production by a 4′-phosphopantetheinyl transferase (PPTase)	Garden soil	*Streptomyces lavendulae*	Functional	[68]

restoring potency and shelf life of the drug [74]. Further, tracking the presence, distribution, and abundance of antibiotic resistance genes in bacterial populations in different ecosystems could be used to determine which might be easily transferred to human pathogens in drug development programs [75].

A popular approach described by Brady and Clardy [76] is the top agar overlay assay using *E. coli* as the host to screen libraries for clones showing antimicrobial activity. Using this method they constructed a cosmid library of eDNA from soil and isolated and identified for the first time the genes for biosynthesis of *N*-acyl-1-tyrosine antibiotic, a new natural long-chain bacterial product, and also the genes for synthesis of the broad-spectrum antibiotic violacein. Lim et al. [54] constructed forest soil metagenomic libraries in fosmids using *Bacillus subtilis* as the target organism and demonstrated that forest soil microbial genes produced a pigment called indirubin that has antimicrobial activity. Using shotgun cloning into *E. coli* of genomic DNA from *Prochloron* sp., a symbiont of the sea squirt *Lissoclinum patella*, Long et al. [62] succeeded in the heterologous expression of the cyclic peptide patellamide. Schipper et al. [77] fused acyl-homoserine lactone synthase promoter to a *lacZ* reporter and were able to identify three eDNA-derived AHL lactonases capable of preventing biofilm formation by *Pseudomonas aeruginosa*. Schmitz et al. [78] discovered several anti-*Bacillus anthracis* lysins cloned from DNA of bacteriophage from bat guano and earthworm gut. In 2014, lysostaphin was discovered by screening a Staphylococcal-derived metagenomic library [79]. Craig et al. [80] found that cosmids conferred the production of novel antibiotics in β-Proteobacterium *Ralstonia metallidurans* but did not work in *E. coli*. Patellamide, a cytotoxic cyclic peptide, and pederin, a novel anticancer agent, were isolated by this new system from a cosmid library constructed using beetle-derived metagenomic DNA. Pederin and its relatives onnamide and psymberin were reported from symbionts of sponges by Piel et al. [58].

The *Streptomyces* symbiont of leaf-cutting ants (*Acromyrmex*) produces the potent antifungal candicidin, which is active against the pathogenic fungus *Escovopsis* [81]. Banik et al. [82] identified two new biosynthetic gene clusters screened and found to be associated with the biosynthesis of teicoplanin- and vancomycin-like glycopeptide antibiotics from a soil DNA cosmid library.

Donia et al. [83] identified more than 14,000 biosynthetic gene clusters, elucidating 3118 small molecule from the genomes of human microbiome, indicating that the human microbiome harbors diverse bioactive compounds such as the thiopeptide antibiotic lactocillin, isolated from the vaginal microbiota, with potent antibacterial activity against many Gram-positive vaginal pathogens.

Enzymes PKS and NRPS are synthesizers of bioactive secondary metabolites such as antibiotics [84,85]. Gillespie et al. [56] reported the broad-spectrum antibacterial turbomycin A and B after direct screening of clones displayed antimicrobial activity, and by screening for polyketides, the broad class of antibiotics that includes epithilone, erythromycin, and rifamycin. Polyketides were first cloned from soil by Seow et al. [86] by designing primers to hybridize with the highly conserved region of polyketide synthase genes and amplifying novel polyketide synthase homologs using PCR directly from soil.

28.5 Conclusion

Metagenomics has opened avenues in bioprospecting the genetic potential of microorganisms by offering access into a world of unexplored microbial diversity from unusual environments. The advent of next-generation sequencing technologies and advances in computational tools has transgressed the traditional barriers to drug discovery. Developments in structural and functional metagenomics have paved the way to discover novel genes and metabolic pathways for obtaining bioactive compounds with better properties and given a ray of hope for addressing the problem of drug resistance. However, the success of these techniques demands the development of approaches like metatranscriptomics and metaphenomics to add to the current status of protein expression and metabolic activity of novel bioactive compounds. Within the next few years, we will see the application of metagenomics or sequence-based technologies for genome-inspired personalized medicine.

Conflict of interests

The authors declare that there are no conflicts of interest regarding the publication of this paper.

References

[1] Mandell GL, Sande MA, Goodman GA, Goodman LS, Gilman A. In: Goodman and Gilman's. The pharmaceutical basis of therapeutics. 8th ed. New York: MacMillan Publishing Company Inc; 1991. p. 1065–98.

[2] Staley JT, Konopka A. Measurement of in situ activities of nonphotosynthetic micro-organisms in aquatic and terrestrial habitats. Annu Rev Microbiol 1985;39:321–46.

[3] Amann RI, Ludwig W, Schleifer KH. Phylogenetic identification and in situ detection of individual microbial cells without cultivation. Microbiol Rev 1995;59:143–69.

[4] Kennedy J, Marchesi JR, Dobson AD. Marine metagenomics: strategies for the discovery of novel enzymes with biotechnological applications from marine environments. Microb Cell Factories 2008;7:27.

[5] Bolhuis H, Cretoiu MS, Stal LJ. Molecular ecology of microbial mats. FEMS Microbiol Ecol 2014;90:335–50.

[6] Cragg GM, Newman DJ. Natural products: a continuing source of novel drug leads. Biochim Biophys Acta 2013;1830:3670–95.

[7] Jensen PR, Moore BS, Fenical W. The marine actinomycete genus Salinispora: a model organism for secondary metabolite discovery. Nat Prod Rep 2015;32:738–51.

[8] Wohlleben W, Mast Y, Stegmann E, Ziemert N. Antibiotic drug discovery. Microb Biotechnol 2016;9(5):541–8.

[9] Pidot S, Ishida K, Cyrulies M, Hertweck C. Discovery of clostrubin, an exceptional poly-phenolic polyketide antibiotic from a strictly anaerobic bacterium. Angew Chem Int Ed Engl 2014;53:7856–9.

[10] Shabuer G, Ishida K, Pidot SJ, Roth M, Dahse HM, Hertweck C. Plant pathogenic anaerobic bacteria use aromatic polyketides to access aerobic territory. Science 2015;350:670–4.

[11] Zengler K, Walcher M, Clark G, Haller I, Toledo G, Holland T, et al. High-throughput cultivation of microorganisms using microcapsules. Methods Enzymol 2005;397:124–30.

[12] Zang E, Brandes S, Tovar M, Martin K, Mech F, Horbert P, et al. Real-time image processing for label-free enrichment of Actinobacteria cultivated in picolitre droplets. Lab Chip 2013;13:3707–13.

[13] Monciardini P, Iorio M, Maffioli S, Sosio M, Donadio S. Discovering new bioactive molecules from microbial sources. Microb Biotechnol 2014;7:209–20.

[14] Kaeberlein T, Lewis KSS, Epstein SS. Isolating "uncultivable" microorganisms in pure culture in a simulated natural environment. Science 2002;296:1127–9.

[15] Berdy B, Spoering AL, Ling LL. Epstein SS in situ cultivation of previously uncultivable microorganisms using the ichip. Nat Protoc 2017;12:2232–42.

[16] Nichols D, Cahoon N, Trakhtenberg EM, Pham L, Mehta A, Belanger A, et al. Use of ichip for high-throughput *in situ* cultivation of "uncultivable" microbial species. Appl Environ Microbiol 2010;76:2445–50.

[17] Ling LL, Schneider T, Peoples AJ, Spoering AL, Engels I, Conlon BP, et al. A new antibiotic kills pathogens without detectable resistance. Nature 2015;517:455–9.

[18] Aoi Y, Kinoshita T, Hata T, Ohta H, Obokata H, Tsuneda S. Hollow-fiber membrane chamber as a device for in situ environmental cultivation. Appl Environ Microbiol 2009;75(11):3826–33.

[19] Jung D, Seo EY, Epstein SS, Joung Y, Han J, Parfenova VV, et al. Application of a new cultivation technology, I-tip, for studying microbial diversity in freshwater sponges of Lake Baikal, Russia. FEMS Microbiol Ecol 2014;90(2):417–23.

[20] Ueda K, Beppu T. Antibiotics in microbial coculture. J Antibiot (Tokyo) 2017;70(4):361–5.

[21] Lodhi AF, Zhang Y, Adil M, Deng Y. Antibiotic discovery: combining isolation chip (iChip) technology and co-culture technique. Appl Microbiol Biotechnol 2018;102(17):7333–41.

[22] Masser DR, Stanford DR, Freeman WM. Targeted DNA methylation analysis by next-generation sequencing. J Vis Exp 2015;96:52488.

[23] DeSantis TZ, Hugenholtz P, Larsen N, Rojas M, Brodie EL, Keller K, et al. Greengenes, a chimera-checked 16S rRNA gene database and workbench compatible with ARB. Appl Environ Microbiol 2006;72:5069–72.

[24] Quast C, Pruesse E, Yilmaz P, Gerken J, Schweer T, Yarza P, et al. The SILVA ribosomal RNA gene database project: improved data processing and web-based tools. Nucleic Acids Res 2013;41:D590–6.

[25] Cole JR, Wang Q, Fish JA, Chai B, McGarrell DM, Sun Y, et al. Ribosomal Database Project: data and tools for high throughput rRNA analysis. Nucleic Acids Res 2013;42:D633–42.

[26] Xu Y, Zhao F. Single-cell metagenomics: challenges and applications. Protein Cell 2018;9(5):501–10.

[27] Wilson MC, Mori T, Rückert C, Uria AR, Helf MJ, Takada K, et al. An environmental bacterial taxon with a large and distinct metabolic repertoire. Nature 2014;506:58–62.

[28] Fitzpatrick D, Walsh F. Antibiotic resistance genes across a wide variety of metagenomes. FEMS Microbiol Ecol 2016;92(2).

[29] Edwards RA, Rodriguez-Brito B, Wegley L, Haynes M, Breitbart M, Peterson DM, et al. Using pyrosequencing to shed light on deep mine microbial ecology. BMC Genomics 2006;7:57.

[30] Couto GH, Glogauer A, Faoro H, Chubatsu LS, Souza EM, Pedrosa FO. Isolation of a novel lipase from a metagenomic library derived from mangrove sediment from the South Brazilian Coast. Genet Mol Res 2010;9:514–23.

[31] Li G, Jiang Y, Fan XJ, Liu YH. Molecular cloning and characterization of a novel beta-glucosidase with high hydrolyzing ability for soybean isoflavone glycosides and glucose-tolerance from soil metagenomic library. Bioresour Technol 2012;123:15–22.

[32] Mai Z, Su H, Yang J, Huang S, Zhang S. Cloning and characterization of a novel GH44 family endoglucanase from mangrove soil metagenomic library. Biotechnol Lett 2014;36:1701–9.

[33] Goncalves AC, dos Santos AC, dos Santos TF, Pessoa TB, Dias JC, Rezende RP. High yield of functional metagenomic library from mangroves constructed in fosmid vector. Genet Mol Res 2015;14:11841–7.

[34] Aakvik T, Lale R, Liles M, Valla S. Metagenomic libraries for functional screening. In: de Bruijn FJ, editor. Handbook of molecular microbial ecology. 1st ed. Hoboken: Wiley-Blackwell; 2011. p. 171–81.

[35] Zhou X, Ren L, Meng Q, Li Y, Yu Y, Yu J. The next-generation sequencing technology and application. Protein Cell 2010;1(6):520–36.

[36] Kanehisa M, Goto S, Sato Y, Kawashima M, Furumichi M, Tanabe M. Data, information, knowledge and principle: back to metabolism in KEGG. Nucleic Acids Res 2014;42:D199–205.

[37] Huerta-Cepas J, Szklarczyk D, Forslund K, Cook H, Heller D, Walter MC, Rattei T, Mende DR, Sunagawa S, Kuhn M, et al. eggNOG 4.5: a hierarchical orthology framework with improved functional annotations for eukaryotic, prokaryotic and viral sequences. Nucleic Acids Res 2016;44:D286–93.

[38] Hunter CI, Mitchell A, Jones P, McAnulla C, Pesseat S, Scheremetjew M, Hunter S. Metagenomic analysis: the challenge of the data bonanza. Brief Bioinform 2012;13:743–6.

[39] Kim M, Lee KH, Yoon SW, Kim BS, Chun J, Yi H. Analytical tools and databases for metagenomics in the next-generation sequencing era. Genomics Inform 2013;11:102–13.
[40] Schloss PD, Handelsman J. Biotechnological prospects from metagenomics. Curr Opin Biotechnol 2003;14:303–10.
[41] Blaby-Haas CE, de Crecy-Lagard V. Mining high-throughput experimental data to link gene and function. Trends Biotechnol 2011;29:174–82.
[42] Ferrer M, Beloqui A, Timmis KN, Golyshin PN. Metagenomics for mining new genetic resources of microbial communities. J Mol Microbiol Biotechnol 2009;16:109–23.
[43] Handelsman J. Metagenomics: application of genomics to uncultured microorganisms. Microbiol Mol Biol Rev 2004;68(4):669–85.
[44] Riesenfeld CS, Schloss PD, Handelsman J. Metagenomics: genomic analysis of microbial communities. Annu Rev Genet 2004;38:525–52.
[45] Ekkers DM, Cretoiu MS, Kielak AM, Elsas JD. The great screen anomaly—a new frontier in product discovery through functional metagenomics. Appl Microbiol Biotechnol 2012;93:1005–20.
[46] Akondi KB, Lakshmi VV. Emerging trends in genomic approaches for microbial bioprospecting. OMICS 2013;17:61–70.
[47] Angelov A, Mientus M, Liebl S, Liebl W. A two-host fosmid system for functional screening of (Meta) genomic libraries from extreme thermophiles. Syst Appl Microbiol 2009;32(3):177–85.
[48] Hain T, Otten S, Both U, Chatterjee SS, Technow U, Billion A, et al. Novel bacterial artificial chromosome vector pUvBBAC for use in studies of the functional genomics of *Listeria* spp. Appl Environ Microbiol 2008;74(6):1892. 01.
[49] Vester JK, Glaring MA, Stougaard P. Improved cultivation and metagenomics as new tools for bioprospecting in cold environments. Extremophiles 2015;19:17–29.
[50] Boulund F, Berglund F, Flach CF, Bengtsson-Palme J, Marathe NP, Larsson DGJ, et al. Computational discovery and functional validation of novel fluoroquinolone resistance genes in public metagenomic data sets. BMC Genomics 2017;18:682.
[51] Hover BM, Kim SH, Katz M, Charlop-Powers Z, Owen JG, Ternei MA, et al. Culture-independent discovery of the malacidins as calcium-dependent antibiotics with activity against multidrug-resistant Gram-positive pathogens. Nat Microbiol 2018;3:415–22.
[52] Allen HK, Moe LA, Rodbumrer J, Gaarder A, Handelsman J. Functional metagenomics reveals diverse β-lactamases in a remote Alaskan soil. ISME J 2009;3(2):243–51.
[53] Feng Z, Chakraborty D, Dewell SB, Reddy BVB, Brady SF. Environmental DNA-encoded antibiotics fasamycins A and B inhibit FabF in type II fatty acid biosynthesis. J Am Chem Soc 2012;134(6):2981–7.
[54] Lim HK, Chung EJ, Kim J, et al. Characterization of a forest soil metagenome clone that confers indirubin and indigo production on *Escherichia coli*. Appl Environ Microbiol 2005;71(12):7768–77.
[55] Wang G, Graziani E, Waters B, et al. Novel natural products from soil DNA libraries in a streptomycete host. Org Lett 2000;2(16):2401–4.
[56] Gillespie DE, Brady SF, Bettermann AD, et al. Isolation of antibiotics turbomycin A and B from a metagenomic library of soil microbial DNA. Appl Environ Microbiol 2002;68(9):4301–6.
[57] Choi SY, Yoon K, Lee JI, Mitchell RJ. Violacein: properties and production of a versatile bacterial pigment. BioMed Res Int 2015;2015:8. Article ID 465056.

[58] Piel J, Hui D, Wen G, Butzke D, Platzer M, Fusetani N, et al. Antitumor polyketide biosynthesis by an uncultivated bacterial symbiont of the marine sponge *Theonella swinhoei*. Proc Natl Acad Sci USA 2004;101:16222–7.

[59] Hildebrand M, Waggoner LE, Liu H, Sudek S, Allen S, Anderson C, et al. bryA:an unusual modular polyketide synthase gene from the uncultivated bacterial symbiont of the marine bryozoan Bugula neritina. Chem Biol 2004;11:1543–52.

[60] Donia MS, Ruffner DE, Cao S, Schmidt EW. Accessing the hidden majority of marine natural products through metagenomics. Chembiochem 2011;12:1230–6.

[61] Grindberg RV, Ishoey T, Brinza D, Esquenazi E, Coates RC, Liu WT, et al. Single cell genome amplification accelerates identification of the apra- toxin biosynthetic pathway from a complex microbial assemblage. PLoS One 2011;6. e18565.

[62] Long PF, Dunlap WC, Battershill CN, Jaspars M. Shotgun cloning and heterologous expression of the patellamide gene cluster as a strategy to achieving sustained metabolite production. Chembiochem 2005;6:1760–5.

[63] Brady SF, Clardy J. Palmitoylputrescine, an antibiotic isolated from the heterologous expression of DNA extracted from bromeliad tank water. J Nat Prod 2004;67:1283–6.

[64] Sekyere JO, Asante J. Emerging mechanisms of antimicrobial resistance in bacteria and fungi: advances in the era of genomics. Future Microbiol 2018;13:241–62.

[65] Chang FY, Brady SF. Discovery of indolotryptoline antiproliferative agents by homology-guided metagenomic screening. Proc Natl Acad Sci USA 2013;110:2478–83.

[66] Feng Z, Kim JH, Brady SF. Fluostatins produced by the heterologous expression of a TAR reassembled environmental DNA derived type II PKS gene cluster. J Am Chem Soc 2010;32:11902–3.

[67] King RW, Bauer JD, Brady SF. An environmental DNA-derived type II polyketide biosynthetic pathway encodes the biosynthesis of the pentacyclic polyketide erdacin. Angew Chem Int Ed Engl 2009;48:6257–61.

[68] Owen JG, Robins KJ, Parachin NS, Ackerley DF. A functional screen for recovery of 4'-phosphopantetheinyl transferase and associated natural product biosynthesis genes from metagenome libraries. Environ Microbiol 2012;14:1198–209.

[69] Barriuso J, Valverde JR, Mellado RP. Estimation of bacterial diversity using next generation sequencing of 16S rDNA: a comparison of different workflows. BMC Bioinformatics 2011. https://doi.org/10.1186/1471-2105-12-473.

[70] Laureti L, Song L, Huangm S, Correm C, Leblond P, Challis GL, Aigle B. Identification of a bioactive 51-membered macrolide complex by activation of a silent polyketide synthase in *Streptomyces ambofaciens*. Proc Natl Acad Sci USA 2011;108:6258–63.

[71] Charlop-Powers Z, Milshteyn A, Brady SF. Metagenomic small molecule discovery methods. Curr Opin Microbiol 2014;19:70–5. https://doi.org/10.1016/j.mib.2014.05.021.

[72] Hemala L, Zhang D, Margesin R. Cold-active antibacterial and antifungal activities and antibiotic resistance of bacteria isolated from an alpine hydrocarbon-contaminated industrial site. Microbiol Res 2014;165. https://doi.org/10.1016/j.resmic.2014.05.035.

[73] Lang KS, Anderson JM, Schwarz S, Williamson L, Handelsman, Singer RS. Novel florfenicol and chloramphenicol resistance gene discovered in Alaskan soil by using functional metagenomics. Appl Environ Microbiol 2010;76(15):5321–6.

[74] Okandeji BO, Greenwald DM, Wroten J, Sello JK. Synthesis and evaluation of inhibitors of bacterial drug efflux pumps of the major facilitator superfamily. Bioorg Med Chem 2011;19:7679–89.

[75] Vecchione JJ, Sello JK. A novel tryptophanyl-tRNA synthetase gene confers high-level resistance to indolmycin. Antimicrob Agents Chemother 2009;53:3972–80.

[76] Brady SF, Clardy J. Long-chain N-acyl amino acid antibiotics isolated from heterologously expressed environmental DNA. J Am Chem Soc 2000;122:12903–4.

[77] Schipper C, Hornung C, Bijtenhoorn P, Quitschau M, Grond S, Streit WR. Metagenome-derived clones encoding two novel lactonase family proteins involved in biofilm inhibition in *Pseudomonas aeruginosa*. Appl Environ Microbiol 2009;75:224–33.

[78] Schmitz JE, Daniel A, Collin M, Schuch R, Fischetti VA. Rapid DNA library construction for functional genomic and metagenomic screening. Appl Environ Microbiol 2008;74(5):1649–52.

[79] Chua KYL, Howden BP, Jiang J-H, et al. Population genetics and the evolution of virulence in *Staphylococcus aureus*. Infection. J Genet 2014;21:554–62.

[80] Craig JW, Chang FY, Brady SF. Natural products from environmental DNA hosted in *Ralstonia metallidurans*. ACS Chem Biol 2009;4:23–8.

[81] Haeder S, Wirth R, Herz H, Spiteller D. Candicidin-producing Streptomyces support leaf-cutting ants to protect their fungus garden against the pathogenic fungus *Escovopsis*. Proc Natl Acad Sci USA 2009;106:4742–6.

[82] Banik JJ, Brady SF. Cloning and characterization of new glycopeptide gene clusters found in an environmental DNA megalibrary. Proc Natl Acad Sci USA 2008;105:17273–7.

[83] Donia MS, Cimermancic P, Schulze CJ, Wieland Brown LC, Martin J, et al. A systematic analysis of biosynthetic gene clusters in the human microbiome reveals a common family of antibiotics. Cell 2014;158:1402–14.

[84] Gallo A, Ferrara M, Perrone G. Phylogenetic study of polyketide synthases and nonribosomal peptide synthetases involved in the biosynthesis of mycotoxins. Toxins (Basel) 2013;5(4):717–42. https://doi.org/10.3390/toxins5040717.

[85] Manivasagan P, Venkatesan J, Sivakumar K, Kim S-K. Pharmaceutically active secondary metabolites of marine actinobacteria. Microbiol Res 2014;169:262–78.

[86] Seow KT, Meurer G, Gerlitz M, Wendt-Pienkowski E, Hutchinson CR, Davies J. A study of iterative type II polyketide synthases, using bacterial genes cloned from soil DNA: a means to access and use genes from uncultured microorganisms. J Bacteriol 1997;179:7360–8.

Chapter 29

Application of 3D cell culture techniques in cosmeceutical research

Surya Nandan Meena, Chellandi Mohandass
Biological Oceanography Division, National Institute of Oceanography, Dona Paula, Goa, India

29.1 Introduction

The global cosmetics market was US$ 460 billion in 2014 and is estimated to reach US$ 675 billion by 2020, growing at a rate of 6.4% [1]. Cosmeceuticals are the next-generation skin care products, and the growing interest of people in cosmeceutical products worldwide has encouraged various research groups involved in cosmeceutical research to invent more efficient and novel cosmeceutical bioactive compounds. Drug discovery programs require accurate in vitro systems for the drug screening and testing. Human skin is the first site of exposure to any cosmetic constituent, so cosmeceutical compounds either may get entered into the system through stratum corneum or may get metabolized in the skin only (Fig. 29.1). Therefore, human skin is the most preferred model for testing the chemical compounds of cosmeceutical interest. Perpetual availability and use of human skin for ex vivo experiments is hardly possible and beyond existing ethics.

29.2 Two-dimensional cell system in cosmeceutical research

Traditional cell culture makes use of two-dimensional (2D) surfaces for ex vivo cell growth. In such environments, cells are forced to adopt unnatural characteristics, including aberrant flattened morphologies. Therefore, there is a strong demand for new cell culture platforms that allow cells to grow and respond to their environment in a more realistic manner. The other main demerit of 2D monolayer skin cell culture system is a myriad of important signals, key regulators, and tissue phenotypes that make it unfit to mimic the complexity of in vivo microenvironment. The cultured cells in the 2D

Advances in Biological Science Research. https://doi.org/10.1016/B978-0-12-817497-5.00029-X

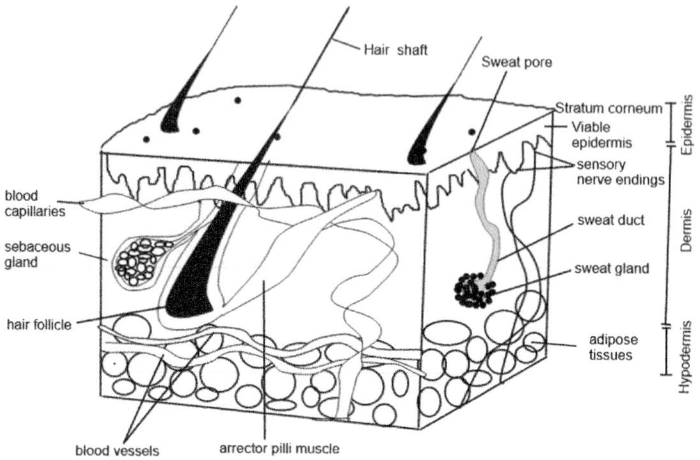

FIGURE 29.1 A cross-sectional view of human skin. *Source: Wikipedia.*

monolayer system adopt an artificial, rigid, and flat surface that sometimes leads to false-positive results of the experiments.

29.3 Role of three-dimensional cell culture system in cosmeceutical research

Alternative to 2D cell culture, three-dimensional (3D) reconstructed human skin (RHS) models, prepared using primary keratinocytes, are now being used. These 3D RHS models have the metabolic capacity [2–5], structure [6] exactly parallel to native human skin, and these characteristics make them a relevant and useful model to evaluate chemical compounds for their cosmeceutical and pharmaceutical activity. Mazzoleni and coworkers reviewed the key differences such as morphology, cellular signaling, cell movement, and reactions to drugs when the same types of cells are cultured in 2D and 3D growth conditions [7].

29.4 Key features of 3D cell culture

3D RHS model is able to synthesize well-ordered epithelial cells and has been proven as a reliable and sustainable platform for screening the compounds or ingredients in photoaging, cosmetic, cancer, and pharmacological studies. As of now, there are several types of 3D RHS models available commercially that exactly mimic the normal human skin [8–10]. In the literature, these systems have been referred to by many names, including 3-D skin, 3D RHS model, skin substitutes, skin grafts, artificial skin, organotypic skin equivalent culture of skin, and reconstructed skin. In the last few years, the number of commercially available skin models has increased. The Model

Skinathic Engineer (Skinathic, Nice, France), Epiderm TM (Mattek, Ehrland, MA, USA), Episcenatum (Episien, Chaponost, France, L'Oréal, SkinEthic, Nice, France), and Applied (Organogenesis Inc., MA, USA) are the some commonly available skin models [11−13]. The RHS models can be engineered as per the need of experiments. Animals routinely used for skin biology or skin cancer research have different skin architecture than human skin, even skin sensitivity and functionality also different from human skin. For instance, mouse skin tissues very effectively regenerate in response to wounds, whereas, human skin leads to scars or lesions (keloids), which are completely absent in mouse skin [14].

29.5 Diverse application of 3D cell culture

Day by day, artificial skin reconstruction models are flourishing and have been proven for many biological applications such as testing the irritating properties of a test compound using MTT assay [15], the phototoxic potential of various cosmetic compounds or formulations [16,17], and penetration and permeability capacity of various cosmetic compounds [18]. Pfuhler and coworkers successfully used the RHS model for the assessment of genotoxicity and cytotoxicity effects of various chemical compounds [19]. The RHS models are also helpful in microbiology or pathology research, as in understanding skin infection and colonization of microbial agents by regulating both skin as well as pathogen [20−22]. The artificial 3D skin structures are designed to cover a wide range of illnesses, such as microbial infections, skin inflammation disorders, skin cancer, UV-damaged skin, and wound problems [23], psoriasis [24,25], atopic dermatitis [26,27], and eczematous dermatitis [28]. Applications of various commercially available 3D skin models in cosmeceutical research are summarized in Table 29.1.

Despite the continuous efforts by researchers across the globe for the bioprospecting of natural resources for their pharmaceutical and cosmeceutical potential, plentiful natural bioactive compounds still wait for exploration of their cosmeceutical and pharmaceutical properties. Compared to the terrestrial biodiversity, the persistence of marine biodiversity in manifold and always has been overlooked, but it may be a gold mine for bioactive compounds of cosmeceutical and pharmaceutical importance. Presently, screening of natural compounds for their biological potential is common by using spectrophotometric, colorimetric, and 2D cell culture−based in vitro assays. Application of 3D skin model for the bioprospecting of largely untapped natural compounds may boost the pharmaceutical and cosmeceutical research in many research laboratories, especially in developing countries, rich in terrestrial and marine biodiversity. This chapter summarizes the basics of 3D RHS model, types of 3D skin models, and different necessary assays involved in cosmeceutical research that would be helpful to the researchers, especially for the young researchers currently involved in cosmeceutical research.

TABLE 29.1 Various 3D reconstructed human skin (RHS) models for the evaluation of cosmeceutical potential of topically applied ingredients/compounds.

Sl. No.	In vitro assay	Commercially available 3D skin models
1	Skin irritation	EpiSkin^{TM} [29], SkinEthic RHE [29], EpiDerm [30], EST1000 or epiCS [31], Phenion [32], StratiCELL [33], Phenion FT [32]
2	Skin corrosion	EpiSkin [34], SkinEthic RHE [35], EpiDerm [35], EST1000 or epiCS [36],Vitrolife-SKIN [37], LabCyte (EPI-model 24) [38]
3	Phototoxicity	EpiSkin [39], SkinEthic RHE [40], EpiDerm [41], EST1000 or epiCS [42], Phenion FT [43], StratiCELL [44]
4	Skin whitening/ melanin content	EpiSkin [45], SkinEthic RHE [46], MelanoDerm [47], iPSC based models [48], MelanoDerm, Reconstituted human pigmented epidermal (RHE-MEL/001) [49], pigmented epidermis model [50]
5	Skin aging	EpiSkin, SkinEthic RHE, Phenion FT AGED skin model, EpiDermFT, Normal human epidermal keratinocytes (NHEK), reconstructed human epidermal (RHE) and photo-ageing model [51]
6	Genotoxicity	EpiSkin [52], SkinEthic RHE [53], EpiDerm [54], EST1000 or epiCS, Phenion, Phenion FT [54], StratiCELL [45]
7	Skin permeability	EpiSkin [2], SkinEthic RHE [55]
8	Skin penetration	EpiSkin [55], SkinEthic RHE [55], EpiDerm [56]
9	Skin wound healing	Phenion FT [57]
10	Skin allergy	EpiSkin, SkinEthic RHE [58]
11	Skin metabolism	EpiDerm, Phenion [59]
12	Skin sensitization	EST1000 or epiCS [60]
13	Skin absorption	Phenion [59]

29.6 Preparation of 3D reconstructed human skin model

29.6.1 The traditional approach for 3D skin model preparation

Generally, the preparation of 3D RHS model is carried out using Alvetex$^{@}$ Scaffold technology [61]. Skin reconstructs for cancer research have been

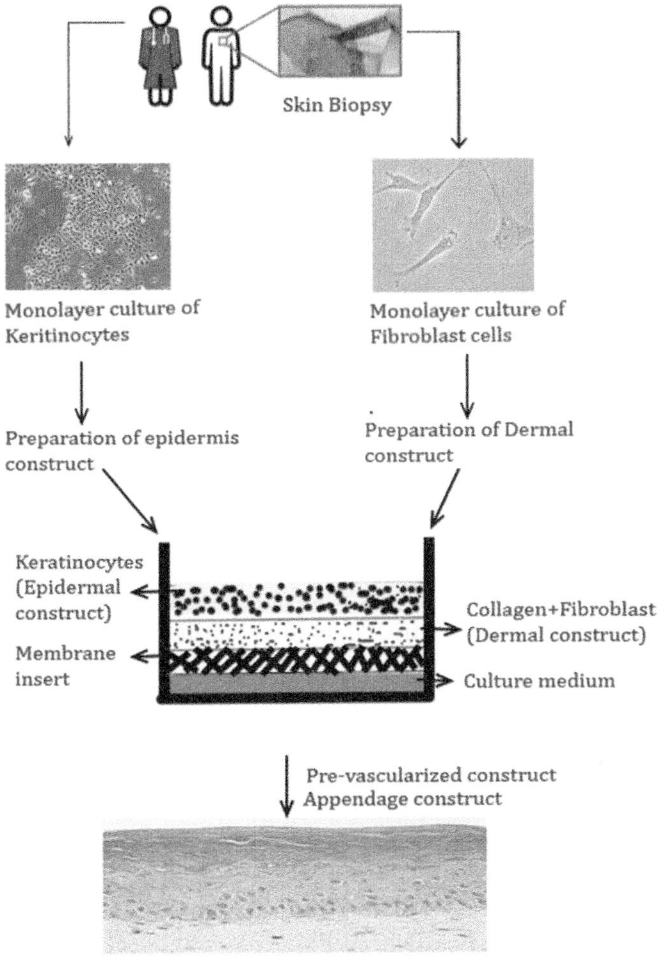

FIGURE 29.2 Construction of 3D reconstructed human skin (RHS) model.

reconstructed by incorporating 3D matrix, melanoma 116 cells, A375 metastatic melanoma cell line [62], and melanoma-tumor globular [63] and squamous carcinoma cell line [64]. In general, the first step in the construction of in vitro 3D skin reconstruct is the construction of the dermal layer, consisting of the fibroblasts and type I collagen. After the polymerization dermal layer, the second step includes the plating of keratinocytes and melanocytes over the dermal layer. After 24 h of plating or after a contraction of collagen gel, the whole structure is allowed to pass through the steel grid to form the air-liquid interface, and finally, after 2 weeks on the interface, artificial skin is obtained (Fig. 29.2).

Generally the full-thickness 3D in vitro skin models that include the epidermis and dermis are useful in analyzing the processes happening in the skin. The ideal RSE models should follow the few basic criteria such as (1) uppermost skin layer (stratum corneum) should possess the lipid layer of viable cells, (2) cell viability should be sufficient to differentiate between positive and negative effects on cells induced by the application of chemical compound, and (3) under defined protocol, results should be consistent. Although the 3D skin model has been successfully proven in various applications, it still does not mimic natural human skin 100 percent in terms of the proper function of the skin barrier, neural connectivity, pigmentation system, sebaceous glands and hair follicles [65].

The recently introduced technique "3D Bioprinting" has resolved some of the drawbacks associated with the currently available 3D RHS model prepared using traditional approach [66].

29.6.2 Bioprinting technology for preparation of 3D skin models

Recently introduced bioprinting technology allows and facilitates the skin structure more effectively with particular architecture [67]. The word "printing" reflects the bioink, which is nothing but a biomaterial and various type of cells used to produce multilayer 3D bioprinting skin model [68]. In general, construction of 3D bioprinting skin is initiated with the collection of various cells (fibroblasts, keratinocytes, and melanocytes) from the patient, which are grown in aseptic conditions using the appropriate medium. The suitable biopolymer is formed and along with cells fed into the bioprinting system.

29.7 Application of 3D skin models in cosmeceutical research

29.7.1 Skin whitening or melanin content

For some of skin problems like hyperpigmentation and inflammatory problems, people generally seek medical help. Currently, there are a number of products available to reduce hyperpigmentation in various ways [69]. Since the enzyme tyrosinase is a well-known rate-limiting factor in melanin biosynthesis, among all the strategies, inhibition of tyrosinase activity has been an effective one for the screening of hyperpigmentation-reducing compounds [70]. Kojic acid, hydroquinone, and arbutin are the most common commercially available hyperpigmentation-reducing compounds screened by targeting tyrosinase enzyme. It is a common observation that compounds screened through enzyme inhibitors using in vitro assays are generally poor in clinical efficacy. Alternatively, a 3D skin model has been proved an effective strategy to screen the compounds for hyperpigmentation-reducing potential. MatTek MelanoDerm is a suitable 3D in vitro model for high-throughput screening (HTS) screening of cosmetic ingredients for skin whitening potential. This model constitutes the normal keratinocytes mixed with melanocytes in the epidermal construct of the model. After good setup of the 3D cell culture using the air-liquid interface, then test compounds applied topically on the skin construct with various doses and incubate for different time intervals. After incubation, melanin content is quantified using a standard protocol. Recently, Mann and screened commercially available skin whitening compounds for their hypopigmentation potential using the MelanoDerm skin model and observed that out of all the tested compounds, thiamidol was the most potent one for the inhibition of melanin biosynthesis [71].

29.7.2 Skin antiaging study using 3D in vitro skin model

In comparison to the skin of a healthy adult human, the asymmetric order and reduced content of elastin, collagen, hyaluronic acid, and Matrix metalloproteinase (MMPs) in the skin are the common symptoms of biologically aged skin [72,73]. The factors generally responsible for skin aging are direct and long exposures to UV radiation, harmful free radicals in many commercially available synthetic cosmetic skin products, and environmental pollution, etc. Overproduction of the reactive oxygen species (ROS) has contributed to the skin-aging symptoms by increasing the expression of collagenase, hyaluronidase, elastase, and MMPs, which degrade the structure of collagen, hyaluronidase, elastin, and MMP, respectively.

Today a range of 3D skin models are available for the evaluation of skin-aging studies and for the HTS of cosmeceutical ingredients. The 3D skin models used in skin whitening studies are EpiSkin, SkinEthic RHE, MelanoDerm, iPSC-based models, MelanoDerm, reconstituted human-pigmented epidermis (RHE-MEL/001), pigmented epidermis model, and epiCS-M. The EpiDermFT skin model is used for the HTS of cosmeceutical ingredients for their skin whitening properties. Using EpiDermFT, it is possible to study the changes in the extracellular matrix of the cells if any are due to the topical application of chemical compounds. Therefore, for the study of antiaging properties of a novel or most potent ingredient, the EpiDermFT-412 skin model is cost-effective and ideal for researchers.

29.7.3 Antioxidant activity

Oxidative stress is the major cause of skin aging that includes wrinkles, pigmentation, and weakened wound healing ability. Application of antioxidants in skin care is well accepted as an effective approach to delay the skin-aging process. Therefore, the antioxidant activity of the samples may be evaluated using a 3D human in vitro skin model as described in the standard protocol [74].

29.7.4 Antiinflammatory activity

Normally the human skin is prone to various skin disorders. Present-day available cosmetic products are blended with ingredients to cure most of skin disorders. But psoriasis, an inflammatory skin disorder, affects approximately 2%–3% of the world population [75]. Screening of natural compounds using 3D skin equivalent model may be helpful in screening antiinflammatory compounds with good clinical efficacy. Chamcheu and coworkers successfully investigated the antiinflammatory effect of delphinidin, a diet-derived anthocyanidin using 3D psoriatic skin equivalent model, and they observed that delphinidin may be a leading candidate for the management of psoriasis [76].

29.7.5 Wound healing assay

Wound healing is a complex process that consists of four main stages: hemostasis, inflammation, proliferation, and remodeling of the tissues. The 3D wound healing model resembles the wound bed during the wound healing process. In order to understand the mechanism of wound healing and the pathogenesis of skin scars, several 3D skin models are available commercially [77]. These models range from a monolayer to organotypic. To understand the wound healing mechanism, a wound is induced mechanically or using chemicals methods to these skin models and then wound propagation induction is initiated to an appropriate stage. Using the high-phase contrast microscope, proliferation and propagation of the cells involved can be seen or studied easily. The wound healing model generally consists of the multilayered epidermal construct of keratinocytes [78] and multilayer dermal construct of fibroblast cells [79,80].

29.7.6 Skin corrosion test

Skin corrosion refers to the production of irreversible tissue damage in the skin following the application of a test compound. The skin corrosion test is mainly based on the principle that after a short-term exposure of the corrosive compounds they are cytotoxic to the stratum corneum of the epidermis if cytotoxicity is immediately determined after chemical exposure. For the skin corrosion test of any cosmetic compound, the test material needs to be applied topically on 3D skin model for a specified time followed by checking the viability of the cells using cell viability test. In a study, Kim and coworkers investigated the skin corrosion effect of various nanoparticles using EpiDerm skin model. After incubating the EpiDerm cells with different types of nanoparticles for 60 min, cells were washed using PBS buffer and subsequently checked for their viability using cell viability test. These findings suggest that FeNPs, AlNPs, TNPs, and AgNPs are "noncorrosive" and "nonirritant" to human skin by a globally harmonized classification system [81].

29.7.7 Skin cell irritation test

The skin irritation properties of chemicals compounds are considered as the way to establish a standard method for the safe use of these potentially harmful chemicals. Standard procedures for the evaluation of chemical compounds for their skin irritation effects are especially needed for cosmeceutical and pharmaceutical ingredients. Skin irritation refers to the production of reversible injury to the skin after using a test substance for up to 4 h. Skin irritation test for cosmetic and pharmaceutical compounds has been conducted successfully using EpiDerm reconstructed human epidermal model [15]. Kim andco-workers performed a skin irritation test on artificial skin cell model (EPI-200; Mattek Corporation, Ashland, MA, USA). They isolated bacteriochlorophyll from Rhodobacter sphaeroides and tested it for cell irritation activity. The test was conducted using the same protocol as the MTT assay used in the cytotoxicity evaluation, and the resultant values were shown in terms of cell viability (%) [82]. Whereas in another study, 3D human skin model (EpiDerm) was used for irritancy/toxicity test of herbal cream derived from *Dodonaea polyandra* and the results showed that the 5% w/w resin cream formulation showed no irritancy in a 3D human skin model [83].

29.7.8 Skin penetration assay

The 3D RHS models have been proven as a useful tool for the evaluation of the skin penetration capacity of topically applied ingredients of cosmetic and pharmaceutical products. Routine methods have been developed to check the penetration level of the new chemical compounds. The routinely used 3D skin models are Episkin^{TM}, SkinEthic RHE, and EpiDerm.

Especially, with skin penetration studies performed using EpiSkin model, there is no need for using the complex device such as Frenz Cells [84]. To test the best absorption of vitamins to the skin, the EpiSkin model was used to the check the absorption of vitamins C and E [55].

29.7.9 Phototoxicity study

Phototoxicity/photo irritation is an acute toxic response that is elicited after the first exposure of skin to any chemical compound followed by subsequent exposure to direct sunlight. Since 2000, the in vitro 3T3 NRU phototoxicity test has been the only in vitro test validated for evaluating the acute phototoxicity of a chemical. However, this monolayer culture of fibroblasts is a simple basic system lacking connections and interactions between different cell types and extracellular matrix (ECM) [85].

Currently many 3D in vitro skin models are available to test the acute phototoxicity effects of ingredients and their photoprotective effects from UVA and UVB. The commonly available models are EpiDerm and EpiSkin, and other models used for phototoxicity studies are summarized in Table 29.1.

Generally, the phototoxic potential of the cosmetic ingredient may not be truly evaluated if applied topically in small quantity, since it does not absorb into the proper cellular system. There is a need for alternative methods than 2D cell-based evaluation of cosmeceutical compound for their phototoxicity evaluation. Presently available 3D skin models have been proven good for predicting the photoirritancy and photoprotection properties of topically applied ingredients in cosmeceutical formulations. The common in vitro 3D skin models used for the evaluation of the phototoxicity of chemical compounds include EpiSkin, SkinEthic RHE, EpiDerm, EST1000 or epiCS, Phenion FT, and StratiCELL.

29.7.10 Genotoxicity assay

Genotoxicity effect of any ingredient in cosmeceutical or pharmaceutical formulations is the alteration in structure, sequence, and segregation of DNA molecules. Therefore, genotoxicity assessment of the chemical compounds of cosmetic potential is a basic need in the early stage of any cosmetic formulation. Comet assay is the standard protocol for the assessment of genotoxicity [19] of any compound, whereas, "micronucleus assay" is a routine use assay for the evaluation of the mutagenicity of any compound. Presently several 3D skin models such as EpisSkin™, SkinEthic RHE, EpiDerm, EST1000 or epiCS, Phenion, Phenion FT, and StratiCELL are routinely used in several laboratories to assess mutagenicity as well genotoxicity in studies.

29.7.11 Skin absorption assay

The first step in evaluating the transdermal or dermal distribution efficiency of any chemical compound is assessment of the percutaneous permeability of the compound.

The materials used in the preparation of 3D in vitro skin models range from silicone membranes, poly(dimethoxysilane), parallel artificial membrane permeability assay (PAMPA), or phospholipid vesicle-based membranes [86]. Later materials have been designed to mimic the stratum corneum. Fast screening of chemical compounds for their passive transport, preferably PAMPA, is the more preferred system [87].

29.8 Conclusion

This chapter summarizes the various 3D reconstructed human skin models, developed as an alternative tool to the traditional in vitro assays. Compared to the 2D cell culture techniques, 3D RHS models are efficient, cost-effective, andand reliable tools for screening the ingredients of pharmaceutical and cosmeceutical importance with more clinical efficacy.

Moreover, these 3D in vitro models have proved to be an alternative to the use of animal models during the drug discovery process. We have provided an overview of the necessity for various assays involved in cosmeceutical research. The 3D RHS models have been proven to be an efficient technology for providing the drug discovery process with higher clinical efficacy. Along with the various benefits of the described 3D skin models, users should always compare the cost and feasibility, reproducibility, etc. of each model for each study.

Acknowledgments

The authors are grateful to the Director, CSIR-NIO, for providing laboratory infrastructure and moral support. We would like to acknowledge the project OLP1707 for consumables and other needed support and to GAP3118 for major financial support. Thanks to SERB-DST, New Delhi, for financial support to SNM through the National Postdoctoral fellowship scheme (PDF/2016/002012).

References

[1] http://www.businesswire.com/news/home/20150727005524/en/Research-Markets-Global-Cosmetics-Market-2015–2020-Market.

[2] Luu-The V, Duche D, Ferraris C, Meunier JR, Leclaire J, Labrie F. Expression profiles of phases 1 and 2 metabolizing enzymes in human skin and the reconstructed skin models Episkin™ and full thickness model from Episkin™. J Steroid Biochem Mol Biol 2009;116(3–5):178–86.

[3] Hu T, Khambatta ZS, Hayden PJ, Bolmarcich J, Binder RL, Robinson MK, Carr GJ, Tiesman JP, Jarrold BB, Osborne R, Reichling TD. Xenobiotic metabolism gene expression in the EpiDerm™ in vitro 3D human epidermis model compared to human skin. Toxicol In Vitro 2010;24(5):1450–63.

[4] Gotz C, Pfeiffer R, Tigges J, Blatz V, Jäckh C, Freytag EM, Fabian E, Landsiedel R, Merk HF, Krutmann J, Edwards RJ. Xenobiotic metabolism capacities of human skin in comparison with a 3D epidermis model and keratinocyte-based cell culture as in vitro alternatives for chemical testing: activating enzymes (Phase I). Exp Dermatol 2012;21(5):358–63.

[5] Van Eijl S, Zhu Z, Cupitt J, Gierula M, Götz C, Fritsche E, Edwards RJ. Elucidation of xenobiotic metabolism pathways in human skin and human skin models by proteomic profiling. PLoS One 2012;26(7). e41721.

[6] Ponec M, Gibbs S, Pilgram G, Boelsma E, Koerten H, Bouwstra J, Mommaas M. Barrier function in reconstructed epidermis and its resemblance to native human skin. Skin Pharmacol Physiol 2001;14:63–71.

[7] Mazzoleni G, Di Lorenzo D, Steimberg N. Modelling tissues in 3D: the next future of pharmaco-toxicology and food research? Genes Nutr 2009;4(1):13.

[8] Dubé J, Rochette-Drouin O, Lévesque P, Gauvin R, Roberge CJ, Auger FA, Goulet D, Bourdages M, Plante M, Germain L, Moulin VJ. Restoration of the transepithelial potential within tissue-engineered human skin in vitro and during the wound healing process in vivo. Tissue Eng 2010;16(10):3055–63.

[9] Souto LR, Vassallo J, Rehder J, Pinto GA, Puzzi MB. Immuno-architectural characterization of a human skin model reconstructed in vitro. Sao Paulo Med J 2009;127(1):28–33.

[10] Sobral CS, Gragnani A, Cao X, Morgan JR, Ferreira LM. Human keratinocytes cultured on collagen matrix used as an experimental burn model. J Burns Wounds 2007;7:e6.

[11] Boelsma E, Gibbs S, Faller C, Ponec M. Characterization and comparison of reconstructed skin models: morphological and immunohistochemical evaluation. Acta Derm Venereol 2000;80:82–8.

[12] Ponec M. Skin constructs for replacement of skin tissues for in vitro testing. Adv Drug Deliv Rev 2002;54:S19–30.

[13] Welss T, Basketter DA, Schroder KR. *In vitro* skin irritation: facts and future. State of the art review of mechanisms and models. Toxicol In Vitro 2004;18(3):231–43.

[14] Khorshid FA. Comparative study of keloid formation in humans and laboratory animals. Med Sci Monit 2005;11(7):212–9.

[15] Kandarova H, Hayden P, Klausner M, Kubilus J, Sheasgreen J. An *in vitro* skin irritation test (SIT) using the EpiDerm reconstructed human epidermal (RHE) model. J Vis Exp 2009;29:1366.

[16] Medina J, Elsaesser C, Picarles V, Grenet O, Kolopp M, Chibout SD, de Brugerolle de Fraissinette A. Assessment of the phototoxic potential of compounds and finished topical products using a human reconstructed epidermis. In Vitr Mol Toxicol 2001;14(3):157–68.

[17] Bernard FX, Barrault C, Deguercy A, De Wever B, Rosdy M. Development of a highly sensitive in vitro phototoxicity assay using the SkinEthicTM reconstructed human epidermis. Cell Biol Toxicol 2000;16(6):391–400.

[18] Garcia N, Doucet O, Bayer M, Fouchard D, Zastrow L, Marty JP. Characterization of the barrier functions in a reconstituted human epidermis cultivated in chemically defined medium. Int J Cosmet Sci 2002;24(1):25–34.

[19] Pfuhler S, Fellows M, Van Benthem J, Corvi R, Curren R, Dearfield K, Fowler P, Frötschl R, Elhajouji A, Le Hégarat L, Kasamatsu T. *In vitro* genotoxicity test approaches with better predictivity: summary of an IWGT workshop. Mutat Res 2011;723(2):101–7.

[20] Holland DB, Bojar RA, Farrar MD, Holland KT. Differential innate immune responses of a living skin equivalent model colonized by *Staphylococcus epidermidis* or *Staphylococcus aureus*. FEMS Microbiol Lett 2009;290(2):149–55.

[21] Corriveau MN, Zhang N, Holtappels G, Van Roy N, Bachert C. Detection of *Staphylococcus aureus* in nasal tissue with peptide nucleic acid–fluorescence in situ hybridization. Am J Rhinol Allergy 2009;23(5):461–5.

[22] Ohnemus U, Kohrmeyer K, Houdek P, Rohde H, Wladykowski E, Vidal S, Horstkotte MA, Aepfelbacher M, Kirschner N, Behne MJ, Moll I. Regulation of epidermal tight-junctions (TJ) during infection with exfoliative toxin-negative Staphylococcus strains. J Invest Dermatol 2008;128(4):906–16.

[23] Barbero AM, Frasch HF. Effect of frozen human epidermis storage duration and cryoprotectant on barrier function using two model compounds. Skin Pharmacol Physiol 2016;29(1):31–40.

[24] Chiricozzi A, Nograles KE, Johnson-Huang LM, Fuentes-Duculan J, Cardinale I, Bonifacio KM, Gulati N, Mitsui H, Guttman-Yassky E, Suárez-Fariñas M, Krueger JG. IL-17 induces an expanded range of downstream genes in reconstituted human epidermis model. PLoS One 2014;9(2). e90284.

[25] Tjabringa G, Bergers M, van Rens D, de Boer R, Lamme E, Schalkwijk J. Development and validation of human psoriatic skin equivalents. Am J Pathol 2008;173(3):815–23.

[26] Pendaries V, Malaisse J, Pellerin L, Le Lamer M, Nachat R, Kezic S, Schmitt AM, Paul C, Poumay Y, Serre G, Simon M. Knockdown of filaggrin in a three-dimensional reconstructed human epidermis impairs keratinocyte differentiation. J Invest Dermatol 2014;134(12):2938–46.

[27] Rouaud-Tinguely P, Boudier D, Marchand L, Barruche V, Bordes S, Coppin H, Roth MP, Closs B. From the morphological to the transcriptomic characterization of a compromised three-dimensional in vitro model mimicking atopic dermatitis. Br J Dermatol 2015;173(4):1006–14.

[28] Engelhart K, El Hindi T, Biesalski HK, Pfitzner I. In vitro reproduction of clinical hallmarks of eczematous dermatitis in organotypic skin models. Arch Dermatol Res 2005;297(1):1–9.

[29] de Fraissinette AD, Picarles V, Chibout S, Kolopp M, Medina J, Burtin P, Ebelin ME, Osborne S, Mayer FK, Spake A, Rosdy M. Predictivity of an in vitro model for acute and chronic skin irritation (SkinEthic) applied to the testing of topical vehicles. Cell Biol Toxicol 1999;15(2):121–35.

[30] Kandárová H, Hayden P, Klausner M, Kubilus J, Kearney P, Sheasgreen J. *In vitro* skin irritation testing: improving the sensitivity of the EpiDerm skin irritation test protocol. ATLA-Alternatives to Laboratory Animals 2009;37(6):671.

[31] http://www.sterlab.com; http://www.straticell.com; http://reconstructed-human-epidermis.com.

[32] Zoller NN, Kippenberger S, Thaçi D, Mewes K, Spiegel M, Sättler A, Schultz M, Bereiter-Hahn J, Kaufmann R, Bernd A. Evaluation of beneficial and adverse effects of glucocorticoids on a newly developed full-thickness skin model. Toxicol Vitro 2008;22(3):747–59.

[33] Eeman M, Bastin C, Leroy M, Belot N, Salmon M. Prediction of skin irritation potential of chemicals using StratiCELL reconstituted human epidermal model. In: ISUP 2010. Innovation for sustainable production, Bruges, Belgium; 2010.

[34] EC-ECVAM. Statement on the scientific validity of the EpiSkinTM test (an in vitro test for skin corrosivity). In: Issued by the ECVAM Scientific Advisory Committee (ESAC10), 1998; 1998.

[35] Kandarova H, Liebsch M, Spielmann H, Genschow E, Schmidt E, Traue D, Guest R, Whittingham A, Warren N, Gamer AO, Remmele M. Assessment of the human epidermis model SkinEthic RHE for in vitro skin corrosion testing of chemicals according to new OECD TG 431. Toxicol Vitro 2006;20:547–59.

[36] EC-ECVAM. ESAC statement on the scientific validity of an in-vitro test method for skin corrosivity testing: the EST-1000. In: Issued by the ECVAM Scientific Advisory Committee (ESAC30), 2009; 2009.

[37] Kojima H, Ando T, Inagaki K, Ohhira M, Kosaka T, Nakamura Y, Torishima H, Morikawa N, Kanno J, Kuboki M, Genno M. Validation of human skin models for skin corrosivity tests in Japan. AATEX 2008;13(1):36–44.

[38] Katoh M, Hamajima F, Ogasawara T, Hata KI. Assessment of the human epidermal model LabCyte EPI-MODEL for in vitro skin corrosion testing according to the OECD test guideline 431. J Toxicol Sci 2010;35(3):411–7.

[39] Lelièvre D, Justine P, Christiaens F, Bonaventure N, Coutet J, Marrot L, Cotovio J. The EpiSkin phototoxicity assay (EPA): development of an in vitro tiered strategy using 17 reference chemicals to predict phototoxic potency. Toxicol Vitro 2007;21(6):977–95.

[40] Bacqueville D, Mavon A. Comparative analysis of solar radiation-induced cellular damage between ex vivo porcine skin organ culture and in vitro reconstructed human epidermis. Int J Cosmet Sci 2009;31(4):293–302.

[41] Liebsch M. Prevalidation of the EpiDerm phototoxicity test InAlternatives to animal testing II. In: Proceedings of the second international scientific conference organised by the European Cosmetic Industry, Brussels, Belgium; 1999. p. 160–6.

[42] https://reconstructed-human-epidermis.com/epics/phototox/.

[43] Meloni M, Farina A, de Servi B. Molecular modifications of dermal and epidermal biomarkers following UVA exposures on reconstructed full-thickness human skin. Photochem Photobiol Sci 2010;9(4):439–47.

[44] http://www.in-cosmetics.com/__novadocuments/8291.

[45] Qiu J, Chen M, Liu J, Huang X, Chen J, Zhou L, Ma J, Sextius P, Pena AM, Cai Z, Jeulin S. The skin-depigmenting potential of Paeonialactiflora root extract and paeoniflorin: in vitro evaluation using reconstructed pigmented human epidermis. Int J Cosmet Sci 2016;8(5):444–51.

[46] Donat V. Skin lightening through a new natural oily composition from botanical origin. Household Personal Care 2017;12:82.

[47] Lee CS, Nam G, Bae IH, Park J. Whitening efficacy of ginsenoside F1 through inhibition of melanin transfer in cocultured human melanocytes–keratinocytes and three-dimensional human skin equivalent. J Ginseng Res 2018, 1–5.

[48] Schafer-Korting M, Bock U, Diembeck W, Düsing HJ, Gamer A, Haltner-Ukomadu E, Hoffmann C, Kaca M, Kamp H, Kersen S, Kietzmann M. The use of reconstructed human epidermis for skin absorption testing: results of the validation study. Altern Lab Anim 2008;36(2):161–87.

[49] https://www.straticell.com/en/In-Vitro-Skin-Models/in-vitro-reconstituted-epidermis.php.

[50] https://www.creative-bioarray.com/skin-whitening-and-pigmentation.htm.

[51] Nakamura M, Haarmann-Stemmann T, Krutmann J, Morita A. Alternative test models for skin ageing research. Exp Dermatol 2018;27:495–500.

[52] Flamand N, Marrot L, Belaidi JP, Bourouf L, Dourille E, Feltes M, Meunier JR. Development of genotoxicity test procedures with Episkin®, a reconstructed human skin model: towards new tools for in vitro risk assessment of dermally applied compounds? Mutat Res 2006;606(1):39–51.

[53] Kandarova H, Liebsch M, Schmidt E, Genschow E, Traue D, Spielmann H, Meyer K, Steinhoff C, Tornier C, De Wever B, Rosdy M. Assessment of the skin irritation potential of chemicals by using the SkinEthic reconstructed human epidermal model and the common skin irritation protocol evaluated in the ECVAM skin irritation validation study. Altern Lab Anim 2006;34(4):393.

[54] Reisinger K, Blatz V, Brinkmann J, Downs TR, Fischer A, Henkler F, Hoffmann S, Krul C, Liebsch M, Luch A, Pirow R. Validation of the 3D Skin Comet assay using full thickness skin models: transferability and reproducibility. Mutat Res 2018;827:27–41.

[55] Rozman B, Gasperlin M, Tinois-Tessoneaud E, Pirot F, Falson F. Simultaneous absorption of vitamins C and E from topical microemulsions using reconstructed human epidermis as a skin model. Eur J Pharm Biopharm 2009;72(1):69–75.

[56] Hu T, Bailey RE, Morrall SW, Aardema MJ, Stanley LA, Skare JA. Dermal penetration and metabolism of p-aminophenol and p-phenylenediamine: application of the EpiDerm™ human reconstructed epidermis model. Toxicol Lett 2009;188(2):119–29.

[57] https://www.phenion.com/products/reconstructed-tissues#Tab-754224_1.

[58] Gazel A, Rosdy M, Tornier C, De Fraissinette AD, Blumenberg M. Transcriptional profiling defines the effects of nickel in human epidermal keratinocytes. J Cell Physiol 2008;217(3):686–92.

[59] Ackermann K, Borgia SL, Korting HC, Mewes KR, Schafer-Korting M. The Phenion® full-thickness skin model for percutaneous absorption testing. Skin Pharmacol Physiol 2010;23(2):105−12.

[60] Koeper LM, Schulz A, Ahr HJ, Vohr HW. In vitro differentiation of skin sensitizers by cell signaling pathways. Toxicology 2007;242(1−3):144−52.

[61] Maltman DJ, Przyborski SA. Developments in three-dimensional cell culture technology aimed at improving the accuracy of in vitro analyses. Biochem Soc Trans 2010;38:1072−5.

[62] Mohapatra S, Coppola D, Riker AI, Pledger WJ. Roscovitine inhibits differentiation and invasion in a three-dimensional skin reconstruction model of metastatic melanoma. Mol Canc Res 2007;5(2):145−51.

[63] Vorsmann H, Groeber F, Walles H, Busch S, Beissert S, Walczak H, Kulms D. Development of a human three-dimensional organotypic skin-melanoma spheroid model for in vitro drug testing. Cell Death Dis 2013;4(7):e719.

[64] Commandeur S, van Drongelen V, de Gruijl FR, El Ghalbzouri A. Epidermal growth factor receptor activation and inhibition in 3D in vitro models of normal skin and human cutaneous squamous cell carcinoma. Cancer Sci 2012;103(12):2120−6.

[65] De Wever B, Kurdykowski S, Descargues P. Descargues, Human skin models for research applications in pharmacology and toxicology: Introducing NativeSkin®, the “missing link” bridging cell culture and/or reconstructed skin models and human clinical testing. Appl In Vitro Toxicol 2015;1(1):26−32.

[66] Jia W, Gungor-Ozkerim PS, Zhang YS, Yue K, Zhu K, Liu W, Pi Q, Byambaa B, Dokmeci MR, Shin SR, Khademhosseini A. Direct 3D bioprinting of perfusable vascular constructs using a blend bioink. Biomaterials 2016;106:58−68.

[67] Richards D, Jia J, Yost M, Markwald R, Mei Y. 3D bioprinting for vascularized tissue fabrication. Ann Biomed Eng 2017;45(1):132−47.

[68] Hospodiuk M, Dey M, Sosnoski D, Ozbolat IT. The bioink: a comprehensive review on bioprintable materials. Biotechnol Adv 2017;35(2):217−39.

[69] Briganti S, Camera E, Picardo M. Chemical and instrumental approaches to treat hyperpigmentation. Pigm Cell Res 2003;16(2):101−10.

[70] Lee SY, Baek N, Nam TG. Natural, semisynthetic and synthetic tyrosinase inhibitors. J Enzym Inhib Med Chem 2016;31(1):1−3.

[71] Mann T, Gerwat W, Batzer J, Eggers K, Scherner C, Wenck H, Stäb F, Hearing VJ, Röhm KH, Kolbe L. Inhibition of human tyrosinase requires molecular motifs distinctively different from mushroom tyrosinase. J Invest Dermatol 2018;138:1601−8.

[72] Binic I, Lazarevic V, Ljubenovic M, Mojsa J, Sokolovic D. Skin ageing: natural weapons and strategies. Evid Based Complement Alternat Med 2013;2013.

[73] Jadoon S, Karim S, Asad MH, Akram MR, Kalsoom Khan A, Malik A, Chen C, Murtaza G. Anti-aging potential of phytoextract loaded-pharmaceutical creams for human skin cell longetivity. Oxid Med Cell Longev 2015;2015.

[74] Grazul-Bilska AT, Bilski JJ, Redmer DA, Reynolds LP, Abdullah KM, Abdullah A. Antioxidant capacity of 3D human skin EpiDermTM model: effects of skin moisturizers. Int J Cosmet Sci 2009;31(3):201−8.

[75] Nestle FO, Kaplan DH, Barker J, Psoriasis. N Engl J Med 2009;361:496−509.

[76] Chamcheu JC, Pal HC, Siddiqui IA, Adhami VM, Ayehunie S, Boylan BT, Noubissi FK, Khan N, Syed DN, Elmets CA, Wood GS. Prodifferentiation, anti-inflammatory and antiproliferative effects of delphinidin, a dietary anthocyanidin, in a full-thickness three-dimensional reconstituted human skin model of psoriasis. Skin Pharmacol Physiol 2015;28(4):177−88.

[77] van den Broek LJ, Limandjaja GC, Niessen FB, Gibbs S. Human hypertrophic and keloid scar models: principles, limitations and future challenges from a tissue engineering perspective. Exp Dermatol 2014;23(6):382–6.

[78] Henemyre-Harris CL, Adkins AL, Chuang AH, Graham JS. Addition of epidermal growth factor improves the rate of sulfur mustard wound healing in an *in vitro* model. Eplasty 2008;8:e16.

[79] Calderon M, Lawrence WT, Banes AJ. Increased proliferation in keloid fibroblasts woundedin vitro. J Surg Res 1996;61(2):343–7.

[80] Kim WS, Lee JS, Bae GY, Kim JJ, Chin YW, Bahk YY, Min HG, Cha HJ. Extract of a neilemakeisak inhibits transforming growth factor-β-dependent signalling by inducing S mad2 downregulation in keloid fibroblasts. Exp Dermatol 2013;22(1):69–71.

[81] Kim H, Choi J, Lee H, Park J, Yoon BI, Jin SM, Park K. Skin corrosion and irritation test of nanoparticles using reconstructed three-dimensional human skin model, EpiDermTM. J Toxicol 2016;32(4):311.

[82] Kim NY, Yim TB, Lee HY. Skin anti-aging activities of bacteriochlorophyll a from photosynthetic bacteria, Rhodobacter sphaeroides. J Microbiol Biotechnol 2015;25(10):1589–98.

[83] Simpson BS, Luo X, Wang J, Song Y, Claudie D, Garg S, Smith N, McKinnon R, Semple S. Development and evaluation of a topical anti-inflammatory preparation containing Dodonaeapolyandra extract. Int J Pharm Pharmaceut Res 2015;18(4):578–99.

[84] Gregoire S, Patouillet C, Noe C, Fossa I, Kieffer FB, Ribaud C. Improvement of the experimental setup for skin absorption screening studies with reconstructed skin EPISKIN®. Skin Pharmacol Physiol 2008;21(2):89–97.

[85] Augustin C, Collombel C, Damour O. Use of dermal equivalent and skin equivalent models for identifying phototoxic compounds in vitro. Photodermatol Photoimmunol Photomed 1997;13(1–2):27–36.

[86] Flaten GE, Palac Z, Engesland A, Filipovic-Grcic J, Vanic Z, Skalko-Basnet N. *In vitro* skin models as a tool in optimization of drug formulation. European J Biomed Pharm 2015;75:10–24.

[87] Kansy M, Senner F, Gubernator K. Physicochemical high throughput screening: parallel artificial membrane permeation assay in the description of passive absorption processes. J Med Chem 1998;41(7):1007–10.

Chapter 30

Advances in isolation and preservation strategies of ecologically important marine protists, the thraustochytrids

Varada S. Damare
Department of Microbiology, Goa University, Taleigao Plateau, Goa, India

30.1 Introduction

Thraustochytrids are unicellular eukaryotic organisms belonging to the Kingdom Stramenopila. This kingdom also includes diatoms, brown algae, the oomycetes fungi, and a variety of flagellates. This eukaryotic group of organisms has been included in the SAR domain under the eukaryotic tree of life [1]. SAR domain represents organisms belonging to Stramenopiles, Alveolates and Rhizaria and have been named as 'SAR' by Adl et al. [1]. The Alveolates consist of dinoflagellates that are well known for the formation of harmful algal blooms (HABs) and the Rhizaria, comprise the foraminifera that help in palaeoceanography owing to their calcareous tests. The thraustochytrids belong to the Phylum Heterokonta and Class Labyrinthulomycota or Labyrinthulomycetes. The Labyrinthulomycetes was earlier comprised of two distinct phylogenetic groups, the Thraustochytrid Phylogenetic group and the Labyrinthula Phylogenetic group [2], but now it comprises three groups, namely the thraustochytrids, aplanochytrids, and labyrinthulids [3]. The aplanochytrids, earlier called the labyrinthuloids, were placed under Labyrinthula Phylogenetic group, but based on the SSU ribosomal sequences, Leander and Porter [4] showed that the genus *Aplanochytrium* was distinct from the labyrinthulids and thraustochytrids. All the three groups are osmoheterotrophic in their mode of nutrition.

Members of Labyrinthulomycetes typically reproduce by means of stramenopilan zoospores. The zoospores are characterized by two flagella of unequal length and are called the heterokont flagella. The anterior whiplash

Advances in Biological Science Research. https://doi.org/10.1016/B978-0-12-817497-5.00030-6

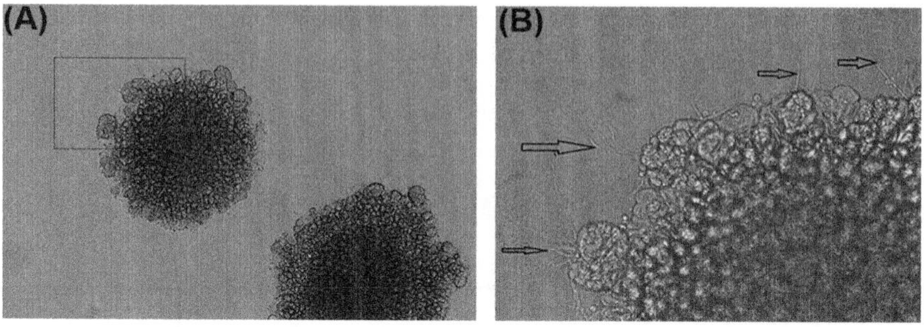

FIGURE 30.1 Photograph of a colony of thraustochytrids as observed under (A) 10× objective of the microscope and (B) 40× objective showing EN elements (arrow). The area marked within the box in (A) is the one magnified in (B).

flagellum is long and covered with tubular tripartite hairs. The posterior tinsel flagellum is shorter than the whiplash. The tripartite tubular hairs are composed of a cone-like base, a tubular shaft, and usually two unequal diverging terminal fibers but only one terminal fiber in case of labyrinthulids and aplanochytrids (see Ref. [5]).

A characteristic feature of the Labyrinthulomycetes is that these organisms produce ectoplasmic net elements (EN elements), which are branched extensions of the plasma membrane and arise from an organelle called the sagenogenetosome or bothrosome [6] (Fig. 30.1). The EN elements not only help these organisms to absorb nutrients from the environment but also help them to attach to a substratum [7]. All thraustochytrids except for the genus *Althornia* produce EN from one or many points in the cell, and unlike labyrinthulids, their cells are not interconnected by EN. Labyrinthulid cells are enrobed by and glide through the ectoplasmic net elements. Aplanochytrid cells are also not completely enrobed by the EN similar to thraustochytrids but are capable of crawling movement by means of EN. This crawling movement is not seen in thraustochytrid cells.

30.2 Occurrence and ecological significance

The Labyrinthulomycetes are exclusively marine in nature and display an absolute requirement for Na^+ ions that cannot be replaced by K^+. However, of late the genus *Diplophrys* has been isolated from freshwater habitats [8]. Molecular phylogenetic studies of this genus demonstrated it to belong to Amphitremida of Labyrinthulomycetes.

The ubiquitous existence of Labyrinthulomycetes in marine environment implies some definite role played by these organisms in the marine ecosystem. Being absorptive or osmoheterotrophic in their mode of nutrition, similar to bacteria and fungi, they obtain their energy for survival via breakdown of various substrates with the help of extracellular enzymes [9,10]. Thus they play a role of decomposers similar to that of bacteria and fungi and that as "left-over scavengers" [11]. In addition to the saprotrophic role, Labyrinthulomycetes are also parasitic in nature. The labyrinthulid *Labyrinthula zosterae* causes the wasting disease of eelgrass, which has caused widespread devastation of large populations of eelgrass in many parts of the globe [12]. Apart from parasitic association of *Labyrinthula* with the seagrasses, there exist parasitic associations with many mollusks [13–16]. *Thraustochytrium caudivorum* parasitizes the flatworm *Macrostomum lignano* [17]. The Quahog-parasite unknown (QPX) is a well-known parasite of the quahogs or hard clams belonging to the Labyrinthulomycetes, which leads to massive killing of the clams and has so far resulted in significant economic losses [18,19].

The Labyrinthulomycetes have also been found to exist in nonparasitic associations with hydroids, corals, and tunicates, but the exact nature of these associations is unknown [7]. These organisms are often encountered as contaminants while growing cell lines of invertebrates and have also been isolated successfully in such conditions [20,21]. In addition to all these, they also serve as a source of nutrition for organisms higher in the trophic level such as the zooplankton. The Labyrinthulomycetes possess docosahexaenoic acid (DHA) as the signature fatty acid. DHA is an omega-3 polyunsaturated fatty acid (ω-3 PUFA) essential for growth and reproduction of zooplankton. Since the zooplankton cannot synthesize DHA, they have to depend on a dietary supplement to obtain this nourishment. Labyrinthulomycetes, therefore, serve as food to the zooplankton [22,23]. Several species of Labyrinthulomycetes often accumulate lipids up to 50% of their body weight, and DHA constitutes 25% or more [24–27]. DHA is an important fatty acid having numerous benefits for human health, too, especially for infant nutrition. Hence these organisms are known to have commercial significance for exploitation of their lipids for nutraceutical benefits [28].

Apart from their existence in association with other organisms, the Labyrinthulomycetes exist in the marine water column mostly in particle-associated form. These organisms are shown to have a positive relationship with the transparent exopolymeric particles or the particulate organic matter in the water column [29–31]. The marine aggregates or marine snow are rich habitats of these organisms and serve as reservoirs for their transfer to different places [32,33].

30.3 Isolation

The Labyrinthulomycetes would henceforth be mentioned as thraustochytrids in the text, except for instances that specifically deal with labyrinthulids or aplanochytrids only and not all the three groups. Thraustochytrids have been isolated from a variety of sources (substrates) and from varied places. The most common substrate is the senescent leaves of mangroves. These organisms have been isolated from mangrove leaves or detritus from various regions across the globe such as Goa (India), Okinawa Prefecture (Japan), subtropical mangroves (25–27°N), Seto Inland Sea (Japan), Futial National Nature Reserve, Shenzhen (China), Hong Kong, Indonesia, and Palombina and Poetto Beach, Goro Lagoon, Ancona Port (Italy) [34–45]. Algae from Maine (United States), Southsea Castle (UK), Hampshire (UK), Dona Paula Bay and Zuari River, Goa (India) and seagrasses from Florida (United States), Puerto Rico, and Eastern Canadian coastal sites from Nova Scotia, Newfoundland, and Labrador have also been explored for isolation of these organisms [34,46–51]. These organisms have also been consistently isolated from the fecal pellets of salp *Pegea confoederata*, coelenteron and hydranth of hydroid, and guts and feces of sea urchin *Lytechinus variegates* [52,53].

A majority of the studies are involved with coastal sites except for isolation of *Aplanochytrium kerguelensis* and four *Thraustochytrium* species from Antarctic waters and that of *Aplanochytrium yorkensis* from zooplankton of the Indian Ocean [9,54]. Table 30.1 describes various substrates and the source of isolation along with the isolation medium used for the same in different studies during the last one decade. Traditionally, the medium described by Porter [6] is being used for isolation of thraustochytrids. But it was seen that in spite of the presence of thraustochytrids in the samples, these were not consistently isolated from the same. Hence many studies were carried out to develop a medium for their isolation.

30.3.1 Isolation of thraustochytrids

Thraustochytrids being zoosporic organisms, the classical method of isolating these organisms involves baiting of various substrates in sterile seawater and sterile pine pollen. The thraustochytrids produce enzymes via the EN elements that help them to break the hard sporopollenin layer of the pollen grain. Hence when baited with pine pollen, these organisms colonize the pollen and get their nutrition from the pollen interior [7]. Thus, pollen baiting is the first step toward isolation of these organisms. Once the samples are incubated along with pine pollen bait, they are monitored every day for the growth of thraustochytrids surrounding the pollen grains. Once the thraustochytrid cells appear under the microscope (Fig. 30.2), an aliquot from the pollen sample is plated on agar medium supplemented with antibiotics. Antibiotics are added to the medium in order to suppress bacterial growth and favor the growth of these protists. Sometimes an antifungal agent is also added to avoid the growth of the fungi since fungi and yeasts are common inhabitants of various substrates and hence easily turn up on the media when antibiotics suppress the bacteria.

Sometimes liquid media is used instead of baiting the sample with pine pollen in sterile seawater, and this media is then plated on the agar. In case of samples such as mangrove leaves or invertebrates, the samples are plated directly on to the agar medium supplemented with antibiotic and/or antifungal

TABLE 30.1 Place of isolation and various substrates used for the isolation of thraustochytrids along with different growth media used in various studies during the last one decade.

Place	Isolation source	Isolation medium (L^{-1})	Reference
Gulf of Eilat, Israel	Surface mucus of hermatypic coral *Fungia granulosa*	B1 liquid media and agar (1 g peptone, 1 g yeast extract and 1.5 g bacteriological agar in 0.22 μm filtered seawater with 500 mg/L streptomycin sulfate and 300 mg/L penicillin G	[72]
Kagoshima Bay, Kagoshima, Japan	Seawater	TPTW medium (yeast extract 1 g, polypeptone 1 g, glucose 1 g, Tween 80 1 g, KH_2PO_4 0.2 g, tomato juice 10 g, agar 15 g, ½ con ASW)	[58]
Hong Kong mangroves	Fallen (black and brown) *Kandelia candel* mangrove leaves	Agar plates consisting of 20 g of agar, 30 g of glucose, 0.2 g of $(NH_4)_2SO_4$, 2 g of monosodium glutamate, 0.3 g of KH_2PO_4, 25 g of NaCl, 1 g of KCl, 5 g of $MgSO_4$ $37H_2O$, 0.3 g of $CaCl_2$ $32H_2O$, 0.1 g of $NaHCO_3$, 2 g of yeast extract, 30 mg of Na_2EDTA, 34.2 mg of H_3BO_3, 2.9 mg of $FeCl_3$ $36H_2O$, 0.02 mg of $CuSO_4$ $35H_2O$, 8.6 mg of $MnCl_2$ 3 $4H_2O$, 0.26 mg of $CoCl_2$ $34H_2O$, and 0.6 mg of $ZnCl_2$	[73]
Rocky shores, coral reefs, estuaries, and salt marshes in Taiwan	Macroalgae, fallen leaves of mangrove trees, sand, and water	0.8% agar medium prepared using full strength seawater, containing 1 g/L peptone, 2 g/L yeast extract, 4 g/L glucose, and antibiotics including ampicillin (sodium form), streptomycin sulfate, and kanamycin sulfate (100 mg/L each)	[74]
Coastal environments of the Argentinean continental shelf	Algal, plant, and animal debris, sediments, sludge with bird feces	Mar Chiquita (MC) isolation medium supplemented with antibiotics (2 g glucose, 1 g peptone, 1 g yeast extract, 1 g monosodium glutamate, 2 g gelatin hydrolysate, 1 g corn steep liquor, 17.5 g artificial sea salt, 20 g agar), MC-BHB medium (MC + brain-heart broth 17.5 g), serum seawater agar medium (SSA) (17.5 g artificial sea salt, 1% horse serum, 12 g agar)	[57]

Continued

TABLE 30.1 Place of isolation and various substrates used for the isolation of thraustochytrids along with different growth media used in various studies during the last one decade.—cont'd

Place	Isolation source	Isolation medium (L^{-1})	Reference
Whangapoua Harbor, North East New Zealand	Estuarine water, sediment and mangrove leaves	Seawater complete (SWC) medium (12 g agar, 3 g yeast extract, 5 g peptone, 3 mL glycerol, 70% artificial seawater), modified Vishniac agar (12 g agar, 1 g glucose, 1 g gelatin hydrolysate, 0.1 g yeast extract, 0.1 g peptone, 70% artificial seawater), By + agar (12 g agar, 5 g glucose, 1 g yeast extract, 1 g peptone, 70% artificial seawater)	[67]
Queenscliff, Victoria, Australia	Seawater samples and sediments containing degraded leaf material, Leaf samples	YP medium (yeast extract 1 g, peptone 1 g, penicillin 500 mg, streptomycin 500 mg, rifampicin 50 mg, and nystatin 10 mg in sterile seawater)	[75]
Philippines mangroves	Hepowit (2006)	Yeast-extract-peptone (composed of 1 g yeast extract, 1 g mycological peptone, 15 g agar, 50% artificial seawater)	[76]
Mangroves around the Mumbai coastline	Estuarine water, decaying mangrove leaves, soil and pneumatophores	MC medium, MC + BHB medium, B1 medium (1 g yeast extract, 1 g peptone, 10 g agar, natural seawater), SWC medium	[68]
Vancouver Island and Port Moody, British Columbia	*Pisaster ochraceus* (sea star)	Serum seawater agar (SSA, 1% agar and 1% horse serum) medium with antibiotics	[56]

West coast of Peninsular Malaysia	Fallen and floating senescent mangrove leaves	GYPsp medium (10 g/L glucose, 2 g/L yeast extract, 1 g/L polypeptone, 0.5 g/L penicillin G, 0.5 g/L streptomycin sulfate, 30 g/L artificial sea salt), GYPc medium (2 g/L glucose, 0.5 g/L yeast extract, 1 g/L polypeptone, 50 mg/L chloramphenicol, 17.5 g/L artificial sea salt)	[77]
Scottish waters	Lesser octopuses (*Eledone cirrhosa*), with a fatal ulcerative dermal necrosis, rainbow trout (*Oncorhynchus mykiss*) with skin lesions	20 g presoaked Fadenagar in seawater, 2 g bacteriological agar, 0.3 g peptone, 0.03 g yeast extract and 3 g malt extract. The plates also contained penicillin (50–100 IU/mL) and streptomycin (100 μg/mL)	[78]
Mangrove forests in Chon Buri province, Thailand	Fallen, senescent mangrove leaves	GYP agar medium containing 1 g each of glucose, yeast extract and peptone with antibiotics	[79]
Coastal habitats of Pearr River Delta region of China	Mangrove leaf samples, seawater samples	Glucose (20 g), peptone (1.5 g), yeast extract (1 g), and artificial seawater (NaCl 25 g, KCl 1 g, KH_2PO_4 0.3 g, $MgSO_4 \cdot 7H_2O$ 5 g, $NaHCO_3$ 0.1 g and $CaCl_2$ 0.3 g)	[80]

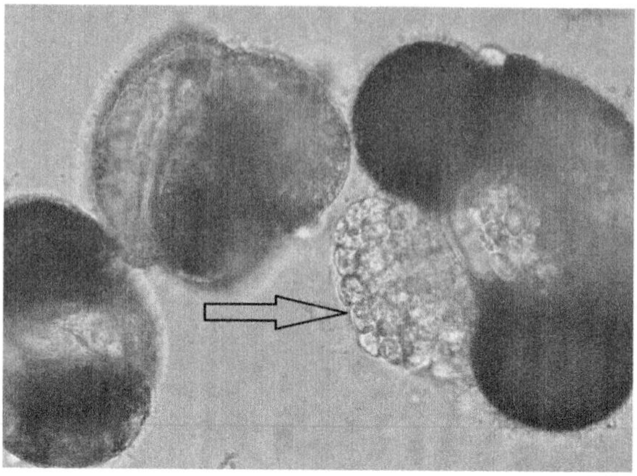

FIGURE 30.2 Photograph of pine pollen with thraustochytrid cluster (arrow) growing over it.

compounds. Before plating any sample as well as before incubating any sample with pine pollen, the sample is washed 4–6 times with sterile seawater or sterile seawater and antibiotics to remove surface debris.

The various media that have been widely used for isolation of thraustochytrids generally consist of glucose as carbon source, yeast extract and peptone as nitrogenous sources. Yeast extract also provides vitamins. The choice of seawater used varied in different studies from natural seawater to artificial seawater and from 100% seawater to 50%–70%. For isolation of thraustochytrids from invertebrates, medium containing serum is generally used. Rabinowitz et al. [21] used diluted (10%) Leibovitz L-15 medium (isoosmotic to seawater) supplemented with 3% heat-inactivated fetal calf serum, 2 mM L-glutamine and 1% mixed antibiotic solution PSA (×100) made of 10,000 units penicillin per ml, 10 mg/mL streptomycin, and 25 μg/mL amphotericin B to isolate thraustochytrids from *Botryllus* (tunicate) primary cultures. Damare and Raghukumar [9] used modified Vishniac (MV) medium containing 0.001% liver infusion broth, 0.01% yeast extract, 0.15% peptone, 0.1% dextrose supplemented with 40,000 U procaine penicillin, 0.075g streptomycin, and 1% fetal bovine serum. The fetal bovine serum is generally used in animal cell cultures to provide essential growth factors [55]. FioRito et al. [56] used horse serum to isolate thraustochytrids from sea star. Rosa et al. [57] tried different media to isolate from animal debris and bird feces, one of which contained horse serum. The serum seawater agar (SSA) medium used in their study demonstrated a low mean growth value (GV) but the highest cultivation efficiency (CE).

The terms GV and CE were proposed by Rosa et al. [57] to characterize the type of thraustochytrid growth obtained after plating different samples. These are measured after 10 days of incubating the agar plates. The cultures are examined by light microscopy, and relative GVs are assigned. These are divided into four categories according to an arbitrary scale based on the characteristics of colonies (see Ref. [57]):

1. GV 0–1: cells that did not proliferate or proliferate defectively only once, respectively;
2. GV 2: small, flat, two-dimensional colonies are formed;
3. GV 3: larger colonies, as a result of cell division on several planes; and
4. GV 4: luxurious, confluent cultures.

For estimating cultivation success, GVs 0–1 are considered bad, GV 2 good, and GV 3 and GV 4 very good. CE for each culture medium is defined as the percentage of strains attaining a GV $>$ 2 on it.

Besides yeast extract, peptone, and glucose, the addition of Tween 80, potassium dihydrogen phosphate, and tomato juice also increases the yield of thraustochytrid isolates after plating the pine pollen–baited samples [58]. Tween 80 is known to enhance the production of exopolysaccharides in medicinal fungi, mushrooms [59,60]. It also increases the concentration of total fatty acids in the cell wall, oleic acid in particular, and maintains the integrity of the cell wall [61]. Generally detergents lyse the cells due to osmotic action, but when protozoa such as *Paramecium* was exposed to toxic doses of Tween 80, which were a little less than LD50 dose, the osmotic mechanism was not seen as only small amount of the compound is in free form and the rest in micellar form [62]. Tween 80 downregulates fatty acid biosynthesis and increases oleic acid and cyclopropane fatty acid levels in lactic acid bacteria [63]. Addition of Tween 80 to the media changes the nutrient availability by decreasing the nutrient particle size and imparting homogeneity to the particles. Thus, at low concentrations Tween 80 will have a growth-promoting effect on a cell by putting stress over it that leads to better uptake of nutrients for survival, whereas at concentrations above critical it is detrimental to the cell. At lower concentrations, Tween 80 slows the growth rate of a few bacteria without causing any effect on the others [64]. Tomato juice has several growth factors such as 4′-o-(β-glucopyranosyl)-DL-pantothenic acid, glutamic acid. Glutamic acid has many roles in metabolism and also during response to various stresses [65]. It was the most preferred amino acid by the aplanochytrids over the others during the study made by Damare and Raghukumar [9]. Pantothenic acid, which is the precursor of CoA, supports the luxurious growth of prokaryotic as well as eukaryotic microorganisms such as yeasts and protozoa [66].

Isolation of thraustochytrids also involves the use of antibiotics similar to the isolation of fungi. Wilkens and Maas [67] tried different antibiotic trials for the isolation from marine environments. Different antibiotics tested were streptomycin, ampicillin, rifampicin, nalidixic acid, tetracycline, and gentamicin. Pandey and Bhathena [68] also tested various other antibiotics such as chloramphenicol and oxytetracycline in addition to the above. Use of antibiotics during isolation leads to better recovery of thraustochytrids in culture. A combination of antibiotics that target different sites of bacterial growth such as those that inhibit cell wall synthesis as well as protein synthesis or DNA synthesis of bacteria are the best choice for removal of bacteria. Rifampicin inhibits DNA-dependent RNA polymerase. Streptomycin, gentamicin, kanamycin, chloramphenicol, and tetracycline inhibit protein synthesis. Ampicillin inhibits transpeptidase needed for cell wall synthesis. Use of an antifungal such as nystatin, amphotericin B, or fluconazole is also preferred to take care of fungal contamination and for increased recovery of thraustochytrids [69]. Nalidixic acid inhibits DNA synthesis. Fluconazole inhibits cytochrome P450 enzyme 14a-demethylase, whereas nystatin and amphotericin B bind to ergosterol present in the fungal cell wall and forms pores in the membrane thus disrupting the membrane integrity and leading to the death of the fungus.

30.3.2 Isolation of labyrinthulids

Labyrinthulids are isolated from marine environment following the same methodology as that for thraustochytrids (Fig. 30.3). However, Yokochi et al. [36] described a new method for isolation involving the bacterium *Psychrobacter phenylpyruvicus*. Pine pollen-baited samples were plated on normal media used for isolation of thraustochytrids, i.e., containing glucose, peptone, and yeast extract, but before plating the samples on the media plates, bacteria were grown onto them. Of the other bacterium (*Shewanella*) and yeast (*Rhodotorula rubra*) studied, *Psychrobacter phenylpyruvicus* gave a better yield of *Labyrinthula* in culture [36]. Further, Yadagiri et al. [70] devised two

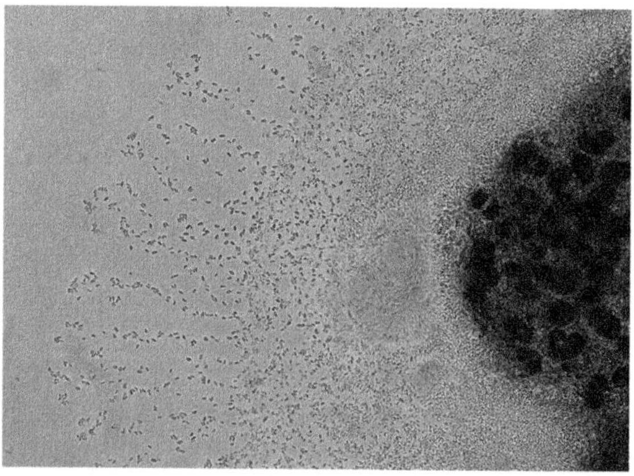

FIGURE 30.3 Microphotograph (under 10× objective) of plating of pine pollen-baited sample on MV agar showing growth of *Labyrinthula*-like cells radiating away from the bacterial colony surrounding the pine pollen.

different media for their isolation without requiring the bacteria. SSA was modified by the addition of glucose, yeast extract, peptone, gelatin hydrolysate, and vegetable extract and called modified SSA. Also, grass extract SSA was prepared by adding perennial ryegrass (*Lolium perenne*) extract. The grass extract was prepared by blending the two-week-old grass in saline to form a crude homogenate, strained through four to six layers of cheesecloth and centrifuged at 10,000 rpm at 4°C for 10 min. The supernatant was centrifuged again and the final supernatant was autoclaved for 10 min at 121°C. This was shaken thoroughly to suspend the precipitate formed after autoclaving and used as grass extract supplement in the SSA.

30.4 Preservation of cultures

Regular subculturing of thraustochytrid cultures on the media of isolation is the preferred choice of maintenance of cultures. Cryopreservation using glycerol is the most common method of storage of these cultures. Cox et al. [71] have studied various ways of preserving these cultures and have found that a mixture of dimethyl sulfide (Me_2SO) and bovine/horse serum is the best choice for preservation of these organisms. The recipe for preservation devised in that study is 10% cell suspension, 10% Me_2SO, 30% serum, and 50% culture media.

30.5 Summary and future prospects

Isolation and maintenance of thraustochytrids in culture is a tedious process since their growth is affected by environmental conditions such as the change in temperature or salinity. Hence it is best to optimize the culture conditions for the isolation of these organisms with respect to the source of the sample, in other words, as per the study area. Baiting with pine pollen is a preferred method before plating on the media as it leads to the luxuriant growth of the protists on the pollen, giving an advantage of obtaining them easily in culture, too, before the evolution of fast-growing contaminants such as fungi after plating in the presence of antibiotics. The polyunsaturated fatty acids produced by thraustochytrid protists make them the potential candidate to harvest for these fatty acids for nutraceutical benefits. These protists can also be used as fish feed to increase the dietary value of the fishes in aquaculture and can serve as good candidates for biodiesel production.

Acknowledgments

The author would like to thank Dr. S. Raghukumar for his introduction to the world of Labyrinthulomycetes protists and for his constant encouragement and guidance. Support by Prof. Sandeep Garg, head of the Department of Microbiology, Goa University—is also highly acknowledged.

References

[1] Adl SM, Simpson AG, Lane CE, Lukeš J, Bass D, Bowser SS, Brown M, Burki F, Dunthorn M, Hampl V, Heiss A, Hoppenrath M, Lara E, leGall L, Lynn DH, McManus H, Mitchell EAD, Mozley-Stanridge SE, Parfrey LW, Pawlowski J, Rueckert S, Shadwick L, Schoch C, Smirnov A, Spiegel FW. The revised classification of eukaryotes. J Eukaryot Microbiol 2012;59(5):429–93.

[2] Honda D, Yokochi T, Nakahara T, Raghukumar S, Nakagiri A, Schaumann K, Higashihara T. Molecular phylogeny of labyrinthulids and thraustochytrids based on the sequencing of 18S ribosomal RNA gene. J Eukaryot Microbiol 1999;46(6):637–47.

[3] Leander CA, Porter D. Redefining the genus *Aplanochytrium* (Phylum Labyrinthulomycota). Mycotaxon 2000;76:439–44.

[4] Leander CA, Porter D. The Labyrinthulomycota is comprised of three distinct lineages. Mycologia 2001;93:459–64.

[5] Dick MW. Straminipilous fungi: systematics of the peronosporomycetes including accounts of the marine straminipilous protists, the plasmodiophorids and similar organisms. Netherlands: Kluwer Academic Publishers; 2001.

[6] Porter D. Labyrinthulomycota. In: Margulis L, Corliss JO, Melkonian M, Chapman D, editors. Handbook of protoctista. Boston, MA: Jones and Bartlett; 1990. p. 388–98.

[7] Raghukumar S. Ecology of the marine protists, the Labyrinthulomycetes (Thraustochytrids and Labyrinthulids). Eur J Protistol 2002;38:127–45.

[8] Takahashi Y, Yoshida M, Inouye I, Watanabe MM. *Diplophrys mutabilis* sp. nov., a new member of Labyrinthulomycetes from freshwater habitats. Protist 2014;165:50–65.

[9] Damare V, Raghukumar S. Morphology and physiology of the marine straminipilan fungi, the aplanochytrids isolated from the equatorial Indian Ocean. Indian J Mar Sci 2006;35(4):326–40.

[10] Bongiorni L, Pusceddu A, Danovaro R. Enzymatic activities of epiphytic and benthic thraustochytrids involved in organic matter degradation. Aquat Microb Ecol 2005;41:299–305.

[11] Raghukumar S, Damare VS. Increasing evidence for the important role of Labyrinthulomycetes in marine ecosystems. Bot Mar 2011;54:3–11.

[12] Sullivan BK, Sherman TD, Damare VS, Lilje O, Gleason FH. Potential roles of *Labyrinthula* species in global seagrass population declines. Fungal Ecol 2013;6:328–38.

[13] Polglase JL. A preliminary report on the thraustochytrid(s) and labyrinthulid(s) associated with a pathological condition in the lesser octopus *Eledone cirrhosa.* Bot Mar 1980;23:699–706.

[14] Jones GM, O'Dor RK. Ultrastructural observations on a thraustochytrid fungus parasitic in the gills of squid (*Ilex illecebrosus* LeSueur). J Parasitol 1983;69:903–11.

[15] Bower SM. *Labyrinthuloides haliotidis* n.sp. (*Protozoa: Labyrinthomorpha*), a pathogenic parasite of small juvenile abalone in a British Columbia mariculture facility. Can J Zool 1987;65:1996–2007.

[16] Bower SM, McLean N, Whitaker DJ. Mechanism of infection by *Labyrinthuloides haliotidis* (*Protozoa: Labyrinthomorpha*), a parasite of abalone (*Haliotis kamtschatkana*) (*Mollusca: Gastropoda*). J Invertebr Pathol 1989;53:401–9.

[17] Schärer L, Knoflach D, Vizoso DB, Rieger G, Peintner U. Thraustochytrids as novel parasitic protists of marine free-living flatworms: *Thraustochytrium caudivorum* sp. nov. parasitizes *Macrostomum lignano.* Mar Biol 2007;152(5):1095–104.

[18] Smolowitz R, Leavitt D. Quahog parasite unknown (QPX): an emerging disease of hard clams. J Shellfish Res 1997;16:335–6.

[19] Whyte SK, Cawthorn RJ, McGladdery SE. QPX (Quahaug parasite X), a pathogen of northern quahaug *Mercenaria mercenaria* from the Gulf of St. Lawrence, Canada. Dis Aquat Org 1994;19:129–36.

[20] Rinkevich B. Cell cultures from marine invertebrates: obstacles, new approaches and recent improvements. J Biotechnol 1999;70:133–53.

[21] Rabinowitz C, Douek J, Weisz R, Shabtay A, Rinkevich B. Isolation and characterization of four novel thraustochytrid strains from a colonial tunicate. Indian J Mar Sci 2006;35(4):341–50.

[22] Damare VS, Damare S, Ramanujam P, Meena RM, Raghukumar S. Preliminary studies on the association between zooplankton and the stramenopilan fungi, aplanochytrids. Microb Ecol 2013;65:955–63.

[23] Damare VS, Raghukumar S. Apparent grazing losses of Labyrinthulomycetes protists in oceanic and coastal waters: an experimental elucidation. Ecol Res 2015;30:403–14.

[24] Bajpai P, Bajpai BP, Ward OP. Production of docosahexaenoic acid by *Thraustochytrium aureum*. Appl Microbiol Biotechnol 1991;35:706–10.

[25] Singh A, Ward OP. Production of high yields of docosahexaenoic acid by *Thraustochytrium roseum* ATCC 28210. J Ind Microbiol 1997;16:370–3.

[26] Yaguchi T, Tanaka S, Yokochi T, Nakahara T, Higashihara T. Production of high yields of docosahexaenoic acid by *Schizochytrium* sp. Strain SR21. J Am Oil Chem Soc 1997;74(11):1431–4.

[27] Yokochi T, Honda D, Higashihara T, Nakahara T. Optimization of docosahexaenoic acid production by *Schizochytrium limacinum* SR21. Appl Microbiol Biotechnol 1998;49:72–6.

[28] Raghukumar S. Thraustochytrid marine protists: production of PUFAs and other emerging technologies. Mar Biotechnol 2008;10:631–40.

[29] Damare V, Raghukumar S. Abundance of thraustochytrids and bacteria in the equatorial Indian ocean, in relation to transparent exopolymeric particles (TEPs). FEMS Microbiol Ecol 2008;65:40–9.

[30] Kimura H, Sato M, Sugiyama C, Naganuma T. Coupling of thraustochytrids and POM, and of bacterio- and phytoplankton in a semi-enclosed coastal area: implication for different substrate preference by the planktonic decomposers. Aquat Microb Ecol 2001;25:293–300.

[31] Raghukumar S, Ramaiah N, Raghukumar C. Dynamics of thraustochytrid protists in the water column of the Arabian Sea. Aquat Microb Ecol 2001;24:175–86.

[32] Damare VS, Raghukumar S. Association of the stramenopilan protists, the aplanochytrids, with zooplankton of the Equatorial Indian Ocean. Mar Ecol Prog Ser 2010;399:53–68.

[33] Lyons MM, Ward JE, Smolowitz R, Uhlinger KR, Gast RJ. Lethal marine snow: pathogen of bivalve mollusc concealed in marine aggregates. Limnol Oceanogr 2005;50(6):1983–8.

[34] Raghukumar S. Detection of the thraustochytrid protist *Ulkenia visurgensis* in a hydroid, using immunofluorescence. Mar Biol 1988;97:253–8.

[35] Perveen Z, Ando H, Ueno A, Ito Y, Yamamoto Y, Yamada Y, Takagi T, Kaneko T, Kogame K, Okuyama H. Isolation and characterization of a novel thraustochytrid-like microorganism that efficiently produces docosahexaenoic acid. Biotechnol Lett 2006;28:197–202.

[36] Yokochi T, Nakahara T, Higashihara T, Yamaoka M, Kurane R. A new isolation method for Labyrinthulids using a bacterium, *Psychrobacter phenylpyruvicus*. Mar Biotechnol 2001;3:68–73.

[37] Bowles RD, Hunt AE, Bremer GB, Duchars MG, Eaton RA. Long-chain ω-3 polyunsaturated fatty acid production by members of the marine protistan group the thraustochytrids: screening of isolates and optimisation of docosahexaenoic acid production. J Biotechnol 1999;70:193–202.

[38] Kumon Y, Yokoyama R, Haque Z, Yokochi T, Honda D, Nakahara T. A new Labyrinthulid isolate that produces only docosahexaenoic acid. Mar Biotechnol 2006;8:170–7.

[39] Wong MKM, Vrijmoed LLP, Au DWT. Abundance of thraustochytrids on fallen decaying leaves of *Kandelia candel* and mangrove sediments in Futian National Nature Reserve, China. Bot Mar 2005;48(5–6):374–8.

[40] Fan KW, Vrijmoed LLP, Jones EBG. Zoospore chemotaxis of mangrove thraustochytrids from Hong Kong. Mycologia 2002;94(4):569–78.

[41] Fan KW, Vrijmoed LLP, Jones EBG. Physiological studies of subtropical mangrove thraustochytrids. Bot Mar 2002;45:50–7.

[42] Fan KW, Jiang Y, Faan YW, Chen F. Lipid characterization of mangrove thraustochytrid-*Schizochytrium mangrovei*. J Agric Food Chem 2007;55:2906–10.

[43] Kamlangdee N, Fan KW. Polyunsaturated fatty acids production by *Schizochytrium* sp. isolated from mangrove. Songklanakarin J Sci Technol 2003;25(5):643–50.

[44] Unagul P, Assantachai C, Phadungruengluij S, Suphantharika M, Verduyn C. Properties of the docosahexaenoic acid-producer *Schizochytrium mangrovei* Sk-02: effects of glucose, temperature and salinity and their interaction. Bot Mar 2005;48(5–6):387–94.

[45] Bongiorni L, Mirto S, Pusceddu A, Danovaro R. Response of benthic protozoa and thraustochytrid protists to fish farm impact in seagrass (*Posidonia oceanica*) and soft-bottom sediments. Microb Ecol 2005;50:268–76.

[46] Booth T, Miller CE. Comparative morphologic and taxonomic studies in the genus *Thraustochytrium*. Mycologia 1968;60:480–95.

[47] Miller JD, Jones EBG. Observations on the association of thraustochytrid marine fungi with decaying seaweed. Bot Mar 1983;26(7):345–51.

[48] Sathe-Pathak V, Raghukumar S, Raghukumar C, Sharma S. Thraustochytrid and fungal component of marine detritus. I. Field studies on decomposition of the brown alga *Sargassum cinereum*. J Ag Indian J Mar Sci 1993;22:159–67.

[49] Raghukumar C, Raghukumar S, Sharma S, Chandramohan D. Endolithic fungi from deep-sea calcareous substrata: isolation and laboratory studies. In: Desai BN, editor. Oceanography of the Indian ocean. New Delhi: Oxford IBH Publication; 1992.

[50] Leander CL, Porter D, Leander BS. Comparative morphology and molecular phylogeny of aplanochytrids (Labyrinthulomycota). Eur J Protistol 2004;40:317–28.

[51] Burja AM, Radianingtyas H, Windust A, Barrow CJ. Isolation and characterization of polyunsaturated fatty acid producing *Thraustochytrium* species: screening of strains and optimization of omega-3 production. Appl Microbiol Biotechnol 2006;72:1161–9.

[52] Raghukumar S, Raghukumar C. Thraustochytrid fungoid protists in faecal pellets of the tunicate *Pegea confoederata*, their tolerance to deep-sea conditions and implication in degradation processes. Mar Ecol Prog Ser 1999;190:133–40.

[53] Wagner-Merner BT, Duncan WR, Lawrence JM. Preliminary comparison of Thraustochytriaceae in the guts of a regular and irregular echinoid. Bot Mar 1980;23(2):95–7.

[54] Bahnweg G, Sparrow FK. *Aplanochytrium kerguelensis* gen. nov. spec. nov., a new phycomycete from subantarctic marine waters. Arch Mikrobiol 1972;81:45–9.

[55] Yao T, Asayama Y. Animal-cell culture media: history, characteristics, and current issues. Reprod Med Biol 2016;16:99–117.

[56] FioRito R, Leander C, Leander B. Characterization of three novel species of Labyrinthulomycota isolated from ochre sea stars (*Pisaster ochraceus*). Mar Biol 2016;163. 170. pp.1–10.

[57] Rosa SM, Galvagno MA, Velez CG. Adjusting culture conditions to isolate thraustochytrids from temperate and cold environments in Southern Argentina. Mycoscience 2011;52:242–52.

[58] Taoka Y, Nagano N, Okita Y, Izumida H, Sugimoto S, Hayashi M. Effect of addition of Tween 80 and potassium dihydrogen phosphate to basal medium on the isolation of marine eukaryotes, thraustochytrids. J Biosci Bioeng 2008;105(5):562–5.

[59] He P, Wu S, Pan L, Sun S, Mao D, Xu C. Effect of Tween 80 and acetone on the secretion, structure and antioxidant activities of exopolysaccharides from *Lentinus tigrinus*. Food Technol Biotechnol 2016;54(3):290–5.

[60] Liu Y-S, Wu J-Y. Effects of Tween 80 and pH on mycelial pellets and exopolysaccharide production in liquid culture of a medicinal fungus. J Ind Microbiol Biotechnol 2012;39:623–8.

[61] Zhang B-B, Cheung PCK. A mechanistic study of the enhancing effect of Tween 80 on the mycelial growth and exopolysaccharide production by *Pleurotus tuber-regium*. Bioresour Technol 2011;102:8323–6.

[62] Dryl S, Mehr K. Cytopathological effects of detergents on *Paramecium caudatum*. Trans Am Microsc Soc 1976;95(4):544–53.

[63] Reitermayer D, Kafka TA, Lenz CA, Vogel RF. Interrelation between Tween and the membrane properties and high pressure tolerance of *Lactobacillus plantarum*. BMC Microbiol 2018;18:72.

[64] Nielsen CK, Kjems J, Mygind T, Snabe T, Meyer RL. Effects of Tween 80 on growth and biofilm formation in laboratory media. Front Microbiol 2016;7:1–10. Article 1878.

[65] Feehily C, Karatzas KAG. Role of glutamate metabolism in bacterial responses towards acid and other stresses. J Appl Microbiol 2013;114:11–24.

[66] Elliott AM. The influence of pantothenic acid on growth of protozoa. Biol Bull 1935;68(1):82–92.

[67] Wilkens SL, Maas EW. Development of a novel technique for axenic isolation and culture of thraustochytrids from New Zealand marine environments. J Appl Microbiol 2012;112:346–52.

[68] Pandey A, Bhathena Z. Prevalence of PUFA rich thraustochytrids sps. along the coast of Mumbai for production of bio oil. J Food Nutr Res 2014;2(12):993–9.

[69] Taoka Y, Nagano N, Okita Y, Izumida H, Sugimoto S, Hayashi M. Use of an antifungal drug, amphotericin B for isolation of thraustochytrids. J Biosci Bioeng 2010;110(6):720–3.

[70] Yadagiri KK, Kerrigan J, Martin SB. Improved methods for axenic culture of Labyrinthula terrestris, causal agent of rapid blight of turfgrasses. Can J Microbiol 2012;58:1230–5.

[71] Cox SL, Hulston D, Maas EW. Cryopreservation of marine thraustochytrids (Labyrinthulomycetes). Cryobiology 2009;59:363–5.

[72] Harel M, Ben-Dov E, Rasoulouniriana D, Siboni N, Kramarsky-Winter E, Loya Y, Barak Z, Wiesman Z, Kushmaro A. A new thraustochytrid, strain Fng1, isolated from the surface mucus of the hermatypic coral *Fungia granulose*. FEMS Microbiol Ecol 2008;64:378–87.

[73] Li Q, Chen G-Q, Fan K-W, Lu F-P, Aki T, Jiang Y. Screening and characterization of squalene-producing thraustochytrids from Hong Kong mangroves. J Agric Food Chem 2009;57:4267–72.

[74] Yang H-L, Lu C-K, Chen S-F, Chen Y-M, Chen Y-M. Isolation and characterization of Taiwanese heterotrophic microalgae: screening of strains for docosahexaenoic acid (DHA) production. Mar Biotechnol 2010;12:173–85.
[75] Gupta A, Wilkens S, Adcock JL, Puri M, Barrow CJ. Pollen baiting facilitates the isolation of marine thraustochytrids with potential in omega-3 and biodiesel production. J Ind Microbiol Biotechnol 2013;40:1231–40.
[76] Ludevese-Pascual G, Pena D, Tornalejo J. Biomass production, proximate composition and fatty acid profile of the local marine thraustochytrid isolate, Schizochytrium sp. LEY7 using low-cost substrates at optimum culture conditions. Aquacult Res 2014:1–11.
[77] Ou M-C, Yeong H-Y, Pang K-L, Phang S-M. Fatty acid production of tropical thraustochytrids from Malaysian mangroves. Bot Mar 2016;59(5):321–38.
[78] Marchan LF, Chang KJL, Nichols PD, Ploglase JL, Mitchell WJ, Gutierrez T. Screening of new British thraustochytrids isolates for docosahexaenoic acid (DHA) production. J Appl Phycol 2017;29:2831–43.
[79] Jaritkhuan S, Suanjit S. Species diversity and polyunsaturated fatty acid content of thraustochytrids from fallen mangrove leaves in Chon Buri province, Thailand. Agri Nat Res 2018;52:24–32.
[80] Song Z, Stajich JE, Xie Y, Liu X, He Y, Chen J, Hicks GR, Wang G. Comparative analysis reveals unexpected genome features of newly isolated thraustochytrids strains: on ecological function and PUFAs biosynthesis. BMC Genomics 2018;19:541.

Chapter 31

Advances in sampling strategies and analysis of phytoplankton

Priya M. D'Costa[1], Ravidas K. Naik[2]
[1]*Department of Microbiology, Goa University, Taleigao Plateau, Goa, India;* [2]*ESSO-National Centre for Polar and Ocean Research, Vasco, Goa, India*

31.1 Introduction

The word *phytoplankton* is derived from the Greek words *phyton* = plant, and *plankton* = wanderer. Phytoplankton are microscopic, single-celled organisms found in all aquatic habitats. They also account for up to 50% of the global primary production [1], and are the base of food webs in aqueous environments. They also play crucial roles in influencing the climate through dimethylsulfoniopropionate (DMSP) production and influences on cloud formation [2], and formation of harmful algal blooms (HABs).

Phytoplankton is composed of both eukaryotic and prokaryotic species. The dominant group in coastal waters is diatoms; other phytoplankton groups are dinoflagellates, coccolithophores, raphidophytes, flagellates, etc. (Fig. 31.1). Diatoms are silica-shelled protists belonging to phylum Bacillariophyta. They are divided into two major orders: order Centrales (centric diatoms having radial symmetry) and order Pennales (pennate diatoms exhibiting bilateral symmetry). Centric diatoms usually inhabit the water column; only some genera grow attached to substrates during their entire life cycle. A few pennate diatoms possess a raphe (long slit along the length of the frustules) that helps in motility and attachment to substrata. Diatoms reproduce asexually by vegetative division and also through sexual means via spores or resting cells. Though the majority of them are photosynthetic, some diatom species also possess a heterotrophic mode of nutrition [3].

Dinoflagellates, another phytoplankton group, are characterized by the presence of cellulose cell walls and two flagella. They possess a range of nutritive strategies (autotrophic, mixotrophic, heterotrophic) and contribute to HABs. They produce resting stages (termed cysts) that help them to survive unfavorable conditions; long-term cysts are reported to provide a seed bank for HABs [4].

Advances in Biological Science Research. https://doi.org/10.1016/B978-0-12-817497-5.00031-8

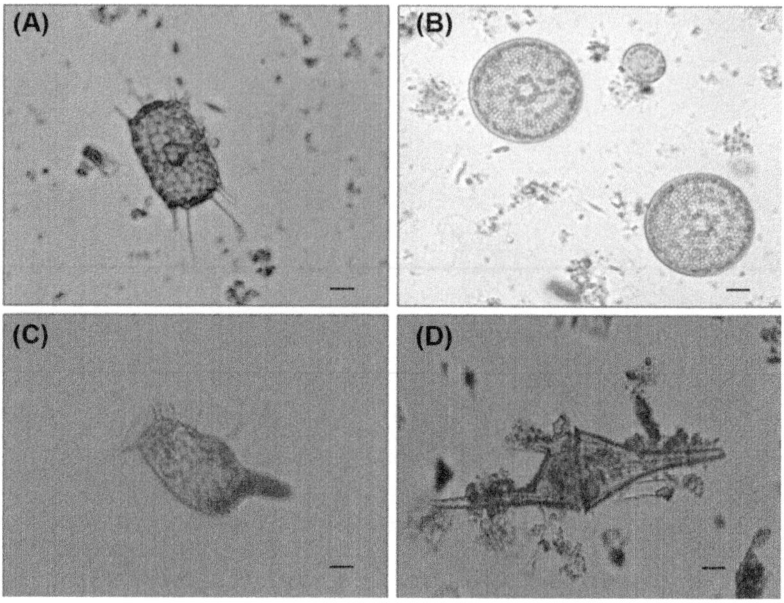

FIGURE 31.1 Diatoms – (A) *Odontella* sp., (B) *Coscinodiscus* sp.; dinoflagellates – (C) *Dinophysis caudata*, (D) *Tripos furca*. (Scale bar in the image = 20 μm).

Picophytoplankton, <3 μm in size [5], are the smallest group of phytoplankton. They have a ubiquitous distribution in both fresh- and marine waters [6], with abundance from 10^4 cells ml^{-1} upwards [7]. They consist of three groups: *Synechococcus* and *Prochlorococcus* (both cyanobacteria) and picoeukaryotes, and contribute a significant proportion of the phytoplankton biomass as well as the primary productivity in various ecosystems [8,9].

31.2 Sampling strategies

Since phytoplankton encompasses a wide range of organisms, both prokaryotic and eukaryotic, with different characteristics, nutritional strategies, varying size ranges, and different econiches, varied strategies have been developed for sampling and analysis of different phytoplankton groups.

31.2.1 Choice of research vessel

The type of cruise vessel used for sampling in coastal waters (the waters above the continental shelf) depends on the depth of the water column and also on the intended number of sampling stations. For near-shore sampling in areas with water depth below 50 m, trawlers of 10–12 m length are used. The sampling duration is restricted to 12 h and, accordingly, few stations are targeted.

In coastal sampling where the water column depth varies between 100 and 500 m, coastal research vessels (CRVs) are used, with a sampling duration of 1–15 days.

Oceanic waters (above the continental slope) are studied for phytoplankton by conducting cruises aboard ocean research vessels (ORVs). The cruise duration may extend up to 60–70 days depending on the endurance of the ship. In addition to this, there is another type of research vessel that is used specifically in polar oceans. They are called polar research vessels (PRVs) and are characterized by their ability to navigate successfully through ice-covered waters. Also known as "icebreakers," these ships are the mainstays of research in polar environments.

Research vessels (CRV, ORV, and PRV) are equipped with basic scientific instruments like conductivity temperature depth/CTD (portable, on CRV), expendable bathy thermograph, expendable conductivity temperature depth; and grabs and corers to sample water and sediment. CTD is mounted on a large round metal frame called a rosette along with water sampling bottles. Different sensors—for chlorophyll, photosynthetic active radiation, turbidity, dissolved oxygen, and nitrate—are mounted on a vertical panel so as to enable collection of water samples and determination of the associated physicochemical parameters from various desired depths. In this manner, depth-wise profiles can also be discerned. Surface samples are generally collected by using a grab sampler or a box corer, whereas, deep cores are collected mainly by using gravity corer by geological/paleontological studies. A multicorer is an instrument that can simultaneously collect several subsamples without disturbing surface sediments.

31.2.2 Sampling in coastal waters

In shallow coastal waters, water samples are collected using either buckets (for surface waters) or water samplers like HYDROBIOS/NISKIN samplers (for near-bottom waters). Sediment samples are collected using either grabs such as the van Veen grab or manually by divers using corers (in shallow waters).

In case of rocky, intertidal areas and beaches, surface water, interstitial water and sediment samples are collected for phytoplankton analysis. Collection of surface waters is done with a bucket. Interstitial water (also called pore water) is collected with a syringe after digging a hole in the sand and allowing the interstitial water to accumulate. However, this method is rather laborious. Sediment, in addition to interstitial water, can be collected from three zones, which are demarcated depending on the tide—low-tide, mid-tide, and high-tide levels. Surface sediment (0–2 cm), when exposed, is collected in zip-lock bags, using a spatula. Vertical sediment cores can be collected for analysis of depth-wise profiles of benthic phytoplankton, and especially for diatom migration investigations, using PVC cores of a suitable diameter (2.5 cm). This enables collection of sediment from depths ranging from surface to up to even 10 cm. Epibiotic phytoplankton are phytoplankton that grow attached to either living hosts (plant/animal) or nonliving substrates (e.g., rocks); these are sampled by transferring the host organism with forceps into polyethylene zip-lock bags. All the samples should be kept in the shade, in cool conditions, and transferred to the laboratory as soon as possible.

31.2.3 Aspects to be considered

One crucial aspect that needs to be considered while planning sampling strategies for phytoplankton is tide timings, especially for intertidal sampling. Secondly, information regarding the depth of the water column, chlorophyll concentration and supporting physicochemical parameters (temperature, salinity, pH, nutrients, suspended particulate matter, etc.) must also be collected. Thirdly, if the purpose of the study is to examine the community structure of benthic phytoplankton, microscopic analysis of a defined number of samples, whether fresh or fixed, will be sufficient. In case of fixed samples, the choice of preservative is of paramount importance, since use of incorrect preservatives will lead to the production of artifacts [10], which will interfere with sample analysis. The most preferred and commonly used fixative for microplankton is Lugol's iodine solution, added to achieve a final concentration ranging from 1% to 10%. Lugol's iodine has several advantages over that of other preservatives, the main one being that flagellates do not lose their flagella when exposed to Lugol's iodine [10]. Also, Lugol's iodine is relatively easy to prepare and is stable for years. However, fixed samples require to be monitored during storage as iodine oxidizes with time.

31.3 Analysis of phytoplankton

Since phytoplankton can be studied for different purposes, the analytical methods depend on the aim of the study and are detailed in the following discussion.

31.3.1 Phytoplankton taxonomy

This study requires observing minute morphological details of individual phytoplankton species to identify and classify them as per taxonomic nomenclature. Conventional light microscopy has been the mainstay of phytoplankton taxonomy and analysis for several decades. However, since the late 1960s, electron microscopy (scanning electron microscopy, SEM; and transmission electron microscopy, TEM) has proved extremely useful in establishing accurate phytoplankton taxonomy using ultra-structural features such as ornamentation of body scales or architecture of flagellar roots and central nodules.

SEM has helped substantially in clarifying the taxonomy of phytoplankton. A case in point is that of *Skeletonema* species. *Skeletonema* species can be differentiated on the basis of certain morphological peculiarities that can be seen only with SEM. *Skeletonema* species, which are abundant in the coastal waters around India, were resolved up to species level using SEM [11]. The intricate frustule designs observed in diatoms using SEM led to them being recognized as a source of nanomaterials [12]. TEM has proved useful for the determination of ultrastructural features in phytoplankton, and the detection of intracellular bacteria in *Pinnularia* [13], a pennate diatom, and in *Alexandrium catenella* and *Protoceratium reticulatum* (toxic dinoflagellates) [14].

31.3.2 Analysis of phytoplankton community structure

Fixed samples, once transported to the laboratory, are kept undisturbed for settling [15], followed by siphoning of the upper water with a 10 μm mesh-covered tube. This is followed by microscopic analysis of aliquots of the concentrated sample, identification based on standard taxonomic keys and quantification.

Given that the above procedure is time-consuming, the FlowCAM (Fig. 31.2), an integrated system for detection and analysis of marine

FIGURE 31.2 The FlowCAM instrument used for analysis of phytoplankton.

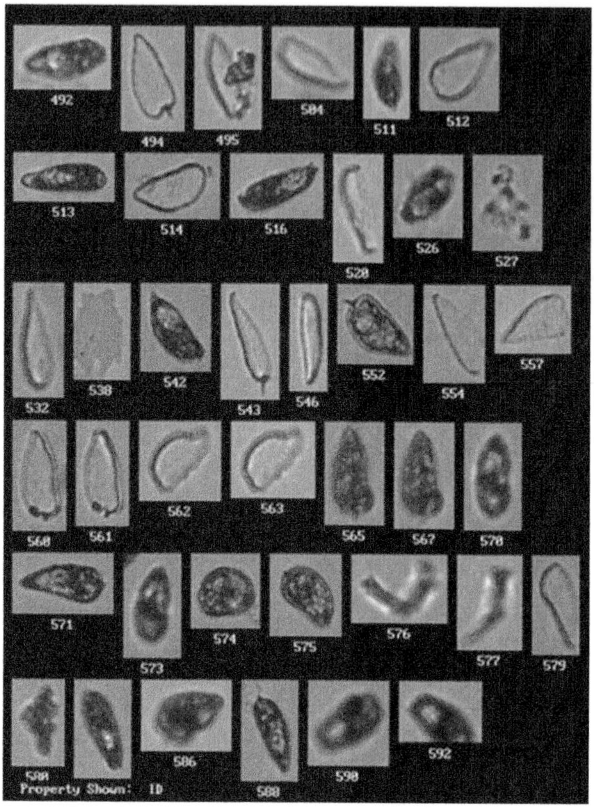

FIGURE 31.3 The image library generated during phytoplankton analysis using FlowCAM.

plankton and particles in a continuous fluid flow, has revolutionized phytoplankton analysis. It was originally developed for investigations of organisms and particles in seawater [16]. It combines the selective capabilities of flow cytometry, microscopy, and fluorescence detection. The main strength of the FlowCAM is the availability of individual images of each particle counted in a library (Fig. 31.3), along with information about its optical properties. By acquiring and storing a digital image of each particle detected, postanalysis identification, differentiation and quantification is made possible. Also, since it acquires high-resolution microscopic images at a very rapid rate, typically up to 10,000 images/minute, rapid estimation of plankton species and biovolume is possible, in addition to information about size, shape, fluorescence, and concentration statistics and, that too, in a fraction of the time required by traditional microscopy. Thus, FlowCAM analysis provides almost real-time information about the dynamics of the plankton community being studied.

Another instrument called the continuous plankton recorder (CPR) has been in use since 1931 for the real-time analysis of phytoplankton in the field [17]. This instrument is towed through the water by a research vessel. In fact, there is a major CPR survey program conducted by the Marine Biology Laboratory, Plymouth, UK. Such surveys have been carried out in various parts of the world's oceans. However, due to the higher mesh size used (>200 μm), it is preferred for zooplankton analysis [17].

31.3.3 Analysis of benthic diatoms

Benthic diatoms (those occurring in sediment) can be studied using various techniques, depending on the aim of the study, as discussed above. It is problematic to discern diatoms microscopically due to the presence of sediment grains, which may hide from view, diatoms attached to sediment grains, on their side away from the light. The extinction-dilution method [18] has been widely employed for study of benthic diatoms (vegetative forms and resting stages). This method involves incubation of the sediment sample in appropriate seawater-based media, after which the diatom growth is monitored microscopically, identified based on standard taxonomic keys, and enumerated. Since it is a culture method, it enables the detection of vegetative cells as well as algal resting stages that tend to be obscured in soil aggregates [19]. The most probable number is then generated based on a statistical table, and then multiplied with the specific gravity of the sediment sample, which is calculated separately. However, this method does not give the exact number of diatom cells but rather enumerates the relative abundance of different taxa. This is because it is based on the presence and/or absence of diatoms and involves a statistical table [18] and probability theory as part of its calculations [19].

31.3.3.1 Modifications of the extinction–dilution method

The extinction–dilution method can be modified to suit the needs of the investigator. For example, artificial seawater can also be used as the basal diluent instead of natural seawater, especially in cases where comparison across seasons or study areas is essential. This will avoid differences in seawater composition from affecting the diatom community under study. Another modification is the use of UV to detect autofluorescence of chlorophyll-containing cells. This helps to detect live diatoms/resting stages even if they are obscured in detritus.

The extinction–dilution method, modified by the incorporation of penicillin (an antibiotic that is effective mainly against actively growing cells), has been used to detect the influence of bacteria on benthic diatom community structure [20]. It has also been used to study the effect of different classes of antibiotics (aminoglycosides – streptomycin, chloramphenicol) on benthic diatom community structure [21].

31.3.4 Analysis of dinoflagellate cysts

Dinoflagellate cysts can be analyzed in several ways depending on the requirements of the study. They are reported to have a long life span; they have been successfully germinated from 32 to 63 cm depth in the sediment, corresponding to a time span of 100 years (based on dating techniques) [22–24]. To check the germination potential of dinoflagellate cysts, sediment suspensions are processed to dislodge sediment particles, and then the concentrated cyst fractions are viewed microscopically to differentiate between entire and ruptured cysts. The entire cysts are picked up and transferred to autoclaved seawater in multiwells (with salinity adjusted to an optimum level, usually 32 psu) and then incubated in cool, bright light (12 h:12 h light: dark conditions) for up to several weeks. Intermittent microscopic observations are required to check whether the cysts have germinated. In this regard, it is important to note that cysts have a mandatory dormancy period during which they will not germinate even when provided with favorable conditions. They will germinate subsequently, only when a "germination window" (favorable conditions) opens up.

Dinoflagellate cysts can also be studied in recent sediments or dated sections from deeper cores by a paleontological method, involving treatment of sediment samples with hydrochloric and hydrofluoric acid, respectively [25]. Hydrochloric and hydrofluoric acids dissolve calcium carbonate and silica of coccolithophores and diatoms, respectively. Dinoflagellate cysts, which have a sporopollenin layer, are resistant to the action of acids. The sodium polytungstate density gradation method, which separates living dinoflagellate cysts from inorganic particles and organic detritus [26], has also been used to study dinoflagellate cyst assemblages [27]. Cysts recovered in this manner can be used for germination experiments [26].

Identification of dinoflagellate cysts can be done based on two approaches: (1) based on the names of the vegetative dinoflagellate forms, and (2) based on paleontological nomenclature. This again depends on the nature of the investigation.

31.3.5 Study of fouling diatoms/biofilms

If fouling diatoms are to be examined, a substrate of choice (glass/fiberglass/polystyrene/stainless steel) is suspended in the water column or anchored in the soil (for benthic diatoms) at the study area for a suitable incubation period. The duration of the incubation period will vary depending on the season, the potential for macrofouling, etc. In a study on the fouling diatom community structure from a monsoon-influenced tropical estuary in Goa, west coast of India, macrofouling interference in biofilm development was observed after

4 days of incubation [28]. Thus, these crucial points must be considered while deliberating on the duration of the incubation period for development of natural marine biofilms.

Biofilms can also be developed on panels suspended in aquaria in the laboratory containing fresh natural seawater that has been transported from the sampling area [29,30]. However, biofilms developed in this manner will differ from those developed in natural environments, mainly in the degree of predation they face, the nutrient pulses and turbulent conditions that they are exposed to, and so on.

Once biofilms are developed, whether in the natural environment or artificially in the laboratory, they can be studied for fouling diatoms either directly, by viewing the substrate microscopically, or by scraping of the biofilm material, fixing it, and then viewing it microscopically to identify and quantify the fouling forms. Brushing and scraping are the conventional means of biofilm removal from solid substrates, especially for fouling diatoms. Since this pivotal step will influence the results of the analysis, it is extremely important to select the right method to dislodge fouling diatoms from the substrate. Patil and Anil [31] compared two methods of biofilm removal (nylon brush and ceramic scraper), with emphasis on diatoms. They reported that the nylon brush showed a higher efficiency in recovering diatoms compared to the ceramic scraper.

31.3.6 Analysis of epibiotic phytoplankton

Epibiotic phytoplankton are usually scraped off the surface of the basibiont host, with a nylon brush into a predetermined amount of filtered seawater. This can be done in case of a hard surface like the chitinous exoskeleton of horseshoe crabs [32]. For epiphytic phytoplankton attached to the surface of seaweeds, they are detached by suspending the seaweeds in a known volume of filtered seawater, mixed manually [33] or by incubating on a shaker (250 rpm) for approximately 30 min. In both the cases mentioned above, the scraped material/seaweed suspension is fixed with Lugol's iodine, concentrated, and enumerated microscopically.

31.3.7 Study of picophytoplankton

Picophytoplankton, due to their small size, are difficult to analyze with traditional methods like epifluorescence microscopy. However, flow cytometry, initially developed for use in the biomedical field, has been successfully applied to picophytoplankton studies since the 1980s. In fact, flow cytometry played a key role in the discovery of picoplankton [34]. Basically, it is a technique for making measurements (-metry) on cells (cyto-), using microscopic methods, performed while the cells flow in a stream past an array of optical detectors. It offers information regarding the pigment content (based on fluorescing pigments—chlorophyll, phycoerythrin, phycocyanin), cell size, scatter properties, and abundance of the major photosynthetic picophytoplankton. It does so at a rapid rate (several thousand cells per minute). It can also be applied for study of larger phytoplankton (100–200 μm and above), with certain modifications incorporated in the instrument [35].

Flow cytometry has been used to study various aspects of picophytoplankton, even in the field. For example, community structure of picophytoplankton by discriminating between phycoerythrin-rich *Synechococcus*, phycocyanin-rich *Synechococcus*, *Prochlorococcus*, and picoeukaryotes. It has also been used to study viral infection of algal cells [36]. Another interesting aspect is that some flow cytometry instruments (sorters) have the ability to physically sort cells into fractions. This has been of immense help in cell cycle studies and discrimination of several morphologically and ultrastructurally variant cells within a single clonal phytoplankton population [37].

For picoplankton analysis, water/biofilm samples (approx. 2 mL) should be collected in cryovials, kept away from sunlight, stored at 4°C without any fixative, and analyzed within 12 h of collection [36]. This can be done using benchtop flow cytometry instruments on board the ship. This is a huge advantage in that samples, in their live state, can be analyzed immediately, and also, cell loss due to the disruptive effects of preservatives is not an issue [38]. If immediate analysis is not possible, samples can be preserved in paraformaldehyde or glutaraldehyde and stored frozen, either at −80°C or in liquid nitrogen, until analysis [39].

31.3.8 Phytoplankton pigment analysis

This study requires filtration of water samples through glass fiber filters, which are then stored at low temperature (−20°C, −40°C, −80°C or liquid nitrogen) until analysis to estimate phytoplankton biomass (chlorophyll *a*) or phytoplankton community structure (pigment indices). Chlorophyll *a* (chl *a*) can be estimated using different methods including spectrophotometry (samples scanned between 750 and 350 nm) and fluorometry (excitation CS-5-60 and emission CS-2-64). Spectrophotometric analysis of pigment extracts can accurately measure chl *a, b*, and *c* (not carotenoids) through the use of trichloromatic equations [40,41]. However, the fluorometric method, with a higher sensitivity (50 times more sensitive for chl *a*) has been widely used in present-day research. In vivo fluorometry is commonly used to detect phytoplankton chlorophyll fluorescence in seawater [42]. It is semiquantitative; the fluorescence of chlorophyll depends on the species present, the time of day, accessory pigments, and the physiological condition of the cells [43]. In extracted fluorometry, the chlorophyll is extracted into methanol or acetone before measurement [44]. The response is quantitative only if chlorophyll *a* is the dominant chlorophyll and there are no chlorophyll *b*, pheophytins, pheophorbides, or chlorophyllides present. Another advantage of fluorometry is that some problems of interference from other chlorophylls and degradation products are reduced [45], since the excitation and emission wavelengths are varied to detect chlorophyll *a*, *b*, and *c* and their degradation products separately.

These methods are now superseded by pigment discrimination and quantification by high performance liquid chromatography (HPLC) directly from seawater samples. HPLC systems were developed for higher plant pigments in the late 1970s. Subsequently, HPLC systems of increasing complexity were developed for chlorophylls and carotenoids from microalgae and natural phytoplankton populations [46–49].

HPLC profiles provide estimates of the oceanic distribution of specific algal classes, based on diagnostic pigments such as fucoxanthin (specific for diatoms), peridinin (dinoflagellates), zeaxanthin (cyanobacteria), and prasinoxanthin (Prasinophyceae) [50,51]. The linear relationship between diagnostic pigment and total chlorophyll *a* is used to calculate the percentage contribution of phytoplankton functional groups [52,53]. It is one of the most prolific separation techniques in use and compares well with other techniques used in phytoplankton analysis (Table 31.1).

A very relevant UNESCO monograph *Phytoplankton Pigments in Oceanography*, published in 1997, covered most of the aspects of HPLC analysis of planktonic pigments, reviewing the application of existing methods to oceanography, and proposing new isocratic and gradient HPLC methods [54]. Generally, reversed-phase HPLC systems are employed rather than normal-phase HPLC systems. C_8 columns are preferred over C_{18} columns due to the ability of C_8 phases to separate isomeric pairs of pigments with slight differences in polarity, such as mono- and divinyl chlorophylls or lutein and zeaxanthin. Isocratic HPLC can rapidly analyze chlorophylls and derivatives from relatively small volumes of seawater. It is especially sensitive when coupled with fluorescence detection. Its limited resolution, however, usually prevents its use for full pigment analysis. Gradient HPLC is the method of choice for full analysis of chlorophylls and carotenoids in oceanography. The equipment is, however, expensive and the analyst requires more time and technical expertise than for nonchromatographic techniques.

31.3.9 Analysis of viability and photosynthetic parameters of phytoplankton populations

Quantification of photosynthetic pigments (mainly chlorophyll) through spectrophotometry, fluorometry, and HPLC does not always accurately reflect the status of the resident phytoplankton community. This is because, firstly, chlorophyll molecules stay intact within a dead cell for up to 2 weeks [55]. Secondly, it is tricky to quantify the number of living cells per volume of water from a biomass index, mainly so when the species and physiological states are

TABLE 31.1 Comparison of different techniques involved in phytoplankton analysis.

S. No.	Property	LM	SEM	FlowCAM	HPLC
1.	Specimen status	Live, preserved	Preserved	Live, preserved	Preserved
2.	Level of preparation of sample	Simple	Expert	Simple	Expert
3.	Magnification extent	Low	High	Low	–
4.	Resolution	Low	High	Low	Resolution in terms of diagnostic pigment indices
5.	Type of output	Colored 2D images	Black-and-white 3D images	Colored 2D images	Chromatogram
6.	Requirement for specific conditions	No vacuum required	Vacuum required	No vacuum required	Solvents required
7.	Cost	Cheap	Expensive	Expensive	Expensive

unknown [56]. Thus, additional techniques that focus on the viability and physiological status of the phytoplankton population are necessary.

The viability of phytoplankton cells can be determined using vital fluorescent stains such as fluorescein diacetate (FDA), 2-(4-pyridyl)-5-{[4-dimethylaminoethyl-aminocarbomoyl-methoxy]phenyl}oxazole (PDMPO) and 5-chloromethylfluorescein diacetate (CMFDA) [56,57]. FDA and CMFDA are membrane-permeable vital stains, which are acted upon by nonspecific esterases, present inside viable, intact cells. The end result is nonpermeable, green fluorescent products, which indicates viable cells [56]. PDMPO has a different mechanism of action; it gets incorporated along with silica precipitation during frustules synthesis in viable, growing cells. Thus, the intensity of PDMPO fluorescence is directly related to the quantity of silica precipitated [57,58].

Another interesting parameter that can be measured is the photosynthetic efficiency of the sample, indicative of the physiological status of the phytoplankton cells. It analyzes the amount and organization of protein molecules around the light-harvesting complex, PSII. It is measured using the fluorescence induction and relaxation (FIRe) technique [59]. The instrument involved is a FIRe fluorometer, fitted with a fiber-optic probe. These measurements are taken from fresh samples, after allowing an adaptation period of 30 min in the darkness. This method is quick and nondisruptive [59].

31.3.10 Toxin analysis

Several phytoplankton species are known to produce toxins, especially those phytoplankton that cause HABs. The presence of toxins is a growing global concern, due to its ill effects on human health and also to the national economy. There are several toxins—domoic acid, saxitoxin, okadaic acid, gymnodimine, nodularin, yessotoxin, brevetoxins, etc.—that are genus/species specific. The reasons for toxin production by phytoplankton could vary from species to species; it could be a part of the defense mechanisms, metabolism, allelopathy, resource competition, and also a result of nutrient stress. Since the mechanisms involved are complex, one needs to use specific experimental approaches with targeted species. However, the most pressing issue that needs to be addressed is the detection and monitoring of coastal waters for toxins from all aquatic bodies and especially their filter-feeding inhabitants. There are different methods by which toxins can be collected, extracted, and analyzed. Toxins can be extracted from muscle tissue of organisms affected during algal blooms or directly from the cell extract of bloom-forming species. However, these are the possibilities only if there is a bloom, whereas, during nonbloom conditions, any given aquatic area can be surveyed for toxins by using grab sediment samples followed by laboratory cleanup and extraction. Such sampling still remains one of the common approaches of researchers. However, a new technique has been introduced by MacKenzie et al. [60] wherein passive sampling of toxin can be done using solid-phase adsorption toxin tracking (SPATT). The SPATT device is a simple instrument made up of a polyester mesh bag containing activated resins of polystyrene-divinylbenzene, which can adsorb lipophilic toxins dissolved in water. The SPATT bags are deployed and retrieved from the study location to get a time-averaged toxin concentration. This technique provides a clean sample matrix, which simplifies the subsequent extraction and analysis using enzyme-linked immunosorbent assay or liquid chromatography-tandem mass spectrometry.

Marine phytoplankton toxins also have medical and commercial significance; dinoflagellate toxins, particularly, are receiving increasing interest [61]. For example, pectenotoxin produced by *Dinophysis* species displays cytotoxic activity against various human cancer cells [62]; goniodomin-A, produced by *Goniodoma pseudogonyaulax*, and gambieric acid, produced by *Gambierdiscus toxicus*, have antifungal properties; zooxanthella toxins produced by *Symbiodinium* sp. exhibit potent vasoconstrictive activity. Similarly, the marine biotoxin okadaic acid is vital in medical research for understanding several cellular processes [63]. Mass culturing of such potent species on a commercial scale in the future could benefit medical research. However, even though toxins have high medical and commercial importance, their inaccessibility, especially in case of dinoflagellates, is the major hindrance in fully exploring their research potential.

31.4 Primary productivity

Primary productivity is the rate of carbon fixation with respect to space and time. There are various methods to estimate the primary production such as oxygen evolution in dark and light incubations, ^{14}C, ^{13}C, ^{18}O, and the fast repetition rate fluorometry (FRRF) method. Comparison of these methods revealed differences of varying magnitude [64,65]. Among these methods, oxygen evolution in dark and light incubation and ^{14}C methods are classical methods that have been in practice for the last 8 decades [66,67]. The comparatively new methods involve the use of ^{13}C, ^{18}O that are stable isotope tracers [68,69].

In the ^{14}C method [70], bicarbonate ion is labeled with a radioactive isotope of carbon. This is done by adding a known quantity of radioactive bicarbonate (HCO_3^-) to two bottles of marine samples containing phytoplankton. Further, these bottles are incubated in two tanks, one of which is covered with a black film to avoid light and the other is without the film. In the bottle with film, only respiration will take place. In the other bottle, which is exposed to light, both photosynthesis and respiration will occur. After the incubation period is over, the samples from bottles are filtered out. The amount of radioactive carbon taken up per unit time is measured using a scintillation counter and the primary productivity calculated (mg C m^{-3} h^{-1}). In the ^{13}C method [68], bicarbonate ion is labeled with natural stable isotope ^{13}C instead of radioactive isotopes, and hence there is no risk factor involved during analysis. In the ^{18}O method [69], samples are enriched with stable isotope ^{18}O as a tracer to measure the gross primary production. This method measures direct gross photosynthetic O_2 production compared to the ^{14}C method. The error in gross primary production estimates from this method is less than 2% [69].

In addition to this, active fluorescence and incubation-free techniques such as FRRF [71] measure the instantaneous depth and time-dependent productivity from active chlorophyll fluorescence at spatial (<1 m) and temporal (1 s) resolutions [72]. This method has better signal-to-noise ratio, which makes possible robust measurements even in oligotrophic regions. This method has the potential to measure physiological parameters instantaneously and also to note the rapid changes that occur in productivity [73]. Yet, this method is also fraught with uncertainties. High concentration of colored-dissolved organic matter (C-DOM) affects the spectral absorption of the FRRF, resulting in errors in estimation [74].

31.4.1 Estimation of primary productivity using remote sensing

Due to the limitation of in situ observations on a regular basis and nonaccessibility of certain study areas, it is difficult to estimate primary productivity on a large scale. These limitations can be overcome by using the remote-sensing approach, wherein satellite-derived chlorophyll can be used to compute the primary productivity of a given area [75]. However, such remote-sensing algorithms need to be calibrated; in a majority of the cases, such calibration has been carried out using the ^{14}C method [76]. The limitations of this method include nonavailability of data during cloud coverage and substantial errors in coastal waters due to C-DOM.

31.4.2 Monitoring of HABs using remote sensing

Generally, HABs are monitored using a two-pronged approach: (1) Physical survey at bloom locations during the potential bloom period, and (2) monitoring of blooms using remote sensing as a tool. The former approach gives more details about the bloom dynamics and species responsible for the bloom. However, practically, it is not possible to be present at every bloom location or to conduct frequent surveys to capture initiation of bloom, its decline, and consequences. In such cases, remote sensing is useful to monitor a large area just by using remote-sensing images obtained from ocean color satellites at daily intervals. By looking at chlorophyll concentration from images, it is easy to identify bloom locations. However, due to the lack of species-specific algorithms, it is difficult to identify the bloom-causing genera/species. Hence, the combined approach can be very useful to conduct a bloom survey at the targeted location.

31.5 Future perspectives

Considering the diversity of phytoplankton parameters and the resulting varied approaches to phytoplankton studies, it is not surprising that several new techniques, individually and in conjunction, are being applied to phytoplankton analysis. Phytoplankton taxonomy has seen several advancements. In recent years, phylogenetic analysis of the 18 and 28S rDNA genes has been increasingly applied to confirm the validity of subgroups that have been defined based on morphological characters. In many instances, new species have been discovered and the number and circumscription of subgroups has been refined, for e.g., *Chaetoceros* [77,78] and *Skeletonema* [79]. This approach has been aided by the incorporation of advanced TEM techniques, which allow far higher resolution compared to earlier techniques.

Phytoplankton diversity at the genomic level has been studied using various metagenomic techniques, which can be divided into two main categories of analysis based on (1) probes and (2) sequencing [80]. In the first category, probes specific for particular phytoplankton genes are designed and fixed onto a suitable array. Subsequently, samples are analyzed using these array-based probes. This technique allows the analysis of expression patterns of particular genes in a mixed phytoplankton community. Several array-based metagenomic techniques have been devised and employed for marine picophytoplankton communities [81,82]. The second metagenomic approach, based on sequencing, has been successfully used to determine phytoplankton diversity [80]. This approach is advantageous since it offers a comparatively unbiased determination of phytoplankton diversity and has, therefore, led to the discovery of several unknown phytoplankton lineages.

DNA metabarcoding is another metagenomic technique that has been applied to unravel diversity within morphologically similar phytoplankton species [83]. It involves combining high-throughput sequencing approaches with sequencing of environmental DNA that has been PCR amplified using primers specific for particular phytoplankton groups. This metagenomic method not only enables the determination of spatial and seasonal distribution patterns without cultivation but also allows the discovery of new taxa linked with new sequences, not assigned to a known organism. Additionally, the use of next-generation sequencing methods has enhanced the number of sequences recovered from individual samples, and overcome the bias against rare taxa [83].

In the context of primary productivity, it is relevant to arrive at a global carbon budget with less error. Thus, it is imperative to set a common global protocol to estimate primary productivity. This can be achieved either by following a common method worldwide or by applying a suitable correction factor to nullify the bias that occurs between different methods [84].

Acknowledgments

PD acknowledges the head of the Department of Microbiology, Goa University, for providing necessary facilities. RN is grateful to the director of NCPOR, MoES, for his encouragement. This is NCPOR contribution no. 59/2018.

References

[1] Falkowski P. The power of plankton. Nature 2012;483(Suppl. 7387):S17.

[2] Laroche D, Vezina AF, Levasseur M, Gosselin M, Stefels J, Keller MD, et al. DMSP synthesis and exudation in phytoplankton: a modeling approach. Mar Ecol Prog Ser 1999;180:37–49.

[3] Admiraal W, Peletier H. Influence of organic compounds and light limitation on the growth rate of estuarine benthic diatoms. Br Phycol J 1979;14(3):197–206.

[4] Smayda TJ, Trainer VL. Dinoflagellate blooms in upwelling systems: seeding, variability, and contrasts with diatom bloom behaviour. Prog Oceanogr 2010;85(1–2):92–107.

[5] Sieburth JM, Smetacek V, Lenz J. Pelagic ecosystem structure: heterotrophic compartments of the plankton and their relationship to plankton size fractions 1. Limnol Oceanogr 1978;23(6):1256–63.

[6] Stockner JG, Antia NJ. Algal picoplankton from marine and freshwater ecosystems: a multidisciplinary perspective. Can J Fisheries Aquat Sci 1986;43(12):2472–503.

[7] Contant J, Pick FR. Picophytoplankton during the ice-free season in five temperate-zone rivers. J Plankton Res 2013;35(3):553–65.

[8] Paerl HW. Ultraphytoplankton biomass and production in some New Zealand lakes. NZ J Mar Freshw Res 1977;11:297–305.

[9] Platt T, Rao DS, Irwin B. Photosynthesis of picoplankton in the oligotrophic ocean. Nature 1983;301:702–4.

[10] Naik RK, Chitari RR, Anil AC. Karlodinium veneficum in India: effect of fixatives on morphology and allelopathy in relation to *Skeletonema costatum*. Curr Sci 2010;25:1112–6.

[11] Naik RK, Sarno D, Kooistra WHCF, D'Costa PM, Anil AC. *Skeletonema* (Bacillariophyceae) in Indian waters: a reappraisal. Indian J Geo-Mar Sci 2010b;39(2):290–3.

[12] Mishra M, Arukha AP, Bashir T, Yadav D, Prasad GB. All new faces of diatoms: potential source of nanomaterials and beyond. Front Microbiol 2017;8:1239.

[13] Schmid AM. Endobacteria in the diatom *Pinnularia* (Bacillariophyceae). I."Scattered ct-nucleoids" explained: DAPI–DNA complexes stem from exoplastidial bacteria boring into the chloroplasts 1, 2. J Phycol 2003;39(1):122–38.

[14] Córdova JL, Escudero C, Bustamante J. Bloom inside the bloom: intracellular bacteria multiplication within toxic dinoflagellates. Rev Biol Mar Oceanogr 2003;38(2):57–67.

[15] Hasle GR. Settling, the inverted microscope method. In: Sournia A, editor. Phytoplankton manual. Paris: Museum National d'Histoire Naturelle; 1978. p. 88–96.

[16] Sieracki CK, Sieracki ME, Yentsch CS. An imaging-in-flow system for automated analysis of marine microplankton. Mar Ecol Prog Ser 1998;168:285–96.

[17] Reid PC, Colebrook JM, Matthews JBL, Aiken J, et al. The continuous plankton recorder: concepts and history, from plankton indicator to undulating recorders. Prog Oceanogr 2003;58(2–4):117–75.

[18] Throndsen J. The dilution-culture method. In: Sournia A, editor. Phytoplankton manual. vol. 6. Paris: UNESCO Monographs on oceanographic methodology; 1978. p. 218–24.

[19] Harris AS, Jones KJ, Lewis J. An assessment of the accuracy and reproducibility of the most probable number (MPN) technique in estimating numbers of nutrient stressed diatoms in sediment samples. J Exp Mar Biol Ecol 1998;231(1):21–30.

[20] D'Costa PM, Anil AC. The effect of bacteria on diatom community structure–the 'antibiotics' approach. Res Microbiol 2011;162(3):292–301.

[21] D'Costa PM, Anil AC. The effect of antibiotics on diatom communities. Curr Sci 2012;102(11):1552–9.

[22] Lundholm N, Ribeiro S, Andersen TJ, Koch T, Godhe A, Ekelund F, et al. Buried alive–germination of up to a century-old marine protist resting stages. Phycologia 2011;50(6):629–40.

[23] Miyazono A, Nagai S, Kudo I, Tanizawa K. Viability of *Alexandrium tamarense* cysts in the sediment of Funka Bay, Hokkaido, Japan: over a hundred year survival times for cysts. Harmful Algae 2012;16:81–8.

[24] Feifel KM, Fletcher SJ, Watson LR, Moore SK, Lessard EJ. *Alexandrium* and *Scrippsiella* cyst viability and cytoplasmic fullness in a 60-cm sediment core from Sequim Bay, WA. Harmful Algae 2015;47:56–65.

[25] Matsuoka K, Fukuyo Y. Technical guide for modern dinoflagellate cyst study. Japan: WESTPAC-HAB/WESTPAC/IOC, Japan Society for the Promotion of Science; 2000. p. 1–29.

[26] Bolch CJ. The use of sodium polytungstate for the separation and concentration of living dinoflagellate cysts from marine sediments. Phycologia 1997;36(6):472–8.

[27] Narale DD, Anil AC. Spatial distribution of dinoflagellates from the tropical coastal waters of the South Andaman, India: implications for coastal pollution monitoring. Mar Pollut Bull 2017;115(1–2):498–506.

[28] Mitbavkar S, Anil AC. Seasonal variations in the fouling diatom community structure from a monsoon influenced tropical estuary. Biofouling 2008;24(6):415–26.

[29] Mitbavkar S, Raghu C, Rajaneesh KM, Pavan D. Picophytoplankton community from tropical marine biofilms. J Exp Mar Biol Ecol 2012;426:88–96.

[30] Khandeparker L, D'Costa PM, Anil AC, Sawant SS. Interactions of bacteria with diatoms: influence on natural marine biofilms. Mar Ecol 2014;35(2):233–48.

[31] Patil JS, Anil AC. Quantification of diatoms in biofilms: standardisation of methods. Biofouling 2005;21(3–4):181–8.

[32] Patil JS, Anil AC. Epibiotic community of the horseshoe crab *Tachypleus gigas*. Mar Biol 2000;136(4):699–713.

[33] Al-Harbi SM. Epiphytic microalgal dynamics and species composition on brown seaweeds (phaeophyceae) on the northern Coast of Jeddah, Saudi Arabia. J Oceanogr Mar Res 2017;5:153. https://doi.org/10.4172/2572-3103.1000153.

[34] Chisholm SW, Olson RJ, Zettler ER, Goericke R, Waterbury JB, Welschmeyer NA. A novel free-living prochlorophyte abundant in the oceanic euphotic zone. Nature 1988;334(6180):340.

[35] Dubelaar GB, Groenewegen AC, Stokdijk W, van Den Engh GJ, Visser JW. Optical plankton analyser: a flow cytometer for plankton analysis, II: specifications. Cytometry. The J Int Soc Anal Cytol 1989;10(5):529–39.

[36] Marie D, Simon N, Vaulot D. Phytoplankton cell counting by flow cytometry. Algal culturing techniques, vol. 1. Academic Press; 2005. p. 253–67.

[37] Arora M, Anil AC, Burgess K, Delany J, Mesbahi E. Asymmetric cell division and its role in cell fate determination in the green alga *Tetraselmis indica*. J Biosci (Bangalore) 2015;40(5):921–7.

[38] Vaulot D, Courties C, Partensky F. A simple method to preserve oceanic phytoplankton for flow cytometric analyses. Cytometry 1989;10:629–35.

[39] Rajaneesh KM, Mitbavkar S. Factors controlling the temporal and spatial variations in *Synechococcus* abundance in a monsoonal estuary. Mar Environ Res 2013;92:133–43.

[40] Jeffrey ST, Humphrey GF. New spectrophotometric equations for determining chlorophylls a, b, c1 and c2 in higher plants, algae and natural phytoplankton. Biochem Physiol Pflanz (BPP) 1975;167(2):191–4.

[41] Lorenzen CJ, Jeffrey SW. Determination of chlorophyll in seawater. UNESCO Tech Pap Mar Sci 1980;35(1):1–20.

[42] Lorenzen CJ. A method for the continuous measurement of in vivo chlorophyll concentration. Deep Sea Res Oceanogr Abstr 1966;13(2):223–7.

[43] Falkowski P, Kiefer DA. Chlorophyll *a* fluorescence in phytoplankton: relationship to photosynthesis and biomass. J Plankton Res 1985;7(5):715–31.

[44] Holm-Hansen O, Lorenzen CJ, Holmes RW, Strickland JD. Fluorometric determination of chlorophyll. ICES J Mar Sci 1965;30(1):3–15.

[45] Neveux J, Lantoine F. Spectrofluorometric assay of chlorophylls and phaeopigments using the least squares approximation technique. Deep-Sea Res Part I Oceanogr Res Pap 1993;40(9):1747–65.

[46] Abaychi JK, Riley JP. The determination of phytoplankton pigments by high-performance liquid chromatography. Anal Chim Acta 1979;107:1–11.

[47] Mantoura RF, Llewellyn CA. The rapid determination of algal chlorophyll and carotenoid pigments and their breakdown products in natural waters by reverse-phase high-performance liquid chromatography. Anal Chim Acta 1983;151:297–314.

[48] Wright SW, Shearer JD. Rapid extraction and high-performance liquid chromatography of chlorophylls and carotenoids from marine phytoplankton. J Chromatogr A 1984;294:281–95.

[49] Gieskes WW, Kraay GW. Analysis of phytoplankton pigments by HPLC before, during and after mass occurrence of the microflagellate *Corymbellus aureus* during the spring bloom in the open northern North Sea in 1983. Mar Biol 1986;92(1):45–52.

[50] Jeffrey SW, Vesk M, Jeffrey SW. Introduction to marine phytoplankton and their pigment signatures. In: Mantoura RFC, Wright SW, editors. Phytoplankton pigments in oceanography. Paris: UNESCO Publishing; 1997. p. 37–84.

[51] Barlow R, Stuart V, Lutz V, Sessions H, Sathyendranath S, Platt T, et al. Seasonal pigment patterns of surface phytoplankton in the subtropical southern hemisphere. Deep Sea Res I 2007;54:1687–703.

[52] Roy R, Pratihary A, Mangesh G, Naqvi SW. Spatial variation of phytoplankton pigments along the southwest coast of India. Estuar Coast Shelf Sci 2006;69(1–2):189–95.

[53] Naik RK, Anil AC, Narale DD, Chitari RR, Kulkarni VV. Primary description of surface water phytoplankton pigment patterns in the Bay of Bengal. J Sea Res 2011;65(4):435–41.

[54] Jeffrey SW. Application of pigment methods to oceanography. In: Jeffrey SW, Mantoura RFC, Wright SW, editors. Phytoplankton pigments in oceanography: guidelines to modern methods. Paris: UNESCO; 1997. p. 127.

[55] Garvey M, Moriceau B, Passow U. Applicability of the FDA assay to determine the viability of marine phytoplankton under different environmental conditions. Mar Ecol Prog Ser 2007;352:17–26.

[56] Steinberg MK, Lemieux EJ, Drake LA. Determining the viability of marine protists using a combination of vital, fluorescent stains. Mar Biol 2011;158(6):1431–7.

[57] Durkin CA, Marchetti A, Bender SJ, Truong T, Morales R, Mock T, et al. Frustule-related gene transcription and the influence of diatom community composition on silica precipitation in an iron-limited environment. Limnol Oceanogr 2012;57(6):1619–33.

[58] Leblanc K, Hutchins DA. New applications of a biogenic silica deposition fluorophore in the study of oceanic diatoms. Limnol Oceanogr Methods 2005;3(10):462–76.

[59] Patil JS, Jagadeesan V. Effect of chlorination on the development of marine biofilms dominated by diatoms. Biofouling 2011;27(3):241–54.

[60] MacKenzie L, Beuzenberg V, Holland P, McNabb P, Selwood A. Solid phase adsorption toxin tracking (SPATT): a new monitoring tool that simulates the biotoxin contamination of filter feeding bivalves. Toxicon 2004;44:901–18.

[61] Camacho FG, Rodríguez JG, Mirón AS, García MC, Belarbi EH, Chisti Y, Grima EM. Biotechnological significance of toxic marine dinoflagellates. Biotechnol Adv 2007;25:176–94.

[62] Zhou ZH, Komiyama M, Terao K, Shimada Y. Effects of pectenotoxin-1 on liver cells in vitro. Nat Toxins 1994;2:132–5.

[63] Bujalance I, Feijoo M, Lopez PM, Munoz MD, Munoz-Castaneda JR, Tunez I, et al. Protective melatonin effect on oxidative stress induced by okadaic acid into rat brain. J Pineal Res 2003;34:265–8.

[64] Quay PD, Peacock C, Björkman K, Karl DM. Measuring primary production rates in the ocean: enigmatic results between incubation and non-incubation methods at Station ALOHA. Glob Biogeochem Cycles 2010;24:GB3014. https://doi.org/10.1029/2009GB003665.

[65] Marra J. Comment on "Measuring primary production rates in the ocean: enigmatic results between incubation and non-incubation methods at Station ALOHA" by Quay PD, et al. Glob Biogeochem Cycles 2012;26:GB2031. https://doi.org/10.1029/2011GB004087.

[66] Gaarder T, Gran HH. Investigations of the production of plankton in the oslo fjord. Rapp Pv Reun Cons Perm Int Explor Mer 1927;42:1–48.

[67] Carpenter JH. The accuracy of the Winkler method for dissolved oxygen analysis. Limnol Oceanogr 1965;10:135–40.

[68] Slawyk G, Collos Y, Auclair JC. The use of the 13C and 15N isotopes for the simultaneous measurement of carbon and nitrogen turnover rates in marine phytoplankton. Limnol Oceanogr 1977;22:925–32.

[69] Bender ML, Grande K, Johnson K, Marra J, Williams PLB, Sieburth J, et al. A comparison of four methods for determining planktonic community production. Limnol Oceanogr 1987;32:1085–98.

[70] Steeman-Nielsen E. The use of radioactive carbon (14C) for measuring production in the sea. J Cons Perm Int Explor Mer 1952;18:117–40.

[71] Kolber Z, Falkowski PG. Use of active fluorescence to estimate phytoplankton photosynthesis in situ. Limnol Oceanogr 1993;38:1646–65. https://doi.org/10.4319/lo.1993.38.8.1646.

[72] Robinson C, Williams PLB. Respiration and its measurement in surface marine waters. In: del Giorgio PA, Williams PLB, editors. Respiration in aquatic ecosystems. UK: Oxford University Press; 2005. p. 147–81.

[73] Sakshaug E, Bricaud A, Dandonneau Y, Falkowski PG, Kiefer DA, Legendre L, et al. Parameters of photosynthesis: definitions, theory and interpretation of results. J Plankton Res 1997;19:1637–70. https://doi.org/10.1093/plankt/19.11.1637.

[74] Laney SR. Assessing the error in photosynthetic properties determined by fast repetition rate fluorometry. Limnol Oceanogr 2003;48:2234–42. https://doi.org/10.4319/lo.2003.48.6.2234.

[75] Kyewalyanga MS, Naik R, Hegde S, Raman M, Barlow R, Roberts M. Phytoplankton biomass and primary production in Delagoa Bight Mozambique: application of remote sensing. Estuar Coast Shelf Sci 2007;74(3):429–36.

[76] Behrenfeld MJ, Falkowski PG. A consumer's guide to phytoplankton primary productivity models. Limnol Oceanogr 1997;42:1479–91.

[77] Gaonkar CC, Kooistra WH, Lange CB, Montresor M, Sarno D. Two new species in the *Chaetoceros socialis* complex (Bacillariophyta): *C. sporotruncatus* and *C. dichatoensis,* and characterization of its relatives, *C. radicans* and *C. cinctus*. J Phycol 2017;53(4):889–907.

[78] Li Y, Boonprakob A, Gaonkar CC, Kooistra WH, Lange CB, Hernández-Becerril D, et al. Diversity in the globally distributed diatom genus *Chaetoceros* (Bacillariophyceae): three new species from warm-temperate waters. PLoS One 2017;12(1):e0168887.

[79] Kooistra WH, Sarno D, Balzano S, Gu H, Andersen RA, Zingone A. Global diversity and biogeography of *Skeletonema* species (Bacillariophyta). Protist 2008;159(2):177–93.

[80] Johnson ZI, Martiny AC. Techniques for quantifying phytoplankton biodiversity. Ann Rev Mar Sci 2015;7:299–324.

[81] Rich VI, Konstantinidis K, DeLong EF. Design and testing of 'genome-proxy'microarrays to profile marine microbial communities. Environ Microbiol 2008;10(2):506–21.

[82] Shilova IN, Robidart JC, Tripp HJ, Turk-Kubo K, Wawrik B, Post AF, Thompson AW, Ward B, Hollibaugh JT, Millard A, Ostrowski M. A microarray for assessing transcription from pelagic marine microbial taxa. ISME J 2014;8(7):1476–91.

[83] Nanjappa D, Audic S, Romac S, Kooistra WH, Zingone A. Assessment of species Ddiversity and distribution of an ancient diatom lineage using a DNA metabarcoding approach. PLoS One 2014;9(8):e103810. https://doi.org/10.1371/journal.pone.0103810.

[84] Regaudie-de-Gioux A, Lasternas S, Agustí S, Duarte CM. Comparing marine primary production estimates through different methods and development of conversion equations. Front Mar Sci 2014;1:19.

Index

'*Note:* Page numbers followed by "f" indicate figures, "t" indicate tables.'

B

C

D

F

G

H

I

J

K

L

M

O

P

Q

R

S

U

CPI Antony Rowe
Eastbourne, UK
December 21, 2019